D1260011

# BIOMES OF EARTH

# BIOMES OF EARTH

## Terrestrial, Aquatic, and Human-Dominated

*Susan L. Woodward*

**GREENWOOD PRESS**
Westport, Connecticut • London

**Library of Congress Cataloging-in-Publication Data**

Woodward, Susan L., 1944 Jan. 20–
  Biomes of earth : terrestrial, aquatic, and human-dominated / Susan L. Woodward.
    p. cm.
  Includes bibliographical references (p. ).
  ISBN 0-313-31977-4 (alk. paper)
  1. Biotic communities. I. Title.
QH541.W633 2003
577—dc21          2003044069

British Library Cataloguing in Publication Data is available.



Library of Congress Catalog Card Number: 2003044069
ISBN: 0-313-31977-4

First published in 2003

Greenwood Press, 88 Post Road West, Westport, CT 06881
An imprint of Greenwood Publishing Group, Inc.
www.greenwood.com

Printed in the United States of America

The paper used in this book complies with the Permanent Paper Standard issued by the National Information Standards Organization (Z39.48–1984).

10 9 8 7 6 5 4 3 2 1

# CONTENTS

# PREFACE

This book brings together in one volume descriptions of the natural vegetation and animal life in the terrestrial and aquatic biomes of Earth, based upon the most recent information generally accessible to American students. The inclusion of human-dominated biomes in a global scheme, as far as I know, has not been attempted before, and just what constitutes such biomes was admittedly a judgment call. Perhaps their inclusion here will serve as a point of departure for finding other ways to define ecological regions that result from the management or transformation of natural ecosystems to such a degree that humans are clearly the dominant or keystone species. It can be argued that the whole Earth is human-dominated and, therefore, so are all conventional biomes, so this might be a timely exercise.

The intended users of the information presented here are high school and undergraduate students in biology and geography, in particular, and any interested layperson, in general. The book is in many ways an outgrowth and expansion of a well-received Web site on terrestrial biomes that I prepared in 1996 as part of the Virtual Geography Department project. The goal has been to sample major expressions of each biome throughout the world, but North America and South America tend to be emphasized whenever a biome occurs on or just offshore from one of those two continents. I have also tried to incorporate some of the excitement of discovery that comes from learning about new and wonderful forms of life and their adaptations to the enormously varied habitats on this planet.

The book is divided into four parts, each with an introduction to explain some of the features that make a set of biomes distinct from those covered in other parts and to introduce some of the terms and concepts that are used throughout a given part. The four parts are Terrestrial Biomes, Freshwater Biomes, Marine Biomes, and Human-Dominated Biomes. The introduction to the book as a whole discusses the biome concept and its development. Within each part are separate chapters on each biome. Most users probably will be looking for information on a single biome, so each chapter can stand on its own, although it would be helpful to read the introduction to the respective part. For each biome, a map shows its global distribution pattern; at least one photograph illustrates typical vegetation if it is a terrestrial biome and typical habitat conditions if it is not. Chapters begin with generalized overviews followed by more specific information from the different regions of the world where the biome occurs, including examples of plant and animal species found there. Brief discussions of the biome's origin in the geologic past, significant and widespread human impacts on its vegetation and animal life, and a brief history of its exploration by scientists are presented, as well as a list of references. A glossary serves the book as a whole, and a cumulative bibliography is also provided.

While I have had experience in many, if not all, of the terrestrial biomes, I am not an expert on any; so I learned many new and interesting things while preparing the manuscript. Writing about aquatic biomes required my exploring entirely new realms, something I always find most enjoyable. Due to the scale of the project and the need to complete it in a limited amount of time, I have relied almost exclusively on secondary sources. Whenever possible, I used books published after 1990. Many excellent works have been written on natural vegetation. It was difficult, however, to find material that included discussions of animals in the framework of biomes, so at times field guides and other identification guides to animals had to be consulted to fill in the gaps. Particularly useful in this regard was *Walker's Mammals of the World.*

# ACKNOWLEDGMENTS

I have been teaching undergraduate courses on biogeography and physical geography for nearly 20 years and have benefited in innumerable ways from working with the students, who brought a variety of backgrounds, interests, and skills to these classes. I would like to single out members of a seminar in applied biogeography who helped me decide upon appropriate biomes for the aquatic and human-dominated sections. The world maps were originally drafted by members of a cartography class at Radford University under the direction of Dr. Lori LeMay and were finalized by Melissa Smith. My appreciation goes out to all these people. Completion of the book would not have been possible without the support of Radford University, especially through its grant of Professional Development Leave for Fall Semester 2001. This book is dedicated to Dr. Jonathan D. Sauer, who provided inspiration that has lasted a lifetime.

# INTRODUCTION

The biome concept developed as a way to classify Earth's terrestrial communities into major ecological units that correlated with regional **climate** types. [Note that terms defined in the glossary are bold highlighted on their first mention throughout the chapters.] Although the term **biome** was first used in 1939 (Clements and Shelford 1939), the roots of the concept go back to late eighteenth century and the voyages of Captain James Cook. The naturalist J.R. Forster accompanied Cook and in 1778, based upon his scientific observations, proposed that the Earth could be subdivided into latitudinal belts of similar climate and vegetation structure. Soon afterward, Alexander von Humboldt, the reputed father of plant geography, traveled through the Americas and noted consistent changes in vegetation associated with changes in elevation. He concluded that the same climate and vegetation changes observed by Forster at different latitudes were replicated at different altitudes, and that climate determined the nature of the plant cover (MacDonald 2003).

A key element in the biome concept is the idea that similar regional climates produce similar morphological adaptations in plants (plant physiognomy) and similar structure in the vegetation, regardless of the actual species present. (Plant physiognomy and vegetation structure receive fuller treatment later in this volume in the "Introduction to Terrestrial Biomes.") It is this emphasis on the characteristic appearance of life instead of its taxonomic relationships that allows the delineation of intercontinental regions of similar climate and corresponding living communities that are the biomes.

A decade or so before Frederic Clements and Victor Shelford talked of biomes, Wladimir Köppen (1923) had developed a classification system for world climate types based on global vegetation patterns. Koeppen, as his name is usually written in English, had focused on annual and seasonal patterns of heat and moisture as the basis for determining different types of climates. He used plant life as a guide to finding significant breaking points in what is in reality a continuous, gradual

**Table I.1**
**Summary of the Koeppen Climate Classification**

---

**Major Climate Groupings**

**A**–tropical climate (all months of the year warm enough for plant growth; enough precipitation to support forests)

**B**–arid and semiarid climates of the subtropics and midlatitudes (insufficient precipitation to support forests; grasslands or desert shrubs prevail)

**C**–subtropical climate (seasonal temperature patterns with warm to hot summers and distinctly cooler winters but without persistent snow cover; enough precipitation to support forests)

**D**–continental climate (strong seasonal variation in temperature with long, cold winters; enough precipitation to support forests

**E**–polar climates (cool to cold year round with very brief growing season; treeless, sometimes called polar deserts)

**Seasonal patterns of precipitation used to differentiate humid climates (A, C, D) only:**

**f**–precipitation fairly evenly distributed throughout the year

**m**–monsoonal/highly seasonal precipitation pattern, but with extremely high amounts of rainfall during rainy season (This pattern usually is applied only to the A climate and in areas where tropical rainforest occurs despite a distinct dry season.)

**w**–winter drought; seasonal precipitation pattern in which the rainy season is associated with the high sun period or summer

**s**–summer drought; seasonal precipitation pattern in which the rainy season is associated with the low sun period or winter; primarily encountered in the subtropics as the mediterranean climate type

**Temperature descriptors added to differentiate subtropical (C) and continental (D) climate types:**

**a**–long, hot summers

**b**–cool summers

**c**–short, cool summers

**d**–extremely short summer

**Precipitation descriptors added to distinguish among dry (B) climates only:**

**S**–semiarid conditions [as a rule of thumb in temperate climates, an annual total of 25–50 cm (10–20 in) of precipitation; enough moisture to support grasslands]

**W**–arid conditions [as a rule of thumb in temperate climates, an annual total of 0–25 cm (0–10 in) of precipitation; only enough moisture to support desert plants]

**Temperature descriptors added to the dry (B) climates:**

**h**–hot deserts (generally frost free)

**k**–cold deserts (winters regularly experience sub freezing temperatures for prolonged periods of time)

**Temperature descriptors added to the cold (E) climates:**

**T**–tundra conditions; a long enough growing season to support a cover of mosses, lichens, sedges and/or dwarfed shrubs

**F**–frozen conditions year round; ice cap

**Table I.1**
**Summary of the Koeppen Climate Classification (continued)**

---

**Some basic climate types and corresponding terrestrial biomes:**

**Af**–tropical wet climate (tropical rainforest biome)

**Am**–tropical monsoon climate (tropical rainforest biome)

**Aw**–tropical wet and dry climate (tropical savanna or tropical dry forest biome)

**BSh, BSk**–hot and cold semiarid climates, respectively (temperate grasslands biome)

**BWh, BWk**–hot and cold desert climates, respectively (desert scrub biome)

**Cfa**–humid subtropical climate (temperate broadleaf deciduous forest biome)

**Cfb**–marine west coast climate (temperate broadleaf deciduous forest biome in Europe; temperate evergreen forests in the Americas)

**Cs**–mediterranean climate (mediterranean woodland and scrub biome)

**Dfa**–humid continental climate with hot summers (temperate broadleaf deciduous forest biome)

**Dfb**–subarctic climate (boreal forest biome)

**Dfc**–subarctic climate (boreal forest biome)

**Dwd**–subarctic climate with dry winters and extreme cold (boreal forest biome)

**ET**–tundra climate (tundra biome)

**EF**–ice climate (no vegetation; ice cap)

---

change in temperature, precipitation amounts, and seasonality over the face of the earth. In other words, he used distinct changes in natural plant cover to indicate a change in climate type. The most basic distinction is made between humid climates, which are indicated by forest vegetation, and those climates that are too dry or too cold to allow for tree growth. A world climate map based on the Koeppen climate classification therefore closely resembles a map of the world's terrestrial biomes. Others have devised other systems of climate classification, often more mathematically sophisticated, but Koeppen's system remains useful as a descriptor of average annual weather patterns and is still presented in most introductory-level textbooks on physical geography. Since it correlates so closely with terrestrial biomes, it helps students develop a basic understanding of the relationship between climate and vegetation. For those readers familiar with the system, Koeppen's symbolic shorthand is given with each terrestrial biome in the pages that follow. For those unfamiliar with the system, a very brief outline is presented later, although an introductory textbook on physical geography or climate should be consulted for more details.

When Clements, a botanist, and Shelford, a zoologist, coined the term "biome," they were both studying **communities**, groups of organisms living together in the same place or habitat. The idea that plants and animals lived together as a self-sustaining unit was formulated by S.A. Forbes (1887) while he was studying lakes. Clements (1916) elaborated upon this concept and produced a major work introducing the notion of ecological succession and the development of **climax communities** in balance with the prevailing regional climate. The way succession was supposed to work was that on any surface initially devoid of life (e.g., a new lava flow, a sand dune, a new glacial lake), a predictable series of plant communities would come to occupy the site, one replacing the other, until finally a stable assemblage of plants would develop and persist until a major change in the climate occurred. Whether the early surface was rock, sand, or water, the site would eventually become covered with the same type of vegetation, and the regional climate would determine the nature of that vegetation. The final persistent plant community was called the climax—or climatic climax—community. Under some circumstances, however, it was recognized that stable communities developed in response to factors other than climate, usually factors related to soil conditions or frequent disturbance. These persistent communities were called subclimaxes, on the belief that if the extraneous conditions were removed, the community would change to one reflecting the regional climate. It is the theoretical climax community that defines most biomes, although the case is made for a few (for example, some tropical savannas and mediterranean scrub) that the vegetation is more truly a subclimax.

Two schools of thought emerged on the nature of a climax community—or any other community for that matter. Clements and his followers viewed a community as a superorganism, and thus as a real and lasting feature on the surface of Earth. Each member of the community was a vital link in the survival of the whole community, and the loss of a member would have repercussions throughout the community and its habitat. An opposite point of view was held by H.A. Gleason (1926) and his followers, who through their observations came to believe that the constituents of a community were a matter of happenstance—that each species reacted independently to environmental change according to its own biological features, including such things as its requirements for specific resources, its tolerance of environmental fluctuations and extremes, and its **dispersal** abilities. In the view of Gleason and his followers, a community was simply a group of organisms with similar requirements and tolerances that had been able to get to and survive in the same site. Vital interactions among the members to sustain the community were not a major factor in the individualistic community concept (MacDonald 2003).

The Clementsian view dominated ecology and **biogeography** for much of the twentieth century and gave rise to a number of terms still used today in describing biomes. However, the individualist concept has come to the forefront in explaining how nature really works, and the emphasis in ecology is no longer on the stability of climax communities but on disturbance and change as universal factors in the dynamic assemblages of plants and animals. Another way to put it is to say that we still talk about climax vegetation and early successional communities as though they were leading to final climax, we talk about successional communities as being in a particular place along the sequence of communities that will take over before the climax is developed, but we don't really believe in a prescribed order of communities of specific composition or of a final, unchanging community in balance with the regional climate. If this sounds crazy, think how we talk about the sun rising and setting each day when we know that it is the earth rotating on its axis that gives us day and night!

Biomes, then, are distinguished according to the structure of their vegetation. Still, different parts of the world are inhabited by different species, and so within a given biome, regional subdivisions are found, each usually restricted to a different continent. A. Grisebach (1872) referred to such regions within a given biome as formations. The tropical rainforest biome, for example, has an African Formation, an American Formation, and an Indo-Malayan Formation because different kinds of trees dominate in each location, but the most conspicuous plants in all are similar in size, shape, leaf habit, and so forth. Influenced by Charles Darwin, Grisebach saw natural selection in a similar climate working on plants of different ancestries and favoring the same adaptations. Despite different evolutionary and geologic histories, morphological outcomes were similar (Kendeigh 1961; MacDonald 2003). In this book, the word "formation" will be replaced by the term "expression" of the biome. This will allow more flexibility in describing vegetation of somewhat different structures (for example, dry tropical forests and woodlands) under the heading of the same biome.

Since regional variation and the distinctiveness of a given expression of a biome are often related to the evolutionary history (phylogeny) of the species present and the geologic history of the landmass involved, it is important for the reader to have a basic understanding of the taxonomic hierarchy, plate tectonics, and a few biogeographic processes. Taxonomy is the science that classifies organisms. Modern taxonomy uses a naming system devised by the Swedish scientist Carl Linneaus in 1758. Each unique form of life has a universally accepted Latin name that consists of two parts. Take us, for example. We are classified by scientists as *Homo sapiens.* Our pet dogs are *Canis familiaris.* The wolf is *Canis lupus* and the coyote *Canis latrans.* Dogs, wolves, coyotes are all *Canis* something. This implies that they are closely related, that they evolved from some common ancestor not too long ago. *Canis* is the name of the **genus** into which they have been placed by taxonomists. The *familiaris*, *lupus*, and *latrans* parts of the names tell specifically which member of the closely related group each is, in other words, which unique **species** it is. Generally speaking, a species is a real or natural unit in that members of a species recognize each other as fellow members of the group and (most of the time) breed only with each other to produce more of the same kind of organism. Rituals, color patterns, reproductive structures, and activity times all evolve to ensure breeding only with members of the same species, or what is called reproductive isolation from other species. Occasionally the safeguards (which may simply be distance between distribution areas) break down and one species breeds with another to produce what is known as a hybrid. Dogs and coyotes can produce coydogs; wolves and coyotes may have produced the red wolf. Most times, however, the hybrid offspring are sterile (as in the case of the mule, the product of a male horse breeding with a female donkey) or they simply do not fare as well in the wild as either of the two parent species and soon disappear.

The categories or levels in the taxonomic hierarchy (Table I.2) other than species are determined by scientists, not nature. They are the pigeonholes into which species are placed so that some order may be created to enable our study and understanding of nature. By examining the fossil record and now the genetic codes of organisms, taxonomists group organisms according to their family tree or evolutionary history. As previously mentioned, a genus is composed of closely related species. Closely related genera (the plural of genus) are placed in the same family. So, to continue with the previous example, the wolf and coyote are also related to foxes (North America alone has several genera of foxes, including *Vulpes*, *Urocyon*, and *Alopex*), the maned wolf (*Chrysocyon brachyurus*) of South America, and the dhole (*Cuon alpinus*)

**Table I.2**
**The Taxonomic Hierarchy as Applied to the Wolf (Canis lupus)**

**Kingdom**: Animalia
**Phylum**: Vertebrata
**Class**: Mammalia
**Order**: Carnivora
**Family**: Canidae
**Genus**: *Canis*
**Species**: *lupus*

of Asia, but less closely related to foxes than to each other. All these animals are basically doglike, so taxonomists have grouped them into the same family, the dog family (Canidae). The dog family, bear family (Ursidae), cat family (Felidae), weasel family (Mustelidae), and some others are sufficiently alike in skeletal and dental features to be placed together in an even broader taxonomic group, the order—in this case the order Carnivora. The carnivores, along with other animals such as rodents, whales, and aardvarks, which all have hair and produce milk from mammary glands to nurse their young, are put into the class Mammalia, the mammals. And so the system progresses through evermore inclusive groups with evermore basic shared characteristics. Similar classes are grouped into a phylum, wherein distinct body plans are recognized (for example, four-limbed, back-boned animals—whether they be fish, amphibian, reptiles, birds, or mammals—are in the phylum Vertebrata, whereas animals such as insects and crayfish with an exoskeleton, segmented body, and jointed appendages attached in pairs to the body segments are placed in the phylum Arthropoda), and phyla are grouped into the highest category in the hierarchy, the kingdom. Any single level or category is called a **taxon** (plural = taxa). Scientists propose different numbers of kingdoms according to the criteria chosen to make this most basic division of life. Only two used to exist: plant and animal. Now most biology textbooks use five: Monera, Protista, Fungi, Plantae, and Animalia.

Taxonomists decide in which pigeonhole—genus, family, etc.—to put an organism according to their best understanding of its origins and evolutionary history. Therefore, the taxonomic names are good clues as to how closely related two species are and thus how similar or dissimilar they are. Members of the same family but different genera are distantly related. What this means is that sometime in the distant past, one species—we might consider it like a great, great, great-grandfather species—through time has diverged into coyotes and wolves, foxes, maned wolf, and so forth. Coyotes and foxes may have the same great, great-grandfather species, but coyotes and wolves have the same father, so to speak.

How new species come into being is not fully understood, but most scientists believe that some type of geographic separation of members of the same species occurs and prevents those on one side of a barrier (a new sea, a rising mountain range, etc.) from breeding with those on the other side. Over a long period of time, the two separated groups will become different as they adapt to the different environmental conditions on either side of the barrier. In short, isolation and time are needed for speciation to occur. The longer and more complete the isolation, the greater will be the proportion of endemic taxa—taxonomic groups that originated in a given area and are still found only there.

The appearance of a new species may signal a change in arrangement of continents and oceans and be intimately involved with earth history and the dynamic geological conditions of our planet. At a global scale, plate tectonics—the movement and deformation of Earth's crust—has been most important. Brief summaries of the relationships between geologic history and the origin of species, climate regions, and particular biomes are given in each of the biome chapters.

Official names, the result of intensive scientific study, are important in the study of biomes because they indicate what some organism is, what it is related to, and what may likely have been its evolutionary history and geographical or distribution history. By convention, these names are given in Latin, and, as a foreign language, the Latin names indicating genus and species are printed in italics. These Latin names are agreed upon in international committees and may be referred to as scientific names or Linnaean names, after the father of modern taxonomy. Of course, we don't refer to the family pet as a *Canis familiaris* but as a dog if we speak English, or *perro* if we speak Spanish, *der Hund* if we speak German, or a cur or mongrel if we wish to be insulting. These are common names. But regardless of what language we speak, it is everywhere *Canis familiaris.* In the rest of this book, common names are given whenever possible, but the important name that will tell you whether the same thing or something very closely related occurs in different biomes or in different expressions of the same biome is the scientific name.

A final word on taxonomy. Taxonomic labels change. With new information, greater insights through studying more specimens, or changes in philosophy, taxonomists may reclassify organisms, and the life-form ends up with a new name. Usually the changes occur at the species or genus level, but sometimes, upon careful reexamination, an organism may be placed in a different family. Newly discovered organisms are compared with species known to science already and may be placed into existing groups or—if extremely different from anything previously described—may require the creation of entirely new orders or even phyla.

Different species, genera, families, and even orders may be found on different landmasses. A number of scientists have undertaken classifying the geographic patterns of life according to taxonomic relationships instead of according to vegetation structure. Two schemes, one for plants and one for animals, have contributed terms commonly used in the description of the taxa comprising biomes. Ronald Good (1947) recognized six Floral Kingdoms: Boreal, Paleotropical, Neotropical, South African, Australian, and Antarctic. Each of these was subdivided into provinces. The most significant province is that of the Cape Province of the South African Kingdom, a tiny part of the Earth's land surface that was singled out because of the large number of plant species and higher taxonomic units found there and nowhere else. Often it is equated with a Floral Kingdom, and the South Africa designation is dropped. Animals are categorized by six Zoogeographic Provinces, originally proposed by the ornithologist Philip Sclater in 1857: Palearctic, Nearctic, Neotropical, Ethiopian, Oriental, and Australian. Sometimes the Palearctic and Nearctic are combined into a single region, the Holarctic. Boundaries of Zoogeographic Provinces differ from those of the Floral Kingdoms because plants and animals have different evolutionary histories and dispersal capabilities.

The biome concept originated as a way to classify terrestrial communities. Later the biome concept embraced the entirety of the ecosystem: the community of living organisms plus the nonliving parts of the environment with which plants and animals and microbes interact to pass the sun's energy through the system and recycle nutrients. Several schemes, employing different names and criteria, have been developed for delineating biomes at both country-specific and global levels. No particular scheme is followed in this book, but it builds upon a general consensus of which vegetation types are significant on a global scale and constitute the foundations of major biomes. The selection of regional expressions is problematic in some instances, perhaps most obviously in deciding what should be called a type of tropical savanna and what should be called tropical dry forest. This issue points to the fact that biomes do not have absolute boundaries; one grades into another, and we merely seek an overall pattern.

The nonterrestrial biomes presented in this book are not well established as such in the literature. The original biome concept depended on natural vegetation, on land plants. Aquatic environments really have nothing comparable, although some coastal and continental shelf marine communities receive their structural foundation from living organisms. Examples are coral reefs, sea-grass meadows, and kelp forests. Freshwater aquatic habitats might be considered a single biome or perhaps three (streams, lakes, and wetlands) but here are grouped into two usually connected biomes—streams and lakes—with wetlands viewed as part of the stream environment. The division is admittedly arbitrary, and the user of the book should feel free to place habitats in whatever and however many biomes seem reasonable to them!

The human-dominated biome categories are also somewhat arbitrary and not truly compatible with the original biome concept. They are mainly the product of the author's interest in cultural biogeography, the distribution of plants and animals resulting from the activities of humans as agents of evolution (e.g., breeding crops and livestock, promoting conditions that encourage the evolution of weeds) and dispersal (the deliberate and accidental transport and introduction of species). The types of plants dominating sites were considered in devising the biomes or major expressions of a biome. The question of degree of human dominance also was considered. The entire Earth is influenced by human activities, so it wouldn't be a great stretch of the imagination to consider all biomes human dominated. Readers will need to judge for themselves upon considering the information presented in each biome chapter.

An attempt has been made to provide the same kinds of information for each biome identified, although this was not always possible or reasonable. Each chapter begins with an overview of basic characteristics, distribution, and climate/environmental controls. The chapters on terrestrial biomes have separate sections on climate, plants, animals, and soil. The aquatic biome chapters describe aspects of the physical environment—in particular substrate but also water temperature and water chemistry—that influence the distribution of life and present challenges to which organisms must adapt to survive. Major regional or environmental expressions of each terrestrial and aquatic biome were selected for further elaboration of the characteristics of the biome and to give an idea of the variation that exists within it. Many more could have been described, and selection was based on available information, continental representation, and, sometimes, the existence of curious anomalies. These chapters conclude with sections on human impacts, the origins in time and space of the particular environmental conditions and species comprising the biome, and a brief history of scientific investigation and the contributions to science that have come from studying the biome.

The human-dominated biomes had to be treated a bit differently than the natural biomes, but the same general chapter structure was maintained. An obvious change was replacing the human-impact section with a description of some of the impacts of the biome itself on natural communities.

## REFERENCES

Clements, Frederic E. 1916. *Plant Succession: An Analysis of the Development of Vegetation.* Publication No. 242, Carnegie Institution of Washington. Washington, DC: Carnegie Institution of Washington.

Clements, Frederic E. and Victor E. Shelford. 1939. *Bio-ecology.* New York: John Wiley.

Forbes, S. A. 1887. "The lake as a microcosm." *Peoria Science Association Bulletin.* Republished in 1920 in *Illinois Natural History Survey Bulletin* 15: 537–550.

Gleason, H. A. 1926. "The individualistic concept of the plant association." *Torrey Botanical Club Bulletin* 53: 7–26.

Good, Ronald. 1947. *The Geography of the Flowering Plants.* London: Longmans, Green and Company.

Grisebach, August. 1872. *Die Vegetation der Erde nach Ihrer Klimatischen Anordnung.* Leipzig: W. Englemann.

Kendeigh, S. C. 1961. *Animal Ecology.* Englewood Cliffs, NJ: Prentice Hall.

Köppen, W. 1923. *Grundiss der Klimakunde: Die Klimate die Erde.* Berlin: DeGruyter.

MacDonald, Glen. 2003. *Biogeography: Introduction to Space, Time and Life.* New York: John Wiley & Sons.

# I TERRESTRIAL BIOMES

## INTRODUCTION

Terrestrial **biomes** are recognized when similar **vegetation** structure has developed under similar regional climatic conditions. The assumption is that—regardless of the actual plant **species** present—the same plant **physiognomy** and the same vertical and horizontal arrangement of the dominant plants on the **landscape** have evolved as adaptations to a given **climate** type wherever it occurs in the world. Species composition is of secondary importance to structure.

The elements of plant physiognomy or **morphology** include such things as size and shape of the whole plant, the nature of the stem—whether woody or nonwoody—the size and shape of leaves, the **deciduous** or **evergreen habit** of the foliage, and the presence or absence of thorns or spines or other ancillary features. Various standardized methods of describing and recording physiognomy have been developed. Two schemes commonly used in the United States are those of C. Raunkiaer (1934) and Pierre Dansereau (1957). Raunkiaer based his system on general size and shape, with particular focus on the type and location of the perennating or renewal bud—the site from which new growth arises each year. Each of Raunkiaer's categories (Table 1.1) is known as a specific **life-form**.

With a few exceptions, Raunkiaer's life-forms do not project a ready image of the appearance of plants and tend to have limited use. The categories most commonly encountered in descriptive accounts of biomes are chamaephytes, hemicryptophytes, and **geophytes**.

Dansereau devised a system of describing plant physiognomy that presents a clearer visualization of what plants actually look like, rather than focusing on characteristics of their life history as Raunkiaer did. Dansereau called his categories **growth-forms** and distinguished among trees, shrubs, **herbs**, **lianas** (woody vines), **epiphytes** (plants that grow on the branches, stems, or leaves of the plants—using their hosts only for support and not for nutrition), and bryophytes (nonvascular plants such as mosses, liverworts, and **lichens**). Each of the categories of woody

**Table 1.1**
**Raunkiaer's Life-forms**

---

**Phanerophtyes**–Tall perennial woody plants; trees, large shrubs, and vines with renewal buds more than 250 mm (10 in) above the ground. Buds are exposed to cold temperatures and drying winds.

**Chamaephytes**–Small perennial woody or herbaceous shrubs with persistent stems and aerial buds up to 250 mm (10 in) above the ground surface. Common in dry and cold climates. In cold climates, the buds lie protected under the snow during winter months.

**Hemicryptophytes**–Small perennial plants with renewal buds at the ground surface or just below in the upper layer of the soil. These plants usually die back to ground layer at the end of each year's growing season.

**Cryptophytes**–Small perennial plants with renewal organs such as bulbs and corms positioned well below the surface. There are three types:

- **Geophytes**: Perennial land plants with renewal organs deep in the soil.
- **Helophytes**: Perennials such as emergent lake plants with renewal buds on rhizomes that are buried in a waterlogged substrate.
- **Hydrophytes**: Perennial water plants with submerged or floating leaves and renewal buds below water.

**Therophytes**–Annual plants that complete their life cycle in one growing season and survive unfavorable conditions as seeds.

**Epiphytes**–Plants that grow on the stems or branches of other plants, their roots hanging in the air. (Dansereau 1957; Lincoln *et al.* 1998; MacDonald 2003.)

---

plants is further elaborated by describing leaf shape, size, texture, and habit. Herbs are subdivided into **forbs** (**broad-leaved annuals**) and graminoids (grasslike plants, **perennial** or annual). Dansereau's method proved easy to use in the field and remains a common way of describing the nature of plants composing a particular biome (Dansereau 1957; MacDonald 2003). It is the main scheme used in the chapters that follow.

**Vegetation** is a descriptive term indicating the overall appearance of the total plant cover, as opposed to **flora**, which refers to all the species of plants inhabiting a given area. Vegetation is divided into structural types that derive from the dominant growth-forms present and the spacing of those growth-forms on the surface and from ground to **canopy**. Among the major structural types of vegetation defining biomes are forests, woodlands, **savannas**, grasslands, **scrub**, and desert. Forests are visually dominated by trees that form a **closed canopy**, that is their **crowns** intermingle and prevent a large percentage of the sunlight that reaches the top of the canopy from filtering through to the forest floor. Forests furthermore have vertical structure since trees shorter than the canopy plants, shrubs, lianas, herbs, and bryophytes create several layers or strata of foliage. A woodland has trees usually of smaller stature than forest trees and has a simpler vertical structure than a forest. The canopy is more open so that enough sunlight reaches the forest floor to allow for a dense, if patchy, shrub or herb layer. Shrublands have a nearly continuous cover of shrubs, while scrub has a discontinuous cover of widely spaced shrubs. Savanna is a parklike vegetation type with widely spaced trees or shrubs standing above a continuous layer of grasses and forbs. Grasslands generally lack trees and shrubs and are characterized by a continuous or nearly continuous cover of grasses and forbs, which may grow in more than one layer. Deserts have a sparse covering of woody shrubs specially adapted to withstand drought. Most of the time, bare ground dominates the landscape, but for very short periods after sufficient rainfall, the leaves and flowers of geophytes may emerge, and the seeds of annuals may germinate to carpet the desert floor.

Terrestrial biomes are distinguished according to vegetation. Accordingly, the biomes described in the following chapters are Tropical Rainforest, Tropical Dry Forest, Tropical Savanna, Desert, **Temperate** Grasslands, Mediterranean Woodland and Scrub, Temperate Broadleaf Deciduous Forest, Boreal Forest, and Tundra.

Any biome description is a generalization of the plant cover dominating a given climate region. Although the focus is on the so-called **climax community**, in actuality, a mosaic of plant communities reflects different **disturbance** regimes and **successional** stages. Each **community** is a separate, local **ecosystem** or a patch. When the patches of the mosaic are repeated again and again over a large expanse of territory, a **landscape** of interacting ecosystems is formed. In the vocabulary of modern **ecology**, then, a biome is a particular landscape.

Each biome occurs in latitudinal swaths, and several are also represented in altitudinal belts on mountains in middle **latitudes**. Alexander von Humboldt first described the altitudinal **zonation** of vegetation as a result of his explorations in the Andes Mountains of South America. He suspected that the belts of vegetation on mountains were similar to the zones described between the equator and the arctic regions of Earth; however, this replication of latitudinal zones of vegetation on mountains is more true for temperate regions of the Northern Hemisphere than in the **tropics**, where Humboldt noticed the phenomenon. Several scientists have since tried to equate altitudinal zones of vegetation with specific temperature patterns. C. Hart Merriam (1894) worked out such a system in the San Francisco Peaks of northern Arizona. Merriam called the vegetation-climate zones (with the animals inhabiting them)—which were expressed both latitudinally and altitudinally—**life zones** and named each after a representative part of the North American continent. The lowest life zone (in both altitude and latitude) was the Lower Sonoran, named after a state in northwest Mexico. Above this were the Upper Sonoran, Transition, Canadian, Hudsonian, and Arctic-Alpine life zones. The climatic boundaries set by Merriam turned out to have no significance, and his system was difficult to apply outside the arid lands of the United States. But, the names he gave the life zones are still used in the mountainous parts of the American West.

The English botanist L. R. Holdridge (1947) proposed a more universal scheme of life zones; he used measures of temperature, precipitation, and potential evapotranspiration to classify vegetation. With the focus on climatic variables as predictors of vegetation type, the Holdridge scheme has remained popular with scientists trying to understand how ongoing and future climate change may alter the distribution patterns of the world's plant formations and the species that comprise them (MacDonald 2003). It also is used often by plant ecologists working in the tropics, where Merriam's life zones bear no resemblance to reality on the ground. Even so, it lacks the descriptive power of Merriam's schemes and therefore is not used in this book.

Animals, of course, are also important components of terrestrial biomes. Animals may clearly exhibit adaptations to both the climate and the vegetation of specific biomes. Adaptations fall under several categories. First are morphological adaptations, such as those epitomized in the nineteenth-century ecogeographic rules of Bergmann and Allen. Bergmann's Rule stated that the colder the climate, the larger the body size of warm-blooded animals (birds and mammals) compared to closely related animals living in warmer regions. Allen's Rule stated that the colder the climate, the shorter the appendages, including legs, ears, and snouts. A comparison between the large, short-eared arctic hare (*Lepus articus*) and the gangly, long-eared antelope jackrabbit (*L. alleni*) of the Sonoran desert bears out both Bergmann and Allen. Other morphological traits that impart survival benefits include such things as a prehensile trail for getting around in a forest or coat color for camouflage. Another group of adaptations are physiological and involve changes in the way an animal body functions. These include the ability to go dormant for periods of time, conserve body moisture in dry environments by concentrating the urine, or schedule reproduction to coincide with times of plentiful food. The third type of adaptation is behavioral and is probably the quickest to evolve in a population. Behavioral adaptations include being a social herd animal in open habitats and benefiting from many eyes and ears alert for predators, or being fossorial and spending the hot daylight hours in burrows below ground during a desert summer. Migrating away from areas when conditions deteriorate may be both a behavioral and physiological response. Common adaptations among animals in each biome are identified in the following chapters.

Viewing biomes as ecosystems demands consideration of the nonliving elements of a given **habitat** type. On land, the substrate—the material that is the ground—is an important influence on both plants and animals. Solid rock is one type of substrate; talus, sand, alluvium, and glacial **drift** are some other types. The different substrates present different challenges to living organisms looking for root holds or shelter and, even more importantly, contribute different minerals as potential plant nutrients. The real interface between the living and nonliving parts of an ecosystem occur in the **soil**, the medium in which plants grow and which seems to defy an exact definition. **Soil** is a product of geological and biological processes and develops over long periods of time. Very briefly, **weathering** breaks rock down into fine particles. Soluble minerals that are released during the weathering process are dissolved in the soil moisture and can be washed down and away from the root zone by a process known as **leaching**. Plants must trap those dissolved minerals necessary for their growth and maintenance (nutrients) before they are leached away. Clay particles and partially decayed plant matter called **humus** can loosely bind nutrients and hold them temporarily in the soil until plant rootlets can retrieve them. Humus is brownish-black in color, and—if it occurs in large amounts in a soil—can stain the soil a dark brown. Soil color is, therefore, a useful indicator of soil fertility: the browner the soil, the more humus, the greater amount of nutrients being stored for ready exchange with plant rootlets, and the more fertile the soil. Different plant species have different nutrient requirements. They will trap only the nutrients they need and incorporate them into plant tissue. Unused nutrients eventually will be leached from the soil. When leaves are shed or plants die, fungi and bacteria decompose the dead material, and the nutrients once held in plant tissues are released back into the soil, where living plants may take them up again. Over time then, plants change the soil chemistry by concentrating the elements and compounds they require and letting the others be removed. Climate plays a major role in determining the rates of weathering, leaching, and decomposition or decay. Generally speaking, rates are fastest where temperatures and precipitation are highest, that is in the humid tropics.

The different combinations of climate, substrate, and vegetation lead to the development of a variety of different soil types. True or mature soils develop as ecological succession proceeds. Each displays layers or horizons of characteristic color, depth, and chemistry. A number of different classification systems are used around the world today. A soil **taxonomy** mimicking the taxonomic hierarchy used for living organism was developed in the United States. In the following chapters, the soil order most representative of each biome is identified, but descriptions and common names—often derived from Russian and British classification systems—are emphasized more.

Human activities have affected all of the terrestrial biomes. In some instances—most notably the mediterranean biome—we aren't even sure what the vegetation was like before the presence of humans, who burned the area, grazed their livestock, and/or removed the natural plant cover to cultivate crops. People living and working nearby even affect preserves, already changed by being but fragments of once more extensive ecosystems. Ecologists and biogeographers who study the distributions of plants and animals seek out the most pristine sites or try to reconstruct the vegetation of the past to give us a better picture of what natural biomes are—or were—like. However, the reality is that contemporary biomes are products of nature and humanity, if indeed those are two different things.

## REFERENCES

Dansereau, Pierre. 1957. *Biogeography: An Ecological Perspective.* New York: Ronald Press.

Holdridge, L. R. 1947. "Determination of world plant formations from simple climatic data." *Science* 105: 367–368.

Lincoln, Roger, Geoff Boxshall, and Paul Clark. 1998. *A Dictionary of Ecology, Evolution and Systematics.* 2nd edition. Cambridge: Cambridge University Press.

MacDonald, Glen. 2003. *Biogeography: Introduction to Space, Time and Life.* New York: Wiley & Sons.

Merriam, C. Hart. 1894. "Laws of temperature control of the geographic distribution of terrestrial animals and plants." *National Geographic* 6: 229–238.

Raunkiaer, C. 1934. *The Life Forms of Plants and Statistical Plant Geography.* Oxford: Clarendon Press.

# 1 TROPICAL RAINFORESTS

## OVERVIEW

The tropical rainforests of the world are famous for containing the greatest number of plants and animals of any biome on Earth, yet common misconceptions and many unknowns surround these rich communities. The tropical rainforest is a broad-leaved evergreen forest with a complex, multistoried structure and several interesting growth-forms, which if are not entirely restricted to the biome are at least conspicuously abundant here. It is estimated that 50 percent of the species living on Earth are found in the 7 percent of the land covered by tropical forests of all types. The greatest proportion of these are in tropical rainforests. In general, rainforest is found at elevations below 1,000 m (3,000 ft) between the latitudes of 10° N and 10° S. Major regional expressions of the biome are found in tropical America, in West Africa and the Congo Basin of Africa, and in the Indo-Malaysian region of Asia. An outlier of the last occurs in a narrow strip along the coast of northeast Australia. The climate of these regions has year-round warm temperatures and abundant precipitation, with no month receiving less than 100 mm (4 in) of rainfall.

Most plants growing in tropical rainforests are woody-stemmed trees and climbers. The crowns of the trees appear to form three layers in the canopy and cast a deep shade over the forest floor broken only by moving flecks of sunlight. Contrary to some popular depictions, the forest floor is not an impenetrable mass of vegetation but has a light cover of tree seedlings and saplings and is rather easy to walk through. (The impenetrable jungle of vines and dense undergrowth is a feature of disturbed sites in the forests, such as along the banks of large rivers, where sunlight is able to penetrate to the forest floor.) Furthermore, the trunks of most trees are very slender with circumferences of less than a meter (3 ft), so the visitor more familiar with mature temperate forests gets the feeling of being in young growth. The average height of the tallest trees is less than 55 m (150–180 ft), which is larger than mature trees of the broad-leaved forests of the eastern United States or western Europe but not nearly as tall as the giant redwoods of California or the

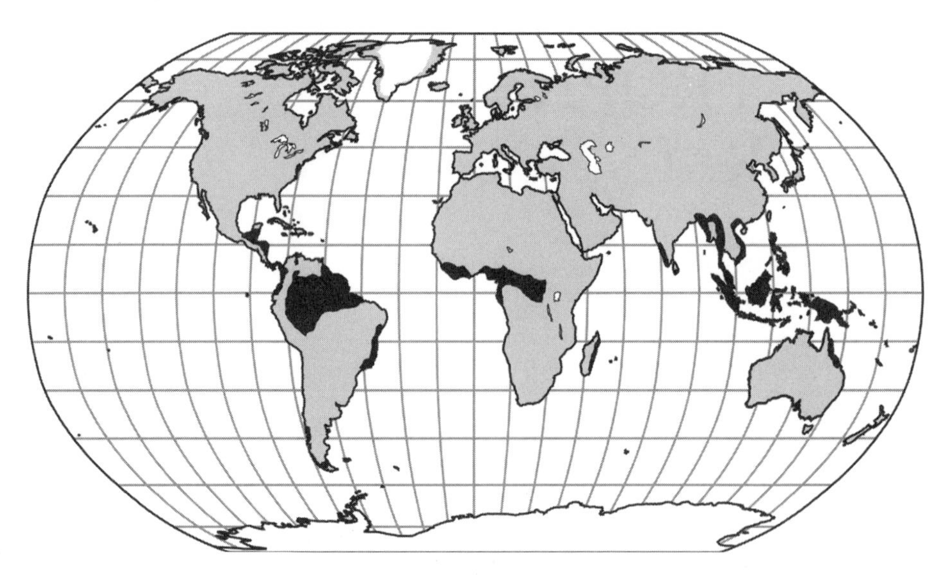

Map 1.1 World distribution of the Tropical Rainforest Biome.

tallest gums of Australia. However, some remarkably large trees are scattered among the smaller trees in tropical rainforests, and their crowns emerge above the general canopy. Nonwoody perennials grow mostly as epiphytes or air plants in the treetops, where they can obtain sunlight; relatively few grow on the dark forest floor. The perennial forbs so common in temperate forests are nearly absent from tropical rainforests, as are annual plants.

The typical tropical rainforest has no single kind of tree occurring in significantly greater numbers than any other. This contrasts with temperate broad-leaved deciduous forests and boreal forests, in which typically two or three kinds dominate. In tropical rainforests, 40 to 100 different trees with a diameter of 10 cm (4 in) or greater can be found in a given **hectare** (2.5 acres), and individuals of the same kind may be far apart. Several plant families are well represented in all three major tropical rainforest regions; these include custard apples (Annonaceae), spurges (Euphorbaceae), figs (Moraceae), myrtles (Myrtaceae) and madders (Rubiaceae). These families are not common as trees in temperate forests. Of those families common to the forests of both the tropics and midlatitudes, many are represented primarily by forbs in the **temperate zone** and trees in the tropical forests. Examples include the rose family (Rosaceae), sunflower family (Compositae) and the **legume** family (Leguminoseae). Legumes are the dominant tree family in tropical America and Africa but have a lesser presence in Asian rainforests, where the dipterocarps (Dipterocarpaceae) are dominant. At lower taxonomic levels, distribution areas are more restricted: many genera and most species are confined to just one tropical rainforest region.

Rainforest plants, with some exceptions, depend upon animals for pollination of their flowers and dispersal of their seeds. Since much of the flowering and fruiting occurs high above the forest floor, many of the animals are found there too. Birds, bats, and insects are important pollinators, as are some small, nonflying, tree-dwelling mammals such as monkeys. Flowers attract them using color, scent, nec-

Photo 1.1 A kapok (*Ceiba pentandra)* beside a road cut through tropical rainforest near Tikal, Guatemala. The umbrella-shaped crown is typical of emergent layer trees. Oropendula nests hang from its branches. (Photo by SLW.)

tar, or oils as signals and rewards. Particular plants and animals have developed close relationships to assure that pollen is carried from one plant to another distant one of the same kind. A large proportion of trees and vines have seeds and fruits attractive to birds and mammals. The plants bribe the animals with carbohydrates or oils so they will collect and consume the fruit. The seed passes through the gut unharmed and is carried away from the parent plant for germination. Fruits of some type are available year round, so some animals are able to depend upon fruit and seeds almost exclusively. Parrots and monkeys harvest fruit in the treetops; rodents wait for it to drop to the forest floor. Where major rivers pass through rainforest, fishes also wait for seeds to drop and are important dispersers for some plants.

Wind dispersal is possible for plants that grow above the general canopy of the forest. The kapok or silk-cotton tree (*Ceiba pentandra*), native to both America and Africa and introduced in Asia, is one of the large trees whose crown emerges above the forest (Photo 1.1). Its seeds are carried on a parachute of fluff, much like thistledown. Orchids, common epiphytes in all tropical forests, produce huge numbers of tiny seeds that are blown away by the air currents in and above the highest levels of the canopy.

As is true for all biomes, the rainforest exists as a mosaic of plant and animal communities. Based on conditions of soil and drainage, three basic types of forest are recognized. Most widespread is that developed on well-drained areas above the annual floodwaters of rivers and known as *terra firme.* This is the typical or mixed rainforest with a great variety of woody plants and no single dominant kind. Other communities have developed in areas that are submerged by floodwaters for several months a year and are dominated by one or a few kinds of trees. These are especially important along the Amazon River and its great tributaries, where they are known as *várzea* or *igapó*, depending on soil nutrients and resulting vegetation (see descriptions following under American rainforests). A third major plant community is the **heath**-forest that develops on nutrient-poor sandy soils and is known from the Americas and Borneo. In addition to these major pieces in the mosaic, small ever-changing components occur when the canopy is broken by the death or windthrow of a tall tree. The gap created is quickly overgrown with vines and fast-growing, short-lived trees such as the cecropias (*Cecropia* spp.) of tropical America, umbrella trees (*Musanga cercropioides*) of Africa, and the macarangas (*Macaranga* spp.) of Asia until slower-growing saplings of upper story trees reclaim the canopy. Cecropias and macarangas have interesting partnerships with ants that protect the trees by eliminating plant-eating insects and vines that might overtop the tree and cast a deadly shade over it.

Humans have lived in the rainforests of the world as hunter-gatherers and as shifting agriculturalists for thousands of years. They, too, created gaps in the canopy as they cleared sites for villages and cropland. It has recently come to light that some areas supported large populations. Once these sites were abandoned, the forest grew back. Traces of these people in the contemporary vegetation may be as subtle as some trees they favored being more abundant than otherwise expected, but it is now well accepted that these forests were not pristine (that is, untouched by humans) until recently. Rather they were probably cut over repeatedly in the past, and the forest was able to recover each time. Nonetheless, the impacts of contemporary land uses such as timbering, mining, ranching, urban development, and road building are on a much greater scale than anything seen previously and are cause for concern. Forests have been cleared from much of the Asian region since the late nineteenth century, and significant clearing is occurring on the margins of the great Amazonian forest today. The rainforests are the storehouses of much of earth's biodiversity and appear to play an important role in global climate patterns. The vast assemblage of plants and animals is still poorly known in most regions. Research is only beginning to uncover the complex relationships between plants and animals and between the living and nonliving parts of the forest ecosystems and how they may affect us all.

## CLIMATE

The tropical wet **climate** (Koeppen's Af) associated with tropical rainforest is characterized by mean monthly temperatures equal to or greater than 18° C (64° F) every month of the year. The warmest and coolest months differ only slightly, usually less than 5° C (9° F). The difference between the highest and lowest temperatures each day is actually greater than the annual range. Frequently it amounts to 6–10° C (10–18° F). Although the angle of the sun is high, frequent cloudiness means that equatorial regions are not the hottest places on Earth. The highest temperatures recorded are rarely higher than 35° C (95° F), while the lowest are usually above 15° C (60° F). Only in southern Amazonia can really low temperatures occur. Invasions of cold air from the south, known as "friagems," though uncommon, can drop temperatures to near freezing.

Total annual precipitation is usually more than 2,000 mm (80 in). Amounts vary from 1,700 mm (65 in) to over 10,000 mm (400 in) in a few locations. This precipitation is not actually distributed evenly through the year, as distinctly wetter and drier periods occur. However, no month receives less than 100 mm (4 in). Dry seasons last fewer than four months and usually fewer than two months. Precipitation is generally associated with the passage of the equatorial low and **intertropical convergence zone (ITC)**, except in the Indo-Malaysian area, where the Asian monsoons determine rain patterns. Poleward of 10° latitude, a single peak of precipitation occurs, but most areas covered with tropical rainforest experience two periods of peak precipitation a year. Much of the rain comes in convectional showers and often at night when land breezes converge with the trade winds. Humidity remains high throughout the year. During the daytime, it generally averages 60–80 percent; at night it is more likely 95–100 percent (Richards 1996).

## VEGETATION

The struggle for light and adaptations to low light intensities beneath the canopy are reflected in the vertical structure and the occurrence of specialized growth-forms in tropical rainforests. The crowns of the broad-leaved evergreen trees form three layers or stories in the forest canopy usually designated as the A, B, and C layers. (These layers are not clearly separated from each other, but nonetheless are distinct enough to serve as descriptive devices for identifying different components of the forest.) The A layer is made up of the crowns of the tallest trees towering 30–60 m (100–120+ ft) above the forest floor, the so-called *emergents* because they emerge or break through the top of the closed forest canopy. The crowns of adjacent A-layer trees do not touch, so this is an **open canopy** through which air moves freely, and sunlight at some time during the day can reach most leaves. The crowns of A-layer trees have an umbrella shape. Below them is the B layer, a closed canopy of spherically crowned trees reaching heights of approximately 24 m (80 ft). This part of the canopy has much less wind, and sunlight reaching lower branches decreases significantly. In the well-lighted zone of the upper B layer, winds affect daily temperatures and humidity. The C layer is composed of trees about 18 m (60 ft) tall and with more conical crowns that interlace to form a closed cover. Very little air movement in this layer or beneath it occurs, relative humidity is constantly high, and temperatures vary little. It is also dark. Less than 3 percent of the light that reaches the top of the forest penetrates through to the C layer (Richards 1996).

Rainforest trees, although of many different kinds, all tend to look alike. The trunks are straight and slender with no branches until near the top. The bark is often very thin (a few mm thick) and smooth, although the trunk itself may be grooved. Leaves tend to be large, somewhat leathery and dark green, and nearly all lack distinguishing indentations or serrations but are elliptical and have what botanists call entire margins. The upper surfaces tend to be shiny. It is common for leaves, particularly among the trees of the B and C layers and on saplings beneath the canopy, to have drip tips, long points or extensions from the central vein of the leaf. Drip tips may be important in draining water off the leaf surface. A dry surface will allow essential transpiration to occur, thereby letting the plant draw water and nutrients up from the roots. It also makes a less inviting site for the growth of algae or mosses that could block all light from the leaf. On the other hand, drip tips,

Photo 1.2 Rainforest tree in Queensland, Australia, with buttressed trunk. Numerous lianas drape nearby. (Photo © Paul A. Souders/CORBIS.)

though characteristic, may not have any adaptive function but merely be the result of fast growth of the leaf tip during the early stages of development in a humid tropical environment (Richards 1996).

Although the forest as a whole is green all year, individual trees replace their leaves more or less on their own schedules. In many trees, the leaves are replaced either all at once over the entire tree or branch by branch. Typically, old leaves drop off after new ones have expanded, but on some trees the crown may be bare for a week or two. Bare branches and a flush of young leaves break the monotony of the dark canopy. The new leaves often are without chlorophyll and hang limp for a week or more after reaching full size. At this time, other pigments are dominate in the canopy. These leaves will slowly acquire chlorophyll and, when fully green, will stiffen.

Many trees have small, inconspicuous whitish or greenish flowers. It is not uncommon for these flowers to grow directly off the tree trunk, a pattern known as **cauliflory**. Flowering, like leaf replacement, is not synchronous among all the trees of a given area or even the same species in a given area, but each tree seems to have its own schedule. Some trees will be in bloom and some trees will be in fruit at any time during the year, but few trees bloom or are in fruit continuously.

Aerial roots and plank buttresses are distinguishing characteristics of many rainforest trees. Aerial roots take different forms; some come off the branches, but many grow from the base of the trunk and form prop roots or stilt roots. Buttresses are flat, woody growths extending from the trunk of the largest trees (Photo 1.2); as they begin growth, they are composed of tissue from roots near the surface, but later stem tissue also contributes to their formation. The purpose of prop roots and buttresses has been debated. They may provide crucial support to tall trees with shallow root systems. And buttresses may be important in channeling water and dissolved nutrients that run down the trunk (stemflow) to the roots below.

Two other layers lie below the trees. The D layer is composed of saplings of B- and C-layer trees rather than different plants altogether. These shade-tolerant young trees grow very slowly. The E layer consists of small plants less than a meter high. These may be seedlings in an arrested state of growth, perennial forbs, saprophytes and root parasites, or ferns. A large proportion of the forest floor has no plant cover at all.

Photo 1.3 Epiphytes, mostly bromeliads and orchids, on the branches of trees in the Mata Atlantica of eastern Brazil. Large-leaved climbers can be seen on the tree trunks on the foreground. (Photo by SLW.)

Nontree growth-forms of tropical rainforests impart a unique appearance to these forests. *Epiphytes* or airplants are the main herbaceous plants of the rainforest (although some are woody). These plants grow in abundance on the branches or trunks of the trees (Photo 1.3), using the tree as a platform for growth but receiving no nutrients from the host. A number of different plants have developed this way of life, including mosses, liverworts, ferns, orchids, bromeliads, and even some cacti. Orchids are most numerous and are found in all rainforests. Hundreds of genera and thousands of species are recognized. Orchids are more numerous in the Indo-Malayasian and American rainforests than in those of Africa. Bromeliads and cacti are New World plants (only one cactus is known from West Africa and Sri Lanka, *Rhipsalis baccifera*—an epiphyte). All told, more than 15,000 different epiphytes are known from the American rainforests, the richest section of the biome; the African forests host the least, an estimated 12,560 (Richards 1996). Other small plants in the canopy are semiparasites, which gain a foundation from the tree branches but also penetrate the host's tissues to withdraw moisture and nutrients. Mistletoes (Loranthaceae and Viscaceae families) are semiparasites and are well represented in Africa and Asia (Richards 1996; Whitmore 1990).

Epiphytes trap water and nutrients in their aerial roots and dead leaves and collect debris and humus high above the forest floor. Most are high in the canopy on A layer or upper parts of B-layer trees where they find abundant sunlight. Epiphytic plants can provide important habitat for a variety of animals including ants, frogs, and snakes.

The tropical rainforest also has a large number of climbers. Some have tendrils that are sensitive to contact with stems and branches of trees and curl around them for support. Others twine their whole stems around hosts, while yet others produce aerial roots that grasp a host plant. As they age, the early attachments may break or die to leave a long, hanging stem that loops like a rope from ground to canopy, where the top of the climber remains clasped to its host tree. Two basic types of climber can be distinguished. Shade-tolerant plants are mainly green-stemmed and attach by way of roots. Philodendrons, cheeseplant (*Monstera deliciosa*), and other members of the Araceae family are common examples. Light-demanding forms, at least as adults, need their foliage in the upper canopy. These are woody plants referred to as lianas. Many families have members that are lianas in the tropical forest. Somewhere between green-stemmed climbers and woody lianas are the rattan or climbing palms of the Indo-Malaysian forests.

Stranglers are a growth-form with no counterpart in temperate forests. As young plants, they are epiphytes high in the forest canopy. However, they send roots down to the soil and in time extract moisture and nutrients from the ground, not the air. The roots initially may follow along the trunk of the host tree but with age branch out and encase the trunk in a latticework of thick roots. If the host tree dies, as it well may when the foliage of the strangler shades its crown, the strangler is able to support itself on these long, strong roots and become a hollow giant in the forest. Most stranglers belong to the genera *Ficus*, *Clusia*, and *Shefflera*. The strangler figs (*Ficus* spp.) are particularly abundant.

Some higher plants have adapted to live in the deep shade of the forest floor. They do not photosynthesize but derive nutrients either by decomposing dead organic material (saprophytes) or by tapping into the roots of trees and lianas that have **photosynthesizing** leaves up in the sunlight (parasites). Saprophytes, since they have no green chlorophyll, are often white, yellow, or bright pink or violet. They are associated with, perhaps actually parasitic upon, abundant **myccorhizal fungi** that scavenge nutrients from the litter and soil. Some terrestrial orchids are saprophytes. True parasites, on the other hand, are rare and localized. They belong to one of two families, Balanophoraceae or Rafflesiaceae. A prime example is *Rafflesia arnoldi*, famous for producing the world's largest flower (Photo 1.4). Its red and orange cabbagelike flower is a meter (3 ft) in diameter and the only above-ground part of the plant. The rest is a mass of roots embedded in the roots of certain lianas of the family Vitaceae.

## SOILS

Many kinds of soils are found in the humid tropics; most would be grouped as oxisols or kaolisols in the U.S. soil taxonomy. They are characterized by advanced weathering and so have a high clay content and by excessive leaching that leaves few mineral nutrients. They tend to be particularly deficient in calcium and phosphorus. With little organic content to stain them brown, tropical soils tend to be bright red or yellowish in both A and B horizons. Color comes from oxides of iron, which are concentrated along with oxides of aluminum and manganese (Richards 1996). These soils are acidic and naturally infertile. As a result, plants tend to conserve and recycle nutrients to the greatest degree possible. At any given moment, most nutrients in the ecosystem are bound in above-ground plant material; stems,

Photo 1.4 A *Rafflesia* flower. The bloom produces a fetid odor to attract pollinators. A root parasite, it is able to grow on the dark forest floor, photographed here in Sabah, Malaysia. (Photo © John Holmes; Frank Lane Picture Agency/CORBIS.)

trunks, leaves, flowers, and fruits. They are lost from the forest plants through leaf fall and as **leachate** that seeps out of leaves and falls to the ground as stem flow or throughfall. Litter is rapidly mineralized by organisms of decay, primarily microbes, and readily taken up again by dense networks of roots close to the surface. Most rainforest trees have shallow lateral roots that extend beyond the periphery of the crown to enable them to capture nutrients from a wide area of the forest floor. (Vertical roots or "sinkers" descend from the lateral roots and may help provide the stability needed especially by tall trees whose crowns are buffeted by winds in the upper canopy.) The activity of roots is enhanced by relationships with myccorhizal fungi that are particularly efficient in collecting phosphorus. Stem flow may be intercepted by epiphytes or by the aerial roots of the trees themselves, or it may be channeled to the subsurface roots through grooves in tree trunks or by the buttresses. Nutrients are added to the system in the form of dust and aerosols blown in from distant areas, by nitrogen-fixing **rhizobial bacteria** in the root nodules of legumes and other plants, and by **cyanobacteria**. Nonetheless, nutrients are limiting factors in rainforests, and removal of the native forest can rapidly deplete what stores are available.

It should be noted that not all soils in the biome are as nutrient poor as the typical oxisols. Soils derived from alluvium carried down from geologically young mountains and soils derived from recent volcanic deposits can be quite fertile and support permanent agriculture and dense human populations. Humans themselves have created fertile soils in some locations during millennia of occupation and agricultural land use.

## ANIMAL LIFE

Tropical rainforests host untold numbers of animals, particularly insects. Just how many is unknown, and why there are so many is still one of the major questions of science. Many vertebrates are adapted to living in the treetops—they climb, jump, swing, glide, or fly through the canopy. Adaptations to a life spent climbing around in the trees includes modified toes, such as in tree frogs for clasping; claws for climbing and hanging, such as those of the three-toed sloth (*Bradypus variegatus*); prehensile tails—tails that can grasp a branch like an extra hand—like those of many New World monkeys; or long forelimbs for swinging from branch to branch, such as gibbons (*Hylobates* spp.) in the Indo-Malaysian forests possess. Vertebrates may feed on the abundant and varied insects. Several mammals have become specialized to feed on ants and termites; for example, lesser anteaters or tamanduas (*Tamandua tetradactyla*) and giant armadillos (*Priodontes maximus*) are found in the American tropics, while pangolins or scaly anteaters (*Manis* spp.) live in Africa and parts of Asia.

Because of the year-round availability of fruits and seeds, a large number of animals are able to eat fruit primarily or exclusively—an uncommon trait in other, more seasonal biomes. These animals are called frugivores and include birds such as parrots, some bats, monkeys and other mammals. Not all frugivores collect fruits in the canopy. Some depend on a constant rain of fruit and seed onto the forest floor, where other edible plant material is not readily available. In South America, the agouti (*Dasyprocta* spp.) and paca (*Agouti paca*) are two rodents that eat fallen fruit and are important dispersers of seeds. This role is filled by the chevrotain or mouse deer (*Tragulus* spp.) in Asian rainforests and the tiny royal antelope (*Neotragus pygmaeus*) in the lowland forests of West Africa. Each region also has wild pigs or pig-like mammals that forage for fruits and other food matter on the forest floor. In South America, peccaries, most commonly the collared peccary (*Tayassu tajuca*), fill this role; West Africa has true pigs such as the giant forest hog (*Hylochoeras meinertzhageni*) and bush pig (*Potamochoerus porcus*); and Southeast Asia has bearded pigs (*Sus barbata*) and baribusa (*Babyrousa babyroussa*). Diets are also similar among large ground-feeding birds such as the South American great tinamou (*Tinamus major*), the Asian ground hornbill (*Bucorvus leadbeateri*), and the New Guinean cassowary (*Casuarius* spp.) (Wolfe and Prance 1997).

Predators of the canopy include hawks, eagles, and owls. Weasellike mammals, such as the South American tayra (*Eira barbara*) and Borneo's binturong (*Arctictus binturong*), move through the branches in search of birds and small mammals. Wild cats such as the ocelot (*Felis pardalis*) of the Americas, the golden cat (*F. aurata*) of West Africa, and the clouded leopard (*Neofelis nebulosa*) of South East Asia and southern China also hunt in the trees (Wolfe and Prance 1997).

In the Amazon Basin and the Congo Basin, large rivers are important parts of the rainforest ecosystem, and adaptations to an aquatic environment are notable among various types of vertebrates. The South American capybara (*Hydrochoerus hydrochoerus*) and West African pigmy hippopotamus (*Choeropsis liberiensis*) are truly semiaquatic mammals, but many others, including wild cats, tapirs, and anteaters, are good swimmers. Uniquely and famously adapted to a flooded habitat is the basilisk lizard (*Basiliscus basiliscus*) of Central and South America. With powerful hind legs and webbed feet, this reptile is able to run across water on bubbles of air trapped momentarily between its toes. For this ability, it has been popularly referred to as the Jesus Christ lizard. Equally fascinating is the turkey-sized hoatzin (*Opisthocomus hoazin*), a strange bird that nests on branches overhanging flooded portions of the Amazon forest. When startled by potential predators, the nestlings jump into the water and then swim back to shore, where they can climb into trees and shrubs using the claws that appear on the wings of juveniles (Sick 1993).

Communication among members of the same species in the dense forest canopy is challenging. A loud voice is one way to overcome this (Kingdon 1989). Parrots, peacocks, and guinea fowl are good examples of loud rainforest birds; the howler monkeys (*Alouatta* spp.) of Central and South America and gibbons (*Hylobates* spp.) of Southeast Asia are mammals that contribute to the din of the rainforest. Other means of communication are bright color and distinctive color patterns (Kingdon 1989). Color and pattern may aid recognition of members of the same species; it can also serve as a warning to potential predators that the prey is poisonous. Some animals fake it and look like some other poisonous animal but are not poisonous themselves. This resemblance is known as Batesian mimicry, a phenomenon first described for butterflies in the Amazon forest. Bright reds, blues, and yellows, in particular, are found among insects, frogs, snakes, and birds. Complex mating rituals are other ways to signal who you are—and that you are ready to breed—in the gloom of the forests. In South America, small birds called manakins have elaborate dances on parts of the forest floor they have cleared of debris just for the purpose. Sunflecks spotlight their plumage. In New Guinea, birds-of-paradise show off their magnificent plumes in acrobatic displays. And on Borneo and Sumatra, as well as in Australia, bowerbirds adorn stick-built structures with flower petals, shiny stones and even colorful, glittering beetle carapaces.

## MAJOR REGIONAL EXPRESSIONS

### American Tropical Rainforests

Three separate rainforests or *selvas* exist in tropical America. The largest of all rainforests occurs in Amazonia and the upper basin of the Orinoco River. This great Amazonian forest extends from the eastern slopes of the Andes to the Guianas and as far south as 15° S in western Brazil and northern Bolivia. Isolated from the Amazon forest are two narrow coastal rainforests. To the south and east is the Atlantic Rainforest, which once covered parts of coastal Brazil from 7° S near Recife to 28° S in the Brazilian state of Rio Grande do Sul. West of the Andes, from just south of the equator in Ecuador northward through Colombia into Panama, is the Chocó, one of the world's wettest rainforests. Rainforest continues northward through Central America reaching its northern limit at 17° N in the Mexican state of Veracruz. At one time it covered many of the islands of the Caribbean as well. Each of the three is a storehouse of rare and unique species and great genetic diversity.

The *terra firme* forest of the Amazon forest is classic mixed rainforest with the tree canopy composed of three intermingling layers and a great diversity of trees, no single one gaining numerical dominance. Among the emergent trees of the A layer are kapok tree (*Ceiba pentandra*), a legume (*Dinizia excelsa*), and Brazilnut (*Bertholletia excelsa*). Strangler figs and woody lianas of many kinds are characteristic. The epiphytes are at least visually dominated by bromeliads. Of particular ecological interest are the tank bromeliads in which the inner leaves of the **rosette** form a cup and hold rainwater. They may hold a gallon or more of water and act as tree-top aquariums—complete with algae and bladderworts, insect larvae, and often frogs and tadpoles. Brightly colored poison arrow frogs (*Dendrobates* spp.) raise their young in these pools. The tadpoles may actually hatch in ground-level pools and hitch a ride to the tree tops one or two at a time on the backs of their parents. Each bromeliad supports one tadpole which feeds on mosquito larvae or, in the case of some frogs in Peru, infertile frog eggs that are laid in the bromeliad by the mother simply to serve as food for her single offspring (Wolfe and Prance 1997).

South America is known as the bird continent, and a large number of the birds live in the Amazon rainforest. The long, hanging nests of oropendula (*Psarocolius decumans*) drape from the branches of kapok trees. Brightly colored parrots, parakeets, and macaws fly in noisy groups above the canopy in search of ripe fruit. Toucans with their bizarre oversized bills also hunt for fruits. Lower levels in the canopy are home to woodpeckers, treecreepers, trogons, jacamars, and many other birds searching for insects; and still lower, the D layer provides habitat for tanagers, flycatchers, antbirds, and manakins. On the ground itself are tinamous, ant pittas, wrens, and ground doves. Those in the canopy tend to be brightly covered but nearly invisible to the viewer on the ground because of the crowns of the C- and B-layer trees. Those that live on or near the ground are often cryptically patterned with stripes, bars, and spots on a brown background, which also makes them difficult to see.

The mammals of the rainforest also tend to be most numerous in the canopy where more food is available to them. They include ant- and termite-eating tamanduas and silky anteaters (*Cyclopes didactylus*); fruit- and insect-eating kinkajous, and many different primates including howler monkeys (*Alouatta* spp.), long-limbed spider monkeys (*Ateles* spp.), mischievous capuchins (*Cebus* spp.), and tiny, squirrel-like marmosets and tamarins. The smallest primate is the pygmy marmoset (*Cebuella pygmaea*), which weighs a mere 85 g (3 oz). A number of marsupials in South America are arboreal, including the woolly opossum (*Caluromys* spp.) and mouse opossums (*Marmosa* spp.). Gaps in the canopy created by the fall of a large tree are colonized by fast-growing trees such as cecropia; jacaranda (*Jacaranda copaia*); and balsa (*Ochroma lagopus*), the light, low density of wood used for building model airplanes and such. The three-toed sloth (*Bradypus variegatus*), though it feeds elsewhere, is most visible in the gaps, where it feeds on the leaves and fruits of cecropias. The forest floor is the domain of tapirs, peccaries [both collared peccary (*Tayassu tajuca*) and the larger, more aggressive white-lipped peccary (*T. pecari*)], agouti and pacas, as well as the jaguar (Wolfe and Prance 1997). It also is home to a large number of frogs—more than 400 in the **genus** *Eleuthreodactylus* alone, which live in the litter and are camouflaged to look like dead leaves (Wolfe and Prance 1997).

On the floodplains of Amazonian whitewater streams (see Chapter 10), the rain forest is a type known as *várzea*. Whitewater streams are those such as the Napo and Madeira, which have a heavy sediment load that makes them look milky or muddy. The sediments derive from the Andes, a geologically young mountain range in which rock materials still have not been leached of their mineral nutrients.

The annual floods of the Amazon and its tributaries deposit nutrient-rich sediment on the floodplains and create a relatively fertile soil. The area on which várzea is found is only under water during the peak of the flood season, when fish swim among the tree trunks. Fish are important consumers of fruit and dispersers of seed (Goulding 1980). Fewer species of trees grow in várzea than in terra firme forest, and the general canopy height is 15–30 m (50–100 ft). However, huge kapok trees emerge above the canopy. Other trees associated with várzea include wild rubber trees (*Hevea brasiliensis*, *H. guyanensis*) and many legumes. Palms are frequent. When várzea is flooded, plants become dormant, and some trees drop their leaves. The so-called **gallery forests** that line permanent streams in the tropical savannas of South America are composed of many of the same trees as found in várzea (Richards 1996).

The other kind of forest found in seasonally flooded areas is *igapó*. This vegetation type occurs in backwater swamps and on the floodplains of blackwater and clearwater streams, both nutrient-poor sites. Blackwater streams flow from the heavily leached, acidic sands of the Guyana Shield north of the Amazon river that is part of the ancient foundation of the South American continent. The blackwater streams carry little by way of sediments. The dark color of the water derives from the plants growing on the Shield that tend to have high amounts of tannic acid as a way of defending their leaves against predators. As the leaves are shed, the dark tannins dissolve in the groundwater and drain into the rivers to turn them the color of strong tea. The Rio Negro, a major tributary of the Amazon, is a prime example of a blackwater stream. Clearwater streams, on the other hand, flow from the Brazilian Shield south of the Amazon, another highland built on ancient, heavily leached bedrock. These streams, such as the Rio Tápajós, are crystal clear with neither sediments nor tannins. The waters flooding igapó are thus low in nutrients and acidic (**pH** = 4.0–4.5) (Wolfe and Prance 1997). During the highwater period at Manaus, Brazil, (April to August) the Rio Negro may be 12 m (40 ft) above November's low water level; elsewhere flood levels may be as much as 20 m (65 ft) higher. The plants that grow in these conditions must be able to tolerate being submerged for 40–60 percent of the year. This limits the variety of plants growing in igapó. The forest consists of often gnarled trees 25–28 m (80–90 ft) tall, shrubs, a few lianas, and no herbs other than epiphytes. Trees are evergreen and have leathery leaves. Though they maintain their foliage throughout the year, they are dormant during flood season and, unlike most tropical plants, produce annual growth rings. One tends to find but one dominant tree and one dominant shrub at any single location. In areas underwater the longest time (approximately 270 days), the dominant tree is *Eugenia inundata* and the dominant shrub is *Symmeria paniculata*. At slightly higher elevations, where floodwaters cover the land for shorter periods of time, more trees, including cecropias and rubber trees, are found. The dominant tree, however, is *Aldina latifolia*, and the dominant shrub is camu camu (*Myrciaria dubia*). Camu camu has fruits that are high in vitamin C and that ripen just as the water level rises. Many fish such as catfishes, the tambaqui (*Collossoma macropomum*), and black piranha (*Serrasalmus rhombeus*) eat fruits from plants such as jauari palms (*Astrocaryum jauary*), camu camu, and rubber trees and disperse their seeds. Another fish, the arauna (*Osteoglossum bicirrhosum*), will leap a meter (3 ft) or more into the air to snatch insects from the leaves overhanging the water. Igapó is also home to freshwater sponges that grow on submerged tree branches. They produce toxins that can cause allergic reactions in swimmers. In the dry season, when water levels have fallen and the sponges are exposed to air, they become dormant (Goulding 1980; Richards 1996; Wolfe and Prance 1997).

In the Guianas and Amazonia north of the Amazon, another distinct forest has developed on the coarse-grained, nutrient-poor white sands of the Guiana Shield that belongs to the general category of tropical heath forests but known in Brazil sometimes by the indigenous name "caatinga-gapo" (white forest—not to be confused with the thornscrub vegetation, which is simply caatinga) or by the more botanical term "campinarana." Soils in the campinarana are nearly always wet, even in the dry season, due to high water tables. The vegetation consists of shrubs, small trees, and grasses with a canopy typically 20–30 m (65–100 ft) above the ground. Woody plants of the legume family are dominant; most widespread is soft wallaba (*Eperua falcata*), sometimes in association with the multistemmed clump wallabas (*Dicymbe corymbosa* and *D. altsoni*) (Richards 1996).

## African Rainforests

Botanists recognize two rainforests in Africa, although on a map they appear to form one continuous rainforest area that covers about 10 percent of the continent. The West African rainforest occurs in the coastal lowlands at elevations below 300 m (1,000 ft) along the Gulf of Guinea and extends from about 10° N at Guinea Bissau to the Nigeria/Cameroon border. The two sections are divided by a 300 km (185 mi) stretch of savanna known as the Dahomey Gap. The Central African rainforest, about 60 percent of the total rainforest in Africa, occurs in the basin of the Congo River from Cameroon inland to Uganda at about 30° E longitude. The West Africa forest has been more accessible to western foresters and other scientists; and, although only about 28 percent of it survives, it tends to be better known than the forest of the Congo Basin. Four general types of rainforest (defined by African researchers as all tropical moist forests with broad-leaved trees and closed canopies) are recognized in Africa and reflect differences in total precipitation. The Wet Evergreen Rainforest occurs in areas receiving more than 1,750 mm (70 in) of rain a year. It contains few important timber trees, and its canopy is generally lower than 40 m (130 ft). One of the ebonies (*Diospyros sanza-minika*) can be distinguished from other trees by its hard black bark with many lengthwise cracks. Locally it is known as "elephant comb" because elephants like to rub against its rough trunk (Martin 1991).

Moist Evergreen Rainforest is found where annual precipitation is between 1,500–1,750 mm (60–70 in) and has more commercially important trees, including African mahogany (*Khaya ivorensis*), sapele (*Entandrophragma cylindricum*), and makore (*Tieghemella heckelli*); its canopy can be 43 m (140 ft) above the ground. Some of the trees shed their leaves for a short time (Martin 1991).

Moist Semi-deciduous Rainforest has the tallest trees and is the most widespread type in West Africa. Average annual rainfall is 1,250–1,750 mm (50–70 in) with the lesser amounts generally found in the northwestern parts of the forest. The upper canopy has both evergreen and deciduous trees in roughly equal proportion, while the lower canopy is formed almost exclusively by evergreen ones. The tallest tress may reach 50–60 m (160–200 ft) in height. Many of the trees are hardwoods harvested commercially for furniture making; especially favored are those with reddish wood such as African mahogany, sapele, makore, and utile (*Entandrophragma utile*) (Martin 1991).

Dry Semi-deciduous Rainforest occurs at the edge of the rainforest. Annual rainfall in these forests is usually 1,250–1,500 mm (50–60 in) a year, and fires occur after long dry spells. The canopy is 30–45 m (100–150 ft) high and contains few commercially important trees. These dry margins of the rainforest are transitions to tropical savanna (Martin 1991).

The trees of the African rainforests have the typical appearance and bear many of the same adaptations as trees in other tropical rainforests. A few can be distinguished by the shapes of trunks or buttresses. For example, the medium-sized tree *Balanaites wilsoniana* has a fluted trunk; atui or dubema (*Piptadeniastium africanum*) has large above-ground roots that run crookedly out from the base of the trunk. It has been estimated that one-quarter of the **vascular plants** in the West African rainforest are lianas, and some are among the largest in the world. The strychnine vine (*Strychnos* spp.) can have base diameter of 30 cm (12 in); the sea bean (*Entada pursaetha*) has a stem 15 cm (6 in) in diameter. Strangler figs are also found. The tangle of vines provide living spaces for squirrels and monkeys (Martin 1991).

It is often said that African rainforests have fewer species than American or Asian, but this claim may be the product of a lack of good information. Some areas have much higher diversity than others, and the region as a whole is poorly known. New plants and animals are continually being discovered. [For example, in 1988 a new monkey, a guenon, was seen for the first time by scientists in Gabon and named *Cercopithecus solatus*. The species name refers to the sun-bleached appearance of its hair (Martin 1991).] It appears that while small areas have high numbers of **endemic** species, many plants and animals are widely distributed throughout the rainforest region. Africa appears to have fewer kinds of some plants that are abundantly represented in other rainforest regions, plants such as orchids and palms (Richards 1996). Some of the African palms, however, have major economic importance. The oil palm (*Elaeis guineensis*), for example, now grown in plantations throughout the tropics, is native to West and Central African rainforests. Raffia palm (*Raphia hookeri*), a swamp plant, has many uses. Fiber from its leaves can be woven into hats, mats, and textiles (raffia), the ribs of the leaves make poles similar to bamboo, and the sap can be made into palm wine (Martin 1991).

The animal life of the African rainforests consists of many small mammals that are seldom seen. More hoofed animals are found there than in other rainforests, and this results in a smaller percent (approximately 30 percent) of mammals confined to the canopy. Among the canopy dwellers, however, are mice (*Thamnomys* spp.) that are preyed upon by the mongoose-like African linsang (*Poiana richardsoni*), scaly anteaters or pangolins (tree pangolin and long-tailed pangolin) with prehensile tails, four kinds of flying squirrels, and the tree hyrax (*Dendrohyrax arboreus*). The gregarious monkeys appear to have divided the canopy into layered living spaces so that different kinds are found at different heights above the forest floor. The red colobus (*Colobus badius*), for example, occupies the highest parts, where it consumes the leaves and fruits of trees and lianas; the black and white colobus (*C. polykomos*) is found at midlevels in the canopy, where it feeds on the leaves of some 120 different plants; diana monkeys (*Cercopthicus diana*) occupy the lower canopy and eat leaves, fruits, oil-rich seeds, and arthropods. Chimpanzees (*Pan troglodytes*) nest in the trees and feed on the ground (Martin 1991).

Some of the larger mammals are confined to the forest floor. These include the forest elephant (*Loxodonta* sp.), a form smaller than the elephant of the tropical savannas and recently recognized as a separate species. Average shoulder height of the forest elephant is 240 cm (about 8 ft), and it has smaller, rounder ears than its savanna relatives, a flatter forehead, and longer and more slender tusks. The forest elephants are usually solitary, but they can also be found in small groups of two to four individuals. They eat the leaves of a variety of plants, especially lianas, and the bark of trees. They also seek out certain fruits, such as those of the makore tree, the strychnine vine, and guinea plum (*Parinari excelsa*), and well-worn trails lead to productive trees. The seeds of these plants pass through the gut unharmed, and though rodents may eat many from the dung, elephants seem to be the major dispersing agent for several plants with large fruits and seeds. Many plants bear their fruit in the dry season (December to March); so in the rainy season food in the forest may be scarce, and it is then that elephants raid farmers' fields (Martin 1991).

Other large mammals include the endemic pigmy hippo (*Choeropsis liberiensis*), now an endangered species, and the forest buffalo (*Syncerus caffer nanus*). Like the forest elephant, the forest buffalo is smaller than its relatives on the open savannas. It has smaller horns and a reddish coat. The forest buffalo requires grass, and thus dwells on the forest edge. A number of other hoofed mammals of various sizes are found, including bongo (*Boocerus euryceros*), bushbuck (*Tragelaphus scriptus*) and the tiny royal antelope (*Neotragus pygmaeus*). The several duikers are especially important in terms of seed dispersal (Martin 1991).

Africa's rainforests have fewer birds than the American ones. Hornbills are diverse and among the easier to see. The eight or nine kinds have differing bill sizes, although all seem to feed on whatever fruit is available. They are most attracted to small red, purple, and black fruits such as produced by the cardboard tree (*Pycnanthus angolensis*) and leave yellow, orange, or green ones for bats (Martin 1991).

Over a hundred kinds of snake inhabit the African rainforests. They include poisonous ones such as the Gabon viper (*Bitis gabonica*), rhinoceros viper (*B. nasicornis*), green mamba (*Dendroaspis viridis*), and several cobras. The world's largest snake, the African or rock python (*Python sebae*), also resides here. [Lengths up to 9.96 m (33 ft) have been reported.] So do some of the world's smallest snakes, the worm snakes (Leptotyphlopidae). The tiniest gets only 15 cm (6 in) long. Another interesting reptile is the hinged tortoise (*Kinixys erosa*). Its shell is 30 cm (23 in) or less long and hinged at the rear to allow room for its relatively long legs. This tortoise can move with considerable speed (Martin 1991).

Tree frogs (family Rhacophoridae) with adhesive disks on their toes are the counterparts of the poison arrow frogs of the American Tropical Rainforest. Like poison arrow frogs, they are brightly colored. They lay their eggs in nests of foam hung between the branches of trees or in tall grass that grows above water (Martin 1991).

As in other tropical forests, plants and animals, especially ants, have formed some close and fascinating relationships. The small tree known as oko (*Barteria fistulosa*) provides housing for ants in hollow, slightly swollen branches. Inside, black ants (*Pachysima aethiops*) raise clearwing bugs (Homoptera), sap-sucking insects that produce "honeydew" from anal glands. The ants eat the honeydew. The ants benefit the tree because they fiercely defend it by stinging all intruders, be they plant-eating insect or unwary scientist. They may even rain down upon a passerby (Janzen 1972). A more bizarre relationship is that between a fungus and an ant. The fungus (*Cordyceps myrmecophila*) uses a primitive ant (*Paltothyreus tarsatus*) as both host and transportation. The fungal spores become embedded in the ant and grow. When the fungus reaches the ant's brain, it "reprograms" the ant to cease being a ground-dweller and climb a tree instead. When the ant reaches the tip of a twig, it bites into it and becomes permanently fastened. The fungus continues to feed on the ant's body and kills it. When the fungus is ready, it sends out a stalk some 2 cm (3/4 in) long with a fruiting body at the end. From this structure, spores are released to float down from the perch in the tree to the ground, where new ants will be infected and the cycle repeated (Martin 1991).

## Indo-Malaysian Rainforests

The rainforests of Indo-Malaysia are highly fragmented since they occur on a number of peninsulas and islands. At the westernmost limits, rainforest occurs along the coast of southwestern India and on Sri Lanka. Rainforest is also found in Thailand and southern China, through the Malaysian Peninsula and the Indonesian islands to the Philippines, New Guinea, and the Bismarck Archipelago just east of Papua New Guinea. Many of the islands of Melanesia, Micronesia, and Polynesia also have rainforest, as do islands in the Indian Ocean including Madagascar, Mauritius, Reunion, Rodriguez, and the Seychelles. An outlier of the forest can be found in the wettest parts of northeastern Australia. The core of the region—from the Kra Isthmus of Thailand and the Philippines in the north south to New Guinea and east to the Bismarck Archipelago—contains a distinct group of plants and is known by plant geographers as *Malesia.* The Indo-Malaysian forests are split into two sections by deep water between the Sunda Shelf and the Sahul Shelf, areas of relatively shallow seas. The Sunda Shelf—on which the Malay Peninsula, Sumatra and Borneo sit—was part of the northern supercontinent Laurasia before continental drift broke it apart. More recently, during the Pleistocene Epoch, the islands of the Sunda Shelf were attached by land to the Asian continent during periods of lowered sea level. The lands on the Sahul continental shelf, including Australia and New Guinea, were part of ancient Gondwana and have never been directly attached to the Eurasian continent. In the deep waters between the two continental shelves are islands such as the Philippines, Sulawesi, and the Lesser Sunda Islands that have been isolated from both continental areas and have received only those plants and animals that were able to overcome the water gap. Animal life shows a particularly sharp divide between the Indonesian islands of Bali and Lombok, with animals in Bali and islands to the west being closely related to Asian forms and those on Lombok and islands lying to its east related to Australian forms. This is particularly true for freshwater fishes and mammals; birds and reptiles have more overlap. The British naturalist Alfred Russel Wallace first described the differences in animal life, and an imaginary line between Bali and Lombok is still referred to as Wallace's Line. Although a number of genera—such as 10 genera of rattan palms—are limited to lands on the Sunda Shelf, plants on the islands are generally more mixed in origins than are animals, and a transition zone rather than a sharp line is recognized by botanists working in Malesia. They call this transition zone Wallacea (Whitmore 1984).

The plants of the Indo-Malaysian forests are fairly well known due to their exploitation by European colonial powers for tropical hardwoods and other export commodities. The largest tracts of rainforest still intact are on the Malay Peninsula, Sumatra, and Borneo (Whitmore 1990; Richards 1996).

The plants of the Malesian rainforest are distinct from those of other rainforests and also from adjacent dry forests. Forty percent of the plant genera are found nowhere else. The largest family is the orchid family, which is represented by somewhere between 3,000 and 4,000 different species. Among trees, dipterocarps dominate the upper canopy of forests on the Sunda Shelf. This family is the most diverse (350 species) and abundant family of large trees in all of the world's tropical rainforests. It also contains the tallest trees of the world's rainforests. Other families abundantly represented include myrtles (Myrtaceae: the genus *Eugenia* has more than 500 species), spurges (Euphorbiaceae: the genus *Ficus* has more than 500 species), and heaths (Ericaeae: the genus *Rhododendron* has 287 species and *Vaccinium* 239). Typically a large number of species occupy the same small area, often exhibiting a local diversity greater than found in most of the Amazon forest (Whitmore 1984).

The most widespread type of rain forest in the Indo-Malaysian region is a lowland evergreen forest characterized by a large number of different kinds of trees growing together, none of them dominant. No tree of the upper canopy represents even 1 percent of the total number of trees present. Two trees of the same kind are usually widely separated. The emergents, frequently dipterocarps and legumes, stand 45 m (150 ft) or higher. Buttresses are common, as is cauliflory. The B-layer trees forming the main canopy are 24–36 m (80–120 ft) tall. Below them are smaller shade-loving trees. The ground cover is sparse and composed mainly of small trees; herbaceous plants growing on the ground are rare. The biggest trees may have clear trunks reaching 30 m (100 ft) into the canopy and diameters of 4.5 m (15 ft). Some are deciduous or semideciduous. Large dangling lianas are abundant, and sometimes climbers also are found on the tree trunks. Epiphytes are rare to frequent. Palms may constitute the undergrowth (Whitmore 1984).

A narrow belt of tropical semi-evergreen rainforest exists between the tropical evergreen forests and tropical seasonal forests (see Chapter 2), which experiences a regular dry period of a few weeks each year. This is the forest type found in the Australian outlier, but it is relatively rare elsewhere. In this forest, about one-third of the emergent trees are deciduous, although they are not leafless at the same time. The other tall trees are evergreen. Unlike the trees of the emergent layers of rainforests in America and Africa that occur as isolated individuals, the dipterocarps in this forest occur in small groups of the same kind, the crowns of adjacent individuals touching. Buttresses are common on both evergreen and deciduous trees, and the canopy is slightly lower than in the evergreen rain forests. Tree bark tends to be thicker and rougher than on trees of the tropical evergreen rain forest; and, probably related to this fact, cauliflory is rarer. The number of trees is somewhat less than in the tropical evergreen rain forests of Indo-Malaysia. Large woody lianas are very abundant, and bamboos are present. Epiphytes are moderately common; many are orchids or ferns. This forest type is prone to fire, and, when burned frequently, it changes to an open grassland (Whitmore 1984).

In the tropical rainforests of Southeast Asia, heath forests have developed on sandy soils comparable to the campinarana of the Amazon. The soils are shallow, acidic, and nutrient-poor, and woody vegetation is low, from less than a meter (3 ft) to about 12 m (40 ft) tall. The vegetation is known as *keranga* ("place where rice won't grow") and exhibits adaptations to drought. Leaves on the shrubs and small, slender trees are shiny to reflect sunlight and reduce heat. They are also relatively small, thick, and leathery, and are held obliquely so as to intercept less of the sun's light. Myrtles with Australian affinities (e.g., the genera *Eugenia* and *Tristana*), are abundant. Many herbs, including sedges (Cyperaceae) and an orchid (*Bromheadia finlaysoniana*), are found, but what may be most interesting is the large number of insectivorous plants. In areas where nitrogen in particular is deficient, one adaptation found among plants—whether in muskeg in the boreal forest or in kerangas in the tropical rainforest—is the ability to extract nitrogen from insects captured by the plant. Bladderworts (*Utricularia* spp.), a sundew (*Drosera spathulata*), and five pitcher plants (*Nepenthes* spp.) do this in the kerangas of Borneo. The pitchers of the *Nepenthes* are microecosystems comparable to those of the tank bromeliads of the American tropics. More than 30 insects and 3 spiders are known to live and breed only in these plants. Other plants have different strategies for acquiring necessary nutrients. A number are associated with myccorhizal fungi. *Casuarina nobilis* has nitrogen-fixing bacteria in root nodules. And a number of plants provide habitation for ants; presumably the droppings and debris from ant harvests contribute nutrients to the host plant. Epiphytic plants from two genera in the family Rubiaceae, *Hydnophytum* and *Myrmecodia*, have swollen stems honeycombed with cavities and inhabited by ants. Kerangas degrade into padangs (a Malay word referring to open places), open savannas of shrubs, small trees, and a sparse ground cover of grasses and sedges (Whitmore 1984; Richards 1996).

The Indo-Malaysian rain forest also has a unique peat swamp forest, known from western Borneo and northern Sumatra. Dome-shaped accumulations of peat are covered by tropical evergreen trees as much as 60 m (200 ft) tall. One of the more common trees is alan (*Shorea albida*), a dipterocarp. The underlying peat deposit is more or less circular; its raised center is flat. The peat is at least 0.5 m (1.5 ft) thick and may be 15 m (50 ft) or more deep in the center. The peat consists of slowly decaying wood, roots, and leaves in a reddish-brown soupy mix. The tall trees that dominate the site produce such a thick mat of roots that it is possible to walk on the surface. The bases of the trees are surrounded with looped or peglike root structures or "knees" called pneumatophores, that allow them to obtain oxygen in a waterlogged situation. The pneumatophores may be 30 cm (1 ft) tall. Undergrowth is composed of sago palm (*Metroxylon sagu*), screw pines (*Pandanus* spp.), pitcher plants, sedges, and young trees (Whitmore 1984; Richards 1996).

Mammal distribution in the Indo-Malaysian rainforests clearly demonstrates the effect of Wallace's Line. Mammals with relatives in Asia and Africa inhabit the western region. Primates are well represented in the canopy by gibbons (*Hylobates* spp.), slow loris (*Nycticebus coucang*), orangutan (*Pongo pygmaeus*), and especially langurs (*Presbytis* spp). The prehensile-tailed pangolin or scaly anteater (*Manis javaneica*) is another mammal of the canopy. Predators include mongooses and their relatives such as palm civets (several genera), Oriental linsang (*Prionodon linsang*), and binturong (*Arctitis binturong*). The clouded leopard (*Neofelis nebulosa*) hunts in the trees and on the ground. Other ground dwellers include bearded pig (*Sus barbata*) and, on Sulawesi and some small neighboring islands, the babirusa (*Babyrousa babyrousa*). East of Wallace's Line, mammals are more apt to be related to groups inhabiting Australia. Those previously listed—all placental mammals—are absent. Instead, New Guinea, for example, has marsupials: cuscuses (*Phalanger* spp.), pygmy possums such as the feather-tailed possum (*Distoechurus pennatus*), gliding possums (*Petaurus* spp.), ring-tailed possums (*Pseudocheirus* spp.), and spiny bandicoots (*Echymipera* spp.). Cassowaries (*Casuarius* spp.), large flightless birds, may take the place of some ground-dwelling mammals (Wolfe and Prance 1997).

## Tropical Rainforests at Higher Elevations

Where mountains rise above humid tropical lowlands, a distinct zonation of vegetation is associated with the climatic changes that occur with increasing elevation, and one encounters montane rainforests. Generally speaking, these higher elevation forests have different plants than those found in the lowland rainforests described above, the trees are not as tall, the structure of the vegetation is simpler, and fewer plant species overall are found. Some of the plants have subtropical and temperate origins; this becomes increasingly so as elevation increases. In most cases, three vegetation zones are recognized: a lower montane rainforest, an upper montane forest, and an alpine zone. The lower montane forest in appearance is similar to the adjacent lowland. The upper montane zone corresponds with the elevation at which fog frequently forms, so plants that draw moisture from the air can be plentiful. These forests are also referred to as cloud forests and are characterized by an exuberance of epiphytes. At its upper limits, this zone consists of dwarfed, crooked trees covered with mosses and lichens. The tropical alpine zone occurs above treeline and is a vegetation type with no exact counterpart outside the tropics. (The alpine zone in wetter parts of the Andes, the paramo, is described with the tundra biome, as is the Afro-alpine zone of East Africa. The drier parts of the Andes, at high elevations, are grasslands called puna and are described in the temperate grassland biome.) At no set elevations do different zones of mountain vegetation occur, and considerable variation exists. To some extent, the elevations depend on the size of the mountain mass. Zones are lower on small isolated mountains than on large mountain chains. They are also lower on coastal mountains than on interior ones (Richards 1996).

In Tropical South America, the western edge of the Amazon forest abuts the Andes. On the eastern flanks of the mountains, three forest zones are commonly recognized. The lowest, the *montaña*, is very similar in appearance to the Amazon forest but different in composition. Montaña gives way to *yungas* between 1,200 and 1,500 m (4,000–5,000 ft). This zone corresponds to the lower montane rainforest. The third zone is known as *ceja de la montaña* (brow of the mountain) or cloud forest and represents the upper montane zone. It occurs from 1,800–2,400 m (6,000–8,000 ft) upwards to 3,400–3,900 m (11,000–13,000 ft), the highest elevations being reached in Peru. The trees of the cloud forest are laden with bromeliads and orchids. The uppermost parts of the zone in the Andes were elfin forests dominated by small *Polylepis* trees (Photo 1.5). (Much of the *Polylepis* forest was converted to grassland or paramo even before the arrival of Europeans, so today remnant forest can be found at elevations above the treeless vegetation of the alpine zone.) Above the *ceja de la montaña*, beginning around 4,500 m (14,750 ft), is the alpine zone—paramo in moister regions and puna in drier areas (Richards 1996).

In Africa, it is difficult to find a complete transition from lowland rain forest to montane rainforest and an alpine zone because many of the highest mountains rise above savannas and because much of the lower forests has been destroyed. However, in the savanna regions of East Africa, at elevations of 1,750–2,500 m (5,700–8,000 ft) there is often an evergreen forest with a variety of species. The leaves of the trees are small. Epiphytes, including the lichen *Usnea*, are conspicuous. Many plants are of subtropical and temperate affinity. Next is a narrow zone

Photo 1.5 *Polylepis* forest near the treeline on eastern range of the Andes, near Papallacta, Ecuador. (Photo by SLW.)

dominated by bamboo (*Arundinaria alpina*). At 2,600 m (8,500 ft), an ericaeous zone has tree-sized heathers (*Erica* spp.) or dwarf *Hagenia abysinnica* trees draped with lichens and mosses. This continues to treeline and the Afro-alpine zone at 3,700–3,800 m (12,000–12,500 ft). Permanent snow is encountered around 4,500 m (14,750 ft) (Richards 1996).

In the Malesian mountains, most plants of the lower montane forests are closely related to plants of Indo-Malaysia or the Old World Tropics as a whole. However, many plants of the upper montane rain forest are related to Australian forms. In New Guinea, the lower montane forest looks like lowland forest but is shorter in stature, emergent dipterocarps are scarce, and buttressing and cauliflory are absent. Fewer kinds of trees are found there than in the lowland forest, and the presence of the southern beeches (*Nothofagus*) is pronounced. Tree leaves tend to be medium to small in size. Lianas are essentially nonexistent, but epiphytes are more abundant than in the lowland forest. What is unique to these forests are huge **conifers** found only on the southern continents. Towering above the general canopy to heights greater than 60 m (200 ft) and sometimes even more than 80 m (260 ft) are hoop pine (*Araucaria cunninghamii*) and klinkii pine (*A. hunsteinii*). The upper montane rain forest has little in common with the lowland forests. It is characterized by broad-crowned trees less than 20 m (65 ft) in height. Leaves are small, thick, and leathery. The trees are primarily from two families, the heaths (Ericaceae) and myrtles (Mrytceae), with some southern conifers. Branches and trunks are covered with mosses, lichens and other epiphytes (Richards 1996).

## ORIGINS

A tropical rainforest existed by 70 million years ago at the beginning of the Tertiary Period, but its location and extent has been constantly changing in response to climate change, continental drift, sea level changes, and episodes of mountain building. Evidence for tropical forest has been found in high latitudes, for example in Alaska, early in the Tertiary Period, when the Earth's climate as a whole was warmer than today. The high latitude tropical rainforest was replaced first by a temperate deciduous forest (in Alaska, by the Oligocene Epoch); and then during the Miocene and Pliocene epochs by a coniferous forest. Continued cooling led to the confinement of the tropical rainforest to the tropics. However, the distribution and composition of the world's tropical rainforests has not been stable since the Pliocene. During the Quaternary Period, as global climate change brought alternating glacial and interglacial conditions to the high latitudes and high elevations, the tropical rainforests may have contracted and expanded several times as the tropics experienced alternating cool, dry and moist, hot periods. Although now questioned, the leading theory has held that during the cool, dry periods the world's tropical rainforests were fragmented into isolated patches surrounded by expanded tropical savannas. In these island habitat patches, old species could survive and new ones could evolve, which may be part of the explanation for the extraordinary diversity of species found in the tropical rainforest biome today. Based on this model, the current rainforests are believed to be rather young, having last expanded from these patches or refugia only 20,000 to 10,000 years ago (Richards 1996). Even if the rainforests were more stable during the Pleistocene than previously believed, new information suggests that human activities greatly altered them during the last 10,000 years, and, by that measure, today's forests are even younger.

## HUMAN IMPACTS

People have been living in tropical rainforests for thousands of years. Hunting and gathering societies may have had subtle effects on the distribution and abundance of plants in the forest. In at least some places, fires were used in the dry season to drive game. Clearings were opened along trails and at campsites. Useful plants spread along trails and were deposited at latrines and middens near campsites and villages, and valued trees and other plants were protected. The nomadic Kukak people of the Colombian Amazonia inadvertently create wild orchards of fruit trees at their campsites today (Denevan 2001). A few hunting and gathering people still survive, people such as the Mbuti and Baka of the Central African Forest and the Penan of Borneo, although they are in contact and trade with agricultural peoples (Hanbury-Tenison 1989).

People have long depended on shifting agriculture throughout the forested regions. Traditional peoples still grow their crops mixed together like the forest plants in temporary plots cleared by girdling trees and burning the dead slash. They also use a variety of wild plants for building material, ropes, mats, food, and medicines and leave patches of the forest intact to preserve the plants they need. It is likely that even in the past they planted useful trees and vines and shrubs as they moved from place to place in the forest and thus altered the natural distribution of some rainforest plants. In the Amazon, for example, babussa palm (*Orbignyia phalerala*), a multiuse plant that survives fire, today occurs in extensive colonies in the rain forest, a likely testimony to previous human settlement in those areas. Other evidence indicates that dense human settlements occurred in the past. On bluffs along the Amazon and its tributaries are frequent patches of *terra preta*, black soil. These are man-made soils up to 2 m (6 ft) thick and enriched with human wastes, the bones of fish and game and people, and ash and other debris from agricultural villages. These patches average about 20 ha (50 ac) in area, although one of the largest known, beneath Santarém at the junction of the Tapajós and Amazon rivers, covers 500 ha (1200 ac). They attest to continuous or recurrent use of the várzea areas by dense agricultural settlements and are still preferred areas for farming, since the black soils—high in organic matter, phosphorus, and calcium—are more fertile than soils under the *terra firme* forests. Other evidence of long-ago human impacts comes from Central America, where early Spaniards in the area described open fields where today dense forest is found. And in Nigeria, many village sites have been unearthed in the Okomu forest. The indications of human use are also evidence of the resilience of the rain forest—its ability to recover from recurrent small scale disturbances and reclaim the humid tropical lowlands. Much of what people until recently have proclaimed pristine forest has probably been cleared patch by patch many times over in the past (Denevan 2001; Smith 1999; Whitmore 1990).

Exploitation of the forests by outsiders also has a long history. Trade in resins and animal parts between South East Asia and China has been active for 2,000 years. The Arab traders who crossed the Indian Ocean to reach the Spice Islands (the Moluccas) exported nutmeg (*Myristica fragrans*), cloves (*Syzygium aromaticum*), and black pepper (*Piper nigrum*) to Europe. The wealth they derived from the spice trade was the key stimulus to Europe's Age of Discovery, as Europeans sought to take over the business for themselves (Whitmore 1990). West African forests lay closest to the ports of Europe, and trade in gold, ivory, cola nuts, and slaves provided funds for the Portuguese era of exploration and early domination of European trade in tropical products (Martin 1991). In 1500, the Portuguese accidentally came upon the Atlantic Forest of Brazil. A trade in brazilwood (*Caesalpinia echinata*), a dyewood that produced a red coloring for textiles, quickly developed. (The wood of this tree replaced a wood from a close relative in the Asian forest that was obtained through intermediaries in the eastern Mediterranean and in Portuguese was called *pau brasil.* The name transferred not only to the Atlantic Forest plant but also to the country in which it was found.) Brazilwood is nearly extinct in the wild today (Dean 1995). As in West Africa, the European presence was largely restricted to a series of fortified storehouses along the coast. Wood and other products from the rainforest, including live animals and pelts, were brought by natives from the interior to these so-called "factories" for shipment to Europe.

By the late nineteenth century, plantations were established to produce rubber, cocoa, and oil palm in locations often distant from the places of origin of these tropical plants. Destruction of the forest for commercial agriculture and timber harvest—and in Amazonia, the introduction of European and African diseases—led to the displacement and annihilation of whole groups of native peoples. Departments of forestry and forestry reserves were created in the colonies in the early twentieth century but did little to slow the clearing of the rain forest. Clearing of the forest indeed accelerated after 1950 due to a combination of mechanized timbering, population growth, and the construction of logging roads that provided access to the interior forests to migrants from outside the rainforest zone. Some governments, such as Brazil's, for a time sponsored projects to settle people in the interior; other immigration was spontaneous. (Most forest clearing in Amazonia occurred in the 1970s and 1980s in the state of Rondonia on the southern edge of the Amazon forest. The rate of deforestation has been declining since, as people prefer settlement in towns and cities with amenities such as electricity and running water not found in rural pioneer settlements.) Conversion of forest to pasture has had a significant impact in both Central and South America. Cattle were the primary livestock produced; however, recently the raising of water buffalo has been expanding upstream on the várzea areas of Amazonia (Smith 1999). Other direct but more localized impacts on tropical forests have come from large hydroelectric projects such as those on the Tocantins River (Brazil) or Batang Ali in Sarawak, that have flooded many hectares, destroying forest and displacing native people living along the rivers. Mining for gold, aluminum ore, and iron also has devastated local areas (Smith 1999; Whitmore 1990).

On a more positive note, the rainforest has become the focus of local, national, and international conservation efforts led by both grassroots and nongovernmental organizations. Rainforests have nearly disappeared from places such as Costa Rica, Sumatra, and West Africa. The Atlantic Forest of Brazil has been reduced to mere fragments. Vast areas of forest, however, still exist in the Amazon and the Congo basins, and the rate of deforestation appears to be slowing. The growing awareness of people all over the globe of the valuable store of biological diversity and the potential medicinal and industrial resources it represents are increasing efforts to preserve what is left through developing methods of sustainable use (Whitmore 1990).

## HISTORY OF SCIENTIFIC EXPLORATION

Christopher Columbus's report of his exploration of Hispaniola in 1493 is the earliest description of tropical rainforest in western literature. The collecting and describing of plants and animals from the world's tropical rainforests followed European expansion as naturalists sought new species of potential use and value to the countries of western Europe. The Dutch naturalist G. E. Rumpf was in Indonesia in 1750, and the Swedish botanist Osbeck first described cauliflory in Java in 1752 (Whitmore 1990). British naturalists, inspired by what they saw in tropical forests, made several major contributions to the development of science. Most important were Charles Darwin, who visited the Atlantic forest of Brazil, and Alfred Russel Wallace, who collected for five years in the Amazon and eight years in the Indo-Malaysian region. Together they proposed the theory of evolution by natural selection—the guiding principle of modern biology, ecology, and biogeography. Wallace also became known as the father of animal geographer based on his studies of animal distributions in the Malay Archipelago, and his name is preserved in Wallace's Line—the boundary between two major units of global animal distribution patterns, the Oriental Zoogeographic Province and the Australian Zoogeographic Province.

The term "tropical rainforest" (*tropische Regenswald*) was coined in 1898 by the German plant geographer A.F.W. Schimper in his classic book, *Plant Geography*, which also gave the first general description of the vegetation—a task not repeated until P. W. Richards's book, *The Tropical Rain Forest*, was published in 1952. Alexander von Humboldt had described the South American rainforest earlier in the nineteenth century and called it *Hylaea*, a term still in use for tropical rainforests in some parts of the world (Richards 1996).

The interest of European Colonial powers in tropical hardwoods and other natural products of the forest led to more intensive forestry inventory and the establishment of botanical gardens in various parts of the tropics to test plants for usefulness and suitability for cultivation. Study in the first half of the twentieth century was concentrated in those areas most accessible to Europeans. The Indo-Malaysian forest became one of the better known even as it and the rainforests of West Africa became heavily exploited and degraded or cleared. In the Americas, important research stations were established at Barro Colorado Island, Panama, by the Smithsonian Institution, and at La Selva in the Monteverde Cloud Forest, Costa Rica, by the Organization of Tropical Studies. Brazil established a major ecological and climatological research center in Manaus. The most basic distribution patterns of the forests were not revealed until the development of remote sensing techniques such as radar and Landsat satellite imagery reduced the need for ground studies in the 1970s and early 1980s. Major vegetation mapping projects included Projecto RADAMBRASIL (1975–83) in Amazonia and the 1983 UNESCO/AETFAT/UNSO Vegetation Map of Africa (Richards 1996; Martin 1991).

One of the most stunning characteristics noted by all early naturalists seeing the rainforests for the first time was the enormous variety of living organisms inhabiting them. The total number is still unknown, and in many ways it can be said that research is just beginning, especially in the interior forests of South America and Africa. Perhaps thousands of plants and tens of thousands of arthropods remain to be discovered by science. Even larger animals have been found in just the last decade or so. Beyond mere inventory, the major research questions of why there are so many species and how they manage to live together are still unanswered. And the major issue remains of how to preserve as much of this forest and as many of its inhabitants—including human beings—as possible.

## REFERENCES


Dean, Warren. 1995. *With Broadax and Firebrand, The Destruction of the Brazilian Atlantic Forest.* Berkeley: University of California Press.

Denevan, William. 2001. *Cultivated Landscapes of Native Amazonia and the Andes.* New York: Oxford University Press.

Goulding, Michael. 1980. *The Fishes and the Forest.* Berkeley: University of California Press.

Hanbury-Tenison, Robin. 1989. "People of the forest." Pp. 189–213 in Lisa Silcock, ed., *The Rainforests: A Celebration.* San Francisco: Chronicle Books.

Janzen, D. H. 1972. "Protection of *Barteria* (Passifloraceae) by *Pachysima* ants (Pseudomyrmecine) in a Nigerian rain forest." *Ecology* 53 (5): 885–892.

Kingdon, Jonathan. 1989. "Bigger, brighter, louder—Signals through the leaves." Pp. 127–151 in Lisa Silcock, ed., *The Rainforests: A Celebration.* San Francisco: Chronicle Books.

Martin, Claude. 1991. *The Rainforests of West Africa, Ecology—Threats—Conservation.* Boston: Birkhäuser Verlag.

Richards, P. W. 1996. *The Tropical Rain Forest.* 2nd Edition. Cambridge: Cambridge University Press.

Sick, Helmut. 1993. *Birds in Brazil, A Natural History.* Princeton, NJ: Princeton University Press.

Smith, Nigel J. F. 1999. *The Amazon River Forest, A Natural History of Plants, Animals, and People.* New York: Oxford University Press.

Whitmore, T. C. 1984. *Tropical Rain Forests of the Far East.* 2nd edition. Oxford: Clarendon Press.

Whitmore, T. C. 1990. *An Introduction to Tropical Rain Forests.* New York: Oxford University Press.

Wolfe, Art and Ghillean T. Prance. 1998. *Rainforests of the World: Water, Fire, Earth, and Air.* New York: Crown Publishers, Inc.

# 2 TROPICAL DRY FORESTS

## OVERVIEW

The term tropical dry forest as used here includes a number of forest and woodland types that otherwise might be referred to as semievergreen, dry **evergreen**, semi-deciduous, **deciduous** or seasonal forests, woodlands, and thornscrub. This is essentially a series of plant communities along an environmental gradient toward increasing aridity and in which the major response to drought is a deciduous **habit**. As total rainfall diminishes and the length of the dry season increases, the degree of deciduousness among the plants becomes more and more pronounced. However, at the driest limits of the **biome**—in the thornscrub—the evergreen habit again becomes common, and **succulents** (plants that store water in their tissues) may become prevalent.

Aspects of **vegetation** structure other than leaf habit also reflect the moisture gradient. Vegetation structure becomes simpler as fewer tree layers are encountered in the drier environments. The height of the **canopy** trees decreases, and fewer **epiphytes** and **lianas** occur. In the drier extremes of the biome, many woody plants bear thorns or spines. Grasses and other **herbs** tend to increase as the canopy becomes less dense in increasingly drier locations.

Tropical dry forests of all types are found in **climate** regions that are warm year round and experience several months of severe drought during the low sun period. This is the same climate region in which tropical **savannas**, occur and frequently these wooded **habitats** are considered part of the tropical savanna biome (Chapter 3). They are usually described, however, in comparison to tropical wet forests. Often neglected as attention focuses on the wet evergreen forests (rainforests) of the **tropics**, dry forest vegetation warrants treatment as a separate biome because it often hosts a different group of plants and animals than do either the neighboring tropical savannas or rainforests. Daniel Janzen, one of the foremost researchers of dry forests in Central America, says this biome is the most threatened of all lowland tropical forest types (Janzen 1988). Dry forests once accounted for 42 percent of all

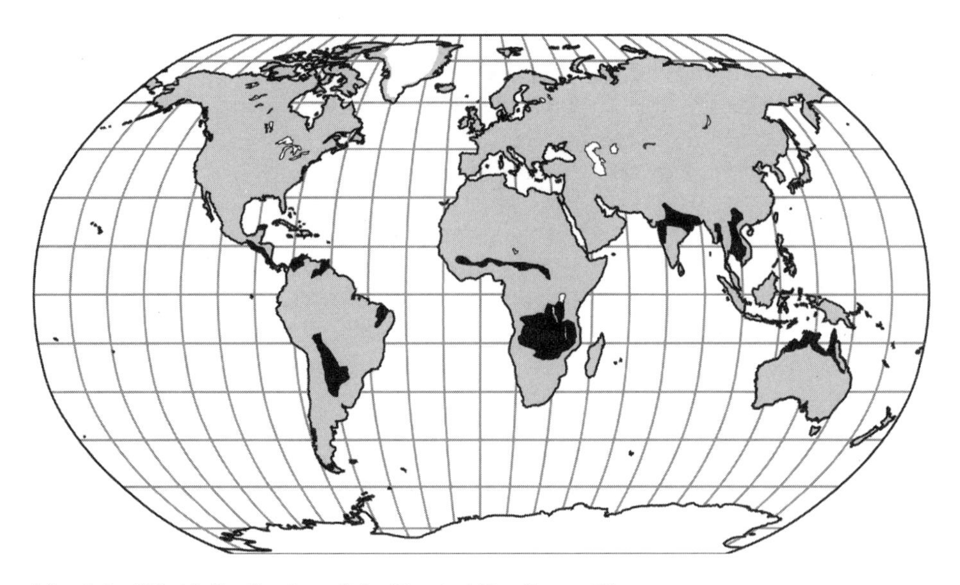

Map 2.1 World distribution of the Tropical Dry Forest Biome.

tropical vegetation—more area than rainforest covered. Now—degraded; transformed; or eliminated by logging, cultivation, and grazing—very few large tracts survive (Mooney et al. 1995; Janzen 1988).

Tropical dry forests are found in Mesoamerica (Mexico and Central America), South America, South and Southeast Asia, Africa—both north and south of the equatorial rainforest—and Australia.

## CLIMATE

The tropical wet and dry climates (Koeppen's Aw) of the Americas and Africa are affected by the seasonal migration of the **Intertropical Convergence Zone (ITC)** north and south of the equator. The rainy season coincides with the high sun period when the ITC is nearby. Depending upon **latitude**, one or two dry periods of varying length may occur when the ITC is removed well to the north or south. In summer, one or two months are dry; and in winter two to six months may be dry. The exceptions to this pattern are close to the Tropics themselves, where a single dry season lasts up to eight months. Total annual precipitation averages 250–2,400 mm (10–95 in). Temperatures show less fluctuation than moisture and remain warm to hot year round.

In Asia, regions of tropical dry forest are influenced by the monsoon, which draws the ITC to latitudes of 20°–25° N in summer. One prolonged dry season extends from November to April, when high atmospheric pressure dominates over the Tibetan Plateau and forces dry continental air southward over India, Myanmar, Thailand, Cambodia, Laos, and Vietnam.

## VEGETATION

The dry forests and the mosaic of communities within them on each continent developed under different climatic conditions, **disturbance** patterns, and geological conditions. The result is great geographic variability in the vegetation, which makes generalization difficult. Examples from each major region of dry forest will be described separately later, but, first, a general description.

Forests categorized as dry evergreen or semievergreen have three tree layers, only the uppermost of which exhibits drought deciduousness. These forests differ structurally from tropical wet evergreen forests in that trees of the emergent or A layer are more widely spaced, which makes the forest canopy in reality two storied. Few trees have buttresses or stilt roots. The evergreen **species** show adaptation to drought in their leaves, which are smaller and thicker than those of rainforest species and have thick cuticles and leathery texture. Bark may be thicker than that of rainforest trees, and there is less **cauliflory**—the formation of flowers, and hence fruits, right on the tree trunk—a feature common in rainforest plants (Furley and Newey 1983).

Forests classified as semideciduous or deciduous tend to have only two tree layers. In a semideciduous forest, the **closed canopy** of the A layer is composed of deciduous species, but the lower layer has many evergreen species. In a fully deciduous forest, both canopy layers contain only deciduous trees. However, it should be noted that deciduousness widely varies from year to year, site to site, and species to species. The leafless season and the usually thin **crowns** of deciduous species allow sunlight to penetrate to the forest floor, which typically allows at least a discontinuous ground layer of grasses and **forbs**. The drier the environment, the more continuous the grass cover and the more probable that fire will be a frequent natural disturbance.

Tropical dry forests in general have fewer epiphytes than tropical wet forests but a large number of climbers. The trees tend to have more conspicuous flowers, and more trees have wind-dispersed seeds than in the wet forests. American tropical dry forests have fewer plant species than rainforests and appear to contain a subset of the genera from a relatively few of the families found in the region's rainforests (Gentry 1995). This seems to hold true in other parts of the world as well.

## SOILS

No one **soil** type correlates with the distribution of tropical dry forests and woodlands. In Africa, they usually occur on a variety of acidic, infertile soils, and—most often, perhaps—on the red, iron-rich, nutrient-poor lateritic soils (oxisols) commonly developed on the ancient plateau surfaces. In Mesoamerica and South America, dry forests are associated with soils more fertile than those occupied by savannas.

## ANIMAL LIFE

The animals living in dry forests are poorly known, as evidenced by the discovery as recently as 1974 of a relatively large mammal, the chacoan peccary (*Catagonus wagneri*), in Paraguay. In general, dry forests have a lower diversity than moist forests due to the absence of frugivores, specialized carnivores, and semiaquatic mammals (Ceballos 1995). In general, unique assemblages of animals are not a characteristic of tropical dry forests. Some **endemic** vertebrates are found, but more commonly, the same animals found in dry forests are also found in nearby rainforests or savannas. However, more ground-dwelling mammals live in the dry forests of South America than in the rainforests (Mares 1992). The greater the isolation of dry forests from other wooded tropical biomes—as, for example, is the case in western Mexico and the South American chaco—the greater the number of unique and restricted species (Ceballos 1995).

In the dry season, the sun dries out the litter on the forest floor, streams may dry up or be greatly reduced in their flow, and plants shed their leaves and cease to grow. However, many of the woody plants flower at this time, have maturing fruits, and are dispersing their seeds (Janzen 1988). Animals respond to strong seasonal changes in their habitat in several ways. Many synchronize their breeding to take advantage of high food abundance. Some migrate to moist **microhabitats** within the dry forest—such as the vegetation growing along streams, hollow logs, or north-facing canyon walls—or to moister forest regions. Hummingbirds in Mexico and howler monkeys in Costa Rica, for example, undergo such local migrations. Animals that remain in the dry forest year round have other strategies. Amphibians and reptiles may become inactive during the dry season. The New World rodents, on the other hand, may become more active at that time. Agouti (*Dasyprocta punctata*) and paca (*Cuniculus paca*) forage for fruits and seeds over lengthier time periods than during the rainy season. The agouti depends upon caches of seeds it has buried, while the paca depends upon body fat stored during times of plenty. Other animals reportedly change their diets seasonally. When flowers are scarce, broad-billed hummingbirds (*Cyananthus latirostris*) and nectar-feeding bats (*Glossophaga* spp.) switch to insects. The tamandua (*Tamandua mexicana*) switches from ants to more succulent termites during the dry season. At least a few mammals are physiologically able to withstand drought. The spiny pocket mouse (*Liomys pictus*) of western Mexico eats seeds that it hoards for the dry season and can survive for months without drinking water (Ceballos 1995).

In the dry forests of both the Americas and Africa, invertebrates tend to be the major plant eaters. Explosions of caterpillars and other insects can defoliate large areas.

## MAJOR REGIONAL EXPRESSIONS

### Africa

Dry forests and woodlands occur in two distinct regions in Africa: the Sudanian region north of the equator and the Zambezian region, south of the equator. In both, severe drought conditions exist for three or more months during the respective dry season. The dry forests are distinguished by their closed canopies and multiple layers of woody plants. The uppermost layer is composed of a variety of deciduous trees, and the understory may contain either deciduous or evergreen plants or both. A grass layer is too sparse and broken to support regular fires. Dry woodlands, on the other hand, are marked by an upper canopy of only one or a very few kinds of small- to medium-sized trees 8–20 m (25–65 ft) tall. Woody plants make up a sparse understory, but grasses and forbs form a continuous cover dense enough to support annual burning. The grasses are not the same species found on the adjacent savannas. The trees of both dry forests and dry woodlands also differ from those that dominate on the nearby savannas. Indeed, many of the tree genera are also represented in African rainforests, although dry forests are always separated from rainforest by belts of savanna (Menaut et al. 1995).

In the Sudanian woodlands that stretch across Africa south of the Sahara from Nigeria to the Sudan, the main trees are the **legumes** *Isoberlinia doka* and *I. tomentosa*. On poorly drained clays, *Terminalia macroptera* replaces them. The woody understory is very sparse, although this may be the consequence of former agricultural practices rather than an adaptation to the prevailing rainfall and fire regimes. Northward in the valleys of the Sahel, small acacias less than 10 m (30 ft) high dominate a drier woodland. *Acacia seyal* grows on clays, *A. raddiana* on sandy soils, and *A. nilotica* on floodplains (Menaut et al. 1995).

The Zambezian region is characterized by miombo or *Brachystegia* woodland. Miombo covers much of the gently rolling Central African Plateau from Angola in the west to Mozambique in the east and from southern Tanzania south to northern Zimbabwe. The plateau surface lies at an elevation of 900–1,500 m (3,000–5,000 ft) and during its four- to six-month rainy season receives 600–1,300 mm (24–50 in) of precipitation. Miombo usually occurs on lateritic soils (oxisols). The species of *Brachystegia* present varies according to soil condition. *B. spiciformis* favors deep soils, *B. longifolia* sandy soils, *B. utilis* sandy-clayey soils, *B. boehmi* heavy clay soils, and *B. floribunda* coarsely textured soils. *Julbernardia paniculata* and *Isoberlinia angolensis* occur on all sites (Menaut et al. 1995). Miombo woodlands and dry forests cover 2.8 million km$^2$ (1 million mi$^2$) and constitute the largest expanse of the tropical dry forest biome in the world. On the margins of the miombo—on the deep Kalahari sands to the south—is a similar woodland dominated by *Baikiaea plurijuga* with other trees such as *Burkea africana*, *Pterocarpus angolensis*, and *P. stevensoni*. Farther south on deep black clays at lower elevations that receive less rainfall, mopane (*Colophospermum mopane*) woodland can be found (Photo 2.1) (Menaut et al. 1997; Desanker et al. 1997).

Herbivores of the miombo are primarily large-bodied mammals including elephant (*Loxodonta africana*); buffalo (*Syncerus caffer*); and selective grazers such as

Photo 2.1 Mopane dry forest. Elephants are a significant disturbance factor in African dry forests because they push over trees and create clearings into which grasses invade. (Photo © Peter Johnson/CORBIS.)

Photo 2.2 Termite mound, Botswana. (Photo © 318; Gallo Images/CORBIS.)

Liechtenstein's hartebeest (*Sigmoceros lichtensteinii*), sable antelope (*Hippotragus niger*), and roan antelope (*H. equinus*). Mammalian browsers are rare. Caterpillars, which undergo periodic outbreaks and defoliate large areas (Desanker et al. 1997), are the consumers of tree leaves. Other important insects in the miombo are the fungus-growing termites, *Macroterma falciger.* Their huge termitaria, 8 m (25 ft) tall and 14–15 m (45–50 ft) wide, are built of clays collected deep in the soil and combined with organic matter (Photo 2.2). Termites gather plant litter and store and process it in their mounds to create a nutrient-rich microhabitat that a variety of trees colonize when the nest is abandoned. Indeed, termite-formed soils characterize much of the miombo area. Without termites, nutrients would be rapidly leached, and soils might become too nutrient poor to allow the regeneration of dry forest trees (Menaut et al. 1995).

Birds of the miombo in Zimbabwe include a number of woodpeckers, including Golden-tailed Woodpecker (*Campethera abingoni*), Bennett's Woodpecker (*C. bennetti*), Cardinal Woodpecker (*Dendropicos fuscescens*), and Bearded Woodpecker (*D. namaquus*); cuckoos such as Klaas's Cuckoo (*Chrysococcyx klaas*), Diederik Cuckoo (*C. caprius*), and African Cuckoo (*Cuculus gularis*); Lesser Blue-eared and Greater Blue-eared starlings (*Lamprotornis chloropterus* and *L. chalybaeus*, respectively); and sunbirds such as the Miombo Double-collared Sunbird (*Nectarinia manoernsis*) and the Yellow-bellied Sunbird (*N. venusta*)—counterparts of New World hummingbirds. Mixed species flocks are common in winter (Solomon 1997).

## Indo-Malaysia

Three major types of seasonal forest are recognized in a zone that extends from India to Vietnam: a dry evergreen forest where annual precipitation averages

1,200–1,500 mm (48–60 in); a mixed deciduous forest dominated by teak (*Tectona grandis*) where precipitation is 1,500–2,300 mm (60–90 in); and a deciduous diptocarp forest, the most widespread of all, where annual precipitation ranges between 1,000 and 2,750 mm (40–70 in). The Asian monsoon influences life in this part of the world. A five- to six-month dry season occurs in the winter. In most cases, seasonal forests have shallow, infertile soils through which water readily passes and is therefore lost soon after it rains (Rundel and Boonpragob 1995). Most of the following more detailed information comes from research in Thailand.

The dry evergreen forest is in some ways an extension of the Malaysian rainforest, although it possesses some deciduous species. Most important are trees of the family Diptocarparpaceae, many found only in this forest type. They form a closed canopy at a height of 25–30 m (80–100 ft) above the forest floor. Beneath them is a sapling layer 5–17 m (15–55 ft) high. One of the distinguishing characteristics of the dry evergreen forest is the presence of many woody vines or lianas that may be dense enough to obstruct one's vision (Rundel and Boonpragob 1995).

The mixed deciduous or teak forest has a more complex structure and composition. Logging has greatly affected this forest, so it is difficult to find natural conditions; however, what remains can be characterized as follows. The closed canopy is 30 m (100 ft) or more above the ground and is composed of teak and the pyinkado (*Xylia xylocarpa*), a deciduous member of the pea family (Leguminoseae). Teak, too, is deciduous, but its time of leaflessness is not perfectly coordinated with the precipitation pattern. Instead, it maintains foliage for a month or two into the dry season and then sprouts new leaves a month or more before the beginning of the rainy season (Eyre 1968). Beneath this is an open, middle story at 10–20 m (30–65 ft) composed of small, often evergeen trees from several families—perhaps most important are trees of the **genus** *Terminalia* (family Combretaceae). Two bamboos dominate the shrub layer, below which is an open ground cover. Lianas occur but are not nearly as abundant as in the dry evergreen forest. Epiphytes are rare (Rundel and Boonpragob 1995).

The deciduous diptocarp forest also contains teak, but the other trees of the canopy are different from those accompanying teak in the mixed deciduous forest type. Four species of diptocarps are prevalent; most common are the eng tree (*Diptocarpus tuberculatus*) and sal (*Shorea obtusa*). All canopy layer trees have relatively large, thick leaves and are deciduous. They do not all lose their leaves at the same time, however; depending upon the species, leaf fall may occur anytime between December and April. Most species are leafless from February to April. A discontinuous shrub layer comprises largely evergreen species, such as dwarf palms (*Phoenix acaulis*), cycads (*Cycas siamensis*), and bamboo (*Dendrocalamus strictus*). **Perennial** grasses and other herbs occur as an open ground layer but are scarce in undisturbed forest. This is a vegetation adapted to withstand fire, and fires occur frequently—roughly every one to three years. Trees, including teak, have thick, rough bark and many are able to resprout from root crowns. A sharp boundary often occurs between deciduous diptocarp forests and the fire-sensitive mixed deciduous of forests of tropical Asia (Rundel and Boonpragob 1995).

The dry forests and associated upland and wetland habitats of Indo-Malaysia are home to a number of rare mammals. Perhaps best known is the Asian elephant (*Elephas maximus*); tamed ones are still used in some places to log teak. Less well known are several members of the cattle family: guar (*Bos guarus*), kouprey (*Bos sauveli*), wild water buffalo (*Bubalus bubalis*), serow (*Capricornis sumatraensis*), and goral (*Nemorhaedus caudatus*). Tigers (*Panthera tigris*) and leopards (*P. pardus*) are the large carnivores of the region (Rundel and Boonpragob 1995).

## Mesoamerica (Mexico and Central America)

Tropical dry forest occurs on the northern Yucatan peninsula and on the rugged foothills and gorges of the western flanks of the Sierra Madre Occidental from southern Sonora south to Chiapis. In Central America, a now highly fragmented dry forest extends along the Pacific Coast from Guatemala to Costa Rica. The Costa Rican forest was among the first described. The vegetation consists of two tree layers. The canopy layer, dominated by leguminous trees of the Mimosaceae and Caesalpinaceae families, stands 20–30 m (65–100 ft) tall. The trees have short, thick trunks and flat-topped crowns that form an **open canopy**. The foliage is thin and shed in the dry season. Beneath this layer is an understory of trees 10–20 m (30–65 ft) high. The trunks of these trees are spindly and crooked; the crowns are small and open. Some of the understory trees are evergreen and many are in the madder family (Rubiaceae). A 2–5 m (6–16 ft) shrub layer consists of multistemmed woody plants armed with spines or thorns. Ground cover is sparse. Woody vines are common, but epiphytes—mostly bromeliads—are uncommon (Hartshorn 1983). The division between thornscrub and dry forest is made according to the height relationships between woody trees and shrubs and tall columnar cacti (Photo 2.3). In thornscrub, the cacti are taller than the shrubs; in dry forest the cacti do not penetrate through the canopy layer of trees (Martin and Yetman 2000).

Photo 2.3 Partially cleared dry forest near Alamos, Sonora, Mexico, used as pastureland. (Photo by SLW.)

Photo 2.4 Thornscrub near Alamos, Sonora, at end of dry season. (Photo by SLW.)

Recent studies have focused on dry forest at its northwestern limits in the Americas (Robichaux and Yetman 2000). This isolated section of dry forest near the Mexican town of Alamos, Sonora, occurs at elevations of 300–800 m (980–2,600 ft) in canyons and up to 1,000 m (3,300 ft) on rocky mountain sides. Below lies thornscrub (Photo 2.4) and above in the mountains is an oak or oak-pine forest. Most woody plants are deciduous, and their leaves tend to be somewhat larger than those of thornscrub plants. Some, such as palo mulato (*Bursera grandiflora*), have drip tips more characteristic of rainforest trees; others, such as kapok (*Ceiba acuminata*), have thorns or spines—more common among woody plants of the thornscrub. Although this forest lies north of the Tropic of Cancer, it harbors many tropical species, including such plants as figs (*Ficus* spp.), kapok, brasil (*Haematoxylum brasiletto*), and tank bromeliads; such birds as trogons, chachalacas, and motmots; and such mammals as mouse opossums (*Marmosa canescens*), vampire bats (*Desmodus rotundus*), and jaguar (*Panthera onca*). They can survive in this location because the Mexican monsoon reliably brings summer rainfall and the Sierra Madre usually blocks "blue northers," the polar air masses that penetrate interior Mexico in winter and cause killing frosts (Martin and Yetman 2000). The Alamos dry forest is multilayered and relatively rich in plant species. A continuous canopy is formed 10–15 m (30–50 ft) above the ground. The tallest trees, plants such as rock fig (*Ficus petiolaris*) and palo joso de la sierra (*Conzattia multiflora*) emerge through the canopy to attain heights greater than 20 m (65 ft). The most common tree of the canopy layer is mauto (*Lysiloma divaricatum*). It is joined by, among other species, kapok, brasil, eight species of *Bursera*, palo santo or morning glory tree (*Ipomoea arborescens*), and four columnar cacti: pitahaya or organ pipe (*Stenocereus thurberi*), sahuia (*S. montanus*), old man cactus (*Pilocereus alensis*), and *Pachycerues pecten-aboriginum*. The understory consists of three layers. Tall shrubs and small trees such as vara blanca (*Croton fantzianus*) and *Jatropha malacophylla* create the highest layer at about 5 m (15 ft). Below this is a 1 m (3 ft) high shrub layer dominated by Sonoran bursage (*Ambrosia cordifolia*). Flor de iguana (*Senna pallida*) is also common in this layer. The ground layer consists of a large variety of perennial and **annual** herbs that sprout mostly during the summer rainy season. Lianas such as jícama (*Ipomoea bracteata*), rosary bean (*Rhynchosia precatoria*), and huirote de violin (*Gouania rosei*) trail up through the trees. Only two epiphytes have been reported—the ball moss (*Tillandsia recurvata*) and rattail orchid or cuerno de chivita (*Oncidium cebollata*) (van Devender et al. 2000; Martin and Yetman 2000).

In addition to the mouse opossum and vampire bat, which reach their northern limits near Alamos, a number of other animals are indicators of the biome. Where conditions permit, amphibians such as Mexican leaf frogs (*Pachymedusa dacnicolor*) and Mexican tree frogs (*Smilisca baudini*) are found. Among the numerous reptiles are parrot snake (*Leptophis diplotropis*), clouded anole (*Norops nebulosus*), Mexican spiny lizard (*Sceloporus horridus*), tropical tree lizard (*Urosaurus bicarinatus*), and Mexican beaded lizard (*Heloderma horridum*) (Schwable and Lowe 2000). The Black-throated Magpie Jay (*Calocitta colliei*) is an indicator bird species, and some 18 bird species are considered characteristic. (At least 150 birds have been recorded from the dry forest, but most are not restricted to it.) Among these are the Green or Military Macaw (*Ara militaris*), Lilac-crowned Parrot (*Amazona finschi*), Lesser Roadrunner (*Geococcyx velos*), Russet-crowned Motmot (*Momotus mexicanus*) and Lineated Woodpecker (*Dryocopus lineatus*). Mexican dry forests are important wintering grounds for migratory birds from western North America (Russell 2000).

## South America

Dry forests north of the Amazon occur on extremely droughty soils in northern Venezuela and on Trinidad, interior Guyana, and northeastern Colombia. South of the Amazon, thornscrub covers a large part of Northeast Brazil as well as the Gran Chaco of South America. The climates of these last two regions are classified as steppe or semiarid (Koeppen's BS).

The northern dry forest receives 760–1,270 mm (30–50 in) of precipitation a year; during the dry season there may be 25–125 mm (1–5 in) a month. The dry forest of Trinidad has been described with two stories, the upper deciduous and the lower typically evergreen, although some understory plants will shed their tiny leaves in years of extreme drought. The crowns of canopy trees are compact and conical or round; few have buttresses. Epiphytes are rare, but lianas are abundant (Eyre 1966).

In the backlands or *sertão* of Northeast Brazil, dry forest and thornscrub is known as **caatinga**. This region receives 300–1,000 mm (12–40 in) of precipitation over a three- to five-month rainy season. Severe droughts that last three to five years occur once every 30 or 40 years and used to force now-legendary migrations of people out of the backlands. The surface is one of uplifted crystalline basement rocks and Paleozoic sedimentary rocks that have weathered into a complex mosaic of soils, most of which are shallow. Caatinga reflects the variety of soil environments and is hence quite variable in structure and species composition. At one end of the spectrum is a dense woodland of mostly deciduous trees reaching heights of

Photo 2.5 Brazilian bottletrees *(Cavanillesia arborea)* in southwestern Bahia, Brazil. (Photo by SLW.)

25–30 m (80–100 ft). Three or more tree and shrub layers can be recognized; the magnificent Brazilian bottletrees or barrigudas (*Cavanillesia arborea*) store water in their trunks and tower above the general canopy (Photo 2.5) (Sampaio 1995). In other areas, caatinga forms a less dense cover of shorter trees 7–15 m (20–50 ft) high; many are thorny (Photo 2.6). Cacti and umbuzeiro (*Spondias tuberosa*) (family Anacardacieae), which store water in underground structures, are conspicuous members of the **community**. By far the most common type of caatinga, though, is a low, dense, largely deciduous shrub community developed on the soils formed from ancient crystalline rock. Cactus, terrestrial bromeliads, and occasionally a small bottletree (*Chorizia glaziovii*) are part of the low caatinga community (Sampaio 1995). Within the vastness of the drylands are some moist habitats. On heavy alluvial soils, small palm forests (*Copernica* spp.) thrive. And on higher massifs—the *brejos*—where **orographic** uplift increases annual rainfall amounts, forest species with affinities to the Atlantic rainforest of coastal Brazil persist.

The families with the largest number of species in caatinga are legumes (Caesalpinaceae and Mimosoidae), Euphorbaciae, Papionoideae, and Cactacae. The most widespread and common species is caatinguiera (*Caesalpinia pyramidalis*).

The mammals of the caatinga reputedly comprise the least diverse community in the neotropics and primarily are similar to those found in either cerrado or rainforests. Of the 86 recorded mammals all are small, 50 are bats and 18 are rodents. The only endemic mammal in the caatinga is a rock-dwelling cavimorph rodent, *Kerodon rupestris* (Ceballos 1995).

The **Gran Chaco** is a patchwork of seasonally flooded grassland, mesquite (*Prosopis* spp.) and tropical dry forest on a flat plain located between the Brazilian shield to the east and the Andes to the west. It covers parts of northern Argentina, northwestern Paraguay, and southeastern Bolivia with a minor extension into

Photo 2.6 Caatinga in eastern Bahia, Brazil. The patch of light-colored plant growth in the center of the photo consists of terrestrial bromeliads. (Photo by SLW.)

southwestern Brazil. The dry forest is dominated by quebracho trees (*Schinopsis quebracho-colorado* and *Aspidosperma quebracho-blanc*). Also common are *Ziziphus mistol* and mesquites. Small trees and shrubs such as *Caesalpinia paraguariensis*, *Cercidium australis*, and *Ruprechtia triflora* can be found in the understory. Terrestrial bromeliads (*Bromelia serra* and *B. hieronymi*) are conspicuous elements of the ground layer. The quebracho trees are harvested for tannin and exported to the world market (Walter 1985; Galera and Ramella n.d.). Relatively few mammals are found in the chacoan dry forest. Anteaters (*Myrmecophaga tridactyla*) and tamanduas (*Tamandua tetradactyla*) feed on the abundant termites. The tree sloth (*Bradypus boliviensis*), tree porcupine (*Coenda spinosus*), and three monkeys (Cebidae) feed in the trees. The vampire bat (*Desmodus rotundus*) and fruit- and nectar-eating bats also occur. The flightless rhea (*Rhea americana*) is perhaps the most conspicuous bird. Several venomous snakes are among the 25 species reported. Some 30 species of frogs and toads are also known from the region (Walter 1985).

## ORIGINS

The history of the development of tropical dry forests is little known. Investigators face a number of problems. Sites for preservation of macrofossils and pollen are few in number. Pollen analysis is hampered by the fact that trees of the dry forest are often distinct from those of wet forests at the species level but not at the generic level, while most fossil pollen can be identified only to genus. This makes pollen data inconclusive as to whether they pertain to dry forest or rainforest. Additionally, many of the flowers of important dry forest trees are pollinated by insects and leave many fewer pollen grains than wind-pollinated plants that are therefore often overrepresented in samples. A scenario for the development of American tropical dry forests is constructed on indirect evidence. Ancestors of dry forest trees had appeared by at least the late Eocene in localized dry sites created by excessively well-drained soils or rainshadow effects. The oldest occurrence of genera such as *Ceiba*, *Acacia*, and *Jatropha* come from the Gatuncillo Formation of Panamá and date from the Mid- to Late Eocene Period. Other genera, such as *Bursera*, are known from the western United States and northwestern Mexico from about the same time. Conditions for the development of a true dry forest community were established with uplift of the Sierra Madre Occidental and the formation of a continuous landbridge (the Isthmus of Panamá ) by the Mid-Pliocene Period (4 million years ago), when dry seasons five to six months long would first have been experienced on the Pacific side of southern Central America and western Mexico. That various scattered elements came together to form a plant community recognizable as dry forest in the Pliocene Period is evidenced in the Gatun formation of Panamá. Since then, the frequent climatic changes of the Quaternary Period led to changes in distribution and composition of the forests and woodlands. Each of the 18–20 glacial periods probably produced conditions suitable for this biome (Graham and Dilcher 1995).

## HUMAN IMPACTS

Indo-Malaysia is one of the areas where agriculture first developed, and this has meant a long history of forest clearing and fire for shifting and permanent cultivation. Growing human populations also required firewood and clearing, which put additional pressures on the dry forests. The British first commercially logged the Indo-Malaysian forests on a grand scale in the late nineteenth century. Before that, shifting agricultural practices of the hill tribes had undoubtedly affected the forests. Clearing of upland forests accelerated with growth of the indigenous human population, and—in Thailand in particular—with the twentieth-century influx of refugees from surrounding war-torn areas. Increased frequency of fire has changed forest to savanna. Most attempts to restore a cover of trees have used **alien** species, especially *Eucalyptus camuldulensis* and *Casuarina junghuhniana*. Native teak is grown in **monocultures** on plantations (Rundel and Boonpragob 1995). Legal and illegal traffic (especially to China) of animal parts believed of medicinal value; the pet trade; and the collection of orchids, beetles, and butterflies all have had harmful effects on wild species of the area.

The American Tropical Dry Forest **ecosystems** have also been widely transformed by deforestation to convert land to cultivation and cattle pasture. Habitat destruction has had negative impacts on animal life already limited in species. In addition, a number of animals are exploited in the wildlife trade. These include three species of peccary, small spotted cats (*Felis wiedii* and *F. pardalis*), parrots and macaws (*Amazona orathrix*, *A. finschi*, *Ara macao*, *A. militaris*, *A. ambigua*), and iguanas (*Iguana* spp. and *Ctenosaura* spp.). Among endangered animals are endemics such as the chacoan peccary, fairy armadillo (*Chlamyphorus retusus*), Chamela wood rat (*Xenomys nelsonii*), and Yellow-headed Parrot (*Amazona orathrix*) (Ceballos 1995).

Clearing of the caatinga is a continuing process. The trees are widely used for firewood and for the charcoal used in the brickyards and steel mills of the region. On St. John's Day (June 23), Brazilians celebrate with bonfires built in front of each dwelling. Often another huge bonfire is built in the town center. This annual event consumes a lot of wood and contributes to the deforestation of the caatinga. Other areas, especially near the São Francisco River, have been cleared for cattle pastures and irrigation agriculture. Caatinga plants originally supplied the poles for grapevines in this wine-producing area. Today poles more often come from tree plantations of Monterey pine (*Pinus radiata*) and eucalyptus. Overgrazing by cattle and goats has also contributed to the demise of caatinga, especially in the drier regions. EMBRAPA, the Brazilian agricultural research agency, is experimenting with ways to improve the economic value of native caatinga plants to ensure their preservation and encourage restoration of caatinga vegetation. One of the plants receiving attention is umbuzeiro (*Spondias tuberosa*), which bears an edible fruit known as umbú. These are currently collected from wild plants and sold mainly in markets in the state of Bahia.

## HISTORY OF SCIENTIFIC EXPLORATION AND RESEARCH

Tropical dry forests and woodlands have been largely neglected in studies of tropical vegetation and are still not well understood. Furthermore, much of the work that has been done in Africa, Asia, and even Latin America is inaccessible to the English-speaking world, as evidenced by the recent compilation of knowledge, *Seasonally Dry Tropical Forests* (Bullock et al. 1995), in which there is a strong bias toward neotropical forests.

The very identification of different and complex vegetation types ranging from semideciduous and dry evergreen tropical forests to dry deciduous forest, woodland, thornscrub, and tropical savanna is unresolved and an impediment to synthesizing information collected around the globe. In the Americas, Beard (1955) recognized a continuum in tropical vegetation and identified three seasonally dry types often used by American researchers: semievergreen seasonal forest, deciduous seasonal forest, and thorn woodland (Murphy and Lugo 1995). Ariel Lugo and his colleagues pioneered ecological studies in the dry forests of Mesoamerica with intensive research in the Guánica Forest on the south coast of Puerto Rico (e.g., Lugo et al. 1978; Murphy and Lugo 1986a, 1986b). Daniel Janzen has studied the ecology of dry forests in Guanacaste National Park, Costa Rica, since the early 1970s and has become a leading proponent for the restoration and conservation of tropical dry forests.

Dárdano de Andrade Lima (1981) was an important pioneer in the study of the plants of the caatinga. The Brazilian agency RADAMBRASIL (1983) mapped and classified caatinga vegetation types. Animals have received less attention, but Michael Mares and colleagues (1981) conducted a scientific inventory of mammals at the same time Andrade Lima was working on plants (Sampaio 1995).

In Africa, a formal gathering of international experts at Yangambi, Zaire, in 1956 established a nomenclature for African vegetation types that is still used. As in colonies everywhere, Europeans did the first work. Among significant early studies were those of Aubréville (1949), Boughey (1957a, 1957b), and Monod (1963). Frank White (1983) provided a definitive classification and map of the vegetation of the entire continent that provided an understanding of dry forest types still current today (Menaut et al. 1995).

Early mapping and classification of Asian forests resulted from British foresters working in India and Myanmar (Champion 1936; Edwards 1950). However, interest in the dry forests of Southeast Asia waned among western scientists until the

1980s because much of the growing body of literature was in languages in which North American and European researchers were not usually fluent. Growing concerns over the potential impacts of global climate change, however, sparked new interest in these ecosystems (e.g., Elliott et al. 1989; Round 1988). Asian scientists, however, continue to conduct much of the research and publish in their native languages (Rundel and Boonpragob 1995).

## REFERENCES

Andrade Lima, D. 1981. "The caatinga dominium." *Revista Brasileira de Botânica* 4: 149–1963.

Aubréville, A. 1949. *Climats, Forêts et Désertification de l'Afrique Tropicale.* Paris: Larose.

Beard, J. S. 1955. "The classification of tropical American vegetation-types." *Ecology* 36: 89–100.

Boughey, A. S. 1957a. "The physiognomic delimitation of West African vegetation types." *Journal of the West African Science Association* 3: 148–163.

Boughey, A. S. 1957b. "The vegetation types of the Federation." *Proceedings and Transactions of the Rhodesian Science Association* 45: 73–91.

Bullock, Stephen H., Harold A. Mooney, and Ernesto Medina, eds., 1995. *Seasonally Dry Tropical Forests.* Cambridge: Cambridge University Press.

Ceballos, Gerardo 1995. "Vertebrate diversity, ecology, and conservation in neotropical dry forests." Pp. 195–220 in Bullock, Stephen H., Harold A. Mooney, and Ernesto Medina, eds., *Seasonally Dry Tropical Forests.* Cambridge: Cambridge University Press.

Champion, H. G. 1936. "A preliminary survey of the forest types of India and Burma." *Indian Forest Records, New Series* 1: 1–286.

Desanker, P. V., P. G. H. Frost, C. O. Frost, C. O. Justice, and R. J. Scholes, eds., 1997. The Miombo Network: Framework for a Terrestrial Transect Study of Land-Use and Land-Cover Change in the Miombo Ecosystems of Central Africa, IGBP Report 41. Stockholm: The International Geosphere-Biosphere programme. Also available at http://miombo.gecp.virginia.edu/igbp/index.html.

Edwards, M. V. 1950. "Burma forest types according to Champion's classification." *Indian Forest Records, New Series* 7: 135–173.

Elliott, S., J. F. Maxwell, and O. P. Beaver. 1989. "A transect survey of monsoon forest in Doi Suthep-Pui National Park." *Natural History Bulletin of the Siam Society* 37: 137–141.

Eyre, S. R. 1968. *Vegetation and Soils.* 2nd edition. Chicago: Aldine Publishing Company.

Furley, Peter A. and Walter A. Newey. 1983. *Geography of the Biosphere, An Introduction to the Nature, Distribution and Evolution of the World's Life Zones.* London: Buttersworths.

Galera, Francisca M. and Lorenzo Ramella. n.d. "Gran Chaco data sheet." *Centres of Plant Diversity,* Vol. 3. *The Americas.* Washington, DC: National Museum of Natural History, Smithsonian Institution. http://www.nmnh.si.edu.

Gentry, Alwyn H. 1995. "Diversity and floristic composition of neotropical dry forests." Pp. 146–194 in Bullock, Stephen H., Harold A. Mooney, and Ernesto Medina, eds., *Seasonally Dry Tropical Forests.* Cambridge: Cambridge University Press.

Graham, Alan and David Dilcher. 1995. "The Cenozoic record of tropical dry forest in northern Latin America and the southern United States." Pp. 124–145 in Bullock, Stephen H., Harold A. Mooney, and Ernesto Medina, eds., *Seasonally Dry Tropical Forests.* Cambridge: Cambridge University Press.

Hartshorn, G. S. 1983. "Plants." Pp. 118–157 in D. H. Janzen, ed., *Costa Rican Natural History.* Chicago: University of Chicago Press.

Janzen, D. H. 1988. "Tropical dry forests, the most endangered major tropical ecosystem." Pp. 130–137 in E. O. Wilson, ed., *Biodiversity.* Washington, DC: National Academy Press.

Lugo, A. E., J. A. Gonzales-Libby, B. Cintrón, and K. Dugger. 1978. "Structure, productivity, and transpiration in a subtropical dry forest in Puerto Rico." *Biotropica* 10: 278–291.

Mares, M. A. 1992. "Neotropical mammals and the myth of Amazonian biodiversity." *Science* 255: 976–979.

Mares, M. A., M. R., Willig, K. E. Steinlein, and T. E. Lacher, Jr. 1981. "The mammals of Northeastern Brazil: a preliminary assessment." *Annals of the Carnegie Museum* 50: 81–137.

Martin, P. S. and David A. Yetman. 2000. "Introduction and prospect: Secrets of a tropical deciduous forest." Pp. 3–18 in Robichaux, Robert H. and David A. Yetman, eds., *The Tropical Deciduous Forest of Alamos: Biodiversity of a Threatened Ecosystem in Mexico.* Tucson: The University of Arizona Press.

Menaut, Jean-Claude, Michel Lepage, and Luc Abbadie. 1995. "Savannas, woodlands, and dry forests in Africa." Pp. 64–92 in Bullock, Stephen H., Harold A. Mooney, and Ernesto Medina, eds., *Seasonally Dry Tropical Forests.* Cambridge: Cambridge University Press.

Monod, T. 1963. "Après Yangambi (1956): notes de phytogéographie africaine." *Bulletin de l'IFAN, Série A* 24: 594–657.

Mooney, Harold A., Stephen H. Bullock, and Ernesto Medina. 1995. "Introduction." Pp. 1–8 in Bullock, Stephen H., Harold A. Mooney, and Ernesto Medina, eds., *Seasonally Dry Tropical Forests.* Cambridge: Cambridge University Press.

Murphy, P. G. and A. E. Lugo. 1986a. "Ecology of tropical dry forest." *Annual Review of Ecology and Systematics* 17: 67–88.

Murphy, P. G. and A. E. Lugo. 1986b. "Structure and biomass of a subtropical dry forest in Puerto Rico." *Biotropica* 18: 89–96.

Murphy, Peter G. and Ariel E. Lugo. 1995. "Dry forests of Central America and the Caribbean." Pp. 9–34 in Bullock, Stephen H., Harold A. Mooney, and Ernesto Medina, eds., *Seasonally Dry Tropical Forests.* Cambridge: Cambridge University Press.

RADAMBRASIL. 1983. Projecto Radambrasil: Levantamento dos Recursos Naturais, Vol 32 (23/24). Rio de Janeiro: IBGE.

Round, P. D. 1988. *Resident Forest Birds in Thailand: their status and conservation.* Monograph No. 2. Cambridge: International Council for Bird Preservation.

Rundel, Philip W. and Kansri Boonpragob. 1995. "Dry forest ecosystems of Thailand." Pp. 93–123 in Bullock, Stephen H., Harold A. Mooney, and Ernesto Medina, eds., *Seasonally Dry Tropical Forests.* Cambridge: Cambridge University Press.

Russell, Stephen M. 2000. "Birds of the tropical deciduous forest of the Alamos, Sonora, area." Pp. 200–244 in Robichaux, Robert H. and David A. Yetman, eds., *The Tropical Deciduous Forest of Alamos: Biodiversity of a Threatened Ecosystem in Mexico.* Tucson: The University of Arizona Press.

Sampaio, Everardo V. S. B. 1995. "Overview of the Brazilian caatinga." Pp. 35–63 in Bullock, Stephen H., Harold A. Mooney, and Ernesto Medina, eds., *Seasonally Dry Tropical Forests.* Cambridge: Cambridge University Press.

Schwable, Cecil R. and Charles H. Lowe. 2000. "Amphibians and reptiles of the Sierra de Alamos." Pp. 172–199 in Robichaux, Robert H. and David A. Yetman, eds., *The Tropical Deciduous Forest of Alamos: Biodiversity of a Threatened Ecosystem in Mexico.* Tucson: The University of Arizona Press.

Solomon, Derek. 1997. "Harare." Durban, South Africa: Southern African Birding. http://www.sabirding.co.za/birdspot/120408.asp.

Van Devender, Thomas R., Andrew C. Sanders, Rebecca K. Wilson, and Stephanie A. Meyer. 2000. "Vegetation, flora, and seasons of the Rio Cuchujaqui, a tropical deciduous forest near Alamos, Sonora." Pp. 36–101 in Robichaux, Robert H. and David A.

Yetman, eds., *The Tropical Deciduous Forest of Alamos: Biodiversity of a Threatened Ecosystem in Mexico*. Tucson: The University of Arizona Press.

Walter, Heinrich. 1985. *Vegetation of the Earth and Ecological Systems of the Geo-biosphere*. 3rd edition. Berlin: Springer-Verlag.

White, F. 1983. *The Vegetation of Africa: A Descriptive Memoir to Accompany the UNESCO/AEFAT/UNSO Vegetation Map of Africa*. Paris: UNESCO.

# 3 TROPICAL SAVANNAS

## OVERVIEW

The tropical **savanna biome** encompasses a variety of **vegetation** types that occur in the tropics where rainfall has a strong seasonal pattern. All types are characterized by a continuous cover of tall grasses with or without a discontinuous upperstory of trees, shrubs, or palms. The term savanna is said to derive from a Native American (Carib) word designating a treeless plain with much grass. The word first appeared in Europe in Spanish (*sabana*) in 1535 and has been known in English since 1555 (Bourliere and Hadley 1983).

Tropical savannas cover some 20 percent of Earth's land surface; major expanses are found in Africa, Australia, and South America (where they are known as *cerrado*). Small areas of savanna occur on the mainland and islands of Middle America. In all regions, savannas tend to be associated with low-nutrient **soils** and frequent fires. It should be noted that India and Southeast Asia also have savannas, but these are considered secondary—that is, they are the product of relatively recent human burning of the original dry forests of those regions, and therefore will not be further discussed in this chapter.

Tropical savannas are often categorized according to the nature of the shrub/tree layer and the spacing of woody plants. A number of classification schemes have been developed, but all incorporate, in one way or another, the following types:

1. Savanna woodlands, in which trees and shrubs are close enough together to form a light **canopy** that still allows for a continuous ground cover of grasses;
2. Park savannas, in which trees are grouped together in clusters with stretches of grassland between adjacent clusters;
3. Tree savannas, in which the upper story consists of scattered trees;
4. Shrub savannas, in which the upper story consists of scattered shrubs; and
5. Grass savannas, in which woody plants are absent altogether.

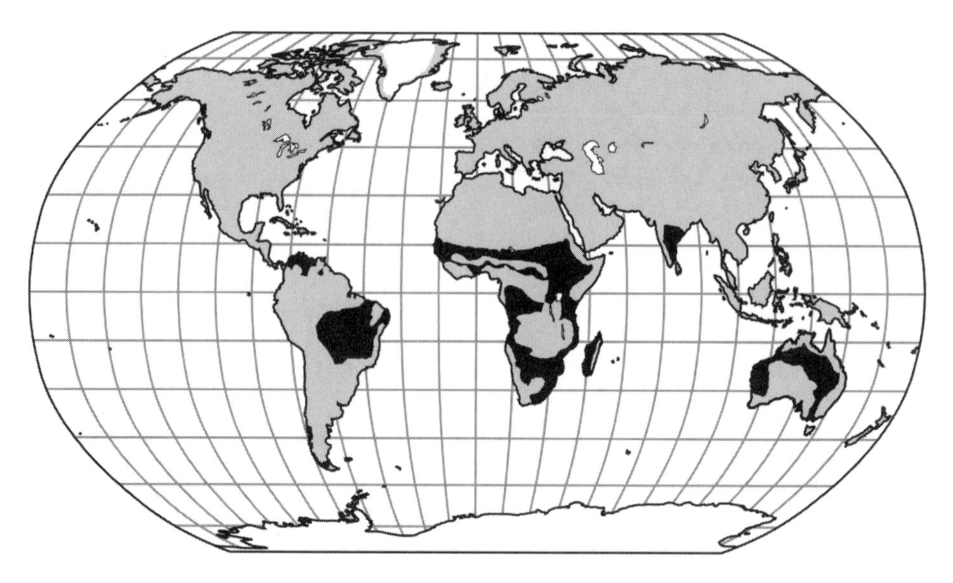

Map 3.1 World distribution of the Tropical Savanna Biome.

Furthermore, savannas may be identified according to a single dominant kind of woody plant, such as the palm (*Paurotis wrightii*) savannas in parts of Mexico or the pine (*Pinus caribea*) savannas of Belize (Bourliere and Hadley 1983; Sarmiento 1983; Cole 1986; Solbrig 1996; Mistry 2000).

Probably no single factor causes tropical savannas, and scientists have proposed several hypotheses to explain their existence. Although associated with tropical wet and dry **climate** regions, the precipitation pattern alone does not explain the occurrence of savannas because seasonal tropical forests (see Chapter 2) are also found under this climate type. Additional environmental constraints thus need to be invoked, and these have included poor soil-drainage conditions (Photo 3.1), soil infertility, geomorphic processes, frequent fire, and impacts of grazing animals. Different savanna regions have different reasons for being, which will be discussed later.

While the herbaceous layer everywhere is dominated by **species** from the same two families—the grasses (Gramineae) and sedges (Cyperaceae)—the woody plants of each continent differ in the species, genera, and families represented. They do have some commonalities, however. Three Gondwanan families (Protaceae, Bombacaceae, and Combretacease) are represented on all three continents, as are widespread families such as Leguminoseae and Myrtaceae. The South American cerrados have, in addition, two families (Velloziaceae and Vochysiaceae) unique to that continent (Solbrig 1996). The cerrados have the greatest number of species, with an estimated 10,000 **vascular plants** occurring in them. Only the Amazon rainforest has more species. Australian savanna has the fewest vascular plants of the three main continental areas of savanna but more species than are found in Australia's rainforest. Africa is intermediate in number of savanna species, with about the same number as found in African rainforest (Solbrig 1996).

Photo 3.1 Palm savanna in waterlogged soil conditions, Rio Grande Do Norte, Brazil. (Photo by SLW.)

Among the herbaceous plants, grasses are most abundantly represented. Any given savanna often has 30–60 species with only 6–10 dominant. The dominant grasses of tropical savannas are distinguished from grasses in dry seasonal forests of the tropics and from most **temperate zone** grasses by the chemical pathway through which carbon is assimilated into their tissue, which will be further explained later. They are referred to as $C_4$ grasses, as opposed to the $C_3$ grasses (and woody plants) elsewhere. In addition to grasses, any given savanna may have 15–30 species of sedges. **Legumes** are the third most important element in the **herb** layer (Solbrig 1996).

The savannas of East and South Africa are famous for their abundance of large mammals (Photo 3.2) including vast herds of grazing and browsing species and a large number of predators. This characteristic, however, does not apply to West African, Australian, or South American savannas, where large animals are noticeable by their near absence.

Humans are believed to have evolved on the tropical savannas of East Africa, and they may have used fire as an environmental management tool for a million years. Much more recently in geologic time did people arrive in Australia and the Americas. It is difficult to ascertain the impacts of the earliest peoples on the tropical savannas. Major transformations in the New World savannas are associated with the introduction of cattle raising to the Americas in the sixteenth century and to Australia in the nineteenth century (Solbrig 1996). Today, the species-rich cerrados are rapidly being cleared for modern, mechanized agriculture and the production of major export crops, especially soybeans.

Photo 3.2 Herd of water buffalo (*Syncerus caffer*) at water hole in Tsavo West National Park, Kenya. The landscape mosaic of trees, shrubs and grasslands are characteristic of African Savannas. (Photo © The Purcell Team/CORBIS.)

## VEGETATION

The vegetation of the tropical savanna biome is actually rather diverse in structure and composition, whether one is comparing the different continental expressions or examining its occurrence on a single continent. (These differences are highlighted in the regional descriptions following.) However, the main element of the plant cover is the same throughout; **tussock** or bunch grasses 50–120 cm (20–50 in) tall, and a few sedges, form a continuous ground cover. They may or may not be accompanied by woody species in an open-canopied shrub or tree layer. Gallery or riparian forests along watercourses and swamps and marshes are often part of the overall savanna **landscape**.

The apparent balance between grasses and woody plants poses an interesting problem for ecologists, since usually a tension exists between herbaceous and woody species that results in the replacement of one by the other depending upon climatic and soil conditions. Some type of regular **disturbance**, such as recurring drought, flood, fire, and/or the consumption of plants by insects and other animals, may be necessary to allow both types of plants to coexist (Bourliere and Hadley 1983; Mistry 2000).

One thing that all researchers comment on is the abrupt boundary that occurs between forest and savanna. Savanna is not an ecotone, not a mix of forest or woodland and grassland species. The trees on the savannas are different species from those found in nearby forests and have different morphologies and different physiologies. The grasses are also different from those growing beneath the canopy of a tropical seasonal forest. Furthermore, in Africa and Australia, different species of herbaceous and woody plants occur in each type of savanna. This is not so much the case in the Brazilian formations, however, where one tends to find the same species in the various types of cerrado (Furley et al. 1992; Cole 1986).

The tropical savanna grasses are predominantly shade-intolerant, **perennial** $C_4$ plants. This means that as carbon dioxide is fixed during the **photosynthesis** process, it is first incorporated into a 4-carbon compound, oxaloacetic acid. The woody species of the **tropics** and most plant species of other **latitudes** are $C_3$ plants. In them, carbon dioxide from the atmosphere is transformed directly into a 3-carbon organic compound, phosphoglyceric acid, in the Calvin cycle. The $C_4$ pathway involves an additional, preliminary step, because specialized cells convert the 4-carbon compound into a 3-carbon compound. $C_4$ photosynthetic reactions are more tolerant of high temperatures and accumulate carbon dioxide more rapidly than $C_3$ photosynthesis—distinct advantages in the hot, seasonally dry tropics. **Stomata** do not have to remain open as long as under $C_3$ photosynthesis, and thus less water is lost from the plant via transpiration (Mistry 2000). Most tussock grasses and sedges have hard leaves with abundant silica bodies. A thick covering of dead leaf sheaths covers the growth buds that lie below ground, close to the surface, and protects them from drought and fire (Sarmiento and Monasterio 1983).

Savanna trees are shorter in stature than forest trees; typically they are 2–6 m (6–20 ft) in height, and even the tallest are less than 12 m (40 ft) high. Trunk diameters are usually small, around 20–30 cm (8–12 in), although the **succulent**-stemmed baobabs (*Adansonia* spp.) of Africa and Australia are spectacular exceptions to this general rule (Photo 3.3). Those trees often have a tortuous branching pattern that gives them a gnarled appearance, frequently described as similar to orchard trees that have been pruned annually. Leaves, whether simple or **compound**, are flat, often leathery, and have entire margins. They may have a thick cuticle and sunken stomata as adaptations to drought. Most species are **deciduous** in that their leaves only live one year, but leaf fall may coincide with the development of new leaves, so the woody layer may appear to be **evergreen** with simply a thinner canopy during the dry season. Many trees have a thick, corky bark that protects them during ground fires, and many have the ability to resprout from roots or deep lateral roots if the above-ground parts are severely damaged. In Africa, at least, spines are common on trees; as reduced leaves these may be an adaptation to drought, but they are also considered a defense against browsing animals. The root systems of many savanna trees combine a deep taproot with a thick system of lateral roots near the surface (Sarmiento and Monasterio 1983).

Another type of woody plant characteristic of, and almost limited to, tropical savannas is what is variously referred to as a **subshrub** (or half-woody plant, suffrutescent shrub, geoxyle, or hemixyle, among other designations). These plants have woody underground water and nutrient storage organs (xylopodia) but only seasonal above-ground shoots. They are represented in African and American savannas by a rich number of species and underground morphologies. Some species always produce annual shoots; others do so only when under stress and otherwise have perennial above-ground growth. Subshrubs of the savannas are generally related to forest trees (Sarmiento and Monasterio 1983).

The time of year that savanna plants actively grow, flower, or fruit varies from species to species. Some produce new plant tissue (leaves, stems, **rhizomes**) year round (although the rate may be very slow during the dry season), others only sea-

Photo 3.3 A baobab (*Adansonia* sp.) towering above acacias on an island in the Okavango Delta, Botswana. Such trees, often hollowed out, can provide shelter, food, and moisture for humans as well as other animals. (Photo © Wolfgang Kaehler/CORBIS.)

sonally. By convention, the beginning of the rainy season is considered the beginning of the growing season. The pattern of flowering is identified according to its relationship to the onset of rains. Species that leaf out and bloom simultaneously at the beginning of the rainy season are said to have precocious flowering. Those that leaf out at the beginning of the rainy season but flower several months later have delayed flowering. Some species have tardy flowering and bloom during the dry season. And a few bloom only after a fire, whenever the burn occurs, and they are said to have opportunistic flowering.

The majority of woody and half-woody plants renew their leaves and bloom during the dry season. The majority of perennial grasses and sedges grow and bloom during the rainy season. Annual grasses and **forbs** are abundant in some but not all savannas; they can be divided into groups that bloom with the onset of the rains (precocious) and those with delayed flowering that bloom toward the middle or end of the rainy season (Sarmiento and Monasterio 1983).

## CLIMATE

The savanna or tropical wet and dry climate region (Koeppen's Aw) lies between the equatorial zone with its year-round precipitation and those subtropical regions with year-round aridity, roughly between 10° and 30° latitude. The most important aspect of the climate for life is the strong seasonality of precipitation. Rains occur during the high sun period; the dry season ranges from three to four months in the areas closest to the equator to eight or nine months at the poleward margins where arid climate regions begin. Total annual precipitation ranges from about 500 mm to 1,500 mm (20–60 in), and occasionally higher, depending primarily upon latitude. The timing of the rains is regulated by the seasonal movement of subtropical high pressure zones and the **intertropical convergence zone (ITC)** in response to the annual migration of the vertical rays of the sun between the Tropic of Cancer and the Tropic of Capricorn. A monsoonal effect is also in play over West and East Africa. Areas close to a Tropic experience one rainy season a year, and areas intermediate between the Tropics and the Equator may have two peaks in precipitation each year.

Temperatures are warm year round, as befits a tropical climate, but there is a greater annual range than in the equatorial zone covered by tropical rainforest. Mean monthly temperatures of the warmest months may be 30–35° C (85–95° F) near the boundary with the rainforest and 25–30° C (75–85° F) at the margins of the deserts. The coolest months average 13–18° C (55–65° F) at the lowest latitudes and 8–13° C (45–55° F) in the drier regions near the Tropics themselves (Cole 1986; Nix 1983).

## SOILS

No single soil type can be considered typical of tropical savannas, yet certain types prevail under certain circumstances of rock type, landform type, drainage conditions, and age. Most are low in plant nutrient content and organic matter, and some actually chemically interfere with nutrient uptake by plants. On ancient upland surfaces, such as ones found in Brazil and western Africa, the **parent material** is deeply weathered, low in clay content, and heavily leached of plant nutrients. Iron compounds are concentrated, which gives these infertile soils a characteristic red or yellow color. In many of the areas of Central Brazil covered by cerrados, the soils have also concentrated aluminum to levels toxic to crop plants. It is possible that these soils formed during hot, wet climates of the Cretaceous and early Tertiary periods—when tropical forests replaced savannas in many places—and are not the product of contemporary tropical savannas (Cole 1986). In U.S. soil taxonomy, these upland soils are classified as oxisols. They are acidic with a mean **pH** of 4.9. In areas where the water table is high during the rainy season but the ground dries out during the dry season, a hardpan called laterite (from the Latin word for brick) forms (Montgomery and Askew 1983).

Soil-forming processes more common to midlatitude regions (i.e., podzolization) have played a role and created soils much less acidic (pH = 6.2) than oxisols in dissected terrain, over crystalline bedrock, and over much of the eastern and southern African savanna region where parent material has been more recently exposed to **weathering** and drainage is better than in the uplands. These types are classified in the U.S. system as ultisols (Montgomery and Askew 1983; Cole 1986).

A third common soil type—associated with grass savannas of low plains and valleys—is black cracking clay soil, a vertisol in the U.S. soil classification. Black cracking soils have a high clay content, which causes shrinking and swelling with changes in the soil water regime. The shrinkage that occurs during the dry season gives the characteristic cracks. These soils develop over base-rich bedrock such as basalt or limestone and hence are high in calcium and magnesium. They have relatively high natural fertility and a pH of 6–7.5 (Montgomery and Askew 1983; Cole 1986).

## ANIMAL LIFE

Herds of many kinds of large grazing and browsing mammals—especially members of the cattle family, such as antelopes, buffalo, and wildebeest, but also elephants, zebras, and rhinoceroses—are what usually come to mind when one thinks of tropical savannas. This picture is true only for East Africa and southern Africa. Smaller mammals, particularly rats and mice, are more typical elsewhere. Squirrels are common in South America and Africa where trees are prevalent. (In Australia, the squirrel niche is filled by arboreal marsupials, possums and gliders.) In addition to rodents, hares (Africa) and rabbits (South America) inhabit savannas. Most small mammals are nocturnal, and many spend the day below ground in burrows. They generally eat a wide variety of plant foods—green foliage, fruits, seeds, roots and tubers—and some also include insects in their diet. The times of stress for these species is toward the end of the dry season and immediately after a fire. Only a few species store food in their burrows, so it is probable that many accumulate fat deposits during the rainy season when food is abundant and then draw upon them for energy as food supplies diminish. Small mammals can survive fast-moving grass fires below ground or, if they are surface-dwellers, can flee the flames. The dangerous period follows a fire when cover and forage has been destroyed. It may be two to three weeks before nutritious new shoots appear (Happold 1983).

Many more mammals in the tropical savannas are carnivorous and insectivorous than are herbivorous. Among the more peculiar are species adapted to extract ants and termites from their nests. South America has anteaters (Order Xenartha), and Africa has pangolins or scaly anteaters (Order Philidota) and aardvark (Order Tubulindentata). From extremely distant lineages, these animals have evolved similar **morphology** and behavior and are prime examples of convergent evolution. A large number of other insect-eating mammals can be found on all continents; many are restricted to a single continent and identified in the regional descriptions following. Common to all three continents are bats that feed on flying insects, although some bats also feed on blood (vampire bats of South America and the false vampires of Africa). With respect to true carnivores (Order Carnivora), Africa has the greatest number of species, which reflects the great variety of herbivores. Dogs, cats, and hyenas and a host of smaller species feed on arthropods, reptiles, birds, and rodents. South America has very few carnivores, and Australia has only one, the dingo (*Canis familiaris dingo*)—probably introduced by people thousands of years ago (Bourliere 1983).

Among true savanna mammals, adaptations to withstand high solar radiation, high temperatures, and limited water are necessary. Strategies such as panting rather than sweating allow the lowering of body temperatures with relatively little water loss. Some animals are able to reduce the output of urine or concentrate their urine as a way of limiting water loss. Reabsorbing water from the feces in the intestinal tract is another way to reduce water loss. Among African hoofed animals, some stand out as particularly well adapted in these physiological ways to withstand drought; in order of efficiency are dikdik (*Modoqua kirki*), oryx (*Oryx* spp.), hartebeest (*Alcelaphus* spp.), impala (*Aepyceros melampus*), and eland (*Taurotragus oryx*) (Sinclair 1983). Other ways to reduce water loss are primarily behavioral. The eland, for example, stands in the shade by day and feeds at night, when the air is more humid and the water content of leaves is higher. It and others select the most succulent leaves to ingest to take in as much water as possible. Another behavioral response is to stand facing away from the sun during the day to expose the least amount of surface area to the sun's direct rays. Coloration also can help. A light-colored coat, or a partly white coat as in the zebra, reflects sunlight and can help prevent overheating. During the dry season, some animals are nomadic and move toward any area that has recently received a dry-season shower; others are migratory and follow the same path each year to reliable water sources. Sedentary species dependent upon free water, animals such as elephant and buffalo, are confined to constricted ranges near water during the dry season (Sinclair 1983).

Morphological and behavioral adaptations allow some species to reach high tree and shrub **crowns**. The giraffe, with its long neck and legs, epitomizes this adaptation. The gerenuk (*Litocranius walleri*) is similarly adapted, though at a smaller scale. Furthermore, the gerenuk has the behavioral adaptation of standing on its hind legs to reach higher, a trait shared with some other gazelles.

Birds are particularly abundant and diverse in tropical savannas. Overall, some 1,500 species in 100 families have been described. Notable are large ground-dwelling birds; each continent has a unique form: the ostrich (*Struthio camelus*) in Africa, the emu (*Dromaius novaehollandiae*) in Australia, and two species of rhea (*Rhea americana* and *Pterocnemia pennata*) in South America. A large number of scavengers and predatory birds such as hawks are represented. Songbirds and small birds are also numerous and generally feed on seeds fallen to the ground or on insects. Fruit eaters in South America and Australia include parrots (Psittacidae) and in Africa, mousebirds (Collidae) and hornbills (Bucerotidae) (Fry 1983).

Lizards and snakes come from the widespread colubrid (Colubridae) and viper and pit viper (Viperidae) families and from tropical families such as geckos (Gekkonidae), skinks (Scincidae), and boas (Boidae). Africa also has chameleons (Chamaeleontidae), and South America has iguanas (Iguanidae).

Amphibians are represented mainly by frogs, toads, and tree frogs that are usually associated with permanent or seasonal waters since water is necessary for their reproduction. They display several adaptations to life in the savanna including camouflage coloration and patterning. Most are active only at night when humidity is highest. Some are active only during the rainy season and burrow underground to moister conditions in the dry season. A few live permanently underground in loose, moist soil (LaMotte 1983).

Of the invertebrates, termites (Isoptera) are a conspicuous and characteristic element of all savannas, since many species build above-ground nests or termitaria. Some of these, especially those of the fungus-growing termites of Africa, are mounds or pinnacles several meters high. Other species make spherical earthen nests in trees and cover their runways on the ground and tree trunks with soil. Termites play a key role in nutrient cycling and in soil aeration. To form their termitaria, they extract fine-grained soil material from depths of as much as 150 cm (5 ft). Depending upon their specific diet, termites bring dead plant matter or **humus** into the nest for food. The fungus-growing types deposit slightly digested plant matter in tiny balls. Fungi in the termitaria digest the cellulose that termites are unable to use and leave an enriched material upon which the termites can feed. Termite mounds become a source of fine-grained, nutrient-rich (especially calcium and nitrogen) soil and may support a distinct group of plants. Leaf-cutting ants (*Atta* spp.) in American savannas play a similar role (Josens 1983; Schultz 1995; Mistry 2000).

## MAJOR REGIONAL EXPRESSIONS

### African Savannas

Three broad geographic subdivisions of the African savannas are often recognized: the West African savannas, the East African savannas, and the southern African savannas. The West African savannas occupy a narrow band between 5° N and 15° N from the Atlantic coast of Senegal and southern Mauritania eastward above the Congo Basin rainforest to the Nile River. They are essentially separated geographically from the savannas of eastern and southern Africa. Within this band are three latitudinal zones with increasing moisture, plant diversity, and height of the vegetation from north to south. In the north, the Sahelian zone is the driest part of the West African savannas; it receives less than 500 mm (20 in) of rainfall a year. This is essentially a dry grassland with widely scattered bushes and small trees 2–5 m (6–15 ft) high. Acacias (*Acacia tortilis, A. laeta, A. ehrenbergiana*) are the most common woody species; annual grasses are dominant in the herb layer.

The intermediate savanna belt, where annual precipitation averages 500–1,500 mm (20–60 in), is called the Sudanian zone. Savanna woodlands composed of *Isoberlinia doka* are common; other woody species such as various acacias (*Acacia dudgeonii, A. albida. A. nilotica*, etc.), *Burkea africana*, and *Terminalia* spp. are also important. Most reach 5–7 m (15–20 ft) in height, although some larger trees, such as baobab (*Adansonia digitata*), reach 7–15 m (20–50 ft). Perennial and annual grasses make up the herb layer (Cole 1986; Mistry 2000).

The southernmost belt of savanna is the Guinean zone, characterized by a moister savanna woodland, where *Anogeissus leiocarpus* and *Lophira lanceolata* are the most common tree species. They attain heights of 12–15 m (40–40 ft). Tall grasses range in height from 0.5 m (1.5 ft) in the north to 1.5 to 3 m (5–15 ft) in the southern part of this zone (Cole 1986).

West African savannas have a high bird diversity but fewer large mammals than the other African savannas. Elephant (*Loxodonta africana*), buffalo (*Syncerus caffer*), several species of antelope, and warthog are present in diminishing numbers. Termites play an essential role in nutrient cycling here as elsewhere in the biome (Mistry 2000).

East African savannas extend from 16° N in Somalia through eastern Ethiopia and Kenya into Tanzania to a latitude of about 9° S. This is a region that typically receives less than 500 mm (20 in) of precipitation a year, usually during two rainy seasons. The first rainy season is the longest and occurs between March and May; the second generally occurs between mid-October and December. In some areas, these two periods merge into a single rainy season; Tanzania, for example, receives rains from November through April. Mean monthly temperatures are from 25° C to 30° C (75–85° F) all year long. The East African savanna is primarily a tree and shrub savanna. Shrubs stand 3–5 m (10–15 m) high with scattered trees emerging above them to heights up to 9 m (30 ft). Plants vary from location to location, but nearly everywhere are acacias and species from the genera *Commiphora* and *Grewia*. Most woody species are multistemmed and deciduous. The few emergent species, such as *Acacia tortilis*, baobab, and *Terminalia spinosa*, have single trunks. The grasses tend to be annuals and short-lived perennials (Mistry 2000).

Within the East African savannas lies the 35,000 km$^2$ (13,500 mi$^2$) Serengeti region, one of the best known sanctuaries for large mammals on Earth and a unique **ecosystem**. The Serengeti is a grass savanna bordered in the north and west by savanna woodlands. The grasslands consist of a mosaic of grass communities, the most prevalent of which is dominated by red oat grass (*Themeda trianda*). Short grasses dominated by dropseed (*Sporobolus* spp.) are abundant in the plains, and tall russet grasses (*Loudetia* spp.) and thatching grass (*Hyperthelia* spp.) are concentrated in the far northwest reaches of the Serengeti. The fringing savanna woodlands have an **open canopy** of thorny acacias and commiphoras, which usually grow no higher than 3–7 m (10–20 ft), and some understory shrubs (e.g., *Grewia* spp. and *Cordia* spp.). The herb layer is 0.5 m to 1.5 m (1.5–5 ft) tall and composed of red oat and other grasses. Twenty-three kinds of wild, hoofed mammals are found in the Serengeti. Over 1.3 million wildebeest, accompanied by plains zebra (*Equus burchelli*) and Thomson's gazelle, migrate across the plains. At the beginning of the rainy season, they move from the northern savanna woodlands to the southern short-grass plains, and spend the wettest part of the year in the driest part of the region. They return to the northern savanna woodlands, the wettest part of the Serengeti, for the dry season. The Serengeti Plains also have resident herds that are nomadic and move from pasture to pasture within the grass savanna. During the wet season, animals such as springbuck (*Antidorcas marsupalis*) and dama gazelle (*Gazella dama*) spend expended periods of time at given locations and then return when the vegetation has recovered. During the dry season, however, they move rapidly between medium and tall grasslands, according to where sporadic showers have occurred (Mistry 2000).

Southern African savannas occur primarily on the plains of the Kalahari Depression from east-central Namibia, across southern Botswana, northern South Africa and southern Zimbabwe into Swaziland and Mozambique with a southward extension along the Indian Ocean coast of South Africa. In South Africa and Namibia, they are commonly referred to as bushveld. In all, southern savannas range from about 18° S in the west to 34° S in the Eastern Cape of South Africa. Distinctions are made between savanna vegetation developed on fertile, though dry, volcanic soils (vertisols) and that developed on wetter, infertile soils (oxisols) derived from the ancient African basement rock. The former is generally confined to the Kalahari in the central part of the region and is characterized by fine-leaved, thorny trees of *Acacia* and *Commiphora* in the upper story. The dominant grasses come from the Eragrostideae, Aristieae and Chlorideae tribes. The ancient rocks of the African shield are exposed north of the Kalahari and along the east coast of South Africa; on their deeply weathered soils grow mostly thornless, **broad-leaved** trees such as *Burkea africana*, many of which contain unpalatable or digestion-inhibiting substances that defend them against predators. The dominant tribe of grasses is Andropogoneae (Cowling et al. 1997; Mistry 2000).

One of the most noticeable aspects of the African savannas, particularly in East and southern Africa, is the great diversity of hoofed mammals. The cattle family (Bovidae) alone has 78 members. The bovids are subdivided into a number of tribes, nine of which are represented on the African savannas in one **habitat** or another. These include members of the Tragelaphini such as eland and kudu, which prefer shrub and thicket habitats in broad-leaved woodlands and the sitatunga that is restricted to swamps. The tribe Bovini is represented by the African buffalo, a grazer. The gray duiker (*Sylvicapra grimmia*) (Cephalophini) prefers savanna thickets, while reedbuck (*Redunca* spp.), waterbuck (*Kobus ellipsiprymunus*), and lechwe (*Kobus leche*) (Reduncini) are restricted to swamps and moist savannas. In dry woodland habitats are members of the Hippotragini, such as roan antelope (*Hippotragus* *equinus*), sable antelope (*H. niger*), oryx (*Oryx* spp.), and an Aepycerotini, the impala. Grazers on the more mesic plains include members of the Alcephini, such as topi (*Damaliscus korrigum*), hartebeest (*Alcephalus* spp.), and wildebeest (*Connochaetes* spp.). The small oribi (*Ourebia ourebi*) and klipspringer (*Oreotragus oreotragus*) (Neotragini) hide in thickets. In addition to the bovids, the large herbivorous mammal group contains one species of elephant, two rhinos [the white rhino (*Ceratotherium simum*), a grazer, and the black rhino (*Diceros bicornis*), a browser], plains zebra, warthog (*Phacochoerus aethiopicus*), bush pig (*Potamochoreus porcus*), hippopotamus (*Hippopotamus amphibius*), water chevron (*Hyemoschus aquaticus*), and giraffe (*Giraffa camelopardalis*) (Sinclair 1983).

Each mammal has a preferred habitat and preferred plants and plant parts it eats. Browsers tend to be more selective than grazers. This divvying up of available food sources may be part of what allows so many different kinds of mammals to coexist. Different species also may graze or browse the same area but at different times of year. Researchers in East Africa, in particular, talk of a grazing succession, in which large animals come into an ungrazed area first. They tend to be less selective than smaller species and eat low quality food such as mature grass stems in large amounts, so they eat the bulk of the tall grass and move on. Smaller species, which require more nutritious food in smaller amounts, then move into the grazed area to feed selectively on smaller plants left by the large grazers or on the new, high-protein shoots that grasses produce in response to the removal of much of their aerial parts by the large animals. Some question just how real this apparent succession of grazers is, but it does appear that Thomson's gazelle follows wildebeest onto an area to take advantage of new growth (Sinclair 1983).

The activities of the abundant large grazing and browsing animals may play a major role in maintaining the open canopy of savanna trees and shrubs that in turn allows the growth of tall $C_4$ grasses. The grasses are the fuel for fires in the savanna, and so herbivores, grasses, and the pattern of fire have a delicate interdependence. Fires stimulate the regrowth of grasses. Too little browsing can lead to grasses being shaded out by the tree or shrub canopy, with a resulting loss of fuel and lowered frequency of fire. On the other hand, overgrazing can thin the grass layer to a level that cannot carry a fire, which thereby promotes the invasion of shrub and trees defended by thorns or unpalatable substances in their leaves (Hopkins 1983). The challenge of wildlife managers is to maintain an ever-changing mosaic of savanna communities that will continue to support a diverse range of mammalian herbivores.

Other mammals worthy of note in the African savannas are the insect and meat eaters. Among the first group are bats, elephant shrew (*Elephatulus rufescens*), and hedgehog (*Erinaceus albiventris*). The previously mentioned pangolins and aardvark also belong in the insect-eating group. Carnivores are diverse in terms of species, size, and food preferences. The hunting dog (*Lysoan pictus*), Abyssinian wolf (*Canis simiensis*), three species of jackal (*Canis* spp.), and striped weasel (*Poecilogale albinucha*) focus on small mammals. The honey badger (*Mellivera capensis*) has a broad diet of scorpions, spiders, snakes, fruits, and carrion, as does the African civet (*Vivvera civetta*). Eight kinds of mongooses specialize on beetles and termites; the aardwolf (*Proteles cristatus*), a type of hyena, specializes on surface-feeding termites. The spotted hyena (*Crocuta crocuta*) is a scavenger, except in the Serengeti where it hunts; the brown hyena (*Hyaena brunnea*) is also a scavenger, but the striped hyena (*H. hyaena*) eats both animals and plants. The cats all take warm-blooded prey. Lion

(*Panthera leo*), leopard (*P. pardus*), and cheetah (*Acinonyx jubatus*) can prey upon the larger grazing and browsing mammals; the caracal (*Felis caracal*) and serval (*F. serval*) concentrate on small rodents (Bourliere 1983).

African savannas host some 700 species of birds belonging to 57 families. In the northern savannas, 180 migrant species from Eurasia are added during the winter dry season. The largest group of native birds are the weavers (Ploceidae) with 61 different species, including bishops and whydahs. One, the quelea (*Quelea quelea*), is superabundant, and breeds at the end of the rainy season in huge colonies of grass nests woven to the branches of trees in the Sahel. All weavers are seed eaters, so when the seed supply in the Sahel is exhausted, the queleas migrate across the equator to the southern savannas to forage as seeds ripen at the end of the Southern Hemisphere rainy season. The weaver finches (Estrilidae) are another large group of seed-eating birds. Old World warblers (Sylviidae) are primarily insect eaters, although members of the large **genus** *Cisticola*—the grass warblers—glean seeds from the grass stalks. The 23 kinds of sunbird (Nectariniidae) are nectar and insect feeders and the 22 barbets (Capitonidae) are berry eaters. Some of the larger birds of African savannas include the secretary bird (*Sagittarius serpentarius*), which preys on snakes; bustards (Otididae), running birds that are omnivores; and, of course, the ostrich. Old World vultures (Aegypiidae)—skillful scavengers of animal carcasses—and falcons (Falconidae) are relatively numerous with 9 and 11 species, respectively (Fry 1983).

## Australian Savannas

The savannas of Australia form an arcing belt that begins in northern Western Australia at about 17° S, runs parallel to the northern margin of the continent and then bends southward along the east coast in Queensland to about 29° S. Rainfall varies from 500–1,500 mm (20–60 in) and is concentrated during the summer months of December through March. The wettest areas lie in the north, and the climate gets progressively drier farther from the coast. Several types of savanna exist, but most have woody plant layers dominated by eucalypts and acacias (Cole 1983). Other prominent woody plants include melaleucas (Myrtaceae), mimosas (Mimosaseae), *Hakea* and *Banksia* (Protaceae), *Bauhinia* (Fabaceae) and *Terminalia* (Combretaceae). Trees and shrubs may be evergreen, brevideciduous (less than 50 percent of the canopy is lost briefly during the dry season); semideciduous (more than 50 percent of the canopy is dropped in dry seasons); or fully deciduous (all leaves are lost during the early to mid-dry season, and plants remain leafless for at least one month (Mistry 2000).

Bunch grasses from several genera dominate the herb layer; the actual species vary according to type of savanna. Different savanna communities occur in more of a mosaic than a strictly zoned pattern and are distinguished primarily on the basis of the composition and height of the grass layer. Most authors recognize six types but do not always agree upon which six. The following are generally agreed upon. Monsoon tallgrass savannas appear on low-fertility soils across the northern part of the continent; dominant grasses include red oat grass (*Themeda triandra*), *Schizachyrium fragile*, *Sorghum* spp., and *Chrysopogon fallax*. Tropical tallgrass savanna and subtropical tallgrass savanna (also known as black spear grass savannas) are found along the east coast with the separation between the two occurring at about the 21° S parallel. *Heteropogon contortus* and *H. triticeau* occur in both, accompanied by *Themeda* spp. and *Bothriochloa* spp. Midgrass savannas dominated by *Aristida* grasses occur on infertile soils in the Northern Territory and Western Australia. On cracking black clays of moderate to high fertility, mostly in Western Australia, are midgrass communities with the grasses *Dichanthium sericeum*, *Bothriochloa decipiens*, *B. bladhii*, and *Chloris* spp. Acacia shrublands, or mulga pastures, occur throughout northern Australia on stony soils of low fertility and low water-holding ability. The woody layer is dominated by mulga (*Acacia aneura*), and the herb layer is composed of *Digitaria* spp., *Monochather paradoxa*, *Eriachne* spp., and *Aristida* spp. (Cole 1986; Solbrig 1996; Mistry 2000).

The larger mammals of the Australian savannas consist of six species of the marsupial family Macropodidae. These are the agile wallaby (*Macropus agilis*), antilopine wallaroo (*M. antilopinus*), northern brown bandicoot (*Isoodon macrurus*), northern quoll (*Dasyurus halluctus*), brushtail possum (*Trichosurus vulpecula*), and sugar glider (*Petaurus breviceps*). The last two are counterparts of the arboreal squirrels in African and South American savannas (Mistry 2000). Small mammals include rodents, the only placental mammals other than bats native to Australia. They are represented by rats (*Rattus* spp.), banana rats (*Melomys* spp.), tree rats (*Colinurus* spp.), and a web-footed water rat (*Hydromys* spp.) (Bourliere 1983).

Australian savannas have a great diversity of birds; common are lorikeets, parrots, and cockatoos—all members of the parrot family. Also found are many honeyeaters (family Meliphagidae), butcherbirds (family Artamidae), finches, and raptors such as falcons and kites. The emu, as noted earlier, is the large flightless bird of Australian grasslands. Among common reptiles are the frill-necked lizard (*Chlamydosaurus kingii*), sand goanna (*Voranus gouldii*), two-lined dragon (*Diporophora billenata*), and numerous skinks and snakes (Mistry 2000).

## South American Savannas

The savannas of South America occur in widely separated regions and include the Llanos del Orinoco or, more often, simply the llanos in the basin of the Orinoco River in Colombia and Venezuela; the Gran Sabana of Venezuela; the Rio Branco-Rupununi savannas just north of the Amazon in the borderlands of Brazil and Venezuela; the Llanos de Mojos, a periodically flooded savanna in the eastern part of Bolivia; the Pantanal, a great flooded area in the drainage basin of the Parana and Paraguay rivers; and the various cerrados of Brazil. Sometimes the Gran Chaco, part of which extends into the midlatitudes, is classified as tropical savanna, and as sometimes is the caatinga, a semiarid thornscrub vegetation type in Northeast Brazil (see Chapter 2). Only the llanos and the cerrados will be described in any detail here, as they are the major expressions of the tropical savanna biome in South America. Discussion of the pantanal, the world's largest wetland, is included in the section on freshwater biomes (Chapter 10).

**Llanos.** The *llanos* (Spanish for plains) is the major savanna region of northern South America and covers some 500,000 km$^2$ (200,000 mi$^2$) from the Guaviare River in Colombia east along the north (left) side of the Orinoco River to its delta. Elevations in the region range from 100–500, (300–1,600 ft) above sea level. Temperatures vary little month to month. The rainy season (April to November) is warm and humid; the dry season is warm and dry. Total annual precipitation ranges from 800 mm (30 in) in the east to 2,500 mm (100 in) in the southwest. Soils are typically highly leached, acidic, and infertile; oxisols dominate in the well-drained areas, vertisols in the poorly drained areas. Due to poor drainage caused by a lateritic hardpan, about 150,000 km$^2$ (60,000 mi$^2$) of low-lying area in the lower Orinico has standing water for months at a time. These seasonal wetlands are known as *esteros*. The vegetation is a grass savanna with **gallery forests** of tropical trees lining the river banks. Moriche palm (*Mauritia minor*) savanna occurs at the outer margins and in other permanently wet situations. In seasonally wet soils, the fire-resistant palm *Copernicia tectorum* grows. A low tree or shrub savanna exists on higher, better drained areas such as the alluvial fans and the upper river terraces in the western piedmont region of the llanos and the tablelands of the high plains region in the south. The woody plants are 2–7 m (6–20 ft) tall and rather spindly. All are evergreen. Leaf replacement and growth and flowering occur during the dry season. During the rainy season, trees add woody tissue to stems and branches. Common trees include *Brysonima crassifolia* and *Curatella americana*, with *Bowdichia virgiloides* dominant at some locations. Hard perennial tussock grasses and sedges and subshrubs dominate the herb layer. The grasses *Trachypogon plumosus* and *Andropogon semiberbis* dominate. Most of the grasses flower during the rainy season. Regardless of whether flowering occurs early or late, the seeds of perennial grasses all germinate at the beginning of the rainy season. Annual grasses are generally restricted to bare ground between tussocks or to areas of wet low-nutrient soils (Sarmiento 1983; Walter 1985; Cole 1986; Mistry 2000).

A high diversity of animal life is associated with the waterways and flooded regions of the llanos. The most well known are the giant rodent, the capybara (*Hydrochoreus hydrochaeris*), and the spectacled caiman (*Caiman crocodilus*). Other animals, such as the crab-eating fox (*Cerdocyon thous*) and a small armadillo, *Dasypus sabanicola*, avoid flooded areas. Occasionally white tailed deer (*Odocoileus virginianus*)—considerably smaller than their North American relatives—may be sighted, but, like most of the many animals found on the llanos, they are not restricted to the savanna habitat. Most mammals, of which 122 species have been recorded, use and even depend upon neighboring forests (Mistry 2000).

**Cerrado.** On the plateaus and high tablelands of the Brazilian Highlands is a unique savanna vegetation that is referred to by the Portuguese term *cerrado*, meaning closed. The cerrado (Photo 3.4) extends from 3° S to 24° S and covers 2 million km$^2$ (770,000 mi$^2$) or 23 percent of the land area of Brazil. Elevations range from 300–1,800 m (1,000–6,000 ft) above sea level. These uplands are formed of ancient shales and crystalline rocks that have been deeply weathered and extremely leached so that acidic red or yellow oxisols are the main soil type. Natural fertility is low; moreover, in many areas aluminum is highly concentrated. Average annual precipitation ranges from 800–2,000 mm (30–80 in), with a strong dry season extending from April through September. Average annual temperature averages 18–28° C (64–82° F). Especially toward the Tropic of Capricorn, summer and winter temperatures can be significantly different (Cole 1986; Ratter et al. 1997).

The diversity of vascular plants found in the cerrado is second only to the Amazon rainforest. Approximately 800 species of trees and shrubs have been recorded in these savannas and many times that number of subshrubs and herbs. Most of these are found only in this vegetation type (Furley and Ratter 1988). The cerrado's species diversity is of the type called beta diversity by ecologists, meaning that if you compare nearby sites you will find many different species on each. Any one site

Photo 3.4 Cerrado, showing the characteristic contorted trunks of trees. This vegetation grows on the upper surfaces of the Brazilian Highlands. (Photo by SLW.)

may not hold a large number of different kinds of plants, but taken together all sites are home to an extraordinary number. The diversity is not evenly dispersed throughout the cerrado but is concentrated toward the central core area and diminished toward the margins. Investigators have identified hotspots of cerrado diversity in the states of Mato Grosso, Goias (in the Alto Araguaia region), Tocantins, and in the Federal District around the Brazilian capital, Brasilia (Ratter et al. 1997).

Brazilians have a system by which variants of the cerrado are identified, and this more or less corresponds to the general categories of tropical savanna used in this chapter. The *campo limpio* (clean field) is an essentially treeless grass savanna; *campo sucio* (dirty field) refers to savanna with widely scattered shrubs and trees; and *campo cerrado* is a tree savanna with total crown cover about 3 percent. True cerrado is a savanna woodland with a total crown cover by woody plants of approximately 20 percent.

The several types of cerrado occur as a mosaic over the landscape, and all contain generally the same plants. The most common trees are 4–8 m (12–25 ft) tall with contorted trunks. The bark is typically thick and corky with many grooves. The branches are also twisted and usually few in number so that the crowns are quite open. Trees tend to have either large, leathery or stiff leaves or finely divided compound ones. Pequi (*Caryocar brasiliensis*) and murici (*Byrsonima coccolobifolia* and *B. crassifolia*) are examples of the former; the leguminous (*Stryphnodendron barbatimao* and *Dimorphiphandra mollis*) are examples of the latter. On many species, the veins of the large leaves may be raised as hard ridges on the undersurface of the leaf. In lixeiro (*Curatella americana*), the accumulation of silica makes the underside of leaves as rough as sandpaper. Some plants, such as pau santo (*Kielmeyera coriacea*) and pau de arara (*Salvertia convallariodora*), have their leaves as rosettes at the tips of branches. The broad leaves are covered with a thick cuticle that gives most trees a gray-green color (Cole 1986).

Trees keep their leaves into the dry season and continue to transpire. Most are brevideciduous (leafless for only a very short time, perhaps a few days) or semideciduous (losing only some of their leaves). New leaves develop as temperatures rise before the summer rainy season. Many woody plants have xylopodia (underground woody structures) in which to store water. The thick bark and hard leaves, thick cuticles, and sunken stomata once viewed as adaptations to fire and drought, respectively, are now believed to be consequences of low nutrient, high aluminum soils. Aluminum interferes with root growth by making phosphorus and calcium insoluble. Under such conditions, plants produce more cellulose and deposit thick cuticle and more sclerenchyma—the structural cells of a leaf (Cole 1986). Many cerrado species also have physiological processes that either exclude aluminum or allow its accumulation in the bark or leaves. Aluminum-exclusion has developed in several families, including the **endemic** pau terra trees (*Qualea grandifolia*, *Q. parvifolia*, and *Voschysia thyrsoidea*) (Vochysiaceae) (Mistry 2000).

The characteristically twisted growth of cerrado trees was once considered the result of fire. It now seems that the terminal (apical) bud is often damaged and lost at the end of the growing season due to drought. Lateral buds have the ability to assume the role of growth bud and form the main stem during the next growing season (Cole 1986; Solbrig 1996).

The herb layer of the cerrado contains mostly perennial bunch grasses; prominent among them are *Tristachya chrysothrix* and *Aristida pallens*, but grasses of other genera are also present. Almost no annual herbs are found (Cole 1986; Mistry 2000). In wetter depressions on the interfluves, buriti (*Mauritia vinifera*) and buritirana (*M. martiana*) palm savannas can be found (Solbrig 1996).

While large native mammalian herbivores are lacking, many forms of animal life are found in the cerrado. Mammals are represented by opossums, tamanduas, armadillos, marmosets and other monkeys, agouti, paca, and smaller rodents. The largest carnivore is the endangered maned wolf or lobo guara (*Chrysocyon brachyurus*). Many of these animals rely upon the gallery forests for their survival. The rodents show the highest proportion of species restricted to the cerrado (Mistry 2000).

Over 400 species of birds are recorded from the cerrado, but they, too, show little specialization on the cerrado environment (Ratter et al. 1997; Mistry 2000). Among birds found in cerrado (and other American savannas) are ground birds, such as the large flightless rhea (Rheidae), screamers (Anhimidae), and tinamous (members of an endemic family, perhaps ancestral to large flightless birds the world over) that eat plant material, and the red-legged seriema (*Cariama cristata*) that eats animal matter. New World or tyrant flycatchers are common; they specialize on flying insects. The finches (Fringillidae) are the only small seed eaters and are comparable to the weavers and weaver-finches of African savannas. Other families with over 20 species include parrots, hummingbirds (Trochilidae), honeycreepers (Thraupidae), and caciques (Icteridae). South America has far fewer ground-dwelling or ground-feeding birds and more arboreal species than in either Africa or Australia (Fry 1983).

Termites and leaf-cutting ants are the main herbivores of the cerrado. They, in turn, are food for other animals such as the tamandua (Pivello 2000).

## ORIGINS

The oldest pollen grains of grass in the tropics date to the middle of the Eocene Epoch, so tropical savannas could not have existed before then. By the Miocene and Pliocene epochs (late Tertiary Period), grass pollen was abundant in the fossil record of northern South America, which suggests that savannas were probably widespread by then. Pollen from two genera of llanos and cerrado woody plants, *Byrsonima* and *Curatella*, are also abundant, which further confirms this assertion. The Quaternary Period, with its changeable climate related to alternating glacial and interglacial periods at higher latitudes, may have seen an expansion of the savanna and contraction of tropical forests into habitat islands several times. It is possible that the llanos and cerrado were connected during intervals of maximum expansion (van der Hammen 1983).

A Tertiary origin for the African savannas is supported by the fossil record of vertebrates, which displays a major diversification of herbivores adapted to consuming grass (van der Hammen 1983). The hoofed animals—first the odd-toed (Perissodactyla) ancestors of rhinoceroses and zebras—appear, their long teeth able to withstand years of wear against the silica-rich grasses they ate. The ruminant stomach of the even-toed (Artiodactyla) ungulates served as an even better adaptation since it, with the help of myriad bacteria dwelling inside, allowed more complete digestion of grasses. Once ruminants such as early members of the cattle family (Bovidae) evolved, they radiated into many, many forms in Africa. The northern and southern savannas may have been connected until the Quaternary, when landform and climatic changes created a geographic barrier in East Africa (Menaut 1983). The role of the ancestors of modern humans (*Homo sapiens*), who appeared 2–3 million years ago on the savannas of east Africa, is still debated. Presumably, once humans learned to control fire, they regularly burned the savanna to make predators more visible, to drive prey in a hunt, and to attract herbivores to the flush of new plant growth that follows a fire. The Quaternary Period in Africa was also a time of expansion and contraction of savannas (with compensating shifts in forest and desert) in response to global climate change associated with glacial and interglacial periods in the midlatitudes and high latitudes (van der Hammen 1983). A similar history probably applies to Australian savannas.

The three continents on which tropical savannas are best developed were part of a single land mass, Gondwana, until the Cretaceous Period. The interior of the supercontinent was dry. Several of the woody plants adapted to savanna drought, therefore, have a common and ancient ancestry. The trees and shrubs of *Acacia*, found in the savannas of all three continents, are an example. Africa and Australia share baobabs (*Adansonia* spp.) and have genera from a very old family of flowering plants, the Protaceae. After the separation of these Gondwanan landmasses from each other, evolution of plants proceeded more or less in isolation. African woody plants mostly derive from drought-adapted ancestors, but South American savanna trees are most closely related to plants of the Amazon and Atlantic rainforests. A few trees in the South American savanna and many of the herbs and subshrubs are peculiar to the savannas, however. The Australian trees are dominated by the genus *Eucalyptus*, which has diversified to occupy all types of environments on that continent (Menaut 1983; Sarmiento 1983; Cole 1986).

## HUMAN IMPACTS

Human use of fire has long been part of the savanna environment. In Africa, hunting, gathering and scavenging peoples have intentionally burned grasses for at least 50,000 years and perhaps much, much longer. Humans occupied Australia 40,000 years ago, and good indications show people manipulated their environment by regularly setting fire to the vegetation. In the Americas, evidence shows man-made fires were occurring 10,000 years ago (and maybe as long ago as 30,000 years), and better evidence shows they were happening frequently by 5,000 years ago (Gillon 1983). Repeated, regular burning opens woodlands and expands the grass layer. It selects for fire-resistant or fire-adapted plants and eliminates those that are unable to survive a burn. Fire can synchronize the flowering in some grasses and thereby aid the exchange of pollen and hence reproduction. Thick bark and the ability to sprout from roots or from lateral aerial buds, as seen in many cerrado woody plants, helps trees and shrubs tolerate fire. We know people had many incentives to burn savannas, and charcoal layers in the ground suggest that they indeed did so. What is not known is just how much this burning influenced the nature of the world's savannas. Some have gone so far as to say that people created many savannas, but this idea has become less popular in recent years as more information on the impacts of soil conditions and periodic drought becomes available.

Early inhabitants of Australia and South America probably played a role in the loss of native large mammals that had existed in the tropical savannas of both continents during the Pleistocene. The argument is made that African mammals evolved in the presence of the early ancestors of man and developed strategies to survive human predation, just as they adapted to predation from the big cats and other carnivores. However, where humans arrived as full blown, skilled hunters, animals had no time to evolve, and the largest—with very slow recruitment rates—were highly susceptible to extinction. Many disappeared.

The next major human impact on tropical savannas came from agriculturalists and herders of domestic livestock—mainly cattle. Again, the impact has been much more longstanding in Africa than elsewhere. Domestic cattle were introduced into Africa about 6,000 years ago, although they did not penetrate into the savannas of southern Africa until about 2,000 years ago. Cattle arrived in the Americas in the sixteenth century and in Australia in the late nineteenth century. Livestock grazing required the burning of the savannas to create new pastures; when herds became too large, overgrazing could destroy the grass cover and lead to desertification. Cattle raising continues to be a major land-use practice in tropical savannas worldwide, but increasingly, cultivation of commercial crops is becoming important.

The savannas of West Africa were strategically placed between the Sahara to the north and the coastal rainforests to the south. As such they became the location of trading centers where goods from the desert and Mediterranean—such as salt, glass, and copper—could be exchanged for products of the rainforest—such as ivory, gold, cola nuts, and slaves. Between the fourth and sixteenth centuries, a series of powerful kingdoms developed in the savanna zone, supported by sedentary agriculture first based on native millets and beans. From the seventh through the eleventh centuries, new crops and domesticated animals represented the Islamic and Arabic influences that reached the zone. These centers of power, wealth, and agriculture would have cleared the savanna for farmland, cut trees for fuelwood and for smelting iron, and removed dangerous large animals or those competing with cattle for forage. The long period of dense human settlement may be one reason for the relative scarcity of large mammals in these savannas today. The European colonial period in West Africa led to depopulation of the savanna zone. The great trading centers and centers of power moved to the coast to interact with the European tradesmen, and the savanna zone became a major source of slaves to export to the Americas. Today, the major land uses are animal husbandry and agriculture. The severe drought in the Sahelian zone in the 1970s forced pastoralists in greater concentrations into the Sudanian zone. At the same time, programs to eradicate the tsetse fly, which carries trypanosomiasis—a disease that weakens and kills cattle—allowed for more intensive use of cattle. Human population growth increased the area used for rain-fed subsistence agriculture and stressed the traditional system of leaving some land in fallow each year to recover soil moisture. Increased human populations also meant increased deforestation. Desertification became a problem at various locations. Commercial agriculture, much of it irrigated, also has expanded. Major products are cotton, peanuts, and maize. Mining is another important extractive industry in the West African savanna, with emphasis on petroleum, iron, diamonds, bauxite, manganese, and limestone. Efforts to restore woody species to the land mainly involved afforestation projects with nonnative eucalypts. Large mammals continue to suffer as their habitat is reduced; black rhino and eland are nearly extinct in many countries, and elephant, topi, and reedbuck are threatened (Mistry 2000).

Land use in precolonial East Africa was based on shifting and sedentary agriculture and nomadic pastoralislm, although the presence of trypanosomiasis limited livestock. European colonization in the nineteenth century and the subsequent development of railroad and communications systems led to the development of commercial agriculture based on coffee, tobacco, sisal, and cotton; and the removal of native vegetation. Wholesale slaughter of wild animals to prevent the spread of rinderpest to domestic livestock, the sport of big game hunting, and the uncontrolled ivory trade led to the decimation of large mammals. In the 1940s, game and safari parks were established to protect and manage wildlife and to promote tourism. Native peoples were removed from the parks, and a pattern of land use that conflicted with that of traditional societies began. Rights to the use of park land continues to be a major issue. Increased agricultural use of the land also has reduced the amount of land available for nomadic pastoralists, and this has led to overgrazing and soil erosion. The provision of permanent waterholes (boreholes) has concentrated domestic herds during the dry season and produced patches of degraded habitat. Tourism, however, is the region's most challenging contemporary management problem (Mistry 2000).

The southern savannas of Africa have a history of human land use similar to the other two regions. Since at least A.D. 300, traditional land use has involved a combination of slash-and-burn agriculture based upon sorghum and millet, hunting, and raising sheep and cattle. The human settlement pattern became one of many small villages. Farming, grazing, collection of fuelwood, and iron smelting all contributed to the clearing of savanna woodlands. By A.D. 1000, regional population centers began to engage in long-distance trading. Cattle became a symbol of wealth and power, and increasing herds led to overgrazing. Cattle continue to be a major part of modern land use. Trophy hunting on commercial game ranches and ecotourism in national parks are also important to the regional economy. Both require continuous management of savanna vegetation and wildlife (Mistry 2000).

Aboriginal use of Australian savannas was limited to hunting and gathering and fire was part of the hunting strategy. In the nineteenth century, European settlers introduced cattle, which became the mainstay of the savanna region's economy. Because native Australian grasses did not tolerate grazing, African grasses such as *Melinis minutiflora* and *Pennesitum polystachyon* were introduced. **Exotic** woody plants such as prickly acacia *(Acacia nilotica)*, parkinsonia (*Parkinsonia aculeata*), mesquite (*Prosopis* spp.), and jujuba (*Ziziphus mauritiana*) were also introduced and have become invasive. Introduced animal species such as starlings and mynas dominate nest sites and threaten native bird species. Today, tourism is promoted as a nonconsumptive use of the savanna. Tourism will be based, in part, upon what can be preserved or restored of aboriginal culture and landscapes. Efforts are under way to reestablish the ancient fire management patterns (Mistry 2000).

The llanos was home to large populations of Native Americans before the colonial period. Concentrated along the waterways, these people depended on crops of manioc and maize, hunting, and gathering. They traded natural products such as tree resins, animal skins, and turtle eggs. However, within a few decades of Spanish settlement (1548), diseases introduced from Europe and Asia decimated the native populations. Diseases such as malaria and Chagas disease also slowed Spanish advances into the llanos; however, between 1640 and 1790, escaped African slaves established communities (*cumbes*) there. European land uses throughout the colonial period remained extensive; cattle were raised on large land holdings (*latifundias*). The llanos remain largely undeveloped and unsettled. Cattle ranching is the primary land use, and most of Venezuela's meat and milk are produced on the grass savannas. To increase pasturage for cattle, **alien** grasses have been introduced. *Panicum maximum* is widespread; *Hyparrhema rufa* has displaced native grasses from the wetter, more fertile parts of the llanos. Flood control programs have also been instituted, along with programs to eradicate native animals thought to compete with or prey upon domestic livestock. However, successful efforts promote the commercial value of capybara for meat and caiman for hides, so some ranchers manage and sustainably harvest these species rather than exterminate them (Mistry 2000).

Before European settlement, the cerrado was inhabited by numerous tribes such as the Kayapo and Guarani, who lived in small villages and depended upon crop cultivation, hunting, and collecting forest products. Their rich vocabulary dealing with fire suggests that burning the cerrado was an important land-management tool. During the colonial period, the cerrado was viewed as a remote backlands used for extensive cattle raising; the first Portuguese settlements were associated with gold mining. Cattle raising and subsistence agriculture characterized the economy until the founding of Brasilia in 1959. Brazil had deliberately moved its capital to this new city to move settlement westward and lay effective claim to its vast backlands. Development followed as planned. Today, increased urbanization and large-scale commercial agriculture is rapidly transforming the cerrado. Large tracts of land are now given over to the highly mechanized cultivation of soybeans, maize, rice, and manioc (Ratter et al. 1997). Through the 1970s, several government agencies worked to develop the technology to overcome aluminum toxicity, apply phosphorus and nitrogen fertilizers, and develop high-yielding varieties of soybean. Now, an extensive, mechanized **monoculture** of soybeans for the export market exists (Mistry 2000). Related to agriculture is environmental degradation from drainage projects, dam construction for irrigation water, the widespread use of herbicides and pesticides, and soil compaction from heavy machinery (EMBRAPA 1996). Cattle ranching has become more intensive than in the past, and large areas of cerrado vegetation have been cleared to create improved pastures. Crossbreeding has led to improvements in the stock and made Brahman cattle the most characteristic animal of the cerrado (Furley and Ratter 1988). Silviculture is also important; pine (*Pinus radiata*) plantations provide pulp for the paper industry, and eucalypt plantations provide charcoal for the state of Minas Gerais's steel industry—the world's largest consumer of charcoal. Mining for minerals such as copper, iron, nickel, manganese, and gold also affects the land and waters of the region. An estimated 37 percent of the species-rich cerrado of Brazil has been transformed, primarily by agriculture, but also by urban and industrial development and mining (Mistry 2000). In efforts to use native animals as a way of conserving them, much like the efforts in the llanos to conserve capybara and caiman, target species include the quaillike tinamous (*Northura* spp.) and pampas deer (*Ozotocerus bezoarcticus*) (Furley and Ratter 1988).

## HISTORY OF SCIENTIFIC EXPLORATION

The investigation of the world's tropical savannas is largely a reflection of the history of European settlement and land use. As mentioned earlier, the word savanna first came into use in a Spanish publication of 1535 (G. F. Oviedo y Valdes's "Historia general y natural de las Indias," vol. 1). Initially the term was used to describe "land without trees but with much grass either tall or short"; this meaning was changed to the current concept of a grassland with trees by the German plant geographers Grisebach (1872), Drude (1890), and Schimper (1898) (Cole 1986). An important paper by Beard (1953) categorized the American tropical savannas into different types and proposed different factors that might account for them. Other biogeographers and botantists have created and/or refined savanna classifications systems.

As for all biomes, the late 1960s saw the beginnings of comparative ecological studies as part of the International Biological Programme. Use of tropical savannas and the ecological constraints and consequences were addressed in 1979 by United Nations Educational, Scientific and Cultural Organization (UNESCO) publication *of Tropical Grazing Land Ecosystems* (Paris: UNESCO) and in 1980 by David R. Harris's *Human Ecology in Savanna Environments* (London: Academic Press).

In Africa, research initially was influenced by colonialism and the search for exploitable natural products and the need or desire to manage European-type agricultural systems. In the twentieth century, the focus switched to wildlife and conservation issues, and pioneering studies on wildlife **ecology** took place on the African savannas. The plants of these environments are now well known, but more ecological studies useful to range management are still needed (Menaut 1983).

The llanos and cerrados of South America were considered remote backlands until the twentieth century, and only in the 1960s did ecological study begin. The Food and Agriculture Organization of the United Nations (FAO) published a work oriented toward livestock production in 1966. A research team from McGill University in Canada conducted major studies of the Rupununi savannas of Guyana in the 1960s. Work by scientists at the Universidad de los Andes, begun in the 1960s and continuing until today, has made the Venezuelan llanos one of the better studied regions of American savannas (Sarmiento 1983).

Some Brazilian ecosystems, such as the Amazon and Atlantic rainforests, are identified in the national constitution as parts of the natural heritage to be conserved, but the cerrado is not one of them. In the late 1980s and 1990s, major projects were initiated to identify potential sites for reserves and to develop strategies for conserving the great **biodiversity** of this biome. A joint British-Brazilian project, "Conservation and Management of the Biodiversity of the Cerrado Biome," includes research teams from the University of Brasilia, CPAC (a cerrado-focused division of Brazil's agricultural research agency, EMBRAPA), and the Royal Botanic Garden Edinburgh. A project on the **biogeography** of the cerrados is centered at the University of Brasilia. Other Brazilian agencies such as IBAMA (Brazil's national environmental agency, roughly comparable to the U.S. Department of the Interior) and CENARGEN (National Center for Genetic Resources and Biotechnology ) are also involved (Ratter et al. 1992).

In Australia, a general failure of agriculture in the tropical savanna zone in the nineteenth and twentieth centuries has recently propelled inhabitants to embrace tourism as a source of income. The main assets of the region were seen to be its natural landscapes and aboriginal peoples—both actively being destroyed until then. The focus on tourism and the need for information on the savanna ecosystem and the role of aboriginal culture in shaping and maintaining it has stimulated important ecological research (Mistry 2000).

## REFERENCES


Beard, J. S. 1953. "The savanna vegetation of northern tropical Amazonas." *Ecological Management* 23: 149–215.

Bourliere, Francois. 1983. "Mammals as secondary consumers in savanna ecosystems." Pp. 463–475 in Francois Bourliere, ed., *Tropical Savannas.* Ecosystems of the World, 13. Amsterdam: Elsevier Scientific Publishing Company.

Bourliere, F. and M. Hadley 1983. "Present day savannas: An overview." Pp. 1–17 in Francois Bourliere, ed., *Tropical Savannas.* Ecosystems of the World, 13. Amsterdam: Elsevier Scientific Publishing Company.

Cole, M. M. 1986. *The Savannas: Biogeography and Geobotany.* New York: Academic Press.

Cowling, R. M., D. M. Richardson, and S. M. Pierce. 1997. *Vegetation of Southern Africa.* Cambridge: Cambridge University Press.

Drude, O. 1890. *Handbuch der Pflanzengeographie.* Stuttgart: Engelhorn.

EMBRAPA. 1996. *Atlas do Meio ambiente do Brasil* (Atlas of the Environment of Brazil). Brasilia, DF: EMBRAPA-SPI.

Fry, C. H. 1983. "Birds in savanna ecosystems." Pp. 337–357 in Francois Bourliere, ed., *Tropical Savannas.* Ecosystems of the World, 13. Amsterdam: Elsevier Scientific Publishing Company.

Furley P. A. and J. A. Ratter. 1988. "Soil resources and plant communities of the Central Brazilian cerrado and their development." *Journal of Biogeography* 15: 97–108.

Furley, P. A., J. Proctor, and J. A. Ratter, eds., 1992. *Nature and Dynamics of Forest-Savanna Boundaries.* London: Chapman and Hall.

Gillon, Dominique. 1983. "The fire problem in tropical savannas." Pp. 617–641 in Francois Bourliere, ed., *Tropical Savannas.* Ecosystems of the World, 13. Amsterdam: Elsevier Scientific Publishing Company.

Grisebach, A. H. R. 1872. *Die Vegetation der Erde nach ihrer klimatischen Anordnung.* 2 vols. Leipzig: Wilhelm Engelmann.

Happold, D. C. D. 1983. "Rodents and lagomorphs." Pp. 363–400 in Francois Bourliere, ed., 1983. *Tropical Savannas.* Ecosystems of the World, 13. Amsterdam: Elsevier Scientific Publishing Company.

Hopkins, Brian. 1983. "Successional processes." Pp. 605–616 in Francois Bourliere, ed., 1983. *Tropical Savannas.* Ecosystems of the World, 13. Amsterdam: Elsevier Scientific Publishing Company.

Josens, Guy. 1983. "The soil fauna of tropical savannas. III. The termites." Pp. 505–524 in Francois Bourliere, ed., *Tropical Savannas.* Ecosystems of the World, 13. Amsterdam: Elsevier Scientific Publishing Company.

LaMotte, Maxime. 1983. "Amphibians in savanna ecosystems." Pp. 313–323 in Francois Bourliere, ed., *Tropical Savannas.* Ecosystems of the World, 13. Amsterdam: Elsevier Scientific Publishing Company.

Menaut, J. C. 1983. "African savannas." Pp. 109–149 in Francois Bourliere, ed., *Tropical Savannas.* Ecosystems of the World, 13. Amsterdam: Elsevier Scientific Publishing Company.

Mistry, J. 2000. *World Savannas: Ecology and Human Use.* Harlow, England: Pearson Education Ltd.

Montgomery, R. F. and G. P. Askew. 1983. "Soils of tropical savannas." Pp. 63–78 in Francois Bourliere, ed., *Tropical Savannas.* Ecosystems of the World, 13. Amsterdam: Elsevier Scientific Publishing Company.

Nix, H. A. 1983. "Climate of tropical savannas." Pp. 37–62 in Francois Bourliere, ed., *Tropical Savannas.* Ecosystems of the World, 13. Amsterdam: Elsevier Scientific Publishing Company.

Pivello, Vania R. 2000. "Cerrado." Brazilian Embassy in London: www.brazil.org.uk.

Ratter, J. A., J. F. Ribeiro, and S. Bridgewater. 1997. "The Brazilian cerrado vegetation and threats to its biodiversity," *Annals of Botany* 80: 223–230.

Sarmiento, G. 1983. "The savannas of tropical America." Pp. 245–288 in Francois Bourliere, ed., *Tropical Savannas.* Ecosystems of the World, 13. Amsterdam: Elsevier Scientific Publishing Company.

Sarmiento, G. and M. Monasterio. 1983. "Life forms and phenology." Pp. 79–108 in Francois Bourliere, ed., *Tropical Savannas.* Ecosystems of the World, 13. Amsterdam: Elsevier Scientific Publishing Company.

Schimper, A. F. W. 1898. *Pflanzen-Geographie auf physiologischer Grundlage.* Jena: Fischer. English translation: 1930 *Plant Geography Under a Physiological Basis.* Oxford: Clarendon Press.

Schultz, J. 1995. *The Ecozones of the World: The Ecological Divisions of the Geosphere.* Berlin: Springer-Verlag.

Sinclair, A. R. E. 1983. "The adaptations of African ungulates and their effects on community function." Pp. 401–426 in Francois Bourliere, ed., *Tropical Savannas.* Ecosystems of the World, 13. Amsterdam: Elsevier Scientific Publishing Company.

Solbrig, O. T. 1996. "The diversity of the savanna ecosystem." Pp. 1–27 in O. T. Solbrig, E. Medina, and J. F. Silva, eds., *Biodiversity and Savanna Ecosystem Processes. A Global Perspective.* Berlin: Springer.

van der Hammen, T. 1983. The paleoecology and paleogeography of savannas." Pp. 19–35 in Francois Bourliere, ed., *Tropical Savannas.* Ecosystems of the World, 13. Amsterdam: Elsevier Scientific Publishing Company.

Walter, H. 1985. *Vegetation of the Earth and Ecological Systems of the Geo-biosphere.* Berlin: Springer-Verlag.

# 4 DESERTS

## OVERVIEW

A sparse **cover** of shrubs or dwarf shrubs typically vegetate the driest regions on earth's land surfaces. Only a very few places, such as large dune fields and **badlands**, are devoid of plant life. Many fascinating lifeforms, both plant and animal, have adapted to the extreme conditions of low and unpredictable moisture and, often, high temperatures during the summer.

Many scientists recognize three kinds of deserts in terms of degree of aridity and **vegetation** cover. Under such a scheme, true deserts usually have less than 120 mm (5 in) of precipitation annually and a very sparse (less than 10 percent) plant cover of primarily dwarf shrubs. Semideserts have between 150 and 300–400 mm (6–15 in) of precipitation and a plant cover of 10–30 percent composed of the dominant dwarf shrubs, **perennial** grasses, shrubs, and/or **succulents** (plants that store water in their tissues). Extreme deserts receive less than 70 mm (3 in) of precipitation a year and are essentially devoid of perennial plants. According to this scheme, most desert regions of the United States would be considered semideserts (Shmida 1985).

Landforms are important controls of plant and animal distribution in desert environments, and a set of landforms peculiar to arid lands will be briefly described here. Moving sand dunes in a sea of sand, actually a minor part of most deserts, form a surface known as *erg*. When a layer impermeable to water lies beneath the sand, moisture can become available for plants. Rainwater percolates easily through the sand and collects on top of the impermeable layer, the overlying sand protects the water from high temperatures and evaporation. Other desert surfaces represent areas where wind action has removed all sand and finer materials. Reg is surface of closely spaced small stones (80 percent of the Sahara is reg). Bare rock surfaces are known as hammadas. In mountainous areas, the intervening basins have many depositional landforms with varying particle sizes and **soil** conditions that influence the types of plants able to grow on them. Streams flowing out of the mountains

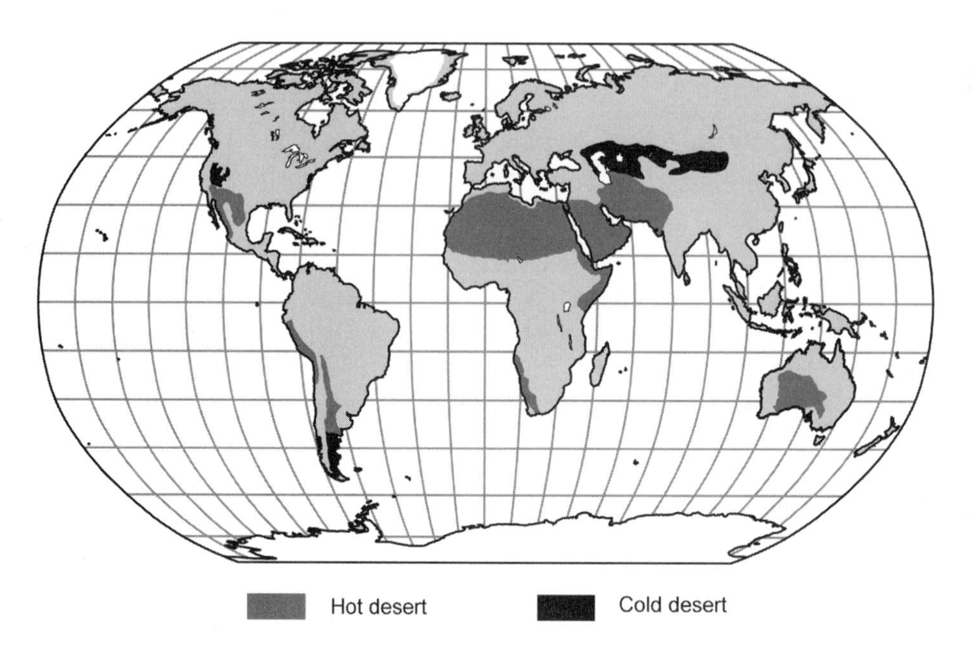


Map 4.1 World distribution of the Desert Biome.

quickly dry up upon reaching the basin floor; the sands and gravels they were carrying are left as alluvial fans at the foot of the mountains. When a group of neighboring fans overlap, they form a surface of unconsolidated materials known as a **bajada**. During rains, small streams may flow across the basin floor, but most of the time only dry stream beds known as washes or wadis are present. When these erode into deep, steep-sided stream beds, the dry channels are called *arroyos*. Most water flowing in desert streams never reaches the ocean but collects in shallow, temporary lakes at the lowest parts of a basin. Such stream systems that never carry water beyond the immediate region are referred to as interior drainage. The dry lake beds, composed of the finest materials washed from surrounding uplands and often high in salts due to the evaporation of streamflow past and present into the lake, are called playas, salares, salines, or salt pans.

Humans have lived in deserts since ancient times. Major early civilizations arose in deserts where permanent rivers—the Tigris and Euphrates, the Nile, and the Indus, for instance—supplied water for irrigated crops. These mighty rivers and innumerable smaller ones around the world have their headwaters in humid **climates**, where they receive enough water to maintain a year-round flow through the deserts they cross. Known as exotic rivers, they present to plants and animals, as well as people, a well-watered environment though in the midst of aridity.

Deserts are too often viewed as wastelands, when, in fact, they are inhabited by fascinating plants, animals, and human cultures that have adapted to cope with drought. Traditional cultures in various parts of the world used their deserts through hunting and gathering, herding livestock, and ingenious systems of irrigation agriculture. Modern use still involves the grazing of livestock. Mining of minerals like phosphates, gypsum, and salt that accumulated via evaporation has been and still is important is several deserts, including the Sahara and Atacama. Other minerals related to other geologic processes, minerals such as copper, gold and diamonds, are also mined. Increasingly, tourism is a major force in the economy of desert regions. People come to take advantage of sunny skies and often spectacular scenery while they play and relax. Ecotourism is growing in response to people's curiosity about the unique forms of life in the desert **biome**. All these activities—hunting and gathering, grazing, mining, tourism—can alter the natural desert communities.

## CLIMATE

The climatic conditions that the deserts have developed varies from place to place. A description of a specific desert's climatic regime will be presented in the regional discussions later. While rainfall is low throughout, the actual amounts that create aridity depend upon temperatures. In all cases, total annual potential evaporation must exceed total potential precipitation. Thus, actual amounts of total annual precipitation in the desert biome as a whole vary from near 0–400 mm (0–15 in) or more. The season of peak precipitation differs from desert to desert. Some have a summer maximum, others have a winter maximum. Some deserts experience two peaks of precipitation annually, while others have small amounts of precipitation distributed rather evenly throughout the year. The main factor in most deserts is the unreliabililty and great variability in rainfall from site to site and from year to year. Summer convectional storms are highly localized. Average total precipitation has little direct meaning other than as an indicator of degree of aridity. Some years may see no precipitation at all, while the next may have two or three times the annual average.

Similarly, temperature patterns vary, although a marked difference always occurs between January and July mean monthly temperatures and often an even greater temperature range between day and night. Cold deserts have prolonged periods of below-freezing temperatures and frequently receive significant amounts of their annual precipitation in the form of snow.

## VEGETATION

The shrub is the dominant and characteristic **growth-form** of both cold and warm deserts (Photo 4.1). Shrubs may be either drought **deciduous** or **evergreen**. Most desert shrubs have very small leaves and either spines or unpalatable, strongly scented oils—presumably as defenses against herbivores. The true xerophytes (dry plants) have a shallow root systems that extends well beyond the **canopy** to capture rainwater from a wide area. Other types of plants, the phreatophytes (groundwater plants), have long tap roots that can reach down several meters to extract water from the nearly saturated zone just above the water table. Phreatophytes are usually confined to washes and dunes or the banks of intermittent and exotic rivers.

Some plants simply avoid drought by going dormant during the driest part of the year. Deciduous shrubs achieve a state of dormancy with the shedding of their leaves. In **subshrubs**, the entire above-ground part of the plant dies back in winter, but woody parts below ground have renewal buds that will send up new shoots with the onset of the rainy season. Another common perennial growth-form that avoids drought is the bulb, an herbaceous plant that stores water and nutrients below ground and sends shoots above ground only after rainfall sufficient to ensure flowering and replenishment of underground stores. **Annuals** are also able to avoid

Photo 4.1 Creosotebush (*Larrea tridentata*), the indicator species of hot deserts in North America and a true xerophyte. Note the even spacing of plants and large proportion of bare ground from which lateral root systems absorb rainwater. (Photo by SLW.)

severe drought if they produce a waterproof coating on their seeds. Desert annuals (Photo 4.2) may have extremely brief life spans as germinated plants and will complete their life cycle in just a few weeks. They are often referred to as **ephemerals** and can remain dormant as seeds for years—even decades—until enough rainwater percolates through the soil to dissolve the seed coating. This also happens to be enough moisture to enable them to germinate, flower, and set seed before going dormant again. Deserts usually have a large number of annuals, but most of the time they are invisible. However, when it does rain and the desert blooms, carpets of brightly colored flowers create one of nature's great spectacles.

Another growth-form commonly found in deserts is the halophyte, a plant adapted to tolerate the high concentrations of salts (sulfate, choride, and carbonate compounds of sodium and magnesium) found in many desert soils. The problem halophytes overcome is that water naturally moves from areas of low salt concentration into areas of high concentration. The salt content of the tissues of most plants is lower than that of saline desert soils, so water is drained out of them and they die. Halophytes have evolved ways to pump salts into their tissues to give them a higher salinity than the soil, so that soil moisture will move into the plants. Since too much salt is toxic to plants, they also have evolved ways to tolerate or rid themselves of excess salts, either by excreting it through specialized salt glands or storing it in leaves and then losing it when the leaves drop off.

Succulents, plants that store water in their tissues, represent a growth-form well adapted to survive extended dry periods. Many plant families have adopted this strategy for desert survival. A multitude of forms have resulted, depending on what part the plant uses as a water storage organ. Succulents are much more common in hot deserts than cold ones.

Photo 4.2 The desert primrose (*Camissonia* sp.), an annual. This species germinates after winter rains in the Sonoran desert and briefly carpets the desert floor. (Photo by SLW.)

The Namib and the Karoo in southwestern Africa harbor some other interesting growth-forms. One is the psammophore, a plant that grows in areas regularly blasted by blowing sand. To prevent the destruction of their tissues, these plants have evolved ways to coat themselves with sand. Some produce a sticky substance that covers the exposed leaves and to which sand adheres. Others have glands or hairs that trap sand. Another growth-form found in the Namib is a window plant, a succulent that is greatly reduced in size and grows embedded in the ground with only the tips of its swollen leaves at the surface. Sunlight penetrates through translucent, unpigmented cells on the leaf tips to reach the chloroplasts buried inside the plant. Many of these plants resemble stones, soil, or dung more than a green plant. Successfully camouflaged to escape predators, they can live for decades. Examples include flowering stones (*Lithops* spp.) (Photo 4.4) and baby toes (*Fenestaria aurantiaca*) (Milton et al. 1997).

The structure of the vegetation of the desert biome varies. Semideserts, such as on bajadas in the Sonoran Desert in Arizona, have multiple layers—with small trees and treelike cactus forming the highest level, and other levels made up of shrubs, dwarf shrubs, and herbs, respectively. In other areas and common in true deserts, a single layer consists of widely spaced shrubs with much bare ground between them. However, research shows that a nonrandom clumping of plants does occur in these situations, as some plants serve as nurse plants for others. For example, a spiny shrub may protect the seedlings of slow-growing succulents from predation, so that a cactus may develop only when clumped with a thorny woody plant.

## SOILS

Soils are usually poorly developed in desert environments due to a lack of **leaching**, the process that contributes to the development of distinct horizons in true soils. Organic matter is minimal because of sparse plant cover and the slow rates of decomposition in arid conditions; for this reason, the soils are usually a light gray color. Extreme calcification is the dominant soil-forming processing; capillary action accumulates calcium carbonate at or near the ground surface, further contributing to the whitish hue of desert soils. At times, a hardpan of calcium carbonate develops as the carbonates become cemented together. This petrocalcic (rock calcium) layer may be several feet thick and, in North America, is called caliche. It is not certain how or when caliche forms; it may well be a product of moister conditions during the Pleistocene rather than current soil-forming processes. Nonetheless, a caliche layer is impermeable to water and acts as a trap to water percolating through the soil after rains. This creates what is known as a perched water table and can produce a source of water within reach of phreatophytes, if not other desert plants. In the U.S. **taxonomy** of soils, desert soils are classified as Aridosols. Brown soils occur in semideserts, especially toward their moister margins, where many intergrade with grasslands. In basins receiving internal drainage, soils high in soluble salts are formed. Moist, salt-rich soils named solonchaks are found where the water table is high, at least periodically, and the top meter of the soil is saturated for a month or more. Halophytes grow on these usually white soils.

## ANIMAL LIFE

Animals, like plants, have to cope with a lack of water and excess heat if they are to live in the desert. Most adaptations in animals involve behavioral responses, but morphological and physiological adaptations are also evident in some forms. Among common behavioral adaptations are nocturnal or crepuscular (being active only near dawn and dusk) activity and burrowing below ground for shelter and nesting sites. Those **species** that are active during the day or that do not burrow regulate body temperature by staying in the shade or orienting their bodies to minimize the area directly exposed to the sun. Morphological adaptations in warm-blooded animals include long appendages (e.g., legs, wings, ears) and relatively small bodies so they readily lose body heat to the environment. A light color that reflects sunlight is advantageous to animals that spend much time above ground during the daylight hours. Physiological adaptations are less common than behavioral and morphological ones, but they are nonetheless significant when they do occur. Some mammals and most desert amphibians are able to aestivate, that is, enter a state of reduced metabolic activity or dormancy during the hottest, driest part of the year. Ways to reduce the loss of body moisture include the absence of sweat glands in desert-adapted mammals and, in a few, the ability to concentrate their urine. Some reptiles, birds, and mammals are able to extract enough water from their food and to conserve enough in their bodies that they never need to drink.

Reptiles tend to be abundant in deserts because the reptilian body is more or less waterproof to begin with. Birds can be abundant in all but the largest deserts, because the adults can fly to sources of free water. The most abundant animals in terms of number and collective weight, however, are ants and termites. Other arthropods that have watertight bodies, such as spiders and scorpions, are not uncommon.

# I. HOT DESERTS

Hot deserts are not hot year round but exhibit large seasonal as well as diurnal ranges in temperature. They are distinguished from cold deserts because, although they may have a few days when minimum temperatures are at or below freezing, in general temperatures are not low enough for long enough to kill cold-sensitive plants.

Four geographic factors are implicated in the development of arid conditions. One or more of these may be influential in any given desert region. First, many hot deserts are located under the subtropical high pressure systems centered at roughly 30° **latitude** in both hemispheres. In these regions, air is being forced down from higher in the atmosphere as part of the global atmospheric circulation pattern, and it is difficult to counteract this movement to produce the upward flow with its associated cooling necessary for precipitation. Major desert regions directly related to the presence of a subtropical high include the world's largest, the Sahara, in the Northern Hemisphere and the deserts of Australia in the Southern Hemisphere. A second location at which hot deserts develop is on the west coasts of continents between the latitudes of 20° and 30°. Aridity in these regions is the product of dry winds blowing off the continent (due to circulation around a subtropical high) and the presence of a cold ocean current offshore. The cold ocean water chills the air above it and condenses any moisture in it to form fog. Local sea breezes during the day push the fog onshore, but because the air mass is cooler, and hence denser than air warmed over land, it cannot rise to produce precipitation. Although very little rain ever falls, moisture is available to plants and animals that can collect droplets from the fog. Some of the world's driest deserts, the Atacama in northern Chile and the Namib in Namibia, are such fog deserts located right next to an ocean. A third location in which deserts are found is in the lee or rainshadow of high mountain ranges. Winds moving across a mountain range force air masses to rise on the windward side, where the temperatures drop and much of the moisture in the air is released. On the opposite side of the range, as the air mass continues, it descends in elevation and is warmed. Under these conditions, its relative humidity decreases—that is, it gets drier—and, as a consequence, moisture is evaporated from the surface rather than released to it in the form of precipitation. The hot deserts of the western United States are essentially rainshadow deserts, as are the Chilean and Peruvian deserts in South America. The fourth and final determinant of dry conditions is an interior location on a large landmass. Distance from a major source of moisture such as the sea—usually aggravated by the rainshadow effects of high mountains or plateaus—means that air masses have lost most of their moisture before they ever reach these locations. The deserts of the United States, the interior Sahara, and parts of the Australian deserts are to one degree or another related to their interior positions on their respective continents, as are the deserts of Asia.

Shrubs are the dominant growth-form in hot deserts, but several other growth-forms also occur there. Certainly, the one most commonly associated with hot deserts is the succulent. Leaf succulents include such plants as agaves and aloes, which both have a **rosette** of basal leaves, and jade plants, which are more like shrubs and have fat, juicy leaves that form on branches. Stem succulents also occur in a variety of shapes. Columnar forms have cylindrical stems; some, such as chollas, have jointed stems; others such as saguaro grow as tall as trees and have a candelabra shape. Globular forms, such as barrel cactus, come in many sizes and with a variety of patterns in the arrangement of spines. Still others have flattened pads; prickly pear cacti are well-known examples of this form. Fruit succulents, such as coyote melon and the desert ancestor of watermelon, have concentrated water in their fruits to promote **dispersal** by thirsty animals. Finally, a group of plants known as caudiciforms, store water and nutrients in swollen stem bases or roots called caudexes that are either below ground or just above the surface. Many of these take on interesting and even bizarre shapes and are popular with plant collectors and hobbyists. Pony-tail palm (*Beaucarnia recurvata)*, actually a member of the agave family, from southeast Mexico is an example often seen as a houseplant.

## MAJOR REGIONAL EXPRESSIONS

### North Africa: The Sahara and the Deserts of Egypt

The Sahara is the world's largest desert and is part of a yet greater dryland zone that stretches from the Atlantic Ocean eastward across northern Africa into Pakistan and India. It also extends from the Mediterranean Sea south to the Niger River at approximately 15° N. Covering some 7 million km$^2$ (3 million mi$^2$), the Sahara is comparable in size to the continental United States, including Alaska. It is generally a level plateau surface interrupted by several large massifs—areas of barren rock hills, reaching no higher than 3,300 m (11,000 ft) (e.g., Hoggar, Tibesti, Adrar, and Aïr). A number of large basins, some sites of oases, others of sand seas (erg) occur. In all, sand covers about 28 percent of the area (Lancaster 1996).

The Sahara lies under the strong influence of a subtropical high, which blocks moisture-bearing winds much of the year. The northern Sahara receives winter rains as an extension of the mediterranean climate regime and has plants and animals related to those of Eurasia and North America. The central Sahara is extremely dry and almost empty of life, and, as such, has long been a barrier to the north-south movement of plants and animals. Therefore, the species of the southern Sahara are related to those of the Old World **tropics**. The southern Sahara experiences summer rainfall peaks. Total precipitation varies from near zero to 400 mm (0–15 in), with aridity increasing from west to east (LeHouérou 1986). Despite the low amount of precipitation, groundwater is plentiful in many regions. Much of it presumably originated during the Pleistocene, when a cooler, moister climate prevailed. Some groundwater reserves, however, are the result of contemporary interior drainage that collects in salt pans or *chotts*, as they are known locally. Oases occur where the water table is close to the surface in depressions.

The highest temperature ever recorded on Earth in the shade—58° C (136° F)—was in the Sahara (at Al'Aziziah, Libya on September 3, 1922). Average temperatures for the warmest month (July) vary from 30° C (86° F) on the Mediterranean Coast to 45° C (113° F) in the interior. Average temperatures for the coldest month (January) range from 10° C (50° F) on the coast to –2° C (28° F) in the massifs of Morocco and Algeria. Hot dry winds known by various names, including sirocco and khamsin, occur 20–90 days a year. At these times, temperatures may be from 35 to 45° C (95–113° F) and relative humidity as low as 5–15 percent. More normal relative humidity is 60–65 percent in winter and 35–40 percent in summer (LeHouérou 1986).

Widely spaced shrubs and subshrubs with varying amounts of bare ground between are the typical vegetation of much of the Sahara. Esparto grass (*Lygeum spartum*) and alfa grass (*Stipa tenacissima*) were once abundant. It is more common today to find low woody plants 0.2 to 0.5 m (0.5–1.5 ft) high, shrubs such as *Artemisia campestri*, *A. monoica*, and *Anthyllis henoniana*. Fleshy, salt-tolerant plants such as *Salicornia fruticosa*, *Salsola* spp., *Atriplex* spp., and *Zygophyllum* spp. grow in saline situations. Cactuslike succulents are found only in the westernmost part of the Sahara along the Atlantic coast, where winter temperatures never fall below 10° C (50° F). Among the succulents are several euphorbias (*Euphorbia balsamifera*, *E. echinus*, *E. regisjubae*, and *E. resinifera*), a group of plants that resemble American cacti. Other succulent plants of the western coasts include *Kalanchoe faustii*, *Senecio anteuphorbium* and *Zygophyllum waterlottii*. Like most deserts, the Sahara is a region of intermittent or ephemeral streamflow; the usually dry streambeds are known as wadis. Since runoff collects in the wadis and percolates into the channel bottoms where it is stored for lengthy periods of time, taller vegetation is associated with them. Shrubs 0.5–3 m (1.5–10 ft) tall dominate. Common are acacias (*Acacia raddiana*), *Calligonum comosum*, *Ephedra alata*, tamarisk (*Tamarix* spp.), and *Ziziphus* spp. Smaller shrubs and annual grasses and **forbs** grow between the shrubs. Trees and perennial grasses may be found in the largest wadis (LeHouérou 1986).

In the driest regions, those with less than 10 mm (0.4 in) of precipitation a year, ephemerals such as *Cleome droserifolia*, *Fagonia* spp., *Plantago ciliata*, and *Stipagrostis plumosa* are the only plants. The Western Desert of Egypt, an area of flat expanses of sand and bare rock (hammada), is nearly devoid of vegetation. Perennial plants can be found only at oases in the numerous closed depressions or along exotic rivers. In such situations one may find acacia trees (*Acacia albida*), date palms (*Phoenix dactylifera*), *Zygophyllum album*, cane (*Phragmites communis*), wild sugarcane (*Saccharum ravennae*), and the grass, *Imperata cylindrica* (LeHouérou 1986; Ayyad and Ghabbour 1986).

Large mammals such as addax (*Addax nasomaculatus*), gemsbok (*Oryx gazella*), bubal hartebeest (*Alcelaphus buselaphus*), and dorcas gazelle (*Gazella dorcas*) once were common in the Sahara, but they pretty much disappeared in the twentieth century. Rodents are still very common, especially in some of the saline depressions and in some sandy areas. The fat sand rat (*Psammomys obesus*) is very common. One also finds jerboas (*Jaculus jacula*, *J. orientalis*, and *Allactaga tretradactylis*) and gerbils (*Gerbillus campestris*, *G. nanus*, *G. pyramidium*, and *G. gerbillus*). Rock dassies or hyrax (Procavia capensis), distant relatives of elephants (although they resemble rodents), live in colonies in the rocky **habitats** of the central massifs. Among carnivorous mammals of the Sahara are jackal (*Canis aureus*), fennec (*Fennecus zerda*), pale fox (*Vulpes pallida*) and Rüppell's fox (*V. rueppelli*) (LeHouérou 1986).

The ostrich (*Struthio camelus*) is no longer common, but quail (*Ternix tachyonurus*) and Barbary Partridge (*Alectoris barbara*) can be found in considerable numbers, as can falcons, hawks, bustards, and storks (LeHouérou 1986).

Among typical reptiles are a land tortoise (*Testudo kleinmanni*), a gecko (*Tropidocalotes tripolitanus*), a chameleon (*Chamaeleon chamaeleon*), a monitor lizard (*Varanus griseus*), and, in sand dune areas, a skink (*Scincus scincus*). The most common snakes are the horned viper (*Cerastes cerastes*) and the very venomous *Vipera lebetina*. Particularly rich assemblages of insects, rodents, lizards, and snakes live together in **nebkas**, accumulations of sand trapped by small shrubs, where one can find the Egyptian cobra (*Naja naja*) (LeHouérou 1986).

The Nile River is the major exotic river crossing the Sahara, and it acts as a biogeographical barrier between the Eastern and Western deserts of Egypt. The animals of the Eastern Desert are more closely related to those of the Arabian Peninsula and western Asia. Among large mammals are Nubian ibex, Beisa oryx, and perhaps a few remaining wild African asses (*Equus africanus*). Birds include the interesting sand grouse (*Pterocles senegaelus*) that nests far from a water source. The male is able to carry water soaked in his breast feathers up to 30 km (20 mi) to dampen the eggs and nest area and provide drinking water for his chicks. Most other birds are found congregated near water. These would include partridge (*Ammoperdix heyi*) and chat (*Oenanthe monache*), and in summer at the oases, turtle-dove (*Streptopelia turtur*), Rufous Warbler (*Erythrpygia galactotes*), and Olivaceous Warbler (*Hippolois pallida*) (Ayyad and Ghabbour 1986).

## Arabian Peninsula

Deserts cover 2.6 million km$^2$ (1 million mi$^2$) on the Arabian Peninsula, but they are not well studied. In the southeast corner is the Rub'Al Khali or Empty Corner, the largest continuous body of sand on the planet. Some of the sands in this 650,000 km$^2$ (250,000 mi$^2$) region are stabilized, others are mobile. Another region of sand dunes is found in the north, the Great Nafud, which is connected to the Rub'Al Khali by a narrow, arcing belt of sand ridges some 1,000 km (600 mi) long. The Arabian deserts have cool to mild winters with rain and very hot, dry summers. In general, average total annual precipitation is less than 100 mm (4 in). The dominant shrubs of the region are *Artemisia monosperma* and *Calligonum comosum*. Characteristic perennial forbs are *Monsonia niva* and *Scrophularia deserti*. After rain, annuals can be abundant and include *Plantago* spp., *Gypsophila cappilaris*, *Senecio* spp., *Rumex pictus*, and *Stipagrostis plumosa* (Abd el Rahman 1986).

Animal life on the Arabian Peninsula is very poorly known. Large mammals are rare but are represented by oryx (*Oryx leucoryx*) and gazelles (*Gazella dorcas*, *G. marica*, and *G. gazella*). Wild hare (*Lepus* sp.) are abundant in southeastern valleys and eastern oases. Among birds living in the Arabian deserts are Houbara bustard (*Chlamydotis undulata*) and sand grouse (*Pterocles* spp.), quail (*Coturnix coturnix*), pigeons (*Columbia livia*), and doves (*Streptopelia* spp.). Smaller birds include bee-eaters, weaverbirds, wagtails, falcons, and hawks (Abd el Rahman 1986).

## India: Thar Desert

The Thar, or Great Indian Desert, lies at the eastern extreme of the great dry belt that stretches from the Atlantic coast of North Africa to this area in northwestern Rajasthan, India. The uplands are rocky plateaus, but the Thar is essentially a region of sandy hills and dunes. Almost all of the rain that falls here comes with the southwest monsoon sometime between June and September. The desert vegetation reflects the length of the dry season. Where it is 11–12 months long, only annuals exist. Subshrubs can survive a dry season 9–10 months long. And in areas where it is essentially rainless only seven months, one finds shrubs and subshrubs dominant. In the foothill regions of granitic uplands, one finds shrub acacia (*Acacia leucophloea*, *A. nilotica*) and *Capparis decidua* with clumps of cactuslike euphorbias (*Euphorbia caducifolia*) in crevices. On rhyolite bedrock, a different acacia (*Acacia senegal*) grows with *Capparis decidua*, *Maytensis emarginata*, and *Salvadora oleodes*. On the lower slopes of the foothills, where runoff infiltrates better, mainly grasses (*Aristida* spp., *Oropetium thomaeum*) and sometimes shrubs grow. The shrub vegetation of the dunes consists of *Aerva* spp., *Calligonum polygonoides*, and *Crotalaria burhia* with clumps of panic grass

(*Panicum turgidum*). On stabilized dunes, acacia (*Acacia senegal*) and mesquite (*Prosopis cineraria*) occur (Gupta 1986).

The mammals of the Thar Desert include desert hare (*Lepus nigricollis*) and such rodents as northern palm squirrel (*Funambulus pennantii*), Indian gerbil (*Tatera indica*), and Indian desert gerbil (*Meriones hurrianae*). Larger mammals are represented by blackbuck (*Antilope cervicapra*), gazelle (*Gazella gazella*), nilgai (*Boselephas tragocamelus*), and ghor-khur or onager (*Equus hemionus*). The hoofed animals move great distances between feeding grounds and watering holes. Among birds are the Rosy Pastor or Rose-colored Starling (*Sturnus roseus*), an insect eater, and seed-eating parakeets (*Psittacula krameria*). Other birds are Jungle Babbler (*Turdoides striatus*), Common Babbler (*T. caudatus*), and Red-vented Bulbul (*Pycnotus cafer*) (Gupta 1986).

## Southern Africa: Karoo-Namib Region

The arid zone of southern Africa in which the Namib Desert and the Karoo are located is believed to have been in existence for 15 million years, making it the oldest desert area on Earth. It is, therefore, not surprising that the region is home to some wonderfully unique species and modes of adaptation. Indeed, 50 percent of the plants are found nowhere else (Kingdon 1989). The **Namib** is a classic example of a fog desert. It lies as a 2,000 km (1,200 mi) strip along the Atlantic Ocean from the San Nicolai River in southern Angola to the Olifants River in Namibia. Offshore is the cold Bengula Current. Coastal stations record an average total annual precipitation of 9–27 mm (0.35–1.0 in), but inland stations near the escarpment that forms the eastern edge of the Namib may record as much as 100 mm (4 in) (Lancaster 1996). As in all deserts, actual amounts vary tremendously from year to year. Walter (1986) gives an example at Swakopmund, where between 1931 and 1933 a total of 4.9 mm (1.2 in) of rain fell. Then in March 1934, 112.9 mm (4.5 in) fell—a once-in-a-century rainfall event. Most of the time, plants and animals get moisture only as fog or dew.

Near the coast—the outer Namib—on the desert pavement are **lichen** fields, dominated by **fruticose** lichens (especially *Teloschistes capensis*). They grow as low, dark orange cushions and cover 40–60 percent of the surface. Other lichens may be interspersed among the cushions, such as *Ramalina maculata* and the nearly black *Xanthoparmelia hyporhytida*. Lichens grow on the lee side of stones or in crevices, where they are protected from salt spray and sand blast. Beneath transparent stones are window algae. Masses of wind-blown "rolling lichen," *Omphalodium convolutum*, accumulate in depressions and runnels. The **thalluses** of these lichens, when dry, roll into dark tubes and are carried away by the wind. Nebkas are a characteristic feature of the outer Namib inland from the lichen fields; they form as sand is trapped by small halophytic shrubs such as *Zygophyllum* and *Salsola* (Walter 1986). Succulent mesembranthemums may grow between nebkas. The most remarkable plant of the Namib, however, is found in the transition zone between outer and inner Namib, an area essentially devoid of perennial vegetation except in the dry stream beds. There, some 50–60 km (30–35 mi) from the coast, one can find *Welwitschia mirabilis*, an ancient evergreen plant with two, continuously growing strap-like leaves (Photo 4.3) (Walter 1986). Plants can live for 1,500 years (Milius 2001), during which time the wind shreds and abrades the leaves so that plant looks like a green heap of refuse. One of the oddities of the plant world, *Welwitschia* is classified

Photo 4.3 *Welwetschia* in its typical dry riverbed habitat, Namibia. (Photo © Michael & Patricia Fogden/CORBIS.)

as a gymnosperm since it produces, on separate individuals, male and female cones. It is a phreatophyte that is able to tap into groundwater supplies.

Sand covers large areas of the Namib. Dunes, if vegetated at all, have perennial grasses (*Cladoraphis*, *Centropodia*, and *Stipagrostis* spp.), often with woody-stemmed leaf succulents such as *Trianthema hereroensis* as codominants (Jürgens et al. 1997).

Animal life in the Namib is surprisingly varied, even in the dunes, where most nutrients are blown in on the winds. Mammals range from elephant (*Loxodonta africana*) and black rhino (*Diceros bicornis*) to the tiny pygmy mouse (*Mus minutioides*). Other large mammals that live there are gemsbok (*Oryx gazella*), springbok (*Antidorcas marsupialis*), and Hartmann's zebra (*Equus zebra hartmannii*). Among smaller mammals are the golden mole (*Eremitalpa granti namibensis*). It is nocturnal and found in the large dune areas, where it swims through the sand to locate the spiders, beetles, and lizards it eats. The vlei rat (*Paratomys littledalei*) is common in the coastal nebkhas (Kingdon 1989).

Ostrich inhabit the Namib; other birds nesting there include Namib lark (*Ammomanes grayi*), desert chat (*Karrucincla schlegelii*), pied crow (*Corvus albus*), and rock kestrel (*Falco tinnuculus*). A number of well-camouflaged **cursorial** bustards (*Eupodotis rueppeli*, *E. caerulescens*, *E. afia*, *E. ludwigi*) of different sizes live in the desert with no need for free water. Reptiles include the fringe-toed, shovel-nose lizard *Aposaurus anchietae*. It stays active during the day until surface temperatures exceed 70° C (158° F), at which time it burrows into the sand. The web-footed gecko (*Palmatogecko* sp.) obtains water by letting fog condense on its head and then wipes the water droplets into its mouth with a long tongue, while another lizard (*Angolosaurus* sp.) obtains water by eating wild melons that grow between the dunes. The side-winding adder (*Bitis peringueyi*) is another characteristic reptile of the dunes; it spends much of its time hidden just under the sand, with only its eyes and the tip of its tail above the surface (Kingdon 1989; Walter 1986).

Photo 4.4 Living stones (*Lithops* sp.) barely distinguishable from pebbles on the desert floor in Namibia. (Photo © Michael & Patricia Fogden/CORBIS.)

Several interesting beetles are among the **endemic** insects inhabiting the dunes of the Namib. The toktokkie beetle (*Onymacris unguicularis*) obtains water from the fog by doing a headstand on the ridge of a dune with its abdomen facing the sea. This is enough of an obstacle to the inland movement of the fog to force condensation. Water droplets collect on the beetle's body and trickle down to its mouth. Other beetles (*Lepidochora discordalis*, *L. kahani*, and *L. porti*) dig shallow trenches in the windward side of dunes and drink the dew that collects in them (Kingdon 1989; Walter 1986).

The **karoo** lies south and east of the Namib in a zone of fairly reliable precipitation from fog and significant amounts of winter rainfall. The growing season is short and cool. The vegetation consists of a huge variety of leaf succulents, both low and dwarf shrubs and rosettes. The most important families are the mesems or ice plants (Mesembryanthemaceae) and the stonecrops (Crassulaceae). A number of nonsucculent shrubs, bulbs, and annuals are also found. In shallow soils, embedded growth-forms are common among such genera as *Haworthia*, *Lithops*, and *Opthalmophyllum* (Photo 4.4). These small leaf succulents are camouflaged as stones, dung, or soil, and their presence is evident only when they put forth showy, colorful flowers. The blooming seasons are predictable and allow for hundreds of species of insect pollinators. Those that bloom in the daytime have yellow, pink, and purple flowers and attract solitary and social bees, wasps, butterflies, and bee-flies. Those plants that bloom at night have white or yellow flowers and fruity scents to attract moths. The stapeliads, succulent members of the milkweed family, have dark red and brown flowers that smell like a dead animal and are pollinated by flies. Most perennials, be they bulbs, succulent, or nonsucculent shrub, grow in autumn and winter and flower in the spring (Milton et al. 1997).

Most of the woody shrubs and dwarf shrubs are evergreen, although the leaves are actually bronze-green, brownish green, or a silvery gray-green. Leaves are hard and often shiny to help reflect light, as in *Maytenus*, *Rhus*, and *Ziziphus*. The gray color of *Olea* serves the same function. Furthermore, the leaves of dwarf shrubs are often rolled up, finely dissected, and covered with hairs or a resin.

Photo 4.5 Halfman (*Pachypodium namaquanum*) with its top knot of leaves, photographed in Goegap Nature Reserve, Republic of South Africa. (Photo © Roger De La Harpe; Gallo Images/CORBIS.)

Some dwarf nonsucculent shrubs undergo a process of stem splitting as they age. The stems break lengthwise from the original plant to create a cluster of individual shrubs, each with its own canopy and roots. Others take on a cushion growth-form. (Werger 1986; Milton et al. 1997).

The karoo is also home to tall plants. Most large shrubs and trees are phreatophytes living in drainage lines. Since they tap into groundwater, they may grow and flower throughout the year. Other trees grow among boulders on shaded hillsides. Some of the stem succulents are among the stranger-looking large plants. The caudiciform quiver tree (*Aloe dichotoma*, *A. ramossissima*) has rosettes of grasslike leaves at the ends of small branches coming off its swollen trunk. Elephant's foot (*Pachypodium namaquanum*), also known as halfman (Photo 4.5), has a single rosette of broad leaves atop its spine-covered succulent stem. The tops or heads tilt to the north, toward the sun. According to a local legend, these are ancestors that God froze when they were migrating south and disobeyed an order not to look back (Kingdon 1989; Milton et al. 1997).

The hoofed mammals of the karoo are the same as those found in the Namib Desert. Among small mammals are burrowing rodents such as ground squirrel (*Xerus inauris*), springhare (*Pedetes capensis*), a mouse (*Malacothrix typica*), gerbil (*Gerbillurus paeba*), and porcupine (*Hystrix africaeaustralis*). By day, burrows may be occupied by bat-eared fox (*Otocyon megalotis*), suricate (*Suricata suricatta*), aardvark (*Orycteropus afer*), aardwolf (*Proteles cristatus*), and brown hyena (*Hyaena brunnea*), all nocturnal hunters. It is not unusual to find several different kinds of animals—including mammals, reptiles, and arthropods—sharing the same burrow during the heat of day (Werger 1986).

## North America: Mojave, Sonoran, and Chihuahuan Deserts

The hot deserts of North America are found in the Basin and Range **Physiographic Province**, including its Mexican Highland section. The major landforms of the province are a swarm of mountain ranges usually trending north-south; many are nearly buried by the valley fill of the intervening basins, which account for more than half of the land surface. Broad bajadas extend over three-fourths of many basins and are important determinants of plant and animal distribution patterns. The upper bajadas are formed of coarse sediments washed out of the adjacent mountain range and usually host more diverse vegetation than lower slopes formed of finer particles. The basins themselves consist of desert pavement; dune areas are infrequent and quite small compared to those of other deserts in the world. Often the stones of the desert surface are darkened with what is called desert varnish, a thin coating of iron and manganese oxides derived from physical and possibly biological processes. In areas of finer particles, crusts are produced by lichens, algae, and **cyanobacteria**. Interior drainage and Pleistocene lakes have led to the formation of more than 200 playas in the lowest parts of the basins, and a number of exotic streams flow through the area although fewer than before a period of arroyo cutting in the late nineteenth century and dam building in the twentieth century. The Colorado River is the largest of these; other examples are the San Pedro, Salt, Gila, and Santa Cruz rivers in southern Arizona. The small size of the desert basins and the proximity of mountains high enough to induce precipitation means that many highly mobile animals are able to move daily to a waterhole. Since caliche is a common occurrence beneath the desert pavement of the basins; perched water tables and stores of groundwater that may last months, or even all year are widespread. With water so readily available at mountain springs or beneath the gravels of desert washes, relatively few animals in the North American have developed physiological adaptations to drought, although behavioral modifications are the same as those found in deserts around the world (MacMahon 2000).

North American hot deserts receive most of their precipitation as rain. What differs among the three deserts is the time of year at which rainfall peaks are most predictable. The Mojave Desert is an inland extension of the mediterranean pattern of winter rainfall; the Chihuahuan Desert receives most of its precipitation from summer convectional storms as a monsoonal effect developed over the continent draws moist air inland from the Gulf of Mexico. The Sonoran Desert, situated between the other two, benefits from both rainfall regimes and experiences a double peak of rainfall with periods of rain in both winter and summer. Summer temperatures may exceed 40° C (100° F), and winter temperatures may drop below freezing in all three deserts. It is rare, however, to have extended periods of time with temperatures at or below 0° C (32° F); indeed, many plants are killed when it is below freezing for longer than 36 hours (MacMahon 2000).

The indicator plant of hot deserts in North America is creosotebush (*Larrea tridentata*). The distribution area of this evergreen shrub coincides pretty closely with that of the hot desert region as a whole. Interestingly, although the same species, the creosotebush in each desert has a different number of chromosomes. The plant found in the Chihuahuan Desert has 13 different chromosomes (that is, n = 13),

Photo 4.6 Joshua tree (*Yucca brevifolia*), the indicator plant of the Mojave Desert. (Photo by SLW.)

each of which is paired with an identical duplicate of itself. Every cell in the plant (except for ova and pollen), thus actually possesses a total of 2n or, in the case of Chihuahuan creosotebush, 26 chromosomes. In the other deserts, creosotebushes possess multiples of 13 chromosomes: for the Sonoran plants n = 26 and for Mojave plants n = 39 (MacMahon 2000). In each desert, a different set of plants joins the creosotebush, and the dominant or characteristic growth-forms vary. The occurrence of key plants distinguishes each desert.

Much of the **Mojave Desert** is dominated by shrubs of creosotebush and white bursage (*Ambrosia dumosa*). The Joshua tree (*Yucca brevifolia*) (Photo 4.6), the distinguishing plant of the Mojave, grows at higher elevations and on upper and middle bajadas. The Joshua tree is visually dominant on the landscape because of its large size, but the shrubs it grows alongside are more numerous and important to the plant and animal communities. At lower elevations, the Joshua tree is associated with creosotebush; elsewhere it occurs with such shrubs as wolfberry (*Lycium* spp.), blackbush (*Coelogyne ramosissima*), bladder-sage (*Salazaria mexicana*), and Mormon tea (*Ephedra nevadensis*). Beavertail (*Opuntia basilaris*), staghorn cholla (*Opuntia acanthocaroa*), and barrel cactus (*Ferocactus* spp.), as well as some perennial grasses (*Hilaria rigida*, *Muhlenbergia porteri*, and *Stipa* spp.) also grow with Joshua trees.

Joshua trees do not grow everywhere in the Mojave, and where it does not grow a diverse **community** of shrubs may appear. Common shrubs on middle bajadas, for example, include wolfberries (*Lycium andersonii* and *L. pallidum*), ratany (*Krameria parvifolia*), Fremont dalea (*Psorothamnus fremonti*), goldenhead (*Acamptopappus schlockeyi*) and yellow paper daisy (*Psilostrophe cooperi*). Small yuccas, such as Mojave yucca (*Y. schidigera*), banana yucca (*Y. baccata*), and Spanish bayonet (*Y. whipplei*), grow with a large number of winter annuals that cover the desert floor in a patchwork of bright color in early spring (MacMahon 2000).

Photo 4.7 The candelabra-shaped saguaro (*Carnegiea gigantea*), signature plant of the Sonoran Desert. Many other cacti, such as the cholla (*Opuntia* sp.) and prickly pear (*Opuntia* sp.) in the left foreground also grow here. The thin, flaring stems of ocotillo (*Fouqueria splendens*) can be seen in the center of the photo. (Photo by SLW.)

The **Sonoran Desert** is a species-rich group of regional plant and animal communities. In the United States, the signature plant is the tall, candelabra-shaped saguaro cactus (*Carnegiea gigantea* = *Cereus giganteus*) (Photo 4.7). The plants tend to be related to tropical forms, and many genera occur both here and in the Monte of Argentina. To deal with the geographic complexity of the vegetation of the Sonoran Desert, scientists sometimes break it into six subdivisions: Vizcaino, Arizona Upland, Magdalena Plain, Plains of Sonora, Central Gulf Coast, and Lower Colorado Valley (MacMahon 2000). Only three will be discussed here. The largest and driest subdivision is the Lower Colorado Valley, which occurs along the Colorado River and extends into lowlands drained by streams tributary to the Colorado River in both Arizona and California. Creosotebush, white bursage, and brittlebush (*Encelia farinosa*) dominate large areas of level land. Small trees, such as blue paloverde (*Cercidium floridum*), honey mesquite (*Prosopis glandulosa*), smoketree (*Dalea spinosa*), and ironwood (*Olneya tesota*), and a variety of shrubs, such as burroweed (*Hymenoclea salsola*), desert lavender (*Hyptis emoryi*), wolfberry (*Lycium andersonii*), and saltbushes (*Atriplex* spp.), grow in and near washes. Where the water table is high along exotic rivers, cottonwood (*Populus* spp.) and willows (*Salix* spp.) can be found, along with dense thickets of phreatophytes such as screwbean mesquite (*Prosopis pubescens*) and **introduced** salt cedar (*Tamarix* spp.). A fringe of arrowweed (*Pulchea sericea*) lies between them and the dry desert (Woodward 1976).

The bajadas host the greatest diversity of both growth-forms and species. Typical larger plants are tall cactus like the saguaro, the green-stemmed foothills paloverde (*Cercidium microphyllum*), catclaw (*Acacia greggii*), and ocotillo or coachwhip (*Fouquieria splendens*). Much of the time, ocotillo looks like a bunch of long, dry, thorny sticks, but it is notable for its ability to leaf out and flower within days of rainfall—regardless of when it rains or how many times a year it rains. Smaller shrubs and stem succulents of the bajadas include bladder-sage (*Salazaria mexicana*), desert senna (*Cassia armata*), indigobush (*Dalea schottii*), hedgehog cactus (*Echinocereus engelmannii*), and barrel cactus (*Ferocactus acanthodes*). A great variety of ephemerals are abundant whenever winter rainfall is adequate. These include woolly plaintain (*Plantago insularis*), an important forage plant for mammalian herbivores; the large, white-flowered evening primroses (*Camissonia* spp.); and blue-flowered phacelias (*Phacelia* spp.).

Mammals of the Lower Colorado Valley include desert bighorn sheep (*Ovis canadensis nelsoni*), mule deer (*Odocoileus hemionus*), coyote (*Canis latrans*), gray fox (*Urocyon cinereoargenteus*), and kit fox (*Vulpes macrotis*). Black-tailed jackrabbits (*Lepus californicus*), cottontails (*Sylvilagus floridanus*), antelope, ground squirrel (*Ammospermophilus harrisi*), kangaroo rats (*Didpodomys deserti*), woodrats (*Neotoma* spp.) and pocket mice (*Perognathus* spp.) are among the more common small mammals. A number of hawks, including the widespread red-tailed hawk (*Buteo jamaicensis*) and Cooper's hawk (*Accipiter cooperii*), prey on the small mammals, as do coyotes and foxes. Other birds of the region include roadrunner (*Geococcyx californianus*), gilded flicker (*Colaptes chrysoides*), phainopepla (*Phainopepla nitens*), cactus wren (*Campylorhynchus brunneicapillus*), verdin (*Auriparus flaviceps*), Costa's hummingbird (*Calypte costae*) and many more. Reptiles include desert tortoises (*Gopherus agassizi*), desert iguana (*Dipsosaurus dorsalis*), chuckwalla (*Sauromalus obesus*), horned lizards (*Phrynosoma* spp.), and rattlesnakes (*Crotalus* spp.).

The only substantial area of large sand dunes in North American hot deserts occurs in the Lower Colorado Valley near Yuma, Arizona. Plant life is sparse and consists of perennial grasses such as *Hilaria rigida* and several shrubs such as Mormon tea (*Ephedra trifurca*), daleas (*Psorothamnus schottii* and *P. emoryi*), wild buckwheat (*Eriogonum deserticola*), and sandpaper plant (*Petalonyx thurberi*) (MacMahon 2000). The fringe-toed lizard (*Uma notata*) and sidewinder (*Crotalus cerastes*) are residents of these dunes (Lowe 1964).

The Arizona Upland subdivision is cooler and moister than the Lower Colorado Valley, but it contains many of the same plants and animals, or their close relatives, plus many others. The valley bottoms and lower bajadas are generally vegetated with creosotebush, frequently in pure stands, and bursage (*Ambrosia deltoidea*). Shrubby acacias are also common. On upper slopes in Arizona, one finds saguaro with foothills palo verde, ocotillo, barrel cactus, and chollas, with a lower shrub layer of plants such as fairy duster (*Calliandra eriophyla*), the **broad-leaved** ***Jatropha cardiophylla***, brittlebush, and jojoba (*Simmondsia chinensis*). Southward in Mexico, elephant tree (*Bursera microphylla*) and two large columnar-type cactus—cardon (*Lophocereus schotti*) and organpipe cactus (*Lamaireocereus thurberi*)—are characteristic (MacMahon 2000). These upper bajadas may host mammals such as white-tail deer (*Odocoileus virginianus*), javelina or collared peccary (*Tayussu tajacu*), hooded skunk (*Mephitis macroura*), and hog-nosed skunk (*Conepatus mesoleucus*)—animals not found in the Lower Colorado subdivision. Common resident birds include Gambel's quail (*Lophortyx gambelii*), roadrunner, cactus wren, curve-billed thrasher (*Toxostoma curvirostre*), and pyrrhuloxia (*Pyrrhuloxia sinuata*). These together with a host of migratory birds make the region a birder's paradise.

Photo 4.8 Chihuahuan Desert scrub. Three shrubs, whitethorn acacia (*Acacia constricta*), creosotebush (*Larrea tridentata*), and tarbush (*Flourensia cernua*), dominate this desert's landscapes, photographed here in an outlier in southeastern Arizona. (Photo by SLW.)

The Vizcaino subdivision of the Sonoran Desert is located in the western part of the central Baja California peninsula. It is essentially a fog desert receiving average annual rainfall of less than 100 mm (4 in). Right along the coast of the Pacific Ocean, in the northern part of the region the maguey (*Agave shawii*), is a conspicuous element on the landscape. It is a leaf succulent, rosette plant with a flower stalk several meters tall. Farther south, a ground cover of fruticose lichens dominates coastal vegetation. Inland, the visual dominant is the strange caudiciform boojum tree (*Fouquieria columnaris*), sometimes described as looking like an upside-down carrot. Boojums are restricted to this part of the Sonoran Desert and the Central Gulf Coast Subdivision, which straddles the Gulf of California. Elephant trees (*Bursera* spp.) also occur in both subdivisions; they become a dominant in the Central Gulf Coast, where they are associated with *Jatropha cuneata* and *J. cinerea*, ocotillo, and several chollas and columnar cacti (MacMahon 2000).

The **Chihuahuan Desert** is less studied than the other two hot deserts of North America. It occurs at generally higher elevations, 400–2,000 m (1,300–6,500 ft), particularly in basins underlain by limestone. A characteristic landform feature is gypsum sands and dunes that host a number of unique plants. The shrub is the dominant growth-form of this desert. Three plants in particular are codominants and form a fairly dense cover: creosotebush, white-thorn (*Acacia constricta*), and tarbush (*Flourensia cernua*) (Photo 4.8). Commonly associated with them are ocotillo, allthorn (*Koebelinia spinosa*), and lechuguilla (*Agave lechuguilla*). Many small to medium-sized cacti grow there, but rosette plants such as agaves, yuccas, sotol (*Dasylirion leiophyllumi*), and beargrass (*Nolina microcarpa*) are much more conspicuous on the landscape. A plant of potential economic importance is guayule (*Parthenium argentatium*), a source of rubber. The desertscrub vegetation is interrupted in bottomlands by grassy swales of tobosa grass (*Hilaria mutica*) or bush muhly (*Muhlenbergi porteri*), sacaton (*Sporobolus airoides*), and the grama grass *Bouteloua eiopoda* (MacMahon 2000).

The snow-white gypsum dunes, such as those at White Sands National Monument, support—especially in the interdune areas—several widespread plants such as soaptree (*Yucca elata*), four-wing saltbush (*Atriplex canescens*), and Indian ricegrass (*Oryzopsis hymenoides*). Of even greater interest to botanists and plant geographers are the approximately 70 perennial herbs and dwarf shrubs restricted to the gypsum sands of the Chihuahuan Desert; among them are members of three endemic genera (*Dicranocarpus*, *Marshalljohnstonia*, and *Strotheria*) (MacMahon 2000).

The Chihuahuan Desert is home to a large number of animals, such as collared peccary, kangaroo rats, roadrunners, and curve-billed thrashers, and many are fairly widespread. The reptiles, however, tend to be more closely confined to the region. Among the restricted forms are the Texas banded gecko (*Coleonyx brevis*), reticulated gecko (*C. reticulatus*), greater earless lizard (*Cophosaurus texanus*), and marbled whiptails (*Cnemidophorus tigris marmoratus*). Two Chihuahuan whiptails (*C. neomexicanus* and *C. tessalatus*) are rare examples of all-female species. Snakes characteristic of the region include the trans-Pecos rat snake (*Elaphe subocularis*) and two whipsnakes (*Masticophis* spp.) (Ricketts et al. 1999).

## South America: Monte and Chilean-Peruvian Fog Deserts

The **Monte Desert** extends from sea level to 2,800 m (9,000 ft) in the Salta Province of Argentina and receives about 60 percent of its rainfall in summer. It has not yet been studied in detail, but its interest stems from its similarities with the Sonoran Desert of North America. It has columnar cactus and many of the same genera of plants, although very few species are identical. Among the genera that occur in these two widely separated deserts are *Larrea* (creosotebush), *Acacia* (acacias), *Cercidium* (palo verde), *Prosopis* (mesquite), *Cassia* (senna) and *Opuntia* (prickly pear and cholla cacti). Tall columnar cacti (*Trichocereus* spp.) are visual dominants on upper slopes; creosotebush (*Larrea cuneifolia*) dominates lower elevations. A component of the Monte absent from the North American deserts is the terrestrial bromeliad, a hard-leaved rosette plant of the pineapple family (Bromeliaceae) represented here by *Deuterocohnia schriteri* and *Dyckia velazcana* (Mares et al. 1985).

Unlike plant life, animals of the two deserts have little in common. The Monte lacks large mammals and seed specialists. The Patagonian hare (*Dolichotis patagonium*) is the largest mammal in this desert. It is quite common, as are mountain cavies (*Microcavia australis*) and tuco-tucos (*Ctenomys fulvus*). The leaf-eared mouse (*Phyllotis darwinii*) lives on rocky hillsides covered with cacti and bromeliads. The Monte armadillo (*Chaetophractes vellerosus*) is another mammal of the desert. Physiological adaptations to aridity are expressed in rodents such as the highland desert mouse (*Eligmodontia typus*), which concentrates its urine and can drink salt water, and the vesper mouse (*Calomys musculinus*), which requires no free water (Mares et al. 1985).

One of the world's driest desert regions lies in a narrow band along the west coast of South America from southern Ecuador to northern Chile. The Peruvian Desert and the **Atacama** of Chile lie parallel to the cold Humboldt Current and in the rainshadow of the high central Andes Mountains. Low coastal ranges 500–800 m high (1,600–2,600 ft) rise up right next to the sea and wring much of the moisture out of the fog as daily sea breezes carry it inland. Total annual precipitation averages less than 50 mm (2 in) in the region as a whole, and some sites get virtually none. Trujillo, Peru, receives on average 3 mm (0.1 in) of rain a year; Iquique in northern Chile can expect at best 0.4–1.9 mm (0.1–0.5 in). North of 8° S, measurable rain and substantial runoff from the Andes occurs every 5 to 12 years as a result of the summer El Niño phenomenon. But south of that latitude, moisture comes more regularly from fogs or **garuas** (known as **camanchacas** in Chile) that are especially frequent in the winter. Relative humidity is high and temperatures are rather low, considering the latitude, as a consequence of the dense fogs. At Iquique, for example, the warmest month (March) has a mean temperature of 21° C (70° F) and the coolest month (September) 14.5° C (58° F). The warmest temperature recorded is 31.3° C (88° F). Garuas can move inland 30 to 50 km (20–30 mi) and fill valleys to elevations of 500–700 m (1,600–2,300 ft). In response to the garuas, a number of unique fog desert communities called *lomas* (Spanish for small hills) have developed. Lomas occur as islands of plant life in otherwise barren landscapes and contain many plants found nowhere else on earth (Dillon n.d.; Rauh 1985).

Closest to the shoreline is a fog loma desert containing a large number of ephemerals. The perennial communities of terrestrial bromeliads of the **genus** *Tillandsia* are a unique aspect of this vegetation zone; they are a feature primarily of the Peruvian desert but can be found from the vicinity of Trujillo, Peru (8° S), south to Iquique, Chile (20° S). These tillandsias, members of a strictly New World family, are the only higher plants able to take up moisture directly from the air through specialized structures called trichomes that appear as a fine, grayish fuzz on the leaves. Their roots have no function relative to water uptake, and on most species are greatly reduced. When developed at all, roots serve merely to anchor the plant to the ground. Tillandsias grow in groups or loose mats in pure stands. The most common species is *T. latifolia;* others are *T. paleaceae, T. purpurea, T. recurvata,* and *T. werdermannii* (Rauh 1985).

Cactus lomas grow in other areas. One common form, *Haagocereus repens,* is a columnar type that lies flat on the ground in the sandy areas of northern Peru; only its tip grows erect. Lichens (*Cladonia* and *Teloschistes*) may cover the stem. In other areas, one can find globular cacti (*Islaga* spp.), short columnar cacti (*Copiapoa cinerea*) that lean northward toward the sun, and tall candelabra-type cacti (*Eulychnia iquiquensis* and *Trichocereus chiloensis*). Soil cacti are another peculiarity of these South American deserts. Soil cacti, such as *Pygmaeocereus rowleyanus* in Peru and *Neochilenia* spp. in Chile, have the greatest proportion of the plant below ground in the form of a thick, succulent tuber. From the tuber, an intricate network of roots spreads out close to the surface to trap the slightest amount of water trickling into the ground. The small, often brown, above-surface shoots blend with the ground to make the cactus nearly invisible except when in flower. Various lichens grow on taller cacti and extract moisture directly from the fog. Rolling lichens (*Rocella cervicornis, Parmeia vagens, Tornabenia ephebaea*) are also part of the vegetation of these cactus lomas (Rauh 1985). Washes may have a sparse cover of shrubs with such plants as *Euphorbia lactiflua, Tetragonia maritima,* and *Nolana mollis.* The nolanas are a diverse group of plants restricted to the region (Dillon n.d.).

A unique wood loma or fog forest occurs above 500 m (1,600 ft) at Fray Jorge National Park in central Chile, where it is humid year round in a narrow elevational zone. Only 20 percent of the moisture derives from rainfall; the rest (1,500 mm) is condensed from the fogs. The dominant plant is an evergreen tree, tique (*Aextoxion punctatum*). With it grow puya (*Puya chilensis*), another kind of terrestrial bromeliad; the columnar cactus *Eulychnia;* and moss- and lichen-covered shrubs such as canelo (*Drymis winteri*) and olivillo (*Myrceugenua corraefolia*). This plant community is considered Chile's northernmost forest (Rauh 1985).

On inland fog-free areas of northern Chile (south to 28° S), the vegetation consists of ephemerals; perennial herbs; and window lichens, which produce their main thallus below ground. This is the so-called flowering desert that most of the time appears completely barren, but in those unusual years when more than 20 mm (0.75 in) of rain falls during June and July (i.e., southern hemisphere winter) it turns into a multicolored carpet of flowers. Some of the ephemerals are *Adesmia* spp. *Cryptantha parviflora*, *Lupinus* spp., *Tetragonia maritima*, and *Viola polypoda*. Perennials include Peruvian lily (*Alstroemeria* spp.), *Argylis radiata*, *Calandrinia grandiflora*, *Ephedra* spp., *Euphorbia copiapina*, *Hippeasterum ananuca*, and nolanas. In southern areas, a sparse cover is composed of small shrubs such as *cacho de cabra* (goat's horn) (*Skytanthus acutus*)—so named because its ripened seed capsule coils into a spiral—*Encelia cansecens*, and *Nolana rostrata* (Rauh 1985; Dillon n.d.).

Despite the sparse plant life, the lomas of the Peruvian desert and the Atacama are not devoid of animal life. Arthropods such as flies, spiders, beetles, and scorpions may occur in large numbers; species vary from one isolated loma to another. At least one bird, a miner (*Geositta peruviana*), has different plumage colors depending upon the background hue of the loma it inhabits. Many of the birds and mammals breeding on the coast and near shore islands are more marine than land animals. For example, guano birds [cormorants (*Phalacrocorax bougainvillei*) and pelicans (*Pelecanus occidentalis*)] may occur in the millions, and southern sea lions (*Otaria flavescens*) can be quite abundant in some localities (Rauh 1985). However, some truly terrestrial forms are found as well. The South American desert fox, or *zorro* (*Pseudalopex sechurae*), and the guanaco (*Lama guanicoe*) are two of the larger mammals.

## Australian Deserts

Australians recognize five major deserts on the continent: the Great Sandy, Gibson, Great Victoria, Simpson (or Arunta), and Sturt deserts. These deserts receive on average 250–380 mm (10–15 in) of precipitation a year, with total amounts increasing northward. Australia has no extremely dry deserts (areas are popularly considered arid when they do not support domestic sheep or cattle). Habitats include flat surfaces on ancient bedrock that have a thick lateritic hardpan formed in moister periods of the geologic past. Salt lakes occur in depressions. Stone deserts exist on tablelands, and about 2 million km$^2$ (770,000 mi$^2$) of sand desert with stabilized dunes is present. Leguminous shrubs, chenopods, annuals of the daisy family, and bluegrasses (family Poaceae) are most common. The two most widespread vegetation types are acacia shrublands and arid hummock grasslands dominated by *Triodia* spp. The acacia shrublands have an **open canopy** of shrubs 3–7 m (10–20 ft) tall and are dominated by 1 of some 11 acacias (*Acacia aneura*, *A. brachystachya*, *A calcicola*, etc.). The ground layer consists of grasses such as *Eragrostis eriopoda*, *Aristida* spp., *Chloris* spp., and *Digitaria* spp., and perennial and annual forbs. Actual composition varies with soil type. The arid hummock grasslands, known as spinifex grasslands in Australia, are dominated by *Triodia* grasses. *Triodia basedowii* is most widespread in the central desert, while *Triodia pungens* becomes important in the north. The hummock grasses can attain heights of 1.5 m (5 ft) and hummock diameters up to 6 m (20 ft). Some, such as *T. basedowii*, die in the center as the clump ages to leave a ring of live grasses around the periphery of the hummock. Individual hummocks are widely spaced with much bare ground between them. This bare ground will be filled by ephemerals after rains. Low trees and shrubs may be scattered over the landscape. Important shrubs include several kinds of senna (*Cassia* spp.), *Dodonaea attenuata*, and *Eremophila* spp. (Williams and Calaby 1985).

Two other vegetation types are found in sizeable areas (approximately 500,000 $km^2$ or 200,000 $mi^2$ each). **Tussock** grasslands dominated by *Astrebla* spp. occur primarily in the eastern and northeastern edges of the arid part of Australia. A chenopod shrubland dominated by widely spaced low bushes of *Atriplex*, *Maireana*, and *Sclerolaena* is found in southern and eastern central areas of the desert region (Williams and Calaby 1985).

The animals of Australian deserts include the tiny insect-eating marsupial mouse (*Planigale ingrami*) and the blind marsupial mole (*Notoryctes typhlops*). The latter has habits and appearance similar to the placental golden moles of arid southwest Africa. Hare wallaby (*Lagorchestes* spp.) are associated with hummock grasslands; other wallabies and the larger kangaroos—the euro (*Macropus robustus*) and the red (*M. rufus*)—occur in localized areas. The broad-ranging echidna (*Tachyglossus aculeatus*) also occurs in the deserts and feeds on the abundant ants and termites. Some 20 native rodents live in arid Australia; these include mice (*Notomys alexis* and *Pseudomys hermannsburgensis*), which are nocturnal, burrowing forms that do not require free water. Most rodents occur in low densities, but a few such as the plague rat (*Rattus fuscipes*) have destructive population explosions when rainfall is above average. Birds are highly nomadic, and zebra finch (*Poephila guttata*) and several parrots [galah (*Cacatua roseicapilla*), little corella (*C. pastinator*), and budgerigar (*Melopsittacus undulatus*)] occur in huge flocks (Williams and Calaby 1985).

These deserts have more reptiles than any other; they include 40 snakes and more than 190 different lizards. Reptiles are particularly diverse and abundant in the *Triodia* hummock grasslands. Among snakes are large pythons (family Boidae); the large, highly venomous mulga snake (*Pseudechis australis*) and fierce snake (*Oxyuranus microlepidotus*); and death adders (*Acanthophis antarcticus* and *A. pyrrhus*). Lizards include two large monitor lizards, perentie (*Varanus giganteus*) and sand goanna (*V. gouldi*); geckos; and skinks. Many lizards are nocturnal, but the dragon lizards (*Amphiboturus* spp.) are diurnal. They perch on shrubs and change the orientation of their bodies to maintain body temperatures below 40° C (100° F). During the hottest part of the day, they seek cover in burrows and remain inactive. Some dragon lizards change body color; they are dark when it is cool and lighter colored when it gets hot (Williams and Calaby 1985).

Australia has a number of desert-adapted frogs. They burrow deep into the soil, where they go dormant (aestivate) to avoid drought and high temperatures. Some spend most of their lives in this inactive state. They survive in different ways. Some absorb soil moisture through their skin even as they aestivate; others make a cocoon of shed skin to waterproof themselves. The water-holding frog (*Cyclorana platycephala*) has a very large bladder and fills it with dilute urine as it goes into dormancy. Still others are able to tolerate extreme dehydration and can lose more than 40 percent of their body weight as they dry out without lethal effect. Frogs such as *Notoden nichollsi* have the ability to rehydrate very quickly when it rains and come to the surface to feed on invertebrates and breed. Their larval stages are brief (two to six weeks) to allow the tadpoles to metamorphose into frogs before pools of water evaporate (Williams and Calaby 1985).

## ORIGINS

The age of the hot desert plant communities is still debated. Depending on the evidence used, some scientists contend that they are very old, others believe they are relatively recent in origin. Contemporary plant distribution patterns that have members of the same family or genus on widely separated landmasses suggest that time measured in geologic periods was needed for plants to migrate from one to the other. The existence of ancient plants such as *Welwitschia mirabilis* point to the possibility that its habitat has been in existence for a very long time—indeed this is the main reason that the Namib Desert is considered the oldest of all. On the other hand, it is common to find desert plants that have most of their closest relatives in nondesert biomes. This suggests that adaptations to drought and heat have evolved relatively recently, and only a few members of a family or a genus have been able to invade arid regions. Fossilized plant remains and pollen support the hypothesis of a recent origin by indicating moister conditions in current desert sites during the Pleistocene Epoch and forests before the Miocene Epoch (Shmida 1985).

In the Sonoran and Chihuahuan deserts of North America, it appears that modern plants derive from tropical **savanna** and dry forest plants of the Late Cretaceous and Tertiary periods. They adapted to increasingly drier conditions during the Tertiary but did not find extremely hot and dry environments to dominate as a real desert vegetation type until the Pleistocene. The Pleistocene, of course, was a time of alternating warm-dry and cool-moist climatic patterns, so the distribution areas of these deserts expanded and contracted. Parts of today's Sonoran and Chihuahua deserts served as refuges for desert plants during the cool-moist periods. All indications are that creosotebush (*Larrea tridentata*), the widespread indicator plant of North America's hot deserts, is a recent immigrant from South America that may have become established only about 10,000 years ago. Plants of the Mojave Desert are immigrants from either the Sonoran Desert or the Great Basin Desert (MacMahon and Wagner 1985).

As for the other hot deserts of the world, the plants of the Monte Desert in Argentina are largely related to plants of the Chaco thornforest (Mares et al. 1985). Middle Eastern deserts have many plants derived from those of the mediterranean woodland and **scrub** biome. An arid climate began to develop in much of the Sahara in the Pliocene Epoch, but true desert climates date to the Pleistocene. However, here too, the Pleistocene had alternating drier and wetter phases. The northern Sahara was still more a savanna like present East Africa than a desert between 5000 B.C. and 500 B.C., when nomadic pastoralists herded cattle in the region and produced rock art that portrayed their herds in addition to elephants, giraffes, giant buffalo (*Homioceros*), and ostrich. Rock art of a similar age suggests that the Eastern Desert of Egypt and Sudan only 2,500 years ago also was, if not bettered watered, at least more densely vegetated than today (LeHouérou 1986; Ayyad and Ghabbour 1986). Areas of the southwestern parts of the Arabian peninsula were covered in acacia savanna and supported herds of gazelles, oryx, zebra, and ostrich together with their predators and scavengers well into the twentieth century (Orshan 1986).

## HUMAN IMPACTS

The great desert belt that stretches from the Atlantic Ocean across North Africa and the Arabian peninsula into India has been home to nomadic and semi-nomadic pastoralists for thousands of years. Livestock grazing, especially by sheep and goats, continues to be the major land-use practice in hot deserts around the world. Even the lomas of Peru were grazed by llamas when Andean peoples descended to the coast to trade for fish and salt. In the valleys of larger rivers and at oases, cultivation of crops is also a long-standing economic activity. Techniques that depend on runoff from current precipitation events and irrigation using groundwater supplies had been perfected in the Americas and the North Africa-Asian drylands long before the modern era. In the Middle East, camels, donkeys, and **breeds** of fat-tailed sheep and goats are all adapted to withstand water losses amounting to a considerable proportion of their body weight. Traditional migration routes took the nomads and their flocks to oases and towns where they could exchange milk, meat, wool, and hides for farm produce and manufactured goods. It is difficult to assess the impacts these people and their animals might have had on the wild plants and animals of the desert.

Most serious degradation seems to stem from more contemporary land uses. Overgrazing became a major problem as the human population increased in North Africa, the Middle East, and India in the twentieth century. Modern commercial agriculture focusing on irrigated cotton, fruits, vegetables, and grains has converted many **hectares** of desert to farmland in all regions. Recreational land uses are a growing problem in North America and a potential problem elsewhere. Hunters, campers, and rockhounds can inadvertently damage the fragile vegetation by trampling plants and the soil-binding crust of lichens and algae. Off-road vehicles are even more destructive because they break up the microphytic crust and increase the likelihood of wind erosion, compact the soil and decrease infiltration rates of water, and send up clouds of dust that coat and suffocate desert plants. They can also collapse the burrows of desert rodents and tortoises, not to mention simply running over the animals themselves. As cities in or near deserts expand, more and more pipelines, electricity transmission lines, highways, and aqueducts crisscross the deserts and result in the removal of plants and interference with the movements of animals. In at least a few areas, overcollecting of succulents and other rare and fascinating growth-forms is a threat to the survival of rare plants. Real problems already occur in the Sonoran desert, and the Namib, Karro, and Atacama face potential problems.

## HISTORY OF SCIENTIFIC EXPLORATION

The Carnegie Institution Desert Laboratory was established on Tumamoc Hill in Tucson, Arixona, in 1903 (Bender 1982). This center of study in the Sonoran Desert was to become a model for desert research stations around the world. One of its most influential researchers in the early days was Forrest Shreve (1910, 1917, 1951). His scheme for distinguishing among the North American deserts (Shreve 1942) was based on lifeforms, vegetation structure, and plant distribution and is still used by many scientists today. The Chihuahuan Desert was examined in a general plant geography study of the continent by Harshberger (1911) and work in Mexico by Muller (1939, 1947) and Leopold (1959). The Structure of Ecosystems

Subprogram of the International Biological Programme (IBP) in the late 1960s and 1970s revitalized interest in hot deserts and resulted in many publications on all three hot deserts (e.g., Solbrig and Orians 1977; Orians and Solbrig 1977) (MacMahon and Wagner 1985).

J. Morello (1958) conducted the seminal work on vegetation of the Monte Desert of Argentina, and the Monte was a focus for comparison with the Sonoran Desert during the IBP (Orians and Solbrig 1977). Very little research beyond basic inventory and taxonomy was conducted on the animals of the Monte until studies such as that by Mares et al. (1977) were initiated by the IBP (Mares et al. 1985). In the Peruvian and Chilean coastal deserts, German scientists such as August Weberbauer and Carl Troll were dominant in early scientific exploration. Troll (1930) first explained the *garua*, the fogs that form over the cold Humboldt Current, itself named for a yet earlier German plant geographer-explorer (Rauh 1985).

M. Zohary (e.g., Zohary 1940; Zohary and Orshan 1956) laid the foundations for desert research in Southwest Asia (Orshan 1986). In India, the earliest botanical investigations are traced to King (1879) and Blatter and Hallberg (1918). By the early 1970s, enough information was available to delineate the desert biome in northwest India (e.g., Joshi and Gupta 1973) (Gupta 1986).

People such as M. Kassas (1953), Theodore Monod (1954) and Henri LeHouérou (1959) pioneered research on the North Africa desert. In southern Africa, the Namib was first explored during the second half of the nineteenth century. Marloth (1908), Bews (1916), and Cannon (1924) provided early descriptions of the karro. However, by the time Acocks (1953) described the karro vegetation types and incorporated them into a vegetation classification scheme for South Africa, the plant community had already been altered by sheep grazing practices of European settlers (Milton et al. 1997; Walter 1986). Heinrich Walter (1936) first described the vegetation of the Namib. Richard Logan (1960) wrote a review of the geography of that desert. In 1963, about 100 km southeast of Walvis Bay at Gobabeb, C. Koch, an entomologist, founded the Desert Research Station, a place that continues to be a center of study of this most interesting desert (Walter 1986). Meanwhile, the buildings of the first desert research station in Tucson have become a National Historic Site (Bender 1982).

## II. COLD DESERTS

Cold deserts (not to be confused with polar deserts that are ice covered) are those that regularly experience prolonged periods with temperatures below freezing. In most, soil moisture becomes frozen and unavailable to plants, which imposes a period of dormancy. Some receive a significant proportion of their annual precipitation as snow. The cold deserts are cold only in winter. High daily temperatures in summer are often over 40° C (100° F).

Almost all cold deserts are interior deserts, with Patagonia in southern Argentina as the main exception (West 1983a). Distance from the sea (a source of moisture)—usually aggravated by the rainshadow effects of high mountains or plateaus—means that air masses lose most of their moisture before they ever reach these locations. Large continental masses in the midlatitudes undergo great annual temperature ranges and have very hot summers and very cold winters, a pattern referred to as continentality. Aridity and cold winters are the conditions leading to cold deserts.

The Gobi Desert of Mongolia and China is a prime example of an interior desert. Other cold deserts are found in Middle and Central Asia, from the Caspian Sea to the great bend of the Huang He (Yellow River) in China. In North America, the Great Basin Desert of Nevada and surrounding states is categorized as a cold desert, as is the windswept plateau at the southernmost reaches of South America—the Patagonian Desert that lies in the rainshadow of the Andes Mountains.

Shrubs and subshrubs dominate the plant life of cold deserts, but perennial grasses and forbs can be important components. Perennial grasses store nutrients in **rhizomes** and die back during cold or severe drought. Another common perennial growth-form that allows avoidance of drought is the bulb, which stores water and nutrients below ground and sends shoots above ground only after rainfall sufficient to ensure flowering and replenishment of underground stores. Cold deserts are home to bulbs such as crocuses, tulips, lilies, and irises, whose relatives have become popular garden plants.

Succulents—plants that store water in their tissues—are not as abundant or diverse in cold deserts as in hot ones, although the Great Basin in North America does have cacti. Most of these are prickly pear cacti—stem succulents with flat, padlike stems. The halophyte is a much more common growth-form in cold deserts than succulents.

The cushion plant is a final growth-form associated with cold deserts. These compact, ground-hugging, multistemmed plants grow in very windy areas where high evaporation rates and wind chill would kill other shrubs. The cushion plant is a characteristic growth-form of the Patagonian Desert but is found locally in other deserts where winds are particularly strong and prevalent.

## MAJOR REGIONAL EXPRESSIONS

### Middle and Central Asia

Deserts stretch across 200° of longitude in Asia. For thousands of years, they were seasonal pasturelands for nomadic herds and their livestock, so it is difficult to know what they might have been without human use. Winters are under the control of the strong high pressure system that forms over eastern Siberia and forces bitterly cold air masses, part of the Asian monsoon system, over much of the continent. The strength of the high pressure system weakens enough over Dzungaria, the Tien Shan, and the Pamirs to create climatic differences and make this area a natural boundary. To the west is Middle Asia, a region able to get winter precipitation from the Atlantic and that extends from the Caspian Lowland to Kazakhstan and southward into Iran and Afghanistan. Precipitation in this region occurs mostly in winter and spring with a peak in March, and it generally decreases in total amount from west to east. Plants that grow and bloom in the spring are important in the Middle Asian deserts (Walter and Box 1983a). One of the largest and best-studied deserts in this region is the Karakum, which lies mainly in Turkmenistan and covers about 350,000 km$^2$ (135,000 mi$^2$) on the Caspian or Turanian Lowland southwest of the Amu-Dar'ya. Eighty percent of the Karakum is erg, an expanse of sand with oases at the bases of surrounding mountains. Small sand hills (nebkas) form where shrubs trap blowing sand. Large accumulations of sand occur as individual, crescent-shaped dunes (barchans) 3–4 m (10–12 ft) high that move back and forth across the landscape according to prevailing wind directions: northwest toward the Pamirs in summer and southeast away from the mountains in winter. Sometimes the barchans are grouped in fields or chains. Lenses of groundwater are available below the dunes. Stable sand ridges are also larger features of the Karakum. These may be several kilometers long, up to 100 m (300 ft) wide and up to 60 m (200 ft) high. Interestingly, trees up to 9 m (30 ft) tall are the dominant plants in the sparse, open vegetation of the sand sea. The trees typically have tiny leaves or none at all; **photosynthesis** takes place in green stems and small shoots that fall off in drought and winter cold. The sand acacia (*Ammodendron conollyi*), a **legume**, is a good example. It grows on slightly mobile sands. Mature specimens are 4–9 m (12–30 ft) high, but the trunk is only 1–3 m (3–10 ft) tall. The upper branches die back each year, and new shoots develop from lateral buds. The leaflets of the **compound leaf** are silver and measure 30 mm (1 in) long and 3 mm (0.1 in) wide. The tip of each leaf ends in a spine. Other small trees include another legume, *Eremosparton flaccidum*, that stands 5 m (15 ft) tall; *Calligonum eripodium* (Polygonaceae), 6 m (20 ft) tall, and black saksa'ul (*Haloxylon ammondendron*), a leafless chenopod that stands 5–9 m (15–30 ft) tall. Black saksa'ul grows in thickets and is economically an important plant since its trunk and branches provide firewood for people, and its young shoots provide fodder for camels and sheep (Walter and Box 1983b).

The white saksa'ul (*H. persicum*) is a shrub and indicator species for the Karakum. It grows with sedges locally called *rang* or *ilak* (*Carex physodes*) on stabilized sands. White saksu'al has long shoots with scalelike leaves and grows 3–5 m (10–15 ft) tall. The outer 10–20 cm (4–8 in) die back annually. It is heavily browsed by sheep but resprouts from its roots. *Calligonum caput-medusae* [height = 2.5 m (8 ft)] and *C. setosum* [height = 1.2 m (4 ft)] are other dominants in the shrub layer. *Salsola richteri* [height = 1 m (3 ft)] and *Ephedra strobilacea* [height = 1–2 m (3–6 ft)] are subdominants. **Semishrubs** include species of *Artemisia* and the endemic legume *Smirnovia turkestana.* Sedges are important sheep forage. *Carex physodes* is a turf-forming variety that sheep eat both green and dried. It sprouts in the autumn as soon as the winter rains begin and reaches a height of 15–40 cm (6–16 in) (Walter and Box 1983b).

The Karakum has about 100 annuals that germinate in winter and bloom in the spring and another 45 or so that germinate in the spring and bloom in summer. Spring-blooming bulbs include tulips (*Tulipa sogdiana*) and irises (*Iris longiscapa* and *I. songarica*) (Walter and Box 1983b).

About 10 percent of the Karakum is covered by saltpans and takyry, clayey depressions that flood in spring. The saltpans are occupied by halophytes, but no higher plants grow on takyry, where the soil fractures into large cracks as it dries out. A thin film of algae, lichens, and cyanobacteria, however, withstand nearly complete dehydration (Walter and Box 1983b). The remaining 10 percent of the desert is moving sand essentially devoid of vegetation.

Before firearms became widespread a century or two ago, several wild large grazing animals inhabited the Karakum, including saiga antelope (*Saiga tatarica*), wild sheep (*Ovis orientalis*), and kulan (*Equus hemionus*). Today, goitered gazelle (*Gazella subgutturosa*) visit the desert seasonally (October to March) to graze the sedges beneath the white saksa'ul; in spring, wild boars (*Sus scrofa*) come long distances from floodplain forests along the Amu-Dar'ya to forage for spring bulbs. The main mammals today are rodents, hares, and hedgehogs. Rodents include the long-clawed ground squirrel (*Spermophilopsis leptodactylus*), a porcupine (*Hystrix hystrix*) that inhabits black saksa'ul thickets, and three jerboas built like mini-kangaroos—the rough-legged jerboa (*Dipus sagitta*), the desert jerboa (*Jaculus lichtensteinii*), and the comb-toed jerboa (*Paradipus ctenodactylus*). The great gerbil (*Rhombomys opimus*) is a common rodent that lives in family groups that congregate in colonies. Active year round, the gerbil stores food in underground chambers in its rather complex burrow system. It feeds particularly on rhizomes of sedges and all parts of the black saksa'ul growing above the sedges. Also common is the jird (*Meriones* spp.), a ratlike rodent that also has burrows. The long-eared desert hedgehog (*Hemiechinus auritis*) inhabits old rodent burrows and feeds on reptiles and arthropods (Walter and Box 1983b).

Red fox (*Vulpes vulpes*) and sand cat (*Felis margaritata*) are among the common predators. The latter is usually found in sand blowout areas in *Haloxylon* stands where birds frequently nest. The birds include the great tit (*Parus major*), one of the few year-round residents; lesser whitethroat (*Sylvia curruca*); scrub robin (*Erythropygia* spp.); black-billed desert finch (*Rhodospiza obsoleta*); streaked scrub warbler (*Scotocerca inquieta*); and turtledove (*Streptoperlia turtur*). Reptiles are represented by *Agama* lizards that sit high in the branches of shrubs but quickly flee into rodent burrows when danger approaches. Interestingly, these lizards can change the color of their skin; they reportedly turn blue when caught. Geckos (*Gymnodactylus russowi* and *G. caspius*); several snakes, including the arrow-snake (*Psammophis lineolatus*) and *Varanus griseus*; and a tortoise (*Testuda horsfeldi*) are also found. The tortoise feeds on young shoots of annuals and is dormant much of the year. Among arthropods, ants and soil cicadas (*Pentasteridius* spp.) are common; termites are relatively rare (Walter and Box 1983b).

In Central Asia—which extends from Dzungaria east to the Huang He and south to the Tibetan Plateau—a strong high pressure system over eastern Siberia blocks moist air masses in winter and spring, and both seasons are exceptionally dry. Plants cannot grow until after the summer rains begin, so, unlike the deserts of Middle Asia, no spring-blooming annuals are found here. Summer rainfall comes from storms moving in from eastern Asia, so total annual precipitation in Central Asia decreases from east to west. The driest part is the Takla Makan desert, where rainfall amounts are near zero.

A series of large and small basins separated by mountains more than 5,000 m (16,000 ft) high characterize Central Asia. The Turpan Depression lies at –155 m (–500 ft), the lowest elevation in China, and is where China's highest temperature, 47.6° C (118° F), was recorded. The lowest temperature, –51.1° C (–60° F), was recorded nearby at Fuyun. Strong winds are an important feature of these deserts due, in winter, to high pressure associated with the Asian monsoon and, in summer, to gravitational flows (katabatic winds) from high mountains and plateaus to the much lower basins (Songqiao 1986). Each basin is distinct in its climate and landforms. Several desert regions are thus recognized, including Dzungaria, the Gobi, the Pei Shan, the Takla Makan, the Gansu Corridor, the Ala Shan desert, and the Ordos Desert. Only three of the larger ones are described here.

In the west, the Dzungarian basin, at an elevation of 500–1,000 m (1,600–3,300 ft), has precipitation evenly distributed throughout the year. At the southern edge of the basin, Ürümqi [elevation 903 m (2,960 ft)], the largest city in the region, receives total annual precipitation of about 250 mm (10 in). The most widespread vegetation type is saksa'ul desert, similar to that described in Middle Asia. The Gobi Desert in Mongolia lies at elevations of 1,000–1,300 m (3,300–4,200 ft) and receives only summer precipitation. ["Gobi" means stony desert (Songqiao 1986).] Average total annual precipitation varies from 100–150 mm (4–6 in) to as low as 40 mm (1.5 in); in the westernmost parts some years have no rain at all. Winters are cold and dry; the January average temperature is –18° C (0° F), yet the soil remains unfrozen since daily highs frequently rise above 0° C (32° F). The vegetation consists of low shrubs and semishrubs, many of which are chenopods such as the characteristic "baglur" (*Anabasis brevifolia*). Others are in genera common throughout the **temperate** areas of Asia (e.g., *Salsola*, *Tamarix*, *Zygophyllum*, *Caragana*, *Calligonum*, and *Artemisia*). The Gobi has served as a refuge for rare large mammals such as kulan, saiga, goitered gazelle, wild camel, and possibly the Przewalski horse. The most common mammals, however, are rodents such as jerboas, gerbils, and dwarf hamsters (*Allocricetulus*, *Cricetulus*, *Phodopus*) (Walter and Box 1983c).

The Tarim Basin is surrounded in the north, west, and south by very high mountain ranges (Tien Shan, Pamirs, and Kunlun Shan, respectively) is the driest region in Central Asia. Yet the basin has abundant groundwater that is replenished annually by meltwaters from mountain glaciers and snowcaps. It also receives water from a major river, the Tarim, which hugs the western, northern, and eastern edges of the depression in a constantly shifting course and empties into the saltpan of Lop Nor. Oasis cities like Kashgar are found at the foot of the mountains; the interior is a huge uninhabited sand desert, the Takla Makan, with dunes as high as 300 m (1,000 ft). In summer, temperatures rise to above 40° C (100° F), and relative humidity may be as low as 2–3 percent. Powerful dust storms may blow for days. In winter, temperatures may fall below –27° C (–17° F) because to the east, the basin is open to the Lop Nor depression and the bitterly cold winter winds blowing out of eastern Siberia. Floodplain forests of poplar (*Populus* spp.) and elm (*Ulmus pumila*) line the Tarim River. The oases are major producers of grapes and other fruits irrigated by ancient underground canals (*karez*) and modern above-ground canals. Elsewhere in the vast sand desert, very little plant life is found except at the surprisingly frequent sites between dunes, where groundwater is within a meter or two of the surface; there tamarix shrubs and some reeds grow (Walter and Box 1983c).

## North America

The only cold desert in North America lies on the high plateaus and basins of the Intermontane West, the land between the Cascades and Sierra Nevada on the west and the Rocky Mountains on the east. It is generally referred to as the Great Basin Desert after the major physiographic unit where it is found. Beyond the basin and range topography of the Great Basin in Nevada, the desert extends onto the deeply dissected Colorado Plateau, the basaltic Columbia–Snake River Plateau, and the Wyoming Basin—built of unconsolidated materials and separating the northern Rockies from the southern part of the range. The desert has few areas of sand dunes, but badlands are an important element of the landscape and have very little plant cover. Desertscrub is found at elevations above 1,200 m (4,000 ft) to about 1,850 m (6,000 ft), where the vegetation changes to pinyon-juniper woodland. Winter snows are much more significant in replenishing the soil moisture than

Photo 4.9 Big Sagebrush (*Artemesia tridentata*). Its gray-green hues color vast areas in the Great Basin Desert on high plateaus in the western United States. (Photo by SLW.)

summer rains (West 1983a). Average total annual precipitation ranges from about 160–420 mm (6–16.5 in). The dominant plants throughout the region are woody *Artemisias*. Big sagebrush (*Artemisia tridentata*) is the characteristic shrub of the Great Basin Desert (Photo 4.9). It stands 1–2 m (3–6 ft) tall and forms an open canopy. A number of dwarf sagebrushes less than 0.5 m (1.5 ft) high are also important. Among them are black sagebrush (*A. nova*), low sagebrush (*A. arbuscula*), silver sagebrush (*A. cana*), and bud sagewort (*A. spinescens*). By their dominance, the sagebrushes impart a grayish-green color to the entire landscape. In the moister areas, perennial bunchgrasses share dominance with the sagebrushes. Bluebunch wheat grass (*Agropyron spicatum*) is most widespread; other grasses more restricted in their distribution include Thurber's needlegrass (*Stipa thurberiana*), Indian ricegrass (*Oryzopsis hymenoides*), Great Basin wildrye (*Elymus cinerus*), and medusa head grass (*Taenithiatherum caput-medusae*). Most of these grasses have greatly decreased in abundance since cattle have been grazing the range. Shrubs such as rabbitbrush (*Chrysothamnus viscidiflorus* and *C. nauseosus*), Mormon tea (*Ephedra nevadensis*), horsebrushes (*Tetrademia* spp.), and broom snakeweed (*Gutierrezia sarothae*) and the prickly pear cactus (*Opuntia polycantha*) have replaced them (West and Young 2000; Knight 1994; West 1983b, 1983c).

The Great Basin has interior drainage. Cooler, moister periods of the Pleistocene Epoch saw the formation of a few huge lakes, such as Lake Bonneville in Utah and Lake Lahontan in Nevada, and many smaller lakes, all of which have since evaporated. The result is many playas and large areas of saline soils at lower elevations. The vegetation in these areas is commonly referred to as saltbush desert shrubland. Common shrubs growing in a fringe around the playas are saltbushes, such as shadscale (*Atriplex confertifolia*) and Gardner's saltbush (*A. gardneri*), greasewood (*Sarcobatus vermiculatus*), spiny hopsage (*Grayia spinosa*) and bud sagewort. Winterfat (*Krascheninnikovia lanata = Ceratoides lanata*) is an important subshrub. In a concentric ring nearer to the playa, where groundwater is closer to the surface, the shrubs and subshrubs give way to saltgrass meadow with salt-tolerant grasses such as inland saltgrass (*Distichlis stricta*), alkaligrass (*Puccinellia nuttaliana*), and alkali sacaton (*Sporobolus airoides*). Inside this ring, groundwater is close enough to the surface to keep soils moist as well as highly saline. The main plants in this innermost zone of vegetation are alkali cordgrass (*Spartina gracilis*), pickelweed (*Suaeda* spp.), and glassworts (*Salicornia* spp.) (West and Young 2000; Knight 1994; West 1983c, 1983d).

Shrubs' strategies for survival in the Great Basin Desert differ. Big sagebrush has deep roots that extend 2 m (6 ft) to retrieve deep soil moisture from winter snowmelt and shallow roots to capture summer rainfall. It has an evergreen **habit** that lets it photosynthesize whenever temperatures and water availability allow. This gives a necessary extension of the growing season to a plant in which photosynthesis is limited in summer, because its **stomata** close during the hottest part of the day to prevent water loss. Sagebrush maximizes photosynthesis in early spring, when soil water is most readily available, by increasing photosynthetic surface area with a new set of ephemeral leaves. These new leaves are larger than the ones maintained over the winter and are shed to conserve water when temperatures rise in early summer. Shadscale is another plant that has two kinds of leaves, one ephemeral and the other overwintering. It, however, is a warm-season $C_4$ plant that is able to fix carbon during high temperatures, so it maintains a constant and moderate rate of photosynthesis throughout the growing season. Winterfat, on the other hand, is a cool-season $C_3$ plant. It photosynthesizes at a rapid rate in spring and then is relatively inactive the remainder of the year (Knight 1994).

Plants have adapted to browsing animals by having spines or producing aromatic compounds that reduce the palability of their leaves. The Latin names of some plants reveal their spiny nature—for example, spiny hopsage (*Grayia spinosa*) and bud sagewort (*Artemisia spinescens*). The scent of the desert air after a summer rain emphatically announces that sagebrush has used the other strategy. Nonetheless, sagebrush is an important winter browse species for native wildlife such as mule deer (*Odocoileus hemionus*), elk (*Cervus elaphus*), and pronghorn (*Antilocapra americana*). Most of the carbohydrates produced in spring and summer are stored in the twigs, and the shrub is tall enough that its twigs are usually exposed above the snow cover, where these mammals can reach them (Knight 1994).

Big sagebrush provides good habitat for larger mammals and a host of other animals. Black-tailed jackrabbit (*Lepus californicus*) is one of the more important herbivores that finds shelter in the tall shrubs. Other small mammals include Townsend's ground squirrel (*Spermophilus townsendii*), sagebrush vole (*Lagurus curtatus*), Great Plains pocket mouse (*Perognathus parvus*), western harvest mouse (*Reithodontomys megalotis*), voles (*Microtus* spp.), and deer mice (*Peromyscus* spp.) (West 1983c). Kangaroo rats (*Dipodomys microps* and *D. ordii*) can be found in saltbush desert vegetation (West 1983d). The principal mammalian predator of jackrabbits and rodents is the coyote (*Canis latrans*); the badger (*Taxidea taxus*) is another important consumer of rodents. Common raptors include golden eagle (*Aquila chrysaetos*), ferruginous hawk (*Buteo regalis*), and western red-tailed hawk (*Buteo jamaicensis*). The sage-grouse (*Centrocercus urophasianus*) is the characteristic ground bird of the cold desert; it uses the sagebrush for shelter and eats its leaves in winter. Reptiles are less diverse than in the hot deserts of North America; amphibians are nearly absent (West 1983a, 1983c).

## South America

Patagonia, lying between 40° and 55° S latitude in southern Argentina, is the only cold desert in South America. It contradicts the usual pattern of a location in the interior of a large continent for cold deserts; it stretches from sea level along the Atlantic Ocean inland to elevations of 1,200 m (4,000 ft) at that point where South America tapers off toward the Antarctic. Patagonia lies in the Southern Hemisphere's zone of strong westerly winds, and hence in the rainshadow of the southern Andes. Precipitation amounts decrease rapidly from as much as 1,080 mm (42 in) in the west to about 160 mm (6 in) a year in the east, a pattern reflected by the vegetation. Temperatures are low throughout the year; freezing temperatures may occur during any month. However, because of the influence of the surrounding oceans, the average minimum daily temperature in the coldest month is only slightly below 0° C (32° F).

The Patagonian Desert is the least studied of the cold deserts and, like the others, has been affected by livestock grazing to the extent that the nature of ungrazed vegetation has been obscured. Although the vegetation has a number of regional variations, two basic categories are described here: a narrow, moister north-south band at the eastern foothills of the Andes and a much broader zone in central Patagonia that covers most of the remaining desert area. Closest to the mountains is a narrow belt of tall tussock grasses, mostly needlegrasses such as *Stipa speciosa*, with spiny shrubs less than 1 m (3 ft) high scattered throughout. *Mulinum spinosum*, *Adesmia campestris*, and *Senecio filaginoides* are the most common of the shrubs. Low perennial herbs, mostly with rhizomes as opposed to bulbs or corms, grow under the tussocks or shrubs. A few tiny annuals also grow. Year round, most plants are yellow or gray or brown. *Mulinium spinosum* keeps its dry, spiny leaves throughout the winter, although it becomes dormant then; *Adesmia campestris* is deciduous. *Senecio filaginoides* has no dormant season but maintains leaves all year. All three bloom at the end of spring and beginning of summer (December and January) (Soriano 1983).

As aridity increases to the east, the grasses lose dominance to cushion plants well adapted to the windy environment. In wetter, western areas, there are large, loose, spiny cushions of *Mulinum spinosum*, but as one goes east, cushion plants become smaller and denser. The dwarf, hard cushions of *Nassauria glomerosa*—a yellow-flowered member of the daisy family that mimics the color and shape of rocks—are the most characteristic and dominant plants over much of Patagonia. At elevations above 400 m (1,200 ft) a larger globular shrub *Chuquiraga avellanedae*, up to about 60 cm (2 ft) high, is a common component of the very sparse vegetation along with *Nassaurias* and needlegrasses. (Near the Atlantic, at latitudes of 46° to 51° S, *Chuquiraga avellanedae* is replaced by *Verbena tridens*.) *Nardophyllum obtusifolium*, a plant that stays bright green even at the driest times of year, and the pink-flowered *Pleurophora patagonia* are other frequently encountered shrubs. In lowlands below 200 m (650 ft), one can find communities of saltbush (*Atriplex lampa* and *A. sagittifolium*) and *Frankenia patagonia* forming a very sparse cover on saline soils (Soriano 1983).

Animals of the Patagonian desert include the widespread guanaco (*Lama guanicoe*). Early European travelers reported herds of a thousand or more of these South American camels; today it is rare to see more than 30 together. The huemul (*Hippocamelus bisculus*), a small deer, once lived in the desert but now has retreated to a few isolated spots in the Andes. The largest mammals remaining are the mara or Patagonian hare (*Dolichotis patagonum*), and the pichi (*Zaedys pichyii*), an armadillo. A few small South American rodents (*Akodon* spp., *Phyllotis* spp., and *Reithrodon* spp.) can be found. Carnivores are relatively diverse, although they occur in low numbers. Among them are South American foxes (*Pseudalopex culpaeus* and *P. griseus*), puma (*Felis concolor*), Pampas cat (*F. colocolo*), Geoffroy's cat (*F. geoffroyi*), Patagonian weasel (*Lyncodon patagonicus*), and hog-nosed skunk (*Conepatus humboldti*). The Patagonian opossum (*Lestodelphis halli*), a small carnivorous opossum, is the southernmost marsupial on the planet. Characteristic birds of Patagonia prefer walking and running to flying and include the lesser rhea (*Pterocnemia pennata*), Patagonian tinamou (*Tinamotis ingoufi*), and a number of shorebirds. Ground-dwelling, seed-eating parrots also live there. Representative reptiles are the yarara ñata (*Bothrops ammodactoides*), chelco (*Homonota darwini*), and several species of *Liolaemus*. Only four kinds of amphibians have been reported from the region, which reflects the low diversity of vertebrates in general (Soraino 1983; West 1983e).

## ORIGINS

In broad outlines, the cold deserts have similar geologic histories. They result from a series of uplifts of basins and, to much greater heights, surrounding mountain ranges during the Tertiary Period. This, along with the poleward drift of North America and Eurasia and the final assembly of the modern large landmasses, lowered temperatures and lowered precipitation. In the Early Tertiary, the Tethys Sea flooded much of what is now Middle Asian desert, and the major centers of aridity lay south and west in the Sahara and north and east in Mongolia. The convergence of the Eurasian and Africa plates essentially eliminated the Tethys by the Late Tertiary. Remnants survive as the Mediterranean, Caspian, and Aral seas. At the same time, the Hindu Kush and the Elburz folded and rose to form the southern mountain barriers of the Middle Asian deserts. Since the Miocene Epoch, the Indo-Australian Plate has been moving underneath the Eurasian Plate, which causes the continuing uplift of the Himalayas and the plateaus and basins to their north. The rise of the Qinghai-Xizang or Tibetan Plateau was most influential in creating the aridity that occurs in northwest China; it reached its contemporary elevation of over 4,000 m (13,000 ft) only in the late Pleistocene Epoch. The rise of the Qinghai-Xizang Plateau contributed to the establishment of the Asian monsoon and the positioning of the strong Siberian-Mongolian High Pressure System at its present winter location near 55° N and caused the rapidly increasing aridity of the Central Asian deserts in its rainshadow. Pleistocene glaciations led to a lowering of temperatures and reduced evaporation rates in the cold deserts. Glaciers formed only on the highest mountains, but streams running from them brought huge loads of gravels, sands, and silts to the desert basins and contributed to the formation of lakes in the centers of the larger depressions. Glacial meltwaters also added to the groundwater supplies of the desert basins. During the warmer, interglacial periods, the lakes evaporated, and strong winds, especially in winter, removed the finer particles carried in earlier by the mountain glaciers' streams. Sand dunes formed in some basins; finer materials were deposited as **loess** on the southern and western foothills. With all these geologic and climatic changes, it is safe to say that the contemporary vegetation of the cold deserts of Asia originated in Late Pleistocene or Early Holocene epochs as current climate conditions developed. The widespread grazing of livestock then greatly modified much of it a few thousand years ago, so that little—if any—natural vegetation actually persists today (Breckle 1983; Walter and Box 1983c, 1983d).

In North America, where the vegetation of the past has been more intensively studied, plants of subhumid climates are first recorded in the central Rocky Mountain region in the Eocene and Oligocene epochs. Uplift of the Cascades–Sierra Nevada began in the Oligocene, and with it began a drying of the interior of the continent that would have favored the evolution of drought-tolerant plants and the extinction of moisture-loving plants. Maximum uplift of the Cascades and Sierra Nevada, together with the Great Basin and Colorado Plateau, did not occur until the late Pliocene Epoch. It was the creation of high mountain ranges to the west that provided the extreme rainshadow effect necessary for a desert and the increased elevation of the intermountain region that established the conditions for a cold desert. The Pleistocene glaciers in the mountains, as in Eurasia, had the indirect effect of reducing evaporation and increasing runoff to the basins below. Over 90 lakes formed in the region now occupied by the Great Basin Desert. The later drying of these lakes resulted in today's salt lakes and playas. Temperature changes during the Pleistocene Epoch and subsequently in the Holocene forced elevational shifts in plant and animal life. Current distributions are still in flux. The mass extinction of large browsing animals such as camels, giant ground sloths, and horses by about 10,000 years ago has had an unknown impact on the desert vegetation (West and Young 2000; West 1983a).

For a long time (and more than once) in the Cenozoic Era, Eurasia and North America were connected. The connections permitted an exchange of plants and animals so that close relatives occur on both of the now-separate continents, as you can tell by looking at the Latin names. A good example is the genus *Artemisia*, an important group in the deserts of both landmasses. Yet the relatives have some interesting differences. Again using *Artemisia* as an example, in North America *Artemisias* are usually woody shrubs—the sagebrushes—but in Eurasia they are usually forbs. (The pattern is reversed in other genera. *Astragalus* and *Salsola* are typically represented by woody plants in Eurasia and forbs—introduced in the case of *Salsola*—in North America.) Among plants, all the migrants between continents were Eurasian in origin (West 1983e). However, among animals, some from North American crossed into Asia (horses are a good example), but many more Eurasian forms successfully moved in the opposite direction.

The Patagonian Desert seems to have originated in the Pleistocene Epoch. Moisture-bearing winds from the Pacific were able to reach the area until the Late Miocene, and a humid subtropical climate may have characterized Patagonia. Increased aridity would have coincided with the rise of the Andes and the blocking of the westerly winds. Pollen records indicate that the Pleistocene was a period of alternating warm-dry and cool-moist conditions. The cooler periods saw expansion of steppe vegetation, the warmer presumably of desert vegetation (Soriano 1983).

## HUMAN IMPACTS

The human impacts on the cold deserts of the world are much the same; what differs is the timing and rate of ecosystem transformation. Domesticated livestock have been the main agents of change. For several thousand years nomadic pastoralists have used the deserts in Middle and Central Asia as pastureland, especially for sheep but also for horses, goats and camels. These animals select the most nutritious and most palatable grasses and shrubs, which slowly changes the composition of the vegetation. In overgrazed areas, the grasses disappear, while the plants that increase are those with bitter or toxic compounds in their leaves or those armed with spines or thorns. Trampling by hoofed animals breaks the crust of lichens and algae that holds the dry ground in place and lets the wind carry the sands away to accumulate in dunes elsewhere. Pastoralists have reportedly responded to degraded pastures by changing the livestock raised from, in this case, first cattle—which require the best grazing conditions—to less demanding sheep, and finally to goats—which are essentially browsers and can consume even thorny plants.

Pastoralism also has affected the animal life of cold deserts. Herding peoples usually try to eliminate any wild animals that may compete with their livestock for forage. In Asia, these would be animals such as the kulan or Persian onager, saiga, and gazelle. They also kill those carnivores that might prey upon the domestic flocks. The impact of the loss or near lost of most native wild animals on the vegetation is unknown.

Cultivation of irrigated crops is widespread in the cold deserts at oases and other favorable locations. The water comes from artesian systems, from above-ground canals that carry water from glacial streams to the desert basin, and in some locations from an intricate system of underground conduits called *karezes* or ganats. The first karezes were constructed in Iran several thousands of years ago and later spread eastward into the drylands of western China. When people congregate in farming villages, they need fuelwood and charcoal, so the removal of larger woody plants has been another consequence of human use of the desert.

Mining can significantly alter the desert ecosystem, but its effects are usually localized. In the Asian deserts, the concentrating and processing of salts of sodium, magnesium, potassium, bromine, and iodine have laid waste to the desert in some places. Recently, discoveries of oil and natural gas have meant the construction of pipelines and some industrial developments that can disturb local ecosystems.

In North America, the Great Basin Desert was home to nomadic hunters and gathers (e.g., Paiute and Shoshoni peoples) until livestock were introduced in the 1800s. The first cattle came in the 1840s with Mormon settlers and Californians raising livestock to support the newly discovered gold mines. Settlers also introduced horses and sheep. Overgrazing led to a decrease in palatable native bunchgrasses and invasions of annuals such as Eurasian cheatgrass (*Bromus tectorum*). Less palatable (to livestock) native shrubs such as rabbitbrush, horsebrush, Mormon tea, and snakeweed increased. The expansion of annuals, particularly cheatgrass, altered aspects of the animal community. Seed-eating rodents such as deer mice, pocket mice, and kangaroo rats, increased in numbers. An introduced Eurasian gamebird, the chukar (*Alectoris chukar*), expanded its range. Black-tailed jack rabbits replaced white-tailed jack rabbits (*Lepus townsendi*), and coyotes increased. A cover of annual grasses provided fuel for frequent fires, something big sagebrush could not tolerate. As the main shrubs were killed, the growth of annual grasses was further promoted, but important forage and shelter for native animals was lost. Mule deer—which had initially increased when livestock were introduced and shrubs expanded at the expense of perennial grasses—went into decline as did other wildlife. By the late 1800s, the land was suitable only as winter pasture for sheep, but sheep continued to be important well into the twentieth century (West and Young 2000; Knight 1994; West 1983a).

In the 1930s, efforts began to improve pastures by removing sagebrush and planting Eurasian grasses that could tolerate heavy grazing and trampling. Desert wheatgrass (*Agropyron desertum*), planted through the 1960s, was of particular importance. Then, concern for declines in native wildlife caused a reevaluation of and shift in land-use practices on these lands, much of which was under the jurisdiction of the Bureau of Land Management. Big sagebrush may be poor forage for sheep and cattle and may compete with the grasses preferred by livestock and ranchers, but it provides good habitat for mule deer, elk, pronghorn antelope, and sagegrouse. Livestock on the public lands were reduced, in part because the sheep industry went into decline after World War II. Increases in pronghorn coincided with the removal of sheep, although they were due in large part to successful game management programs in various states. Today, ranching, along with hunting and mining, are considered the traditional land uses in the Great Basin Desert. More modern land uses center around recreation (West and Young 2000; Knight 1994; West 1983a).

The story is much the same in Patagonia, even if less well documented. Sheep, introduced in the late nineteenth century, have grazed the entire region; wool production continues to be the major industry of the region. Heavy grazing is viewed as a major influence on the vegetation throughout Patagonia. In the early twentieth century, the European hare (*Lepus europeaus*) was introduced and has since become widespread. As a consequence of competition from sheep and hares and of overhunting, the larger native mammals—both herbivores and carnivores—as well as the lesser rhea have greatly declined in numbers (Soriano 1983).

## HISTORY OF SCIENTIFIC EXPLORATION

Such people as Marco Polo brought the first reports of the deserts of Middle and Central Asia to Europe via the Silk Road in the thirteenth and fourteenth centuries. Russians and then other Europeans began purposeful exploration in the second half of the nineteenth century. The expeditions of Semenov in 1857 and those of Przewalski between 1870 and 1878 spanned the Russian period of geographic exploration. Nineteenth-century exploration by other countries included such expeditions as those of the German geographer Richthofen (1868–72) and the Swede Sven Hedin (1894–97, 1899–1902). The travels of these people brought knowledge of geology, climate, and ethnic groups of the deserts to European scientists but lacked a major emphasis on natural history. Ecological and other scientific studies began at Repetek in the Karakum in 1912, when the Sand-Desert Station was established. Studies of desert **ecology** continued with further impetus coming in the mid-1960s from the International Biological Programme. Most of the literature available to Western scientists continues to be in Russian and German. Two important comprehensive works are M. P. Petrov's (1966–67) two-volume work on the deserts of Central Asia and Heinrich Walter's (1974) publication on the vegetation of Eastern Europe, Middle and Central Asia. Much of the information contained in these and other studies was not translated into English until the 1980s (Walter and Box 1983a). It should be noted that scientists in such countries as Iran and China have made significant contributions to knowledge about Asian cold deserts, but these are generally poorly known in Europe and the United States.

The Great Basin was largely unexplored territory in 1844, when John C. Frémont crossed it. Soon after, settlement began: Mormons fleeing persecution came from the east, and Californians seeking new pastures for cattle came from the west. Scientific investigation awaited the twentieth century. Forrest Shreve (1942) described and named the deserts of the western United States and bestowed the name Great Basin Desert on the shrublands of Nevada and the Colorado Plateau. Most ecological work has to one degree or another revolved around the use of this desert as rangeland for sheep and cattle. Another strong focus has been on paleoecology and the study of past vegetation. A fossil record dating back to the Tertiary is available from this region. Quaternary vegetation is reconstructed from pollen records preserved in Pleistocene lake beds and from plant materials embedded in Pleistocene pack rat (*Neotoma* spp.) middens that survive in numerous dry caves.

Naturalists explored Patagonia in the nineteenth century, the most famous was Charles Darwin, who provided general descriptions of the landscape, including vegetation. Scientific investigation did not begin until the twentieth century, when the focus was the economic potential of the region (Soriano 1983). By mid-century, however, such Argentine scientists as Alberto Soriano and A. Cabrera began significant work on the ecology of Patagonian communities. The work of these scholars and their students continues.

## REFERENCES

Abd el Rahman, A. A. 1986. "The deserts of the Arabian peninsula." Pp. 29–54 in Michael Evenari, Imanuel Noy-Meir, and David W. Goodall, eds., *Hot Deserts and Arid Shrublands, B.* Ecosystems of the World, 12B. Amsterdam: Elsevier.

Acocks, J. P. H. 1953. "Veld types of South Africa." *Memoirs of the Botanical Survey of South Africa* 28: 1–192.

Ayyad, Mohamed A. and Samir I. Ghabbour. 1986. "Hot deserts of Egypt and the Sudan." Pp. 149–202 in Michael Evenari, Imanuel Noy-Meir, and David W. Goodall, eds., *Hot Deserts and Arid Shrublands, B.* Ecosystems of the World, 12B. Amsterdam: Elsevier.

Bender, Lionel 1982. *Reference Handbook on the Deserts of North America.* Westport, CT: Greenwood Press.

Bews, J. W. 1916. "An account of the chief types of vegetation in Southern Africa, with notes on plant succession." *Journal of Ecology* 4: 129–159.

Blatter, E. and F. Hallberg. 1918. The flora of the Indian desert (Jodhpur and Jaisaimer)." *Journal of the Bombay Natural History Society* 26: 218–246.

Breckle, S. W. 1983. "Temperate deserts and semideserts of Afghanistan and Iran." Pp. 271–319 in Neil E. West, ed., *Temperate Deserts and Semi-deserts.* Ecosystems of the World, 5. Amsterdam: Elsevier Scientific Publishing Company.

Cannon, W. A. 1924. "General and physiological features of the vegetation of the more arid portions of South Africa with notes on the climatic environment." *Yearbook of the Carnegie Institution of Washington* 8(354), 1–159.

Dillon, Michael O. n.d. "Lomas formations of the Atacama Desert, Northern Chile," *Centres of Plant Diversity, Vol 3: The Americas.* Department of Botany, National Museum of Natural History. http://www.nmnh.si.edu/botany/projects/cpd/sa/sa43.htm.

Gupta, R. K. 1986. "The Thar Desert." Pp. 55–99 in Michael Evenari, Imanuel Noy-Meir, and David W. Goodall, eds., *Hot Deserts and Arid Shrublands, B.* Ecosystems of the World, 12B. Amsterdam: Elsevier.

Harshberger, J. W. 1911. *Phytogeographic Survey of North America.* New York: Hafner Publishers.

Joshi, M. C. and R. K. Gupta. 1973. "Ecology of arid and semiarid zones in India." Pp. 111–153 in *Progress of Plant Ecology in India.* New Dehli: Today & Tomorrow. (Cited in Gupta 1986.)

Jürgens, N., A. Burke, M. K. Seely, and K. M. Jacobson. 1997. "Desert." Pp. 189–214 in R. M. Cowling, D. M. Richardson, and S. M. Pierce, eds., *Vegetation of Southern Africa.* Cambridge: Cambridge University Press.

Kassas, M. 1953. "Landforms and plant cover in the Egyptian desert." *Bull. Soc.Geogr. Egypt* 24: 193–205.

King, G. 1879. "Sketches of the flora of Rajasthan." *Indian Forestry* 4: 226–236.

Kingdon, Jonathan. 1989. *Island Africa: The Evolution of Africa's Rare Animals and Plants.* Princeton, NJ: Princeton University Press.

Knight, Dennis H. 1994. *Mountains and Plains, The Ecology of Wyoming Landscapes.* New Haven: Yale University Press.

Lancaster, Nicholas. 1996. "Desert environments." Pp. 211–237 in William S. Adams, Andrew S. Goudie, and Antony R. Orme, eds., *The Physical Geography of Africa.* Oxford: Oxford University Press.

LeHouérou, H. N. 1959. *Recherces Écologiques et Floristiques sur la Vegetation de la Tunisie Méridionale.* Algiers: Institut des Recherches Sahariennes, University of Algiers.

LeHouérou, Henri Noel. 1986. "The desert and arid zones of northern Africa." Pp. 101–147 in Michael Evenari, Imanuel Noy-Meir, and David W. Goodall, eds., *Hot Deserts and Arid Shrublands, B.* Ecosystems of the World, 12B. Amsterdam: Elsevier.

Leopold, A. S. 1959. *Wildlife of Mexico, the game birds and mammals.* Berkeley: University of California Press.

Logan, R. F. 1960. *The Central Namib Desert, South West Africa.* National Research Council Publication 758. Washington, DC: National Academy of Science.

Lowe, Charles H., ed., 1964. *The Vertebrates of Arizona.* Tucson: The University of Arizona Press.

MacMahon, James A. 2000. "Warm Deserts." Pp. 285–322 in Michael G. Barbour and William Dwight Billings, eds., *North American Terrestrial Vegetation.* 2nd edition. Cambridge: Cambridge University Press.

MacMahon, J. A. and F. H. Wagner. 1985. "The Mojave, Sonoran and Chihuahuan deserts of North America." Pp. 105–202 in Michael Evenari, Imanuel Noy-Meir, and David W. Goodall, eds., *Hot Deserts and Arid Shrublands, A.* Ecosystems of the World, 12A. Amsterdam: Elsevier.

Mares, M. A., W. F. Blair, F. A. Enders, D. Greegor, J. Hunt, A. C. Hulse, D. Otte, R. Sage, and C. Tomoff. 1977. "The strategies and community patterns of desert animals." Pp. 107–163 in G. H. Orians and O. T. Solbrig, eds., *Convergent Evolution in Warm Deserts.* US/IBP Synthesis Series 3. Stroudsburg, PA: Dowden, Hutchinson & Ross.

Mares, M. A., J. Morello, and G. Goldstein. 1985. "The Monte Desert and other subtropical semi-arid biomes of Argentina, with comments on their relation to North American arid areas." Pp. 203–237 Michael Evenari, Imanuel Noy-Meir, and David W. Goodall, eds., *Hot Deserts and Arid Shrublands, A.* Ecosystems of the World, 12A. Amsterdam: Elsevier.

Marloth, R. 1908. *Das Kapland, insoriderheit das Reich der kapflora, das Waldgebeit und die Karrpo, pflanzengeographish dargestellt.* Jena: Gustav Fischer.

Milius, Susan. 2001. "Torn to ribbons in the desert." *Science News*, 160: 266–268.

Milton, S. J., R. I. Yeaton, W. R. J. Dean, and J. H. J. Vlok. 1997. "Succulent Karoo." Pp. 131–166 in R. M. Cowling, D. M. Richardson, and S. M. Pierce, eds., *Vegetation of Southern Africa*. Cambridge: Cambridge University Press.

Monod, T. 1954. "Modes contracté et diffus de la végétation saharienne." Pp. 35–37 in J. L. Cloudsley-Thompson, ed., *Biology of Deserts*. London: Tavistock House.

Morello, J. 1958. "La provincia fitogeografica del monte." *Opera Lilloana* 2: 1–155.

Muller, C. H. 1939. "Relations of the vegetation and climate types in Nuevo Leon, Mexico." *American Midland Naturalist* 2: 687–728.

Muller, C. H. 1947. "Vegetation and climate of Coahuila, Mexico." *Madroño* 9: 33–57.

Orians, G. H. and O. T. Solbrig, eds. 1977. *Convergent Evolution in Warm Deserts*. US/IBP Synthesis Series 3. Stroudsburg, PA: Dowden, Hutchinson & Ross.

Orshan, G. 1985. "The deserts of the Middle East." Pp. 1–26 in Michael Evenari, Imanuel Noy-Meir, and David W. Goodall, eds., *Hot Deserts and Arid Shrublands, B*. Ecosystems of the World, 12A. Amsterdam: Elsevier.

Petrov, M. P. 1966–67. *Deserts of Central Asia*, 2 vols (in Russian). Leningrad: Nauka.

Rauh, Walter. 1985. "The Peruvian-Chilean deserts." Pp. 239–267 in Michael Evenari, Imanuel Noy-Meir, and David W. Goodall, eds., *Hot Deserts and Arid Shrublands, A*. Ecosystems of the World, 12A. Amsterdam: Elsevier.

Ricketts, Taylor H., Eric Dinerstein, David M. Olson, and Colby J. Loucks et al. 1999. *Terrestrial Ecoregions of North America: A Conservation Assessment*. Washington, DC: Island Press.

Schmida, A. 1985. "Biogeography of the desert flora." Pp. 23–77 in Michael Evernari, Imanual Noy-Meir, and David W. Goodall, eds., *Hot Deserts and Arid Shrublands, A*. Ecosystems of the World 12A. Amsterdam: Elsevier.

Shmida, Avi. 1985. "Biogeography of the desert flora." Pp. 23–77 in Michael Evenari, Imanuel Noy-Meir, and David W. Goodall, eds., *Hot Deserts and Arid Shrublands, A*. Ecosystems of the World, 12A. Amsterdam: Elsevier.

Shreve, Forrest. 1910. "The establishment of the giant cactus." *Plant World* 13: 235–240.

Shreve, Forrest. 1917. "The establishment of desert perennials." *Journal of Ecology* 5: 210–216.

Shreve, Forrest. 1942. "The desert vegetation of North America." *Botanical Review* 8(4): 195–246.

Shreve, Forrest. 1951. "Vegetation of the Sonoran Desert." *Carnegie Institution Washington Publication 591*.

Solbrig, O. T. and G. H. Orians. 1977. "The adaptive characteristics of desert plants." *American Scientist* 65: 412–421.

Songqiao, Zhao. 1986. *Physical Geography of China*. New York: Wiley & Sons, Inc.

Soriano, Alberto. 1983. "Deserts and semi-deserts of Patagonia." Pp. 423–460 in Neil E. West, ed., *Temperate Deserts and Semi-deserts*. Ecosystems of the World, 5. Amsterdam: Elsevier Scientific Publishing Company.

Troll, C. 1930. "Die trophischen Andenländer: Bolivien, Peru, Ecuador, Columbien und Venezuela." Pp. 309–490 in F. Klute, ed., *Handbuch der Geographischen Wissenschaften*, Band Südamerika. Potsdam: Athenaion Wildpark.

Walter, H. 1936. "Die ökologischen Verhältnisse in der Namib-Nebelwüste (Südwestafrika) unter Auswertung der Aufzeichnungen des Dr. G. Boss (Swakopmund). *Jahrbuch der Wissenschaften Botanische* 84: 58–222.

Walter, H. 1974. *Die vegetation Osteuropas, Nord- und Zentralasiens*. Band VII, Vegetationsmonographien der einzelnen Grossräume. Stuttgart: Fischer-Verlag.

Walter, H. 1986. "The Namib Desert." Pp. 245–282 in Michael Evenari, Imanuel Noy-Meir, and David W. Goodall, eds., *Hot Deserts and Arid Shrublands, B*. Ecosystems of the World, 12B. Amsterdam: Elsevier.

Walter, H. and E. O. Box. 1983a. "Overview of Eurasian continental deserts and semi-deserts." Pp. 3–7 in Neil E. West, ed., *Temperate Deserts and Semi-deserts*. Ecosystems of the World, 5. Amsterdam: Elsevier Scientific Publishing Company.

Walter, H. and E. O. Box. 1983b. "The Karakum Desert, an example of a well-studied eu-biome." Pp. 105–159 in Neil E. West, ed., *Temperate Deserts and Semi-deserts*. Ecosystems of the World, 5. Amsterdam: Elsevier Scientific Publishing Company.

Walter, H. and E. O. Box. 1983c. "The deserts of Central Asia." Pp. 193–236 in Neil E. West, ed., *Temperate Deserts and Semi-deserts*. Ecosystems of the World, 5. Amsterdam: Elsevier Scientific Publishing Company.

Walter, H. and E. O. Box. 1983d. "Middle Asia deserts." Pp. 79–104 in Neil E. West, ed., *Temperate Deserts and Semi-deserts*. Ecosystems of the World, 5. Amsterdam: Elsevier Scientific Publishing Company.

Weberbauer, A. 1911. "Die Pflanzenwelt der peruanischen Anden." In *Vegetation der Erde*, 12. Leipzig: Englemann (cited in Rauh 1985).

Werger, M. J. A. 1986. "The Karoo and Southern Kalahari." Pp. 283–359 in Michael Evenari, Imanuel Noy-Meir, and David W. Goodall, eds., *Hot Deserts and Arid Shrublands, B*. Ecosystems of the World, 12B. Amsterdam: Elsevier.

West, Neil E. 1983a. "Overview of North American temperate deserts and semi-deserts." Pp. 321–330 in Neil E. West, ed., *Temperate Deserts and Semi-deserts*. Ecosystems of the World, 5. Amsterdam: Elsevier Scientific Publishing Company.

West, Neil E. 1983b. "Great Basin-Colorado sagebrush semi-desert." Pp. 331–349 in Neil E. West, ed., *Temperate Deserts and Semi-deserts*. Ecosystems of the World, 5. Amsterdam: Elsevier Scientific Publishing Company.

West, Neil E. 1983c. "Western intermontane sagebrush steppe." Pp 351–374 in Neil E. West, ed., *Temperate Deserts and Semi-deserts*. Ecosystems of the World, 5. Amsterdam: Elsevier Scientific Publishing Company.

West, Neil E. 1983d. "Intermontane salt-desert shrubland." Pp. 375–397 in Neil E. West, ed., *Temperate Deserts and Semi-deserts*. Ecosystems of the World, 5. Amsterdam: Elsevier Scientific Publishing Company.

West, Neil E. 1983e. "Comparisons and contrasts between the temperate deserts and semi-deserts of three continents." Pp. 461–472 in Neil E. West, ed., *Temperate Deserts and Semi-deserts*. Ecosystems of the World, 5. Amsterdam: Elsevier Scientific Publishing Company.

West, N. E. and James A. Young. 2000. "Intermountain valleys and lower mountain slopes." Pp. 255–284 in Michael G. Barbour and Dwight Billings, eds., *North American Terrestrial Vegetation*. 2nd edition. Cambridge: Cambridge University Press.

Williams, O. B. and J. H. Calaby. 1985. "The hot deserts of Australia." Pp. 269–312 in Michael Evenari, Imanuel Noy-Meir, and David W. Goodall, eds., *Hot Deserts and Arid Shrublands, A*. Ecosystems of the World, 12A. Amsterdam: Elsevier.

Woodward, S. L. 1976. Feral Burros of the Chemehueri Mountains, California: The Biogeography of a Feral Exotic. Ph.D. dissertation, University of California, Los Angeles, 178 p.

Zohary, M. 1940. "Geobotanical analysis of the Syrian Desert." *Palestine Journal of Botany* 2: 46–96.

Zohary, M. and G. Orshan. 1956. "Ecological studies in the vegetation of the Near East deserts II. Wadi Araba." *Vegetatio* 71: 15–37.

# 5 TEMPERATE GRASSLANDS

## OVERVIEW

In the **midlatitudes**, natural grasslands are found in the semiarid interiors of North America and Eurasia and in rainshadow situations in South America and southern Africa. The **vegetation** is dominated by **perennial** grasses—some sod-forming and others bunch or **tussock** forming—and a variety of **forbs**, mainly **legumes** or daisies. The plant **community**'s composition and structure are quite complex. Variation is related to climatic differences, local **edaphic** conditions, fire history, and grazing patterns.

Each of the world's major regions of temperate grassland has a different common name. In North America, the grasslands are referred to as prairies and in Eurasia steppe. In South America, they are called pampas. European and North American scientists call the grasslands of southern Africa the veld, although this is actually a local term referring to vegetation in general. Australia and New Zealand also have temperate grasslands, but these are relatively minor and will not be described here.

Each major expression of the temperate grassland **biome** has zonal subdivisions that reflect decreasing amounts of precipitation. In North America, as one progresses from east to west across the midcontinent from the subhumid **climate** region of the Central Lowlands to the drier margins of the semiarid climate region on the Great Plains, one passes from tallgrass prairie to mixed prairie to shortgrass prairie. In Eurasia, the zones run north-south from the moister meadow steppe that borders the temperate **broadleaf** deciduous forest biome through true or typical steppe to dry steppe and shrub steppe. Along the moisture gradient toward drier conditions, grasses (and forbs) decrease in height, diversity, and percent **cover**. Root development and changes in the **soil** also decrease (see following discussion).

Human activities have greatly altered all temperate grasslands, and few, if any, truly pristine examples survive. Cultivation and intense grazing by domestic live-

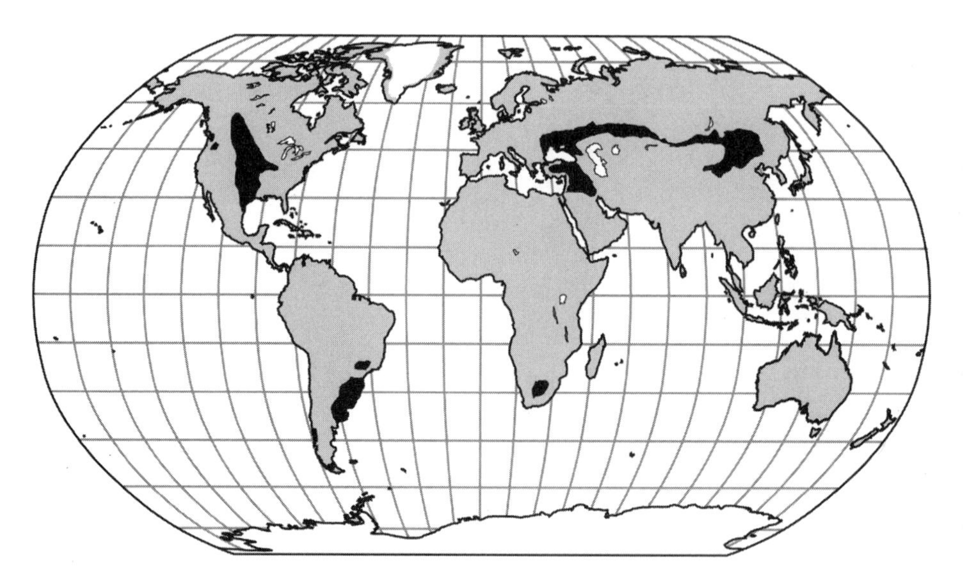

Map 5.1 World distribution of the temperate grassland biome.

stock have destroyed or degraded the biome, including the native animals, to such a degree that reconstruction of the former natural situation is difficult, perhaps impossible. The major crops produced in these regions today are themselves grasses (e.g., wheat, maize) or oilseeds related to wild grassland forbs (e.g., sunflower and rapeseed or canola).

## VEGETATION

Perennial grasses are the characteristic plants of the temperate grasslands. Grasses are well adapted to survive hot summers, fire, and—in many instances—grazing and trampling by large mammals. Narrow, leafy shoots or tillers are held upright to reduce heating in the summer; large fibrous root systems effectively trap moisture and nutrients from the upper horizons of the soil; and the growth buds are at or just below the surface, where they are protected both from the fires that sweep through the **canopy** and from grazing animals. Two basic **growth-forms** of grass occur. The turf- or sod-forming type usually reproduces vegetatively by lateral buds or **rhizomes** and spread to form a dense mat of entwined roots. Tussock or bunchgrasses grow as individual clumps of tillers around a central stem and reproduce by seed (Furley and Newey 1983).

Depending upon the actual climatic conditions of a given region, both $C_4$ (warm-season) and $C_3$ (cool-season) grasses are found in the temperate grasslands. These designations $C_4$ and $C_3$ refer to two different chemical pathways by which carbon dioxide is taken from the atmosphere and converted (i.e., fixed) into carbohydrates during photosynthesis (see also tropical savanna biome, Chapter 3). The $C_4$ pathway allows for photosynthesis to occur at higher temperatures than is possible with $C_3$ photosynthesis, the type typical of most green plants. $C_4$ grasses are thus associated with those parts of the temperate grassland with hot summers. They are also referred to as warm-season grasses because they do not begin to grow

Photo 5.1 Temperate grassland, with many forbs as well as grasses. (Photo by SLW.)

until early summer. Peak growth occurs when temperatures reach about 35° C (95° F). The $C_3$ or cool-season grasses, on the other hand, start growth in early spring and maximize growth at around 20° C (68° F); in North America native perennial $C_3$ grasses are more or less restricted to the shortgrass prairie of the northwestern Great Plains (Brown 1993).

Perennial forbs are abundant in the temperate grasslands (Photo 5.1), and as they bloom in spring and summer, create swaths of changing color across the vast plains on which grasslands are typically found. They die down in the fall and regenerate next spring from underground tubers, bulbs, and rhizomes. Members of the legume, daisy, and mint families are common; many, such as the clovers, provide nutritious forage for wildlife and domestic animals.

Especially in the drier or degraded grasslands that have no bare ground, **annual** grasses and forbs, dwarf shrubs, and **subshrubs** occur.

## CLIMATE

Temperate grasslands are generally associated with semiarid continental climates with 25–75 cm (10–20 in) of precipitation annually (BSk in the Koeppen classification). The seasonal pattern of precipitation differs from region to region, as is discussed in the following regional descriptions. These areas experience cold winters and hot summers. A wide daily range in temperatures is common as well. In both North America and South America, temperate grasslands extend into the drier margins of the humid subtropical climate (Cfa) regions, where the precipitation is sufficient to support broadleaf deciduous forests. These marginal climates are referred to as subhumid.

## SOILS

The soils developed under temperate grasslands are among the most fertile on earth. They are typically deep, high in dissolved nutrients, and high in organic matter. The latter gives them a black or deep brown color. **Parent material** in the Northern Hemisphere is usually unconsolidated materials such as glacial till, glacial **outwash**, or especially **loess** that is often initially high in calcium carbonate. In the South American pampas, the parent material is of volcanic origin and also nutrient rich (Acton 1992).

In a semiarid climate, evaporation and capillary action prevent **leaching** of dissolved plant nutrients from the soil column. The deep mat of grass roots rapidly traps nutrients and also prevents their removal from the **ecosystem**. The soils are neutral or slightly alkaline ($pH > 7$). Calcification is the dominant soil-forming process in this biome. When it rains and water percolates through the soil, calcium carbonate is dissolved and moves downward through the A horizon into the B horizon. Evaporation once the brief rains stop draws soil water with its dissolved minerals upward again until it dries out and the calcium carbonate is precipitated. This leads to a concentration of calcium carbonate in the soil at a depth that varies according to total annual precipitation. In the subhumid and wetter parts of the semiarid regions, a layer rich in carbonates (in the B horizon) may be 2 m (6 ft) deep, and the presence of calcium carbonate may be discernible only by its reaction with an acidic solution. However, the drier the climate, the closer the deposit is to the surface and the calcium carbonate nodules become more visible.

The **humus** or organic material in the soil that facilitates the exchange of mineral nutrients between soil and plant root derives from the annual dying of aerial portions of the grasses and forbs of the grasslands and, just as importantly, from the continual dying and decay of their roots, which in the moister areas may extend to depths of 3.5 m (12 ft).

Grassland soils are classified in the U.S. Soil Taxonomy in the order mollisols. Moll refers to the soft, friable texture of these soils. Within the order, soils range from very black types to lighter brown ones. Russian soil scientists did the early work on temperate grassland soils, and the Russian term for the blackest soils, *chernozem*, prevails even where the American system is used. Chernozems occur in the moister parts of the grassland in the northern Central Lowlands of North America on glacial **till** and loess and across the northern steppe of Eurasia from Romania through Ukraine and into Kazakhstan and western Siberia on loess (Acton 1992). Chestnut-colored soils occur in warmer and more arid regions.

## ANIMAL LIFE

Typical mammals of the temperate grasslands fall into two groups: fleet-footed hoofed mammals (ungulates) that congregate in herds and burrowing rodents that often live in colonies. Bison (*Bison bison*) (Photo 5.2) and pronghorn (*Antilocapra americana*) are the characteristic ungulates of the North American grasslands, while wild horses (*Equus przewalskii* and *E. gmelini*), asses (*E. hemionus* and *E. kiang*), and saiga antelope (*Saiga tatarica*) were once abundant on the grasslands of Eurasia. The black-tailed prairie dog (*Cynomys ludovicianus*) epitomizes the colonial rodent in North America, while Eurasia has mole rats (*Spalax microphtalmus*) and social voles (*Microtus socialis*), among others. Most mammals are active year round, but a

Photo 5.2 Shortgrass prairie with American bison (*Bison bison*). Note the bunchgrass growth-form characteristic of this expression of the temperate grassland biome. (Photo by SLW.)

few of the rodents hibernate in the winter. Birds are necessarily ground nesters such as the horned lark (*Eremophila alpestris*), which is widespread across North America and Eurasia. Snakes and lizards occur but are not very diverse. Amphibians tend to be rare and are represented mostly by toads and frogs. Insects such as grasshoppers are the most important herbivores.

## MAJOR REGIONAL EXPRESSIONS

### North American Prairies

The major grassland region of North America corresponds with the Central Lowlands and Great Plains **physiographic provinces**, the flat to gently rolling midsection of the continent that stretches from the eastern foothills of the Rocky Mountains in the west to approximately the 100° W meridian. An eastern extension crosses Iowa and northern Missouri into northern Illinois and has outliers as far east as Ohio. Other areas of temperate grassland in North America include the Palouse Prairie of eastern Oregon and Washington state, the grasslands of California's Central Valley, and the desert grasslands in the American Southwest and Mexico.

Three distinct plant groupings occur in the major, central grassland region as north-south belts running from the prairie provinces of Canada (Alberta, Saskatchewan, and Manitoba) to Texas. These are, from east to west, the tallgrass prairie, mixed prairie, and shortgrass prairie. Each is described in some detail.

Tallgrass prairie is sometimes referred to as bluestem prairie after its dominant grasses. It is also sometimes called the true prairie, since the term prairie comes from the French word for meadow, and it was this eastern region that French explorers first encountered. It occurs in the subhumid and moister semiarid regions of the eastern prairie belt. This is a multilayered vegetation with tall grasses, mid-sized grasses, and short grasses. Most of the dominants are tall grasses such as big bluestem (*Andropogon gerardii*), little bluestem (*Schizachrium scoparium*), Indian grass (*Sorghastrum nutans*), prairie sandreed (*Calamavilfa longifolia*), switchgrass (*Panicum virgatum*), and sand bluestem (*Andropogon hallii*). Most of these are sod-forming types, but little bluestem is a bunchgrass. All are $C_4$ grasses. In late summer, when the tallest grasses have matured, they may reach heights of 200–300 cm (80–120 in). Sideoats grama (*Bouteloua curtipendula*), another bunchgrass, is one of the shorter grasses [20–100 cm (8–40 in)] and the most drought tolerant. The grasses are accompanied by a variety of perennial forbs including yarrow (*Achillea* spp.); purple coneflowers (*Echinacea angustifolia*); goldenrod (*Solidago* spp.); sunflowers (*Helianthus* spp.) and other members of the daisy family; and legumes such as wild alfalfa (*Psoralea tenuiflora*), prairie clover (*Petalostemon* spp.), and slender lespedeza (*Lespedeza virginica*). Tallgrass prairie is the richest in **species** of the North American midcontinent grasslands and is associated with the richest chernozem soils. Fire appears to be a major factor in maintaining the tallgrass prairie and preventing the invasion of woody species (Brown 1993; Kucera 1992; Vankat 1979). The animals of the tallgrass prairie are most notable for those missing now that most of the region has been converted to agricultural land. Gone are the herds of bison, antelope, and elk (*Elaphus canadensis*), as is the mule deer (*Odocoileus hemionus*). Gone, too, is their chief predator, the wolf (*Canis lupus*). The swift fox (*Vulpes velox*) is no longer found in most areas. Taking their place are coyotes (*Canis latrans*) and red fox (*Vulpes fulva*). Rodents survive in the rare protected areas of tallgrass prairie. These include the thirteen-lined ground squirrel (*Spermophilus tridecemlineatus*), meadow vole (*Microtus ochrogaster*), plains harvest mouse (*Reithrodontomys montanus*), grasshopper mouse (*Onychomys leucogaster*), and pocket gopher (*Geomys bursarius*). The eastern cottontail (*Sylvilagus floridanus*) and white-tailed jackrabbit (*Lepus townsendii*) are other small mammalian herbivores of the tallgrass prairie. Among smaller carnivores are badger (*Taxidea taxus*), least weasel (*Mustela frenata*), and striped skunk (*Mephitis mephitis*). Native birds include Horned Lark (*Eremophila alpestris*), Grasshopper Sparrow (*Ammodramus savannarum*), Savanna Sparrow (*Passerculus sandwichensis*), Dickcissel (*Spiza americana*), Eastern Meadowlark (*Sturnella magna*), Western Meadowlark (*S. neglecta*), Upland Sandpiper (*Bartramia longicauda*), and Burrowing Owl (*Speotyto cunicularia*). The once-abundant Greater Prairie Chicken (*Tympanuchus cupido*) has declined in numbers and is being replaced by the **introduced** Chinese Ring-necked Pheasant (*Phasianus colchicus*) (Kucera 1992).

Mixed prairie is a two-layered grassland vegetation that occurs entirely on the Great Plains from southeastern Alberta, southern Saskatchewan, and southwest Manitoba south to Texas. This is a latitudinal range from 52° N to 29° N and an elevational range from 500 m (1,600 ft) in the east to 1,100 m (3,600 ft) in the west. Annual precipitation in this semiarid climate region varies from 600 mm (24 in) in the east to 350–400 mm (14–16 in) in the west. The soils tend to be brown and chestnut-colored mollisols. The vegetation consists of mixed stands of midsized and short grasses. Dominant grasses include midsized $C_3$ species such as western wheatgrass (*Andropogon smithii*), dropseed (*Sporobolus cryptandrus*), Junegrass (*Koele-*

*ria cristata*), needle-and-thread (*Stipa comata*), and shorter $C_4$ grasses such as blue grama (*Bouteloua gracilis*) and buffalograss (*Buchloë dactykoides*). The midsized grasses at maturity stand 60–120 cm (24–48 in) tall; the short grasses 20–60 cm (8–24 in). Perennial forbs such as western yarrow (*Achillea lanulosa*), heath aster (*Aster ericoides*), and plains beebalm (*Monarda pectinata*) are also prominent members of the plant community in mixed prairie. Also found are halfshrubs such as fringed sagebrush (*Artemisia frigida)* and dwarf shrubs such as wild rose (*Rosa* spp.) and Bolander silver sagebrush (*Artemisia cana*). Brittle prickly pear cactus (*Opuntia fragilis*) and plains prickly pear *(O. polycantha)* also occur. In the northern plains, trembling aspen (*Populus tremuloides*) has invaded the mixed prairie since fires were suppressed after European settlement. In this grassland community, new shoots appear in late March to early April, and green grass is still available to animals until early December. The greatest amount of green grass, however, is present in May and June, since grass blades are continually dying. New growth replaces them throughout the growing season (Coupland 1992a; Brown 1993; Furley and Newey 1983). Animal life of the mixed prairie is quite similar to that of the tallgrass prairie. Other species recorded include, among small mammals, northern pocket mouse (*Thomomys talpoides*) and meadow jumping mouse (*Zapus husonicus*); the Sharp-tailed Grouse (*Pedioectes phasianelles*) among larger birds; and among songbirds Baird's sparrow (*Ammodramus bairdii*), Sprague's Pipit (*Anthus spragueii*), Chestnut-collared Longspur (*Calcarius ornatus*), and McCown's Longspur (*Rhynchophanes mccownii*). Eastern Meadowlark does not usually occur this far west. The list of reptiles for mixed prairie includes the prairie rattlesnake (*Crotalus viridus*), the plains hognose snake (*Opheodrys vernalis*), bull snake (*Pituophis melanoleucus*), and plains garter snake (*Thamnophis radix*). The short-horned lizard (*Phyrnosoma douglassi*) is also present. Only a few amphibians occur; most common are the Great Plains toad (*Bufo cognatus*) and the leopard frog (*Rana pipiens*) (Coupland 1992a).

Shortgrass prairie occupies some 280,000 km$^2$ (175,000 mi$^2$) on the central Great Plains from about the Colorado/Wyoming state line south to the Texas/Mexico international border (41° N to 32° N). Precipitation varies from 300–550 mm (12–22 in), with about 70 percent of it occurring between May and August. Dominant plants are shallow-rooted, sod-forming grasses. Most important are blue grama and buffalograss, both of which attain heights of 10–40 cm (4–16 in) at maturity. Other common grasses include needle-and-thread, western wheat grass, and wire grass (*Aristida longiseta*), which form an open layer above the shortgrasses in ungrazed areas. Small shrubs, subshrubs, and prairie prickly pear also occur. The pronghorn is the only large herbivore that occurs in large numbers today, but the shortgrass prairie once was home to vast herds of bison. Hares and rabbits represent the major herbivores today. Black-tailed jackrabbit (*Lepus californica*), white-tailed jackrabbit (*L. townsendii*), and desert cottontail (*Sylvilagus audubonii*) all occur here. The black-tailed prairie dog (*Cynomys ludovicianus*) is a vital food source for Ferruginous Hawk (*Buteo regalis*), Swainson's Hawk (*B. swainsoni*), the Golden Eagle (*Aquila chrysaetos*), coyotes, and badgers. Other animals as diverse as Burrowing Owls, prairie rattlesnakes, thirteen-lined ground squirrels, and desert cottontails use prairie-dog burrows for shelter and nesting sites. The mounds formed at burrow entrances harbor a unique community of annual forbs, and bison and antelope preferentially graze prairie-dog towns. Other important burrowing animals found in parts of the shortgrass prairie region are two pocket gophers: the plains pocket gopher (*Geomys bursarius*) and northern pocket gopher (*Thomomys talpoides*), both members of a uniquely North American rodent family (Geomydiae). Common sparrows and finches include Brewer's Sparrow (*Spizella breweri*); the Chestnut-collared and McCown's longspurs, also associated with mixed prairie; and the Lark Bunting (*Calamospiza melanocorys*). Horned Lark and Western Meadowlark are other common songbirds. The Mountain Plover (*Charadrius montana*) is a common shorebird nesting in the drier parts of the region. Prairie rattlesnakes, western hog-nosed snakes, and gopher snakes (*Pituophis catenifer*) are particularly abundant in the shortgrass prairie, and the prairie garter snake is locally abundant near ponds. Other common reptiles include skinks (*Eumeces* spp.), horned lizard (*Phrynosoma douglassi*) and a box turtle (*Terrapene ornata*). Common amphibians include Great Plains toad, Woodhouse's toad (*Bufo woodhousei*), and leopard frog (Lauenroth and Milchunas 1992).

The Palouse Prairie lies in an area of winter (mid-September to mid-June) precipitation and summer drought and thus differs from the midcontinent grasslands just described. Total annual precipitation ranges between 330–600 mm (13–24 in). The soils, while derived from Pleistocene loess, have been enriched by volcanic ash from the explosion and collapse of Mt. Mazama, which formed Crater Lake some 6,500 years ago. The dominant grasses of the Palouse before European settlement were bluebunch wheatgrass (*Agropyron spicatum*), Idaho fescue (*Festuca idahoensis*), and sandberg bluegrass (*Poa sandbergii*). They stood 10–30 cm (4–12 in) high at maturity. Perennial forbs associated with this grass community include western yarrow (*Achillea lanulosa*), *Castilleja lutescens*, old man's whiskers (*Geum triflorum*), lupine (*Lupinus sericeus*), brodiaea (*Brodiaea douglasii*), and groundsel (*Senecio integerrimus*). In wetter areas—comparable to the meadow-steppes of Eurasia (discussed later)—dwarf shrubs such as snow berry (*Symphoricarpos albus*) and wild roses (*Rosa nutkana*, and *R. woodsii*) are present, and a thin layer of annual forbs less than 10 cm (4 in) in height occurs, composed of such plants as willow herb (*Epilobium paniculatum*), Indian lettuce (*Montia linearis*), and shining chickweed (*Stellaria nitens*). Perennial grasses and forbs retain green leaves throughout the relatively mild winters and flower between May and September. Annuals germinate in midwinter, bloom in early spring, and spend the rest of the year as dormant seeds. Shrubs tend to be green only in summer. Before they were drained and cultivated, seasonal marshes in the Palouse were notable for the dominance of the bulb camass (*Camassia quamas*), an important food source for Native Americans. The animal life of the Palouse was similar to that of the midcontinent grasslands. Bison and pronghorn, joined by white-tailed deer, constituted the larger grazing animals. Smaller mammals included white-tailed jackrabbits, mountain cottontail (*Sylvilagus nuttallii*), yellow-bellied marmots (*Marmota flaviventris*), Columbia ground squirrel (*Spermophilus columbianus*), mice, and voles. They were preyed upon by coyotes and badgers. Rattlesnakes (*Crotalus viridis*), garter snakes (*Thamnophis ordinales*), and bull snakes (*Pituophis catenifer*) are still common (Daubenmire 1992).

The California Prairie is a mediterranean grassland that, like the Palouse, is in a winter rainfall climate region. It once covered some 10 million ha (25 million ac) in the foothills of the Central Valley and some of the coastal valleys of California. The original vegetation was perennial bunchgrasses dominated by purple needlegrass (*Stipa pulchra*) and Malpais bluegrass (*Poa scabrella*). Of lesser importance were needlegrass (*Aristida hamullos*), wild ryes (*Elymus glaucus* and *E. tritcoides*—the only sod-forming grass), *Koeleria cristata* and *Melica imperfecta*. Annual forbs and grasses (e.g., *Vulpa microstachys* and *V. octoflora*) grew between the bunchgrasses. Perennial bulbs such as brodiaeas and *Chlorogalum* spp. were associated with this grassland. Since 1775, the composition of the California grassland has been permanently changed. Today, it is composed mainly of introduced annual grasses and forbs such as wild oats (*Avena fatua* and *A. barbata*), brome (*Bromus mollis*), and storksbill or filaree (*Erodium botrys*) (Heady et al. 1992).

Desert grasslands occur from southeastern Arizona to western Texas and south into northern Mexico at intermediate elevations of 1,000–1,750 m (3,300–5,700 ft) in the Sonoran and Chihuahuan deserts. Although their composition is variable, perennial $C_4$ (warm season) grama grasses (*Bouteloua* spp.) and tobosa grasses (*Hilaria* spp.) characterize these grasslands. Other grasses include plains lovegrass (*Eragrostis intermedia*) and muhly (*Muhlenbergia* spp.). Leaf-**succulent** shrubs of the lily family are abundant on some sites. These include yuccas (*Yucca* spp.), sotol (*Dasylirion wheeleri*), beargrass (*Nolina microcarpa*), and agaves (*Agave* spp.). Woody shrubs occur in many parts of the desert grasslands. These invaders include **evergreen** species such as junipers (*Juniperus deppeana* and *J. monosperma*); live oaks (*Quercus emoryi* and *Q. grisea*); creosotebush (*Larrea tridentata*); and deciduous species such as ocotillo (*Fouqueria splendens*), mesquite (*Prosopis* spp.), and acacias (*Acacia* spp.). Many of the woody shrubs have increased in abundance since the late nineteenth century. Mammals are a mix of forest, grassland, and desert species made possible by the interspersed mountains and plains of the Basin and Range physiographic province in which these grasslands are located. Large animals include both white-tailed deer and mule deer. The peccary or javelina (*Tayassu tajacu*) frequents foothill regions. Smaller mammals are represented by black-tailed jack rabbit, antelope jack rabbit (*Lepus alleni*), desert cottontail, spotted ground squirrel (*Spermophilus spilosoma*), and kangaroo rats (*Dipodomys ordii* and *D. spectabilis*). Carnivores include coyote, bobcat (*Felis rufus*), badger, gray fox (*Urocyon cinereoagenteus*), kit fox (*Vulpes macrotis*), and striped skunk. The relatively few species of breeding birds include, in often brushy locations, Scaled Quail (*Callipepla squamata*), Gambel's Quail (*C. gambelli*), and Montezuma Quail (*Crytonyx montezumae*) (Schmutz et al. 1992; Vankat 1979; Lowe 1964).

## Eurasian Steppes

The geographic pattern of temperate grasslands in Eurasia is unlike that found in North America. The steppes stretch across the continent in a wide east-west belt and show considerable north-south elongation only in the east, in China. Across much of Russia and Mongolia, the steppe belt is 800–1,000 km (500–600 mi) wide (between 48° and 57° N latitude), but its east-west extent is about 8,000 km (5,000 mi) (between 27° and 128° E longitude). Differentiation of the vegetation along a moisture gradient from moist grasslands to dry grasslands occurs in latitudinal zones from north to south, rather than from east to west, as is the case in the North American prairies. This vast area has a great diversity of plant assemblages. Russia geobotanists have traditionally grouped them into four major types of steppe. From north to south, they are as follows:

1. Meadow steppe or forest steppe, a mosaic of grassland and broad-leaved deciduous forests dominated by the oak *Quercus robur* in a subhumid climate region;
2. True or typical steppe, characterized by bunchgrasses with many forbs in semiarid climate regions and bunchgrasses with few forbs in drier areas;
3. Desertified bunchgrass and dwarf shrub-bunchgrass in arid climate regions; and
4. Desert dwarf subshrub-bunchgrass steppe in the driest areas.

Certain trends in the vegetation are apparent from north to south. These include a decrease in the number of species present; a decrease in the height of the grass layer from 80–100 cm (30–40 in) in the north to 15–20 cm (6–8 in) in the south; and a decrease in the percent cover of grasses from 70–90 percent to from less than 10–20 percent. Throughout, perennial bunchgrasses adapted to withstand cold winters and drought dominate. Needlegrasses or feathergrasses (*Stipa* spp.) are the most characteristic; the actual species vary from region to region. In the meadow steppe, *Stipa pennata* is dominant; in the true steppe, *S. lessingiana*, *S. pulcherrima*, and *S. zaleskii* are found. In the driest desert grasslands, *S. gobica* and *S. glareosa* are among the prominent shortgrasses. Other genera of bunchgrasses represented in the Eurasia steppes are the wheatgrasses (*Agropyron*), *Cleistogenes*, fescues (*Festuca*), oat grasses (*Helictotrichon*), and Junegrass (*Koeleria*). Bunch sedges (*Carex* spp.) also occur. These grasses grow throughout the growing season but show slower growth rates in July and early August. With them are a large number of forbs with colorful flowers (Lavrenko and Karamysheva 1993). These forbs come into bloom at different times and create an ever-changing palette of color across the steppes. Each stage is known as an aspect. For example, during spring in early April, the meadow steppe has a brown aspect when the mauve *Pulsatilla patens* is in flower and the sedge *Carex humilis* is in pollen. A yellow aspect takes place at the end of April when the *Adonis vernalis* flowers. A blue aspect occurs at the end of May, when the flowering of forget-me-nots (*Myosotis sylvatica*) dominates, even though white anenomes (*Anenome sylvestris*) and yellow groundsels (*Senecio campestris*) also bloom then. At the end of June, clover (*Trifolium montanum*), shasta daisies (*Chrysanthemum leucanthemum*), and *Filipendula hexapetala* produce a white aspect (Walter 1985). In Russia's Central Chernozem region, as many as 11 aspects or flowering phases occur each year. The number declines in other regions with only five or six aspects in the driest areas. The number and timing of aspects vary from year to year, according to actual precipitation amounts and temperatures (Lavrenko and Karamysheva 1993).

Thousands of years of human activity have greatly altered the Eurasian steppes. Heavy grazing by domestic livestock, fire, and cultivation have changed vegetation and animal life. The steppes of Mongolia are least changed and offer some idea of the earlier animal life elsewhere. As on the North American prairies, mammals tend to be gregarious and occur in herds or colonies. The wild ungulates, now reduced in numbers and distribution, include the Asiatic wild ass (*Equus hemionus*); takhi or Przewalski's horse (*E. przewalskii*); and saiga (*Saiga tatarica*), a rather strange-looking antelopelike animal with an inflated nose. The now-extinct tarpan (*Equus gmelini*) and aurochs (*Bos taurus*), ancestor of domestic cattle, once occupied western parts of the steppe. Rodents typically dig burrows, and their activities are considered major determinants of the nature of soil, vegetation, and microrelief in the steppe. Among the more common rodents are the steppe lemming (*Lagurus*

*lagurus*), Brandt's vole (*Lasiopodomys brandtii*), Chinese vole (*L. mandarinus*), common vole (*Microtus arvalis*), narrow-skulled vole (*M. gregalis*), social vole (*M. socialis*), and mole vole (*Ellobius talpinus*). Mole rats (*Spalax microphthalmus*) belong to an **endemic** family of rodents, the Spalacinae. The squirrel family is represented on the Eurasian steppes by suslik (*Spermophilus pygmaeus*), steppe marmot or bobak (*Marmota bobak*), and tarbagan (*M. siberica*)—members of genera also found in North American grasslands. The Daurian pika (*Ochotona daurica*) is a prominent relative of hares and rabbits. The small mammals of the steppe do not hibernate but accumulate stores of plant material for winter consumption. During the summer, most eat green plant parts, especially the juicy bases of grass stems, buds, rhizomes, bulbs, and tubers. Like tundra rodents, they undergo population cycles of four to five years (Lavrenko and Karamysheva 1993).

The grasslands of China are affected by extreme continentality (a wide range of temperatures between summer and winter) and the alternating pressure and wind systems of the Asian monsoon. A strong high pressure system over Mongolia during the winter is of particular significance. As a consequence, winters are long, cold, and very dry. Drought continues into spring, also a period of high winds, so no spring flushes of flowering occur comparable to the meadow steppes of Russia. No summer dormancy comparable to the Russian steppe occurs either. Most precipitation comes between June and September, when winds blow into the continent from the southeast, and most plants flower between June and August. Annual precipitation is greatest in the eastern part of China's steppe [500 mm (20 in)] and decreases to the west [200 mm (8 in)], a pattern more resembling North America's grasslands than those of the rest of Eurasia. Chinese scientists recognize three types of steppe, from east to west: meadow steppe, typical steppe, and desert steppe. Feathergrasses (*Stipa* spp.) are dominant in each type. Meadow steppe is a humid grassland occurring at elevations of 120–250 m (400–800 ft) in the Northeast (formerly known as the Manchurian Plain). The soils are chernozems or dark chestnut soils. The dominant grass is the bunchgrass *Stipa baicalensis*, which reaches a height of about 60 cm (24 in). Subdominants are sod-forming grasses *Leymus chinensis* and *Arundinella hirta*. They are accompanied by a large number of perennial forbs, including legumes such as *Lespedeza daurica*, *Astragulus adsurgnes*, and *Medicago ruthenica*, and plants from a number of other families. The tansy (*Filifolium sibiricum*) is dominant; others include *Adenophora stemophylla*, *Thalictrum simplex*, *Hemerocallis minor*, and *Delphinium glandiflorum* (Ting-Chen 1993).

Typical steppe extends from the Northeast and inner Mongolian plateau south onto the Loess Plateau. Elevations in this region range from 800–1,400 m (2,600–4,600 ft), and average annual precipitation is 300–400 mm (12–16 in). The vegetation is uniformly low; grasses stand 30–50 cm (12–20 in) high at maturity and cover 30 to 50 percent of the ground surface. Dominant grasses are tufted species of *Stipa* in association with other tuft grasses such as *Cleistogenes squarrosa*, *Koerelia gracilis*, and *Poa sphondylodes*. Forbs are scarce. *Artemisia frigida* is an abundant dwarf subshrub. In the west, near the transition to desert steppe, dwarf and medium-sized shrubs (*Caragana* spp.) become common (Ting-Chen 1993).

Desert steppe is found in the northern and northwestern parts of the Loess Plateau. Vegetation is low in height [20–30 cm (8–12 in)] and diversity and covers only about 30 percent of the surface. The drought-resistant tuftgrass *Stipa gobica* and *Artemisia xerophytica* are dominants. Drought-adapted shrubs and subshrubs, bulbs (*Allium mongolicum* and *A. polyrrhizum*), and some annual forbs (e.g., *Artemisia pectinata*) also are found in desert steppe (Ting-Chen 1993).

The most abundant native animals of the Chinese steppes are burrowing rodents that live in colonies, as is common among the small mammals of other steppe regions. These include Brandt's voles, narrow-skulled voles, zokor (*Myospalax aspalax*), suslik (*Spermophilus dauricus*), and bobak. In typical steppe areas, the striped, hairy-footed hamster (*Phodopus sungorus*) occurs, while clawed gerbil (*Meriones unguiculatus*) and desert hamster (*Phodopus roborouskii*) are found in desert steppe. Ungulates are represented by the Mongolian gazelle (*Procapra guffurosa*) on typical steppe and goitered gazelle (*Gazelle subgutturosa*) on desert steppe. Near extinction are the takhi and the two-humped or bactrian camel (*Camelus bactrianus*). The Mongolian wild ass or kulan (*Equus heminonus*) is increasingly rare. Few resident birds are present; most common and most widespread are Horned Lark (*Eremophila alpestris*), Skylark (*Alauda arvensis*), Mongolian Lark (*Melanocorypha mongolica*), the Isabelline Wheatear (*Oenanthe isabellina*), and the Common Wheatear (*O. oenanthe*). Similarly, few reptiles exist; common are two lizards (*Eremias argus* and *Phyrnocephalus frontalis*) and a snake (*Elaphe dione*). Amphibians, also rare, are only found in moist areas; the toad *Bufo raddei* is most abundant, followed by the frog *Rana temporaria* (Ting-Chen 1993).

Alpine grasslands in China warrant mention because they have supported pastoral cultures for millennia, and they are home to a unique assemblage of large grazing animals. Alpine steppe reaches its greatest extent on the Tibetan Plateau, at elevations of 3,000–5,300 m (10,000–17,000 ft), the highest as well as largest and geologically youngest plateau on Earth. Other important areas of alpine steppe occur in the mountains surrounding the desert basins of Xinjiang province. In the Altai, steppe occurs at elevations of 900–2,000 m (3,000–6,500 ft). On the north slopes of the Tien Shan, desert gives way to steppe at 1,000 m (3,300 ft); on the range's south slopes, steppe occurs between 1,200–1,600 m (4,000–5,250 ft). In the Kunlun Shan, alpine steppe begins at about 1,700 m (5,600 ft). These mountain pastures, forming mosaics with **needle-leaved** forest at higher elevations, receive 300–500 mm (12–20 in) of precipitation annually. Only the alpine steppe of the Tibetan Plateau will be described in any detail. This high-elevation grassland receives 450–700 mm (18–28 in) of precipitation. Soils are shallow and often underlain by permafrost. Perennial drought-tolerant, cold-adapted plants and alpine **cushion plants** dominate the vegetation. The short grasses [20–30 cm (8–12 in)] produce a tough turf with 50 percent cover. Dominant grasses are *Festuca ovina*, *Poa alpina*, and *Stipa purpurea*. Perennial sedges (*Carex* spp.) are also found. Cushion plants include *Androsae tapete*, *Arenaria musciformis*, and *Thylacospermum caepitosum*. Small forbs (in the genera *Artemisia*, *Ajuga*, and *Corydalisa*) and dwarf shrubs such as cinquefoils (*Potentilla* spp.) are scattered throughout. At higher elevations, *Kobresia* replaces *Stipa*, and cushion plants are more abundant (Ting-Chen 1993; Schaller 1998). The Tibetan steppe is home to a number of rare ungulates. The chiru or Tibetan antelope (*Pantholops hodgsonii*) is probably more of a goat than an antelope and is classified in the only **genus** restricted to the Tibetan Plateau. The Tibetan argali (*Ovis ammon*) is restricted to the level terrain of western Tibet and is the largest of the world's wild sheep. It has massive curled horns. Males have a shoulder height of 118 cm (46.5 in) and weigh 100 kg (220 lbs). The Tibetan gazelle (*Procapra picticaudata*) and wild yaks (*Bos grunniens*) frequent alpine meadows and steppes. The largest of the wild asses, the kiang

(*Equus kiang*), is most abundant in alpine steppes but also occurs at lower elevations in desert steppe. This handsome animal is 142 cm (56 in) at the shoulder and weighs 250–300 kg (550–660 lbs). Among smaller mammals on the Tibetan Plateau are woolly hares (*Lepus oiostolius*), black-lipped pikas (*Ochotonoa curzoniae*), and marmot (*Marmota himalayana*). Predators include red fox (*Vulpes vulpes*), sand fox (*V. ferrilata*), wolf (*Canis lupus*), and the very rare Tibetan brown bear (*Ursos arctos*). Resident birds are few but include the bird that occurs throughout the temperate grasslands of the northern hemisphere, the Horned Lark (*Eremophila alpestris*), as well as Rufous-necked Snowfinch (*Montifringilla ruficollis*) and Blanford's snowfinch (*M. blanfordi*). Among the summer migrants is the very distinctive Common Hoopoe (*Upupa epops*) (Schaller 1998).

## South American Pampas

Temperate grasslands occupy 700,000 km$^2$ (270,000 mi$^2$) of flat to gently rolling terrain surrounding the Rio de la Plata in central-eastern Argentina, Uruguay, and the state of Rio Grande do Sul in southeastern Brazil. They occur under generally subhumid conditions. Precipitation varies from 400 mm (16 in) a year in the southwestern pampas to 1,600 mm (63 in) in the northeastern areas. Two rainy seasons occur: one in the spring and one in the fall. Summers are dry. Temperatures are relatively mild throughout, with colder winters [125 days with temperatures below 0° C (32° F)] in the west and only 20 days a year with such temperatures in the east. It is not entirely clear why the vegetation of this region is grassland rather than forest. Trees do occur in **gallery forests** lining the banks of rivers. The Brazilian pampas have isolated stands of trees, while in Uruguay, trees grow mostly at the edges of marshes (Soriano 1992).

The soils are mollisols developed on deposits of loess and alluvium composed of loessic silts 300–5,000 m (1,000–16,000 ft) deep and overlying ancient crystalline and sedimentary bedrock. The southern pampas are associated with chernozems; the western pampas with chestnut soils. In the humid eastern pampas, alfisols—more typical of temperate deciduous forests and resulting from podzolization processes rather than calcification—are found. Regional variations in the plant composition of the pampas vegetation occur due to differences in geomorphology, soils, and drainage. Different scientists have categorized the regional types in a variety of ways; only general categories are described here. As elsewhere in the world, grazing and cultivation have severely modified these temperate grasslands, and reconstruction of pristine vegetation is problematic (Soriano 1992).

The pampas grasses included many feathergrasses (*Stipa* spp.); 25 species have been reported from the area as a whole. Other important grasses, represented by several species each, are bluegrasses (*Poa* spp.), *Piptochaetum*, three-awns (*Aristida* spp.), and melics (*Melica* spp.). Perennial forbs and subshrubs include members of genera in the daisy family such as *Baccharis*, *Eupatorium*, and *Hypochoeris*. Legumes are not prominent. One type of pampa is the so-called rolling pampa just south of the Rio de la Plata, where the tallest plants are 50–100 cm (20–40 in) high. The grass community is composed of both sod-forming types and bunchgrasses in several layers. The dominant grass, *Bothriochloa laguriodes*, is a sod-forming grass. *Stipa neesiana* is a bunchgrass about 50 cm (20 in) tall that occurs in a middle layer, while small tuftgrasses commonly named *flechillas* or little darts (e.g., *Piptochaetum mon-*

*tevidense*, *Aristida murina*, and *Stipa papposa*) form the lowest layer. Sedges and small forbs grow among the grasses. Grazing has converted this grassland to two layers: a lower layer about 5 cm (4 in) high and a taller layer of bunchgrasses and small woody plants (Soriano 1992).

The pampa just north of the Rio de la Plata was probably structurally similar to rolling pampa but was dominated by different grasses. The most important grasses were tall grasses (*Paspalum notatum*, *Schizachyrium condensatum*, *Andropogon lateralis*, and *Axonopus compressus*). Also typical were *Leptocoryphium lanatum* and *Tridens brasiliensis*. The northerm pampa has fewer species of *flechillas* than to the south. Characteristic forbs include *Gomphrena celosioides*, *Mitracarpus megapotamicus*, *Euphorbia papillosa*, and *E. selloi*. Legumes are well represented by species in the genera *Adesmia*, *Arachis*, *Desmodium*, *Lupinus*, *Medicago*, *Mimosa*, *Phaseolus*, *Trifoium*, and *Vicia*. Grazing has converted most of the area to short grasses (Soriano 1992).

The pampas seem to have lacked the herds of large native grazing mammals so characteristic of Northern Hemisphere temperate grasslands. The only large herbivore was the pampas deer (*Ozotoceros bezoarticus*), now confined to a few wildlife reserves. Among the more abundant mammals today are plains viscacha (*Lagostomus maximus*), large rodents that live in colonies. They dig elaborate burrow systems that may be occupied by a variety of other animals. Opossums (*Didelphis azarae* and *Lutreolina crassicaudata*) and several armadillos (*Chaetophractus villsosus*, *Chlamyphorus truncatus*, and *Dasypus hybridus*) are also common. Many of the small rats and mice are found only in South America and include South American field mice (*Akodon*, spp. and *Bolomys* spp.), vesper mice (*Calomys* spp.), rice rats (*Orysomys* spp.), burrowing mice (*Oxymycterus* spp.), and water rat (*Scapteromys* spp.). Other strictly South American rodents found on the pampas include cavies or wild guinea pigs (*Cavia* spp.), cui (*Galea* spp.), and the pocket gopherlike tuco-tuco (*Ctenmoys* spp.). The Patagonian hare (*Dolichotis patagonum*), a long-legged rodent, is now rare. The largest carnivores, the jaguar (*Panthera onca*) and puma (*Felis concolor*), have been exterminated. Other predators are rare and include pampas fox (*Pseudalopex gymnocercus*); the ferret-like little grisón (*Galictis cuya*); and two small cats, the pampas cat or *gato pajero* (*Felis colcolo*) and Geoffroys's cat or *gato montés* (*Felis geoffroyi*). Birds are very abundant, as is true for South America as a whole. About 390 species have been recorded from the grasslands, many of which are migrants from North America. Most conspicuous is the flightless Rhea (*Rhea americana*). Several tinamous—partridgelike birds—are also present, as are many doves and parrots. In addition, one finds aquatic birds such as ducks, rails, herons, and ibises. The Campo Flicker (*Colaptes campestris*) is a large ground-feeding woodpecker. Among a number of predatory birds are the Ornate Hawk-eagle (*Spizaetus ornatus*) and Crested Caracara (*Polyborus plancus*). Reptiles and amphibians (frogs, toads and treefrogs) are diverse (Soriano 1992).

## Other South American Temperate Grasslands

A cold-temperate grassland exists in South America in Patagonia, an arid to semiarid region east of the southern Andes between roughly 40° and 52° S. These are the latitudes of prevailing westerly winds, so the Andes create a rainshadow, and the region receives less than 250 mm (10 in) of precipitation a year. Due to the tapering of the South American continent, the surrounding seas prevent temperatures from falling far below 0° C (32° F) in winter. One of the most noticeable aspects of the climate is the strong wind that blows much of the time. Low tussock grasses (*Stipa* spp. and *Festuca* spp.) are characteristic and grow in association with drought-adapted cushion plants from several plant families. The cushion growth-form may be a response to high winds (Walter 1985).

Another cold grassland exists on the western slopes and high altiplanos or basins within the Andes from northern Peru to north-central Chile and east onto the Altiplano of Bolivia at elevations comparable to the Tibetan Plateau (see previous description). This is *puna*, a vegetation composed of small drought-adapted bunchgrasses (Photo 5.3). *Stipa jehu* is prominent. Dwarf shrubs of the daisy family also occur (Eyre 1968). The puna is home to both species of South American camel, the more wide-ranging guanaco (*Lama guanicoe*), and the vicuña (*Vicugna vicugna*) (Photo 5.4), a form restricted to high elevations 3,500–5,750 m (11,000–19,000 ft). Guanacos are believed to be the ancestors of the domesticated llamas and alpacas that have been grazed in high Andean pastures for millennia. Some question just

Photo 5.3 Puna at elevations near 4,000 m (13,000 ft) in the Andes near Copiapo, Chile. (Photo by SLW.)

Photo 5.4 Three guanacos (*Lama guanicoe*) and a single vicuna (*Vicugna vicugna*) (far right) at a seep in the puna zone of the Andes, near Copiapo, Chile. (Photo by Regina Antonia and SLW.)

how natural much of the puna is; it may owe its character and extent to ancient land-management practices aimed at improving pasturage for the domesticated forms (Gade 1999).

### African Veld

Temperate grasslands occur in the eastern part of southern Africa on high plateaus and the eastern slopes of mountains at elevations 1,500–5,000 m (5,000–16,000 ft). These grasslands are often referred to in the literature of Europe and America as the veld, although the term veld actually means vegetation, not grassland. (It would be more correct to refer to these grasslands as highveld, since they are the vegetation of higher elevations.) The veld occurs in a subhumid region of summer rainfall, where annual precipitation is greater than 750 mm (30 in). Summer temperatures are high, so evaporation rates are also high. However, the grassland is apparently maintained by grazing and frequent burning, not directly by the climate. At the lower elevational range of the veld, where winters are relatively mild, a tussock grassland is dominated by warm-season grasses, in particular the nutritious redgrass (*Themeda triandra*). Other common grasses include other members of the Andropogonaceae tribe (e.g., *Cymbopogon plurinodis*, *Diheteropogon filifolius*, and *Heteropogon contortus*) and panic grasses (e.g., *Brachiaria senata*, *Digitaria eriantha* and *Setaria flabellata*). Mismanagement by burning at the wrong time of year has changed the composition in many areas to a dominance of wiregrass (*Aristida junciformis*) and blousaadgrass (*Eragrostis* spp.). A number of forbs, typically of the daisy family, such as *Helichrysum* spp. and groundsels (*Senecio* spp.), are found. Cool-season grasses become more abundant with increasing elevation; these include bromes (*Bromus*), Junegrass (*Koeleria*), and *Helictrichon*. At the highest elevations, where winter temperatures are low, the grasses are short [less than 50 cm (20 in)] and dense (Tainton and Walker 1993; O'Connor and Bredenkamp 1997; Eyre 1968).

## ORIGINS

Grasses first appeared in the Eocene Epoch, about 55 million years ago. Their subsequent development and spread is often viewed as the stimulus for the appearance and success of the ungulates: the odd-toed perissodactyls such as members of horse family; and the even-toed artiodactyls, including pigs, camels, and antelopes and other members of the cattle family. The North American shortgrass prairie came into existence in the Oligocene Epoch after a general cooling of the global climate and after the continued uplift of the Rocky Mountains (about 30 million years ago) created a rainshadow to their east. At that time, drought-tolerant grasses and forbs probably replaced a temperate forest. Evidence suggests these plants had both temperate and tropical origins (Lauenroth and Milchunas 1992). The Palouse Prairie formed in response to the rise of the Cascades in the Pliocene Epoch. Plants of the region that were drought tolerant persisted as rainshadow conditions developed, and they were joined by plants moving in from the Great Basin (Kucera 1992). Both midcontinental and western grasslands, however, were disrupted by the repeated continental and mountain glaciations of the Pleistocene Epoch and reconstituted for the last time less than 10,000 years ago. In the midcontinent, a continental ice sheet covered the land at least once as far south as northern Missouri. South of the ice, glacial North America was characterized by an open spruce forest inhabited by a variety of large grazing and browsing mammals such as mammoths, which are now extinct. When the ice retreated, glacial till coated the ground ice had once covered, and glacial outwash and loess were found elsewhere. So, contemporary soils and vegetation both date to post-Pleistocene time. The tallgrass prairie and particularly its eastern extension may be younger still. The existence of this subhumid grassland appears to be related not to drying and warming of the region but to frequent fires—both natural and human set—and perhaps the grazing of vast herds of bison. Charcoal evidence shows fire as a common feature of the environment since at least 9,000 years ago (Kucera 1992). Suppression of prairie fires since European settlement has led to the encroachment of trees, which previously had been confined to fire-safe ravines and along stream banks. The composition of the native prairies is still changing. Tall $C_4$ grasses are moving westward and replacing shortgrasses in a migration that had begun at least by A.D. 500 (Brown 1993).

The plants and some animals of Eurasian steppes are closely related to those of the North American prairies, which suggests a common origin and history. Pleistocene vegetation patterns differed greatly, however. In North America, forested **landscapes** replaced early grasslands; in Eurasia, grass-dominated landscapes expanded. The actual groupings of plants would have been different from today, however; so it is safe to say that contemporary steppes are also Recent in terms of plant composition. Russian plant scientists believe burrowing animals play a major role in the nature of the modern steppe and give little credit to changes by fire or grazing of large mammals.

The Tibetan Plateau may be an exception to the rule of a post-Pleistocene origin for a natural grassland. The plateau was not covered by ice during the Pleistocene, although it was affected by materials washed out of and blown off mountain glaciers that extended onto the margins of the plateau (Schaller 1998). Perhaps its greater antiquity accounts, at least in part, for its greater diversity of large mammals compared to other temperate grasslands.

The origins of the South American pampas are still debated. They seem not to reflect regional climate, particularly in the subhumid areas. The use of fire by aboriginal hunting people is a possible explanation but one without widespread acceptance (Soriano 1992). As previously mentioned, some think the puna has probably expanded under long human use (Gade 1999). Similarly, the highveld of southern Africa is a grassland for unknown reasons. As in the tallgrass prairies of North America, fire is required today to prevent the invasion of woody plants. Human hunters have occupied the area for at least 250,000 years and presumably have used fire as a tool for driving game and improving pasturage that long (Eyre 1968).

## HUMAN IMPACTS

The history of human impacts in all temperate grasslands follows a common theme. Fires set by early hunting and gathering peoples probably influenced the composition and distribution of grasslands for thousands of years. The next impact was that of domesticated livestock, in particular cattle, which through grazing and trampling selected more resistant grasses and forbs. Often this meant a decline in tall grasses in favor of shorter ones. The timing of the introduction of grazing livestock varied according to continent. Pastoralism, developed perhaps 10,000 years ago in the steppes of Eurasia, became significant in the North American grasslands only in the mid-1800s, and did not begin in the pampas of South America until the twentieth century. Domestic livestock resulted in changes in the animal life as well, as cattlemen eliminated large, potentially competing wild grazing animals and larger carnivores. Overgrazing is associated with desertification and the invasion of shrubs. Eurasian grasses and forbs have invaded the Americas as a consequence of opportunistic hitchhiking and the deliberate planting of improved pastures. Crop cultivation in temperate grasslands awaited the development of steel plows that could cut through the heavy sod and make the naturally fertile soils beneath accessible. Dry farming and irrigated farming then consumed vast areas of grassland. These practices, poorly managed, contributed to the erosion of the topsoil (e.g., the Dust Bowl in the southern mixed prairie of the 1930s) and increased concentration of salts in the root zone (salinization). Today, throughout the world, only small patches of relict grassland survive, and, for the most part, the nature of grasslands before agriculture can only be guessed.

In North America, the tallgrass prairie is mostly gone as a consequence of agriculture, as it became an early rangeland and then the breadbasket of the United States and Canada. Grazing, in combination with droughts, contributed to the demise of tallgrasses. They were replaced by cool-season Eurasian grasses such as the Kentucky bluegrass (*Poa pratensis*), redtop (*Agrostis alba*), and tall fescue (*Festuca arundinaceae*). Settlement in the northern mixed prairie began in the 1870s and 1880s, after which time prairie fires were controlled. Aspen (*Populus tremuloides*) has invaded and formed dense woodlands over large areas as a result. Grazing of the rangelands led to the decline of the mid-sized grasses in favor of shorter, more drought-resistant, shallow-rooted species. Crop production today is small grains (spring wheat in the north; winter wheat in the south) and oil seeds (sunflower and canola). The shortgrass prairie is both rangeland and irrigated cropland. Irrigated alfalfa, maize, sugar beet, and cotton are produced. Wheat is grown with dry-farming techniques. The shortgrass prairie has expanded northward from Colorado as a result of grazing (Coupland 1992a; Kucera 1992; Lauenroth and Milchunas 1992).

The native perennial grasses of the Palouse Prairie were unable to withstand heavy grazing and trampling by livestock and have been replaced by introduced annual grasses, such as cheatgrass (*Bromus tectorum*). Game birds such as Chukar (*Alectoris graeca*), Gray Partridge (*Perdix perdix*), and Ring-necked Pheasant (*Phaisanus colcicus*) also have been successfully introduced. In the Palouse, winter wheat is grown today in rotation with peas or lentil (Daubenmire 1992).

Native California perennial grasses also were unable to withstand heavy grazing pressures. They have been replaced largely by Mediterranean annual grasses and forbs, most notable wild oats (*Avena* spp.), brome grasses (*Bromus* spp.), and filaree (*Erodium* spp.). Cattle were first introduced from Mexico at the beginning of the Spanish mission period in 1769. Some of the weeds of the Mediterranean preceded them; their presence is preserved in the adobe bricks of early settler homes. Livestock production increased with the discovery of gold in California in the 1840s and reached its peak between 1850 and 1870. California suffered a severe drought in the midst of this period (1862–64), and the two factors led to a permanent change in the composition of the prairie. Today, much of this area is under irrigation agriculture (Heady et al. 1992).

The Spanish introduced cattle into the desert grasslands of Mexico and the American Southwest in the sixteenth century. Numbers grew rapidly, although Apache raids held them in check in what now is the United States. Major expansion of cattle raising occurred in the late nineteenth century after the Apache were conquered. The 1880s and 1890s saw large numbers of cattle produced on the grasslands of southern Arizona and New Mexico and driven north to railheads for shipment east. This same period was also a time of drought that culminated in a period of arroyo cutting—the downcutting of rivers and streams that led to a lowering of the water table. Experts still do not agree on the cause-and-effect relationships of overgrazing, drought, and arroyo cutting, but an obvious response in the vegetation has been the invasion of shrubs such as mesquite (*Prosopis juliflora*) and prickly pear and other cactus onto the desert grasslands.

The pampas were home to nomadic hunters and gatherers who probably used fire to drive game and increase grasses. In 1527, Spanish introduced horses to the region, where they multiplied with legendary rapidity. By 1585, some 80,000 horses lived in the vicinity of Buenos Aires. Cattle were introduced in 1573, and they, too, increased quickly. By 1609, the first of the famous Argentine cattle hunts was organized to harvest hides for export to Europe. Fire was used to herd both horses and cattle. In the late nineteenth and early twentieth centuries, a meatpacking industry developed that shipped chilled beef and mutton to Europe. A new wave of European immigrants, the construction of railroads, and use of fencing allowed the expansion and intensification of sheep and cattle production on the pampas at that time. The moister eastern grasslands were used to produce calves that were then fattened for market on the drier pampas to the west. The sod was broken to cultivate alfalfa, barley, and oats for livestock feed. Wheat, maize, and most recently soybeans have been grown on the pampas and added to the exports to a world market in the twentieth century (Soriano 1992).

In the Old World, the history of human impacts is more elusive because pastoral societies have used those lands for a much longer time. Essentially, nothing prepastoral exists to compare with contemporary grasslands. More recent changes involve plowing the steppe to convert it to cropland. In the steppes of the Central Chernozem region of Russia, the main crops cultivated are wheat, sugar beet, and buckwheat. The area, which may be closest to pristine is in Mongolia, where burning and mowing are rare and grazing is moderate (Lavrenko and Karamysheva 1993).

Nomadic sheepherding was the dominant economy over much of the Chinese steppes until 1950, when the national government mandated that sedentary agriculture—livestock husbandry and crop cultivation—would replace it. Removal of trees was also part of a plan to increase food production for the growing Chinese population. These practices had disastrous results in terms of increased soil erosion, dust storms, gullying, desertification, and salinization. Government policies changed in 1985, and remedial practices were initiated (Ting-Chen 1993). Most noticeable is the widespread planting of trees.

## HISTORY OF SCIENTIFIC EXPLORATION

Science came to the temperate grasslands after grazing had altered them. For example, the first botanists collecting in California, David Douglas (1831–32) and Karl Hartweg (1846–47), arrived 50 years after irreversible changes in the plant life had occurred (Heady et al. 1992). In the United States and Canada, several researchers have classified and mapped grassland communities and studied community dynamics beginning with Clements's (1920) work on **succession**. Important contributors included Carpenter (1940), who defined the biome but limited it to the midcontinent grasslands; Weaver (1954), who produced a descriptive classic that is still a basic reference; Coupland (1961), who classified Canadian prairie types; and Stewart (1956) and Sauer (1975), both of whom emphasized the role of fire in prairie **ecology**.

Considerable Russian literature addresses grassland ecology and classification, dating from the late nineteenth century into the twentieth—before plows destroyed much of the steppe (beginning in the 1940s)—and a growing body of Chinese literature deals with grasslands. Most scientific exploration and research has been, and continues to be, motivated by practical considerations of the grasslands as cattle range.

## REFERENCES


Acton, D. F. 1992. "Grassland soils." Pp. 25–54 in R. T. Coupland, ed., *Natural Grasslands: Introduction and Western Hemisphere*. Ecosystems of the World, 8A. Amsterdam: Elsevier.

Brown, Dwight A. 1993. "Early Nineteenth Century grasslands of the midcontinent plains," *Annals of the Association of American Geographers* 83(4): 589–612.

Carpenter, J. R. 1940. "The grassland biome." *Ecological Monographs* 10: 617–684.

Clements, F. E. 1920. *Plant Indicators: The Relation of Plant Community to Process and Practice*. Washington, DC: Carnegie Institute of Washington.

Coupland, R. T. 1961. "A reconsideration of grassland classification in the northern Great Plains of North America." *Journal of Ecology* 49: 135–167.

Coupland, R. T. 1992a. "Mixed prairie." Pp. 150–182 in R. T. Coupland, ed., *Natural Grasslands: Introduction and Western Hemisphere*. Ecosystems of the World, 8A. Amsterdam: Elsevier.

Coupland, R. T. 1992b. "Overview of South American grasslands." Pp 363–366 182 in R. T. Coupland, ed., *Natural Grasslands: Introduction and Western Hemisphere*. Ecosystems of the World, 8A. Amsterdam: Elsevier.

Daubenmire, Rexford. 1992. "Palouse prairie." Pp. 297–312 in R. T. Coupland, ed., *Natural Grasslands: Introduction and Western Hemisphere*. Ecosystems of the World, 8A. Amsterdam: Elsevier.

Eyre, S. R. 1968. *Vegetation and Soils: A World Picture*. Revised 2nd edition. Chicago: Aldine Publishing Company.

Furley, Peter A. and Walter W. Newey. 1983. *Geography of the Biosphere: An Introduction to the Nature, Distribution and Evolution of the World's Life Zones*. London: Buttersworth.

Gade, Daniel W. 1999. *Nature and Culture in the Andes*. Madison: University of Wisconsin Press.

Heady, H. F., J. W. Bartolome, M. D. Pitt, G. D. Savelle, and M. C. Stroud. 1992. "California prairie." Pp. 313–335 in R. T. Coupland, ed., *Natural Grasslands: Introduction and Western Hemisphere*. Ecosystems of the World, 8A. Amsterdam: Elsevier.

Kucera, C. L. 1992. "Tall-grass prairie." Pp. 227–268 in R. T. Coupland, ed., *Natural Grasslands: Introduction and Western Hemisphere*. Ecosystems of the World, 8A. Amsterdam: Elsevier.

Lauenroth, W. K. and D. G. Milchunas. 1992. "Short-grass steppe." Pp. 183–226 in R. T. Coupland, ed., *Natural Grasslands: Introduction and Western Hemisphere*. Ecosystems of the World, 8A. Amsterdam: Elsevier.

Lavrenko, E. M. and Z. V. Karamysheva. 1993. "Steppes of the former Soviet Union and Mongolia." Pp. 3–59 in R. T. Coupland, ed., *Natural Grasslands: Eastern Hemisphere and Résumé.* Ecosystems of the World, 8B. Amsterdam: Elsevier.

Lowe, Charles H., ed., 1964. *The Vertebrates of Arizona.* Tucson: The University of Arizona Press.

O'Connor, T. G. and G. J. Bredenkamp. 1997. "Grassland." Pp. 215–257 in R. M. Cowling, D. M. Richardson, and J. M. Pierce, eds., *Vegetation of Southern Africa.* Cambridge: Cambridge University Press.

Sauer, C. O. 1975. "Man's dominance by use of fire." Pp. 1–13 in R. H. Kesel, ed., *Geoscience and Man 10.* Baton Rouge: Louisiana State University.

Schaller, George B. 1998. *Wildlife of the Tibetan Steppe.* Chicago: University of Chicago Press.

Schmutz, E. M., E. L. Smith, P. R. Ogden, M. L. Cox, J. O. Klemmedson, J. J. Norris, and L. C. Fierro. 1992. "Desert grassland" Pp. 337–362 in R. T. Coupland, ed., *Natural Grasslands: Introduction and Western Hemisphere.* Ecosystems of the World, 8A. Amsterdam: Elsevier.

Soriano, Alberto. 1992. "Rio de la Plata grasslands." Pp. 367–407 in R. T. Coupland, ed., *Natural Grasslands: Introduction and Western Hemisphere.* Ecosystems of the World, 8A. Amsterdam: Elsevier.

Stewart, O. C. 1956. "Fire as the first great force employed by man." Pp. 115–133 in W. L. Thomas, ed., *Man's Role in Changing the Face of the Earth.* Chicago: University of Chicago Press.

Tainton, N. M. and B. H. Walker. 1993. "Grasslands of Southern Africa." Pp. 265–290 in R. T. Coupland, ed., *Natural Grasslands: Eastern Hemisphere and Résumé.* Ecosystems of the World, 8B. Amsterdam: Elsevier.

Ting-Chen, Zhu. 1993. "Grasslands of China." Pp. 61–82 in R. T. Coupland, ed., *Natural Grasslands: Eastern Hemisphere and Résumé.* Ecosystems of the World, 8B. Amsterdam: Elsevier.

Vankat, John L. 1979. *The Natural Vegetation of North America, an Introduction.* New York: John Wiley & Sons.

Walter, Heinrich. 1985. *Vegetation of the Earth and Ecological Systems of the Geo-biosphere.* 3rd edition. Berlin: Springer-Verlag.

Weaver, J. E. 1954. *North American Prairie.* Lincoln, NE: Johnsen.

# 6 MEDITERRANEAN WOODLAND AND SCRUB

## OVERVIEW

A distinctive woodland and **scrub vegetation** associated with the dry summer mediterranean **climate** type occurs on the west coasts of continents between 30° and 40° **latitude** and in areas surrounding the Mediterranean Sea. Taking its name from the area where it covers the greatest area, the mediterranean scrub biome occurs in five widely separated regions on six continents: in southern California; central Chile; the Cape Province of the Republic of South Africa; southern and western Australia; and around the Mediterranean Sea in Europe, Southwest Asia, and northern Africa. Typically, shrubs in mediterranean regions display adaptations to drought, nutrient-poor **soils**, and fire. Most **broad-leaved** plants, such as live oaks and **heaths**, are **evergreen** and have hard or leathery leaves. Natural and human-induced **disturbances**—most commonly frequent burning and livestock grazing—have strongly influenced the development of vegetation in all regions of the biome. Mediterranean **ecosystems** are home to about 20 percent of the world's plants (Ricketts et al. 1999). A high proportion of **endemic species** and genera—forms that originated in the area and today are found only there—is a hallmark of mediterranean scrub wherever it occurs, but especially in the Western Cape of South Africa. All five mediterranean regions are included among the world's 25 hotspots of **biodiversity** and endangerment—places that have both exceptionally high numbers of species, many found nowhere else on earth, and immediate threats to their survival from human activity (Mittermeier et al. 1999).

Most regions of mediterranean scrub have at least two distinct shrubland types, one interior and the other coastal. At higher elevations and higher latitudes, as total yearly rainfall amounts increase, scrub may give way to woodland and forest. Since a summer dry period still occurs, scientists classify at least some of these woodlands and forests as parts of the mediterranean biome. However, when the trees do not exhibit the typical adaptations to drought possessed by mediterranean shrubs, most

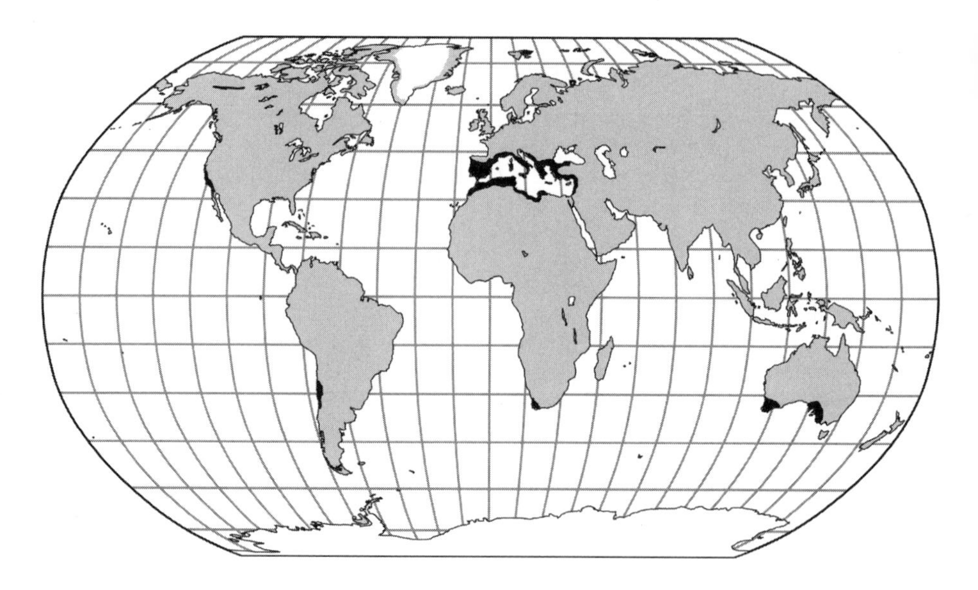

Map 6.1 World distribution of the Mediterranean Woodland and Scrub biome.

believe the forests should be excluded from the biome. In this chapter, such woodlands and forests will be identified, but full descriptions will not be provided.

The North American, South American, and European areas covered by mediterranean scrub are geologically young and have high mountains that offer a diversity of **microhabitats** based on different combinations of **exposure**, elevation, and type of bedrock. In these regions, a complex mosaic of plant communities reflects the topographic variation. In contrast, mediterranean vegetation in South Africa and Australia has developed on ancient, stable surfaces with lesser differences in elevation. Some scientists argue that the long stability of the surface and the long isolation of these regions from other areas of humid subtropical vegetation, account for the exceptionally high number of species in general and endemic species in particular in these two parts of the biome.

Mediterranean scrub vegetation is known by a variety of local names: chaparral and coastal sage in California; matorral and espinal in Chile; tomillares in Spain; maquis and garrigue in France; macchia in Italy; phyrgana in Greece; bath'a and goresh in Israel; fynbos and strandveld in South Africa; and kwongan and mallee in Australia (Conacher and Sala 1998; Dallman 1998).

## CLIMATE

The Mediterranean Climate (Koeppen's Cs) is unique among the global climate types in that the rainy season coincides with the cool or winter period—usually the nongrowing season for plants—and summers are dry. Average total annual precipitation ranges between 380–1,000 mm (15–40 in) per year. Even though the climate is considered humid, the unusual timing of the dry season imposes stresses on plant and animal life similar to those of deserts for at least a few months of the year.

This is a subtropical climate: summers can be hot; winters are cool, but freezing temperatures are rare. In general, temperatures are moderated in winter by the

Photo 6.1 Mediterranean scrub near Los Angeles, California. Fire breaks are cleared along crest lines. (Photo by SLW.)

proximity of the sea, which remains warmer than the continental landmasses, and in summer by fogs generated by cool ocean currents just offshore. The combination of annual precipitation and temperature patterns results in a limited but predictable period when both soil moisture and temperatures are sufficient for active plant growth. Plants and animals must be able to withstand these seasonal rhythms to survive in the mediterranean scrub biome.

## VEGETATION

Today, the natural vegetation of most mediterranean regions is evergreen scrub (Photo 6.1). That scrub represents solely or even primarily a response to the mediterranean climate is questionable. Many scientists believe forests, woodlands, and **savannas** covered most areas before human land-use practices altered the **landscape** and favored those shrub species that, in addition to summer drought, tolerated thin, nutrient-poor soils, frequent burning, and—in some regions—grazing and browsing livestock.

The evergreen **habit** of most woody species is a possible adaptation to the short growing season. When plants maintain foliage throughout the year, they avoid losing precious time producing new leaves at the beginning of the brief growing season. However, it is necessary to protect those leaves from dehydration once summer drought arrives. Among the plants of the mediterranean biome, common adaptations to reduce **evapotranspiration** and prevent dehydration include leathery leaves with waxy cuticles and sunken **stomata**—the pores through which leaf tissue exchanges gases, including water vapor, with the atmosphere. The leaves of some plants assume an upright or vertical orientation to avoid direct exposure of leaf tissue to the damaging heat of the sun. Others have a silvery color and/or hairy leaves to reflect sunlight and reduce leaf surface temperatures and evaporation rates.

A reduction in leaf surface area—resulting in small, sometimes needle-shaped leaves—is yet another way some broad-leaved species reduce water loss. Some plants are drought-dimorphic, that is, they have two types of leaves: the larger, softer leaves of winter and spring are replaced by smaller, harder leaves in summer. Fewer species are drought-**deciduous**, which means they shed their leaves during the dry summer months.

To capture as much soil moisture as possible, mediterranean shrubs characteristically have both long tap roots to obtain water stored at depth and an extensive network of roots near the surface of the ground to intercept rainfall as soon as it percolates into the soil. Water storage tissues, such as bulbs and tubers or **succulent** leaves and stems, are frequent among nonwoody **perennial** plants. In addition, a large number of plants—**annuals** and **ephemerals**—complete their life cycles in a single growing season or less. They survive prolonged drought as seeds buried in the litter or soil.

To cope with low-nutrient soil conditions, many plants are associated with fungi that grow around or in their roots. The fungi send long, thin filaments into the soil to extract important minerals not otherwise available to the plants. The fungi convert these minerals into compounds of nitrogen, phosphorus, and potassium that plants can use and exchange them with the plants' roots for carbon compounds produced in **photosynthesis**. This association between plant and fungus is called a mycorrhiza, and the fungi involved are called mycorrhizal fungi. Mycorrhizae are not limited to the mediterranean scrub biome, but they are extremely important for the survival of many species that compose mediterranean scrub. In Australia and Africa, species of the protea family (Proteaceae) have so-called proteoid roots that accomplish the same thing as mycorrhizal fungi. Proteoid roots are fine, cottony rootlets that plants form near the ground surface at the beginning of the rainy season to extract nutrients leached from the decaying litter. These rootlets wither and die after a few months. Similar structures appear on some grasses, sedges, and restios (reedlike plants of the family Restionidaceae).

Fire is an extremely important part of natural and human-dominated disturbance in the mediterranean scrub biome and eliminates those species that cannot withstand frequent, regular burning. However, some plants have ways to survive fire, and some species even require burning to regenerate. Although surface tissues may be destroyed in a burn, many mediterranean shrubs—the so-called sprouters—are able to resprout from root crowns, thickened woody structures at the base of the trunk. Chamise (*Adenostoma fasciculatum*), the characteristic shrub of the southern California chaparral, is a prime example of a sprouter (Photo 6.2). Other plants—the seeders—can recolonize a burned site rapidly since they produce large quantities of seeds that remain viable in the soil or litter for years, even decades, until a fire stimulates their germination. In the Northern Hemisphere regions of the biome, closed-cone pines, such as the Monterey pine (*Pinus radiata*) native to California, maintain unopened cones on their branches until a fire occurs. The heat of the fire causes the cones to open, and the seeds are released to fall onto ground newly enriched by a layer of ash. In the Southern Hemisphere, the woody seed capsules of some plants in South Africa and Australia work the same way.

Photo 6.2 Chamise (*Adenostoma fasciculatum*) sprouting from its rootcrown shortly after a burn near Malibu, California. (Photo by SLW.)

**Geophytes**, those perennial plants with renewal buds well below ground, are particularly abundant and diverse in the mediterranean scrub biome because the bulbs, corms, or **rhizomes** from which they regenerate are insulated by the soil during an average wildfire. They often produce a colorful wildflower display in the nutrient-enhanced, post-burn **habitat** the spring after a fire. In California, brodiaeas (*Brodiaea* spp.) and mariposa lilies (*Calochortus* spp.) are good examples of this type of plant.

Many mediterranean shrubs also are aromatic. (Culinary and perfumery herbs such as oregano, thyme, sage, and lavender have their origins in the Mediterranean Basin formation.) Aromatic oils in the leaves may offer several advantages. They make the foliage unpalatable to many herbivores and thus reduce predation. They are flammable and may promote the fire that gives many species a competitive edge over those less well adapted to survive fire. And evidence shows that during the heat of the day, these oils evaporate and fill the stomata; they thus may help reduce the loss of water vapor through the leaf pores (Dallman 1998).

## ANIMAL LIFE

No exclusively "mediterranean" adaptations appear among the animals of this biome other than the ability to tolerate or avoid drought and heat. Since animals are mobile, many can simply go to a water source—a spring or permanent stream—every day and drink. They can avoid the hottest part of the day by burrowing below ground where it is cooler and being active only at night. A few animals in mediterranean regions have physiological adaptations to withstand drought comparable to those found among desert animals. Some may enter a state of torpor or dormancy (i.e., aestivation) during the summer; others migrate seasonally to higher elevations or other latitudes. A few species are able to survive on the water they can extract from their food and require no drinking water at all.

The geographic isolation of each of the world's mediterranean regions by oceans, mountains, and deserts has favored the evolution of a number of animal species restricted to single sections of this biome, just as it has done for plants.

## SOILS

No one type of soil is characteristic of the mediterranean climate. Instead, the soils tend to reflect **parent material**. Most are deficient in nitrogen and phosphorus, important plant nutrients, but some soils may be rich in calcium, magnesium, sodium, and iron. In many instances, water infiltration rates are low due to either water repellant properties or compaction and stoniness. One of the more fertile types of soil commonly associated with this biome is known as *terra rossa* (red earth). Terra rossa is not a zonal soil because it does not reflect regional climatic conditions. It is a fine-grained, red material rich in calcium and iron and usually associated with limestone bedrock. Scientists do not agree on the origin of this material. Some think it might have been produced during a former period of tropical climate; others think it might have originated outside of mediterranean areas, and prevailing winds blew it to its present locations.

## MAJOR REGIONAL EXPRESSIONS

### The Mediterranean Basin: Maquis and Garrigue

In this discussion, the French terms *maquis* and *garrigue* will be used for the two main types of scrub vegetation found in the Mediterranean Basin, regardless of the country in which it is growing. The Mediterranean Basin serves as the model against which other regions in the biome are compared, and it is where the biome reaches its maximum extent—covering three times the area of the other four regions combined. The Mediterranean's scrub vegetation is not continuous, however, but fragmented into several sections separated from each other by mountains and seas. Maquis and garrigue usually occur at elevations below 1,000 m (3,000 ft). In Europe, they are found on the Iberian peninsula, in southeastern France, on the Italian and Balkan peninsulas, and on the islands of the Mediterranean. At the eastern end of the Mediterranean Sea, in Asia, mediterranean scrub occurs in Turkey, Lebanon, and Israel; in North Africa the biome is confined to the Maghreb, that is, the northern parts of Morocco, Algeria, and Tunisia. Many genera, species, and subspecies are restricted to only a part of this large region. The actual species present at any given area in the northern Basin varies from east to west, with the most species-rich region being near the center, in the Balkans.

Maquis typically consists of tall shrubs and short trees of varying **canopy** heights. Hard-leaved, evergreen tree species include Kermes oak (*Quercus coccifera*), holm oak (*Q. ilex*), strawberry tree (*Arbutus unedo*), mastic tree or lentisk (*Pistacia lentiscus*), carob (*Ceratonia siliqua*), and wild olive (*Olea europa*). The **needle-leaved** Aleppo pine (*Pinus halepensis*) and junipers (*Juniperus* spp.) are also widely distributed. Tall shrubs include several species of rock rose (*Cistus* spp.) and oleander

(*Nerium oleander*). These shrubs may attain heights of 4–6 m (12–20 ft) to form a dense evergreen woodland unlike any found in other mediterranean regions (Blondel and Aronson 1999).

The maquis is home to many bulbs, annuals, and ephemerals. The large number of annuals and ephemerals may be the product of millennia of interaction with agricultural and pastoral systems. Plows and grazing regularly create short-lived open patches of ground that these plants evolved to take advantage of. Their wind-blown seeds let them easily and quickly invade such sites. However, since they thrive and their seeds germinate only in full sun, the species are forced to disperse to new open sites to survive once mature plants shade the clearing. Many Mediterranean annuals have become highly successful weeds when transported to other parts of the world.

Garrigue, the other common scrub type of the Mediterranean Basin, consists of low shrubs and is characteristic of hotter, drier habitats than those supporting maquis. Garrigue is usually associated with hard limestone bedrock. Typical shrubs are spiny or aromatic. Many of the aromatic shrubs are members of the mint family, among them rosemary (*Rosmarinus officinalis*), lavender (*Lavandula* spp.), thyme (*Thymus vulgaris*), and sage (*Salvia* spp.). Drought-deciduous and drought-dimorphic species also are common in garrigue. The diversity of geophytes is second only to the Western Cape of South Africa. Many species of tulip, narcissus, iris, crocus, and cyclamen are native, as are many terrestrial orchids. This low-shrub vegetation is referred to as *phyrgana* in Greece and *bath'a* in Israel; over much of its range it is considered to be a degraded form of maquis, the result of human misuse of the land (Dallman 1998; Blondel and Aronson 1999).

Large mammals inhabiting this region include herbivores such as roe deer (*Capreolus capreolus*), fallow deer (*Dama dama*), ibex (*Capra hispanica*), and European wild boar (*Sus scrofa*); carnivores such as wolf (*Canis lupus*) and Spanish lynx (*Lynx pardina*) are also native to the region. One primate, the Barbary macaque (*Macaca sykvanus*), can be found in southern Spain and North Africa. Smaller mammals include hedgehogs, moles, and shrews. Spiny mice (*Acomys* spp.) and voles (*Pitymys* spp.) are among the endemic species. The world's smallest mammal, the Etruscan shrew (*Suncus etruscus*), which weighs 1.6–2.4 g (0.05–0.08 oz), is also an endemic (Blondel and Aronson 1999).

More reptiles are represented than amphibians, which is not surprising since reptiles by their very nature are preadapted to drought. Various snakes, lizards, tortoises, and two chameleons are associated with mediterranean scrub in the Basin.

Most resident bird species are widely distributed through the region, unlike many other animal (and plant) species in the Mediterranean Basin; therefore, a relatively low level of diversity of birds and relatively few endemic species occur. However, differentiation into subspecies has occurred in many species, for example, in the jay (*Garrulus glandarius*) and the blue tit (*Parus caeruleus*). The most impressive aspect of birdlife is the large number of nonresident species present much of the year. The lands surrounding the Mediterranean Sea are major stopover and staging sites for vast numbers of migrating birds. An estimated 230–250 Eurasian species (thrushes, wrens, kinglets, etc.) spend their nonbreeding seasons in the Mediterranean Basin, which makes the number of species there higher in winter than in summer. Another 130 species pass through the region twice a year on their migrations between Eurasian breeding grounds and sub-

Saharan wintering grounds. One scientist estimated that some 5 billion birds cross the Mediterranean Sea in spring and again in fall. The bird-eating Eleonora's falcon (*Falco eleonora*) breeds on cliffs along the Mediterranean Sea in late summer—a rare breeding time for birds—when it can rely on the seasonal abundance of migrating birds to feed its young (Blondel and Aronson 1999).

## Western North America: Chaparral and Coastal Sage

Mediterranean-type vegetation occurs along the west coast of North America from San Francisco (37° 45′ N) to northern Baja California, Mexico (31° N). It is more or less restricted to coastal areas by the mountain ranges that parallel the coastline. It has two distinct plant associations: chaparral (from the Spanish *chapa*, scrub oak) and coastal sage.

Chaparral has a fairly rich variety of hard-leaved evergreen shrubs that grow in a mosaic of communities reflecting recent fire history (Photo 6.3). A 12- to 20-year fire cycle perpetuates stands chamise (*Adenostoma fasciculatum*). This shrub has small, needlelike leaves and contains highly flammable oils. Chamise quickly resprouts from root crowns after a fire and tolerates dry, low-nutrient soil conditions. It is one of the most widespread species in the Californian mediterranean formation and can be considered an indicator species for the biome. Chamise frequently occurs in nearly pure stands with a very limited herb or ground layer.

Where fire is less frequent, shrub species diversity becomes much higher. California lilac (*Ceanothus* spp.), mountain mahogany (*Cercocarpus* spp.), sumac (*Rhus ovata*), scrub oak (*Quercus dumosa*), toyon (*Heteromeles arbutifolia*), and manzanita (*Arctostaphylos* spp.) develop a canopy of more or less uniform height [2.5 m–4 m (8–12 ft), depending on site conditions]. Most of these species regenerate from seed rather than by resprouting. Closed-cone pines such as knobcone pine (*Pinus attenuata*) also occur in this association. For a few years immediately after a fire, bulbs and annuals may carpet the ground with wildflowers, among them brodiaeas (*Brodiaea* spp.), larkspurs (*Delphinium* spp.), poppy (*Rhomneys coulteri*), and mariposa lily (*Calochortus* spp.). However, as the shrub component is reestablished, the number of nonwoody species declines. On more humid, north-facing slopes, a forest of the chaparral shrub and tree species just mentioned, along with live (hard-leaved, evergreen) oaks (*Quercus* spp.), California bay or laurel

Photo 6.3 Chaparral in Tuna Canyon near Los Angeles, California. The sign reads "Positively no smoking, no fires." (Photo by SLW.)

Photo 6.4 Elfin forest of scrub oaks (*Quercus* sp.) in area protected from fire and grazing on Santa Cruz Island, California. (Photo by SLW.)

Photo 6.5 Oak savanna. This may have been the characteristic vegetation of mediterranean California before frequent fires set by aboriginal peoples expanded the distribution area of chaparral. (Photo by SLW.)

(*Umbellularia californica*), Pacific madrone (*Arbutus menzeisii*), and chinquipin (*Castanopsis chrysophylla*) will develop.

Where chaparral fires (and grazing and browsing) have been prevented for 50 years or more, such as on parts of Catalina and Santa Cruz islands (in the Channel Islands), an elfin forest of live oaks has developed (Photo 6.4). Some believe that with even more prolonged suppression of fire, an oak savanna—perhaps the climatically determined natural vegetation—would occur. These scientists tend to believe that oak savanna (Photo 6.5) would cover much of inland mediterranean North

Photo 6.6 Coastal sage community along fog-shrouded coast near San Diego, California. (Photo by SLW.)

America if frequent fire did not prevent its establishment. On the mainland, several oak tree species—some evergreen and some deciduous—do occur in open grassland situations.

The presence of soft-leaved low shrubs indicate the presence of coastal sage. Strongly influenced by the high humidity generated from sea fog formed over the cool California Current, coastal sage rarely extends more than a few miles inland (Photo 6.6). It tends to have an **open canopy** that permits an abundance of **forbs** and grasses in the ground layer. Soft-leaved species are less tolerant of drought than the hard-leaved species found inland, so many coastal sage species are drought dimorphic or drought deciduous (Dallman 1998). Dominant species include California sagebrush (*Artemesia californica*), black sage (*Salvia mellifera*), and purple sage (*S. apiana*).

Coastal mediterranean California is home to several ecological islands with endemic **conifer** species of extremely restricted range. One such species is Monterey pine (*Pinus radiata*), a closed-cone pine found naturally at only three sites that together cover 11,000 acres. Ironically, this species is widespread as an **introduced**, fast-growing timber crop species elsewhere in the world and in many places is considered a scourge that is leading to the demise of native trees. Other examples of endemic conifers in coastal California include Torrey pine (*P. torreyana*), Monterey cypress (*Cupressus macrocarpus*), and Gowen cypress (*C. goveniana*).

Some authors include the coastal redwood (*Sequoia sempervirens*) forests as part of the mediterranean biome. These and other giant conifers such as Douglas fir (*Pseudotsuga menzeisii*) and western hemlock (*Tsuga heterophylla*) grow where the uplift of air masses coming off the Pacific Ocean, caused by the presence of the coastal ranges, produces high amounts of rainfall. Along with nearly daily fogs, the precipitation completely eliminates summer drought. These huge needle-leaved trees are accompanied by an understory of heaths (family Ericaceae): rhododendrons, azaleas, huckleberries, and the like.

The animal life of mediterranean North America is relatively diverse. Where habitat is suitable, mule deer (*Odocoileus hemionus*) and elk (*Cervus elaphus*) can be found. Predators include mountain lion (*Felis concolor*), bobcat (*F. rufus*), kit fox (*Vulpes macrotis*), and coyote (*Canis latrans*). Small mammals, such as the endemic white-eared pocket mouse (*Perognathus alticolis*), kangaroo rats (*Dipodomys* spp.), and other rodents are common. Chaparral and coastal sage are home to some 100 species and subspecies of birds [e.g., acorn woodpecker (*Melanerpes formicivorus*) and scrub jay (*Aphelocoa coerulescens*)]. The endangered California gnatcatcher (*Poliotila californica californica*) is confined to coastal sage; the secretive wrentit (*Chamaea fasciata*) prefers dense chaparral. Many reptiles and amphibians are represented; the lizard **genus** *Liolaemus* dominates with 34 species. Five endemic or nearly endemic species of lungless salamanders (Plethodontidae) also occur here. The biome is home to many species of butterflies and a high diversity of native bees (Davis and Richardson 1995; Ricketts et al. 1999).

## Western South America: Matorral and Encinal

Mediterranean scrub vegetation occurs in central Chile as an almost mirror image to North America's mediterranean region. Confined between the Pacific Ocean and the Andean Cordillera, it stretches from La Serene (31° S) south to Conception (37° S). Offshore is the cold Humboldt or Peru Current, which generates coastal fogs known in Chile as *camanchacas*. The mountainous topography and latitudinal extent of the biome create a variety of microhabitats, and matorral (from the Spanish *mata*, shrub) occurs in a patchwork of many plant communities with different structural characteristics and species composition. The Chilean part of the mediterranean scrub biome has many more deciduous plant species than California's chaparral does and many more species with thorns. Prominent among thorny plants are tall columnar cactus (*Trichocereus* spp. and *Eylchnia* spp.), an element virtually absent in other mediterranean vegetation regions. The presence of cactus is evidence that fire is not a significant factor in matorral.

The coastal matorral is comparable to California's coastal sage and somewhat similar to the garrigue of the Mediterranean Basin and strandveld of the Western Cape, South Africa. It is composed of low, soft-leaved shrubs, many of which are drought deciduous. The wild coastal fuchsia (*Fuchsia lycoides*) is one example. Unlike its counterparts, however, coastal matorral contains tall terrestrial bromeliads or puyas (*Puya* spp.) with sharp, serrated leaves as well as columnar cactus (*Trichocereus chilensis*).

The dominant shrub inland is the thorny *espino* (*Acacia* spp.) (Photo 6.7). Mesquite (*Prosopis* spp.) is another common plant in the canopy layer. This vegetation is called espinal to distinguish it from other types of matorral. The acacias, mesquites and other shrubs, such as *trevo* (*Dasyphyllum diacanthoides*), *huañil* (*Proustia pungens*) and *chilca* (*Baccharis glutinosa*), reach heights of 2–6 m (6–20 ft) and grow in an open, savannalike thicket. Puyas and cacti may be present, and perennial bunchgrasses cover much of the ground between shrubs. Vines, bulbs, and a ground layer of spring-blooming ephemerals also occur. On moister, south-

Photo 6.7 Matorral dominated by mesquites and acacias west of Santiago, Chile. (Photo by SLW.)

facing slopes, the shrub canopy becomes closed. Espinal may be a product of overgrazing and wood cutting, the latter of which eliminated the hard-leaved trees [i.e, litre (*Lithraea caustica*) and quillay (*Quillega saponica*)] that once grew here (Bahre 1979).

Hard-leaved evergreen woodlands and forest can be found on the slopes of the coastal ranges of central Chile. One of the most common trees is the *peumo* (*Cryptocarya alba*); it is in the same family (Lauraceae) as the California bay and the bay laurel of the Mediterranean Basin (Dallman 1998). The Chilean palm (*Jubaea chilensis*), the world's southernmost palm, is a disappearing component of this forest. Farther south or at higher elevations, where winter rainfall is more plentiful, southern beeches (*Nothofagus* spp.) come to dominate. Hard-leaved evergreen species are found at lower elevations; deciduous species occur at higher elevations. Bamboo (*Chusquea quila*) is a common understory plant in *Nothofagus* forests. Farther south or at still higher elevations, where temperatures are lower and winter rains may be even greater, stands of an endemic conifer, monkey puzzle tree (*Araucaria araucana*), occur.

Animal life in central Chile is not exceptionally diverse. Amphibians are represented by some 26 species, nearly half of which are unique to the region. The pointy-nosed Darwin's frog (*Rhinoderma darwinii*), one of two species in the family Rhinodermatidae, is an example. Reptiles, mostly lizards, are more diverse with 39 species; 33 percent are found nowhere else on Earth. Native mammals of the matorral are few and include guanaco (*Lama guanicoe*), the Chilean fox (*Pseudalopex* ( = *Dusicyon*) *griseus*), and several rodents. Most small mammals [(e.g., South American field mice (*Akodon* spp.)] are omnivores; only two rodent species [the rice rat (*Oryzomys longicaudatus*) and the leaf-eared mouse (*Phyllotis darwini*)] are primarily seed eaters. The mouse marsupial (*Marmosa elegans*) is insectivorous. About 175 species of birds, not a particularly great diversity, live here. Of significance as seed dispersers for mediterranean shrubs are the fruit eaters or frugivores—native birds such as the tenca (*Mimus thenca*), thrush (*Turdus falklandii*), dove (*Columba araunaca*), and diuca (*Diuca diuca*) (Bahre 1979; Dallman 1998; Mittermeier et al. 1999).

## Western Cape, South Africa: Fynbos and Strandveld

The Western Cape of the Republic of South Africa displays such an extremely high species richness and such a large number of endemic plants that it is placed in a Floral Kingdom (Cape Floristic Province) of its own. The *fynbos*, as mediterranean scrub is called here, has more than 8,000 species in over 200 genera. Sixty-nine percent of the plant species are found nowhere else on earth, and six plant families are endemic. Some of the typical families represented in the fynbos are the proteas or sugarbushes (Family Protaceae) (Photo 6.8) with 69 endemic species; the endemic reedlike restios (family Restionaceae); and the heaths (family Ericaceae), which are worldwide in their distribution. Also present and highly diversified are the daisy family (Asteraceae) and the orchid family (Orchidaceae). Cycads, ancient cone-bearing plants that are distantly related to pines but look superficially like palms, are also part of the formation. Bulbs such as freesia, iris, amaryllis, and gladiolus—familiar as garden and house plants in the northern hemisphere—are native to and abundant in the fynbos, as are the leaf-succulent aloes.

Evergreen shrubs of varying heights characterize the fire-prone fynbos. Several vertical layers can be recognized. Flowering peaks in the spring months of Septem-

Photo 6.8 King protea (*Protea cyanaroides*), one of many plants endemic to the fynbos. (Photo © Nigel J. Dennis; Gallo Images/CORBIS.)

ber and October, but the bloom period extends nearly year round in the moister eastern parts of the region. Some attribute this to the fact that many species are tropical or subtropical in their origins (Dallman 1998). Most plants (some 3,000 species) are low shrubs 0.5–2 m high (2–6 ft) and members of the heath family. Proteas, many with showy flowers and varying in height from 2–4 m (6–12 ft), provide a dense overstory. Most shrub species of the fynbos regenerate from seed, although a few of the proteas will resprout after fire. The single tree species, silvertree (*Leucadendron argenteum*), can be found only on the humid lower slopes of Table Mountain near Capetown (Dallman 1998; Mittermeier et al. 1999).

The coastal mediterranean scrub of South Africa is known as strandveld. It is dominated by evergreen restios, low-growing shrubs, succulents, and bulbs. Indeed, the world's greatest diversity of geophytes occurs in the Cape Floristic Province, and most of them are found in the strandveld.

A third shrub formation in mediterranean South Africa is sometimes considered part of the biome. This is the renosterveld, dominated by low, fire-prone, small- and soft-leaved shrubs of the daisy family—most notably the species for which the vegetation is named, the renosterbos or rhinoceros bush (*Elytropappus rhinocertis*). Renosterveld occurs on relatively fertile, shale-derived soils of the coastal lowlands (Cowling et al. 1997).

Insects are very important elements of the animal life of the Western Cape. Plants of the fynbos are pollinated mostly by flying insects; moths and beetles, in particular, have evolved close relationships with particular plant species. Harvester ants are important seed dispersers; an estimated 1,200 plant species depend upon ants collecting and burying their seeds. Indeed, the seeds of ant-dispersed plants have globules of fat attached to them to attract ants, which then store the seeds in their underground nests, coincidentally protecting them from predators and fire.

The large number of reptiles and amphibians (109 species) includes endemic chameleons (*Bradypodium* spp.); a very large number of tortoises; and the Cape ghost frog (*Heleophryne rosei*), whose large mouth when a tadpole and suckered toes as an adult allow it to cling to rocks in fast-moving streams and graze the algae growing there (Kingdon 1989).

Sugar-birds (*Promerops* spp.) seek nectar in proteas and in the process pollinate them, as do sunbirds (*Nectarinia* spp.), counterparts of New World hummingbirds. Mammals are represented by small antelopes such as the Cape grysbok (*Raphicerus melanotis*), steenbok (*R. campestris*), common duiker (*Sylvicapra grimmia*), and klipspringer (*Oreotragus oreotragus*). Larger antelopes include the rare bontebok (*Damaliscus dorcas dorcas*) and Vaal rhebok (*Pelea capreolus*). Leopards still occasionally prey upon these herbivores (Kingdon 1989; Mittermeir et al. 1999).

Animal life in the Western Cape, like the plant life, includes many forms found nowhere else. Some are evolutionarily quite primitive (for example, the Cape ghost frog mentioned above), which attests to the long isolation of the region that, until recently, discouraged their displacement by more advanced forms from other parts of the world.

### Western and Southern Australia: Kwongan and Mallee

The mediterranean vegetation of Australia occurs in two separate regions, one in the state of Western Australia and the other in the state of South Australia. They are separated by the arid Nullarbor Plain. Species richness is extremely high in Western Australia (indeed, the highest in all Australia); hundreds of tree and shrub species in the genera *Eucalyptus* and *Acacia* alone occur there, while only about 30 species in each of these genera appear in South Australia. *Kwongan* is a southwestern Australia aboriginal word for open, scrubby vegetation on sandy soil (Blondel and Aronson 1999). Mallee actually refers to the **growth-form** of those eucalypts that resprout from root crowns and develop into multistemmed shrubs.

Kwongan resembles the inland scrub formations of the other mediterranean regions, especially South Africa's fynbos. It contains a number of shrubs and small trees of the genus *Banksia* (family Protaceae), some of which may grow to 6 m (20 ft) tall if fires are infrequent enough. Other protea genera are also represented as shrubs. Most kwongan species regenerate by seed after fires. Some (e.g., *Banksia ornata*, *Hakea rostrata*, and *Casuarina muellerana*) store their seeds on their branches in woody capsules that open only in the heat of a fire; others have seeds that fall to the ground where they may remain dormant for years until the heat of a fire removes their tough seed coatings. A species-rich ground layer is present; strawflowers or everlastings (annuals in the daisy family) provide a blanket of color in spring. Sundews (family Droseraceae) are characteristic and diverse (47 species are known from the vicinity of Perth). These are insectivorous plants that overcome low-nutrient soils by digesting insects as a source of nitrogen. Other common components of kwongan are the grass tree (*Xanthorrhoea* spp.), restios, and shrubs of the Australian heath family (Epacridaceae), the citrus family (Rutaceae), the fanflower family (Goodeniaceae), and the kangaroo paw family (Haemodoraceae) (Dallman 1998).

Mallee is dominated by pungent, gray-green evergreen shrubs of the genus *Eucalyptus*. These eucalypts are capable of resprouting after fires, which are frequent. On the poorest soils, mallee has a diverse understory of heathlike shrubs (family Epacridaceae) known as mallee heath. On richer soils, the understory is composed of grasses and forbs. In the drier parts of South Australia, highly flammable porcupine grass (*Triodia irritans*) is a notable member of the understory. This grass grows in hummocks up to three feet across, and its leaves end in sharp spines that can puncture one's skin—hence its common and scientific names (Dallman 1998).

In the wettest areas in the southwestern tip of Western Australia, karri (*Eucalyptus diversicolor*) forests prevail. Karri are enormous, hard-leaved trees that reach 250 feet in height. A multi-layered understory is made of smaller trees, shrubs, and vines in this forest, which is home to large number of terrestrial orchids. On drier sites, a dry forest with nearly pure stands of jarrah (*E. marginata*) grows. Jarrah trees, also hard-leaved, attain heights of about 100 feet. Understory shrubs include banksias, sheoaks (*Casuarina* spp.), and paperbarks (*Melaleuca* spp.) (Dallman 1998; White 1990).

The richness of animal species is not nearly as great as that of the plants, but animals play important roles as pollinators and seed dispersers in kwongan and mallee. Insects, birds, and mammals are involved. Nectar-feeding birds, especially the honey eaters (family Meliphagidae), pollinate proteas. The tiny, endemic honey possum (*Tarsipes rostratus*), a nocturnal marsupial, and bats also pollinate members of the protea family.

Other animals that inhabit kwongan and mallee include the western gray kangaroo (*Macropus fulginosus*); the very rare and unique numbat (*Myrmecobius fasciatus*), a small termite-eating marsupial anteater that is the mammal emblem for the state of Western Australia; and the quokka (*Setonix brachyurus*), another small, rare marsupial. Small wallabies (*Macropus* spp.) and hare wallabies (*Lagostrophus* spp.) are grazers and browsers. Rat kangaroos; bandicoots; and the bush rat (*Rattus fuscipes*), one of Australia's few native, nonmarsupial mammals, feed on bulbs and other geophytes. A most unusual bird of the region is the ground-dwelling mallee fowl (*Leipoa ocellata*), which incubates its eggs in a heap of composting leaf litter. Reptilian diversity is high, as is the case for Australia as a whole. More than 190 species are recognized in the southwestern area; about 26 percent are endemic, among them several tortoises. Amphibians are fewer, but 80 percent of them are restricted to Australian areas of mediterranean scrub (Kalin-Arroyo et al. 1995; Furley and Newey 1983; Mittermeier et al. 1999).

## ORIGINS

The plant species that dominate and characterize the mediterranean scrub biome today trace their ancestries back at least 40 million years, even though the mediterranean climate with its dry summers first appeared only about 3 million years ago at the end of the Miocene Epoch of the Tertiary Period. This was when contemporary atmospheric and oceanic circulation patterns were first established. Movements of the earth's tectonic plates had opened the land barrier between the Arctic and Pacific oceans, created the Central American isthmus separating the Pacific and Atlantic oceans, opened the Drake Passage between Antarctica and South America, and confined oceanic circulation in the Indian Ocean. The repositioning of the major land masses had resulted in the permanence of subtropical high pressure cells over the seas, the development of upwelling and cool surface waters on the eastern borders of the ocean basins, and a cold current circling the Antarctic continent. The world's major mountain systems were also in place, although they were continuing to rise. These factors combined to create increasingly arid conditions in tropical and subtropical latitudes on the west coasts of continents. Between 30° and 40° latitude, aridity became a factor only during the summer months, when winds spiraling out of the eastern side of the subtropical high pressure cells blow parallel to or away from the adjacent coast.

The plant assemblage of each region has a different history, but it appears that most mediterranean species were selected from among plants that had first evolved in humid climates with year-round precipitation. Those species preadapted to drought were able to survive the increased aridity since the late Miocene and the subsequent climate changes of the Quaternary Period, when ice sheets advanced over and retreated from the northern continents several times to affect temperature and precipitation patterns worldwide. In mediterranean regions, summers became progressively drier each time the ice retreated.

In North America and in the Mediterranean Basin, most ancestral species were understory plants of the subtropical broad-leaved evergreen forests that were broadly distributed across North America and Eurasia during the Tertiary. The Mediterranean Basin received plants from other areas as well, lying as it does at the biogeographic crossroads of three continents, whereas North American chaparral plants derive primarily from species that dominated the western interior of that continent before a mediterranean climate existed (Axelrod 1989).

Chile's mediterranean plants are related mostly to tropical and subtropical rainforest species, but a significant number have close relatives in **temperate** vegetation types. Chile, of course, extends into higher latitudes where a temperate rain forest of *Nothofagus* and other species with relatives in the matorral continue to thrive. Subtropical parts of southeast Brazil and northeast Argentina are other regions containing close relatives of matorral plants.

The Western Cape plant species have many of their closest relatives in the high elevation heathlands of tropical Africa and in the kwongan of Western Australia. The origin of a mediterranean vegetation in the Western Cape Province is traced to the development of the cold Benguela Current in the early Pliocene (Cowling et al. 1997). With the associated drying of the climate, fire became a major disturbance factor and selected for those species able to tolerate frequent burns. Fire may also have promoted rapid speciation among the survivors. Fire-prone species generally have short generation times that provide more chances for mutations to occur. Furthermore, fires do not affect a landscape uniformly but in a patchy manner, so fire may have fragmented the vegetation into a series of habitat islands. The isolation so created—combined with the poor **dispersal** capabilities of most fynbos species—prevented gene flow from other populations and thereby enhanced opportunities for new species to form (Mittermeier et al. 1999).

In Australia, the ancestors of modern eucalypts and proteas were members of a Tertiary rainforest. It may be that the early hard-leaved plants were pioneer species on the vast sandy regions exposed whenever sea level fell. Proteoid roots or associations with mycorrhizal fungi would have enabled some species to invade the nutrient-poor sands. Today, species of *Banksia* and *Casuarina*, as well as restios, are colonizers of sand dunes (White 1990). Southwestern Australia was isolated from the southeastern part of the continent during the Cretaceous when a large embayment covered what is now the Nullarbor Plain. The physical barrier between the two Australian regions of mediterranean climate continued when the embayment became—literally—dry land. The long-standing separation of the two regions of mediterranean scrub in Australia has led to the development of distinctive assemblages of plants in each (Dallman 1998).

## HUMAN IMPACTS

Human influences on and alterations to the mediterranean scrub biome are extensive and, in some instances, of great antiquity. Altered frequency of fire, overgrazing, clearing of original woodlands and forests for timber and agriculture, agricultural practices, introduced species, urbanization, and—increasingly—tourism all have played a role in shaping contemporary vegetation and landscapes.

The Mediterranean Basin is one of the most human-affected regions on earth; very little of the natural vegetation remains. Humans have used fire as a tool for manipulating the environment for at least 100,000 years. The eastern part of the Basin was one of the world's major centers of plant (wheat, barley, peas, lentils, grapes, olives, etc.) and animal (sheep, goats, cattle, pigs) domestication beginning about 10,000 years ago. The first evidence of forest clearing to accommodate farming dates to about 8,000 years ago. By 4,000 to 5,000 years ago, agriculture was being practiced across the area. The Basin is also home to many of the early urban cultures of Western Civilization. We know from the works of Homer that the region was forested with live oaks, pines, cedars, wild carob, and wild olive in the Classical Greek Period. The cutting of the famed cedars of Lebanon began around 3000 B.C. First Greeks and then Romans expressed concern over the effects of deforestation as an ever-increasing human population required that more and more trees be removed to clear land for agriculture and to provide construction timbers for growing towns and seaborne enterprises in both trade and warfare. Stripped of their natural vegetation, hillsides became severely eroded, sometimes exposing bedrock or leaving only a veneer of soil too thin to support a dense cover of protective vegetation. Human livelihoods came to depend upon the grazing and browsing of sheep and goats on the open scrub that was able to survive these conditions. The presence of livestock exerted new pressures on the vegetation and shifted the advantage to unpalatable plants and to species able to colonize the dry, open, low-nutrient surface between shrubs—namely annuals or weeds. Much of contemporary maquis and garrigue is the product of human land use. Hard-leaved evergreen scrub probably was originally confined to coastal areas but expanded with the increasing dominance of human-caused disturbances. One should bear in mind, however, that human impacts are aggravated in this region (and in some of the other mediterranean regions) by natural phenomena common to the Basin such as steep slopes, torrential rains, earthquakes, and weak subsurface geology (e.g., soft marls or highly jointed bedrock) (Blondel and Aronson 1999; Conacher and Sala 1998; Mittermeier et al. 1999).

Since the 1960s, the European part of the Basin has lost population in the rural areas because of changing economic conditions brought about by agricultural policies of the European Common Market/European Union. The overgrazing of the past has been replaced by undergrazing, and a dense cover of flammable shrubs has been developing, which increases the frequency and severity of fire. In the Maghreb of North Africa, the situation is reversed; rural population and livestock have increased since the countries there gained independence in the late 1960s. Throughout the Basin, today, expanding cities are destroying the surrounding shrublands—and agricultural lands—for urban development. Coastal habitats are being converted to resorts and their associated landscape elements (roads, parks, marinas, private estates, etc.).

In the mediterranean region of California, burning by aboriginal peoples may have expanded the area originally covered by chaparral, but cattle grazing during the Spanish, Mexican, and early American periods of the settlement history of the area initiated major land degradation. The replacement of native annual and perennial bunchgrasses with European sod-forming annuals (e.g., *Bromus, Avena*) is one consequence. In the modern period, fire suppression policies in California changed the nature of the vegetation and of the fires themselves. When fires are controlled or prevented year after year, species composition alters and fuel (both the living shrubs and the litter) accumulates. Now when fires do occur, they often become infernos that rage out of control and destroy rather than renew chaparral. Cool fires that occur when scrub is less dense and litter has not accumulated do not alter soil chemistry but may actually increase the solubility of various compounds and provide a more nutrient-rich environment for the regrowth of both sprouters and seeders. Hot fires that occur after decades of fire suppression can alter soil structure, reduce porosity, and may create a water-repellant soil. Most seeds stored in the litter cannot survive temperatures above 150° C (300° F), hence, hot fires can prevent the rapid recolonization by seeders that would stabilize slopes; this can lead to accelerated erosion. Slopes denuded by intense fires often suffer mudslides with the onset of the winter rains. Continued population (and wealth) growth in California is expressed in the encroachment of suburbia onto these hazard-prone slopes with resulting high costs in terms of property damage and sometimes loss of life when fires sweep through wooded subdivisions or estates.

Agriculture continues to contribute to the loss of the chaparral. Like all mediterranean regions, California is a major producer of fruit, vegetables, wine and cut flowers. Natural vegetation continues to be removed for expanded agricultural land uses. Long-term consequences to soil structure and chemistry resulting from the use of heavy machinery, fertilizers, pesticides, and irrigation may prevent reestablishment of chaparral when the fields and orchards are abandoned. Salinization is a problem, and, in some places, toxic levels of selenium have accumulated as a by-product of irrigated agriculture.

Chile's aboriginal peoples may have influenced the nature of the matorral. They practiced irrigation agriculture, herded llamas, and cut trees for firewood. Most human impacts, though, are related to European settlement. The Spanish brought cattle and other livestock to Chile. At first, they used natural meadows, but they expanded pastureland by burning surrounding vegetation as their cattle populations increased. As pastures became degraded through overgrazing, cattle were replaced with sheep and finally goats, the animals best able to convert shrubs to milk, protein, fiber, and leather. The presence of grazing and browsing livestock may have been a major selection force for shrubs armed with thorns, the abundance of which distinguishes Chilean matorral from other mediterranean scrub. (This difference may also be related to the low frequency of natural fires in the Chilean region.) Most of the matorral has been heavily grazed—and not only by livestock. More recently, European rabbits (*Oryctolagos cuniculus*), introduced at the beginning of the twentieth century, have caused a precipitous decline in bunchgrasses and the seedlings of woody species except where they are protected under thorny shrubs (Bahre 1979; Dallman 1998; Mittermeir et al. 1999).

The likelihood that espinal derives from human disturbances was mentioned. Woodcutting for fuelwood and charcoal contributed to the demise of trees. The bark of the quillay, known as soapwood in the English-speaking world, forms a lather in water and was exported to cleanse fine fabrics, lenses, and precision instruments. Woodcutting was also an integral part of the region's copper mining economy of the nineteenth and twentieth centuries because charcoal was used for smelting the ore (Bahre 1979).

European agricultural crops and practices were introduced to Chile after the Spanish conquest, and land was cleared for the cultivation of wheat in particular. A shifting, dry-farming technique was practiced, and it destroyed large areas of matorral. The most destructive period was in the nineteenth century, when wheat production rapidly expanded—first to satisfy markets associated with the gold rushes in California and Australia and later a British market. Dry farming involved clearing and plowing large tracts of land to grow wheat, barley, cumin, peas, lentils, and chick peas for three or four years. By then, the soil was depleted of both moisture and nutrients, and the plot was abandoned; this left the soil exposed to wind erosion and created an agri-desert (Bahre 1979; Conacher and Sala 1998). Although some shrub species have since reinvaded, diversity is low and plants grow in straight parallel lines that reveal the furrows and ridges made years ago.

Chile's economy today is highly dependent on the export of agricultural and wood products from its mediterranean climate region. Matorral and forest continue to be converted to orchards, vineyards, and tree plantations. Trees from other mediterranean regions in the world, primarily Monterey pine (*Pinus radiata*) and eucalyptus, especially blue gum (*Eucalyptus globulus*), are preferred over native species since they grow more quickly and are generally disease free outside their natural distribution areas.

In the Western Cape of South Africa, humans and our near ancestors have been part of the ecosystem for a million years or more. For at least 100,000 years, hunter-gatherers, including modern Khoi-San, practiced what has been called fire-stick agriculture: they burned the vegetation to encourage the growth of edible geophytes and also the grasses that attracted game. About 2,000 years ago, a cattle-based pastoral economy entered with the Khoi-Khoi, who pushed the San into more marginal habitats on mountain slopes, which then too were subjected to regular burning. Pollen evidence suggests that both hunter-gathers and herders influenced the relative abundance of certain proteas as well as grasses and geophytes (Conacher and Sala 1998).

European settlement began in 1652, and with it came the introduction of typical European mediterranean agriculture—wheat fields, vineyards, and orchards—and clearing of native vegetation. The Western Cape is still an important and expanding agricultural area with the newest addition to the suite of crops being native proteas for the cut-flower industry. Of all human impacts, cultivation certainly has had the greatest effect on native vegetation by causing its large-scale conversion to cropland. The renosterveld was been almost entirely destroyed. Animals have been affected as well. Hunting led to the extinction of several large mammals, including the endemic bluebuck (*Hippotragus leucophaeus*), the quagga (*Equus quagga*), and the Cape lion (*Panthera leo*) (Conacher and Sala 1998).

Today, what remains of the enormously species-rich fynbos is threatened by population growth and urbanization, afforestation with nonnative trees, and **alien** species invasions—some species being escapes from the tree plantations (Conacher and Sala 1998). Among the invasive nonnative plants are Monterey pine (*Pinus radiata*) from California, maritime pine (*P. pinaster*) and Aleppo pine (*P. halepensis*) from the Mediterranean Basin, and several Australian species including *Hakea sericea*, *Acacia saligna*, and *A. cyclops*. Among problematic introduced animals is the Argentine ant (*Iridomyrmex humilis*), which could replace the native harvester ants upon which so many proteas depend for the dispersal of their seeds. The Argentine ant does not bury seeds but leaves them on the surface, where they are vulnerable to seed-eating animals. Loss of native ant species could cause changes in the abundance of many native shrub species (Conacher and Sala 1998; Mittermeir et al. 1999).

In Australia, the earliest evidence of human use of fire goes back at least 40,000 years. Aboriginal people regularly burned vegetation for hunting; and natural lightning fires occurred also. The impact on species composition and vegetation structure is unknown but must have been significant. There is little question, however, about changes wrought by Europeans on the vegetation—and on the aboriginal peoples, who remained hunters and gatherers until well after contact. The changes to the vegetation were severe and, in this case, recent: most have occurred in the twentieth century. Although European settlement began in 1829, major expansion of European agriculture on the inherently poor soils of western and southern Australia did not occur until after the introduction of superphosphate fertilizers in the 1890s. Extensive areas of natural vegetation were then cleared for wheat and pasture crops.

Cattle, sheep, and goats were introduced and multiplied rapidly. Native grasses were overgrazed and replaced with alien forage species. Desertification and salinization of streams and soils often followed. Europeans also introduced wild animals from Europe, sometimes with dire consequences; best known are the rabbit plagues, originating with an 1859 deliberate introduction of the European rabbit (*Ortylagus cuniculus*) and still not completely under control. Overgrazing by rabbits accelerated erosion and desertification processes. Introduced cats and other small carnivores threaten native birds and marsupials, including important pollinators and seed dispersers of native plants (Conacher and Sala 1998).

A modern plague in Australia is related to the soil fungus, *Phytophthora cinammoni*, apparently introduced from Indonesia in the 1920s. It causes a plant disease known as jarrah dieback. This fungus attacks roots so they are unable to absorb water and nutrients. The most obvious impacts are in the dry forests, which show dieback of the crowns of the tall jarrah trees. However, the fungus affects and kills some 900 other species, primarily natives (Conacher and Sala 1998). It spreads slowly through the soil in undisturbed forests, but its spores are spread rapidly when soil is transported via lumber trucks and road construction. Fire—a common management tool throughout Australia—may also be implicated in its spread: the fungus becomes more prevalent where little or no litter exists, plus burning tends to eliminate prickly moses (*Acacia pulchella*), a plant that repels the fungus and increases the occurrence of banksias that do not. Dieback is considered *the* greatest threat to the native plants of mediterranean Australia (Conacher and Sala 1998).

Among other alien species that are spreading into kwongan and mallee are horticultural escapes including bulbs such as gladiolus, ixia, and freesia—all native to South Africa—and other forbs such as the South African daisy (*Senecio pterophorus*) and the bristly European annual borage known as Paterson's curse (*Echium plantagineum*) (Mittermeier et al. 1999).

## HISTORY OF SCIENTIFIC EXPLORATION

It was in the mediterranean biome that western science first took root. Early Greek naturalists such as Theophrastus and Aristotle observed the workings of nature in the Mediterranean Basin in the fourth and third centuries B.C. With European colonization of the other mediterranean regions came major exploratory expeditions and inventorying of native plants and animals. Similarities in structure, noted especially among the plants of the five mediterranean climate regions, inspired a number of scientific studies on what is known as convergent evolution. This is the hypothesis that species of very different ancestry that occur in similar environments will evolve over time to look and function much the same because they have adapted to similar conditions. Comparative studies of the various regions of the mediterranean scrub biome looked like the perfect way to test this hypothesis. (Interestingly, the consensus that developed as research progressed was that convergent evolution was *not* a major factor explaining the similarities of mediterranean vegetation.) Many modern comparative ecological and biogeographic investigations stem from the Mediterranean Scrub Project of the Structure of Ecosystems Subprogram, Origin and Structure of Ecosystems Integrated Research Program, International Biological Program funded by the National Science Foundation of the United States beginning in 1970. International symposia, collaborative research projects, and major publications synthesizing research results continue to be developed. Among the topics upon which comparative research has focused are ecological function (di Castri and Mooney 1973), biological invasions (Groves and di Castri 1991), fire **ecology** (Moreno and Oechel 1994), plant-animal interactions (Arianoutsou and Groves 1994), **biogeography** (Kalin Arroyo et al. 1995), biodiversity (Davis and Richardson 1995), implications of global climate change (Moreno and Oechel 1995), and implications of landscape disturbance (Rundel et al. 1998).

In addition to comparative studies, a huge body of scientific literature addresses many aspects of each of the five regions within the mediterranean scrub biome. Most studies involve the highly diverse plants and their ecology—past, present, and future.

## REFERENCES


Arianoutsou, M. and R. H. Groves, eds., 1994. *Plant-Animal Interactions in Mediterranean-type Ecosystems.* Dordrecht: Kluwer Academic Publishers.

Axelrod, D. I. 1989. "Age and origin of chaparral." Pp. 7–19 in S.C. Keeley, ed., *The California Chaparral: Paradigms Revisted.* Sci. Ser. No. 34. Los Angeles: Natural History Museum of Los Angeles County.

Bahre, Conrad J. 1979. *Destruction of the Natural Vegetation of North-Central Chile.* Berkeley: University of California Press.

Blondel, Jacques and James Aronson. 1999. *Biology and Wildlife of the Mediterranean Region.* Oxford: Oxford University Press.

Conacher, Arthur J. and Maria Sala, eds., 1998. *Land Degradation in Mediterranean Environments of the World: Nature and Extent, Causes and Solutions.* New York: John Wiley & Sons.

Cowling, R. M., D. M. Richardson, and S. M. Pierce. 1997. *Vegetation of South Africa.* Cambridge: Cambridge University Press.

Dallman, Peter. R. 1998. *Plant Life in the World's Mediterranean Climates.* Oxford: Oxford University Press.

Davis, G. W. and D. M. Richardson, eds., 1995. *Mediterranean-type Ecosystems: The Function of Biodiversity.* New York: Springer-Verlag.

di Castri, F and Harold A. Mooney, eds., 1973. *Mediterranean-type Ecosystems: Origin and Structure.* New York: Springer-Verlag.

Furley, Peter A. and Walter W. Newey. 1983. *Geography of the Biosphere, An Introduction to the Nature, Distribution and Evolution of the World's Life Zones.* London: Butterworths.

Groves, R. H. and F. di Castri, eds., 1991. *Biogeography of Mediterranean Invasions.* Cambridge: Cambridge University Press.

Kalin Arroyo, M. T., P. H. Zedler, M. D. Fox, eds., 1995. *Ecology and Biogeography of Mediterranean Ecosystems in Chile, California, and Australia.* NY: Springer Verlag.

Kingdon, Jonathan. 1989. *Island Africa: The Evolution of Africa's Rare Animals and Plants.* Princeton, NJ: Princeton University Press.

Mittermeier, Russell, Norman Myers, and Cristina Goettsch Mittermeir. 1999. *Hotspots: Earth's Biologically Richest and Most Endangered Terrestrial Ecosystems.* Mexico City: CEMEX, S.A. and Washington, DC: Conservation International.

Moreno, J. M. and W. C. Oechel, eds., 1994. *The Role of Fire in Mediterranean-Type Ecosystems.* New York: Springer Verlag.

Moreno, J. M. and W. C. Oechel, eds., 1995. *Global Change and Mediterranean-Type Ecosystems.* New York: Springer Verlag.

Ricketts, Taylor H., Eric Dinerstein, David M. Olson, Colby J. Loucks et al. 1999. *Terrestrial Ecoregions of North America, A Conservation Assessment.* Washington, DC: Island Press.

Rundel, P. W., G. Montenegro, and F. M. Jaksic, eds., 1998. *Landscape Disturbance and Biodiversity in Mediterranean-type Ecosystems.* Berlin: Springer.

Thrower, N.J.W. and D. E. Bradbury. 1977. *Chile-California Mediterranean Scrub Atlas, A Comparative Analysis.* Stroudsburg, PA: Dowden, Hutchinson, & Ross, Inc.

White, Mary E. 1990. *The Flowering of Gondwana: The 400 Million Year Story of Australian Plants.* Princeton, NJ: Princeton University Press.

# 7 TEMPERATE BROADLEAF DECIDUOUS FORESTS

## OVERVIEW

The **temperate broadleaf deciduous** forest **biome** occurs in the humid middle **latitudes**, where cold winters and warm to hot summers mark distinct seasons. This is essentially a Northern Hemisphere biome, although a few places in southernmost South America have a comparable forest type. Temperate broadleaf deciduous forest is the natural **vegetation** of the eastern United States and was the natural vegetation of western Europe, where it extended from the British Isles eastward onto the continent through much of central Europe to the Ural Mountains. A third major Northern Hemisphere region of the biome is in eastern Asia, north of the Yangtze River (Chang Jiang) (approximately 30° N). The forest reaches north to latitudes of 50° to 60° N in the Northeast of China (Manchuria), the Kamchatka Peninsula, and central Japan. The western limits of the biome in Asia lie between approximately 125° E in the northwest to 115° E in the southwest of China. Smaller, minor outliers of the biome occur in Southwest Asia in the highlands of northern Turkey and western Iran and at lower elevations in the Caucasus Mountains (Röhrig 1991a). In South America, a deciduous forest composed of entirely different trees than those found in the Northern Hemisphere occurs on either side of the Andes in Chile and westernmost Argentina south of about 37° S latitude. The eastern margin in Argentina borders the dry grasslands of Patagonia.

The three major Northern Hemisphere expressions of the biome are geographically isolated from each other but were part of a continuous forest belt in the geologic past. Thus they contain closely related plants today. The most important and typical trees are members of the beech family [Fagaceae, the beeches (*Fagus* spp.) and oaks (*Quercus* spp.)]. Other important trees belong to the birch family (Betulaceae) [e.g., hornbeams (*Carpinus* spp.), hazels (*Corylus* spp.), and hophornbeams (*Ostrya* spp.)], walnut family (Juglandaceae) [especially, in North America, the hickories (*Carya* spp.)], and the maple family (Aceraceae). Due to the deciduous **habit** of the **canopy** trees, sunlight reaches the forest floor at the beginning of the grow-

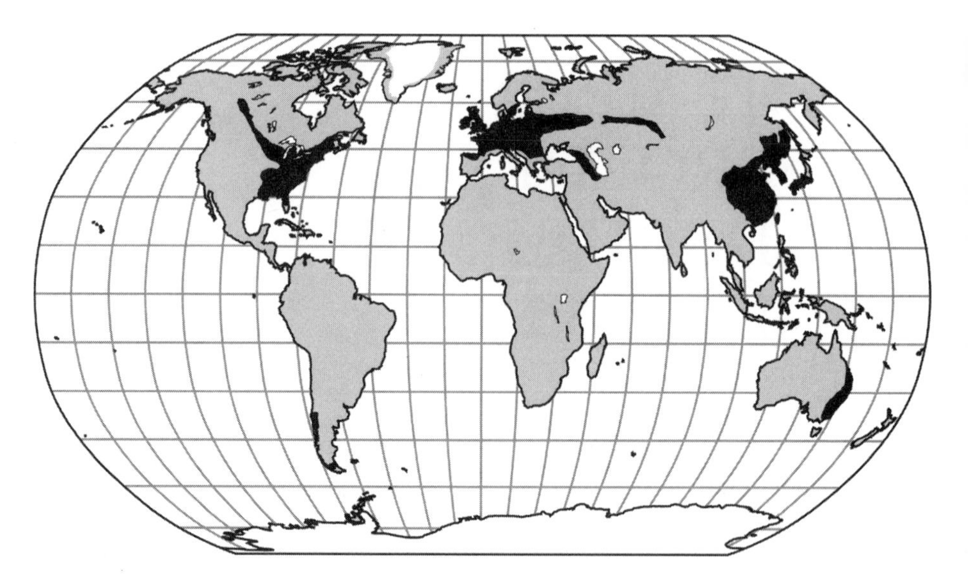

Map 7.1. World distribution of Temperate Broadleaf Deciduous Forest Biome.

ing season, and many **forbs** and shrubs are able to grow and flower before the **crowns** of the tallest trees come into full leaf and cast shade over them. Temperate broadleaf deciduous forests are the most **species**-rich vegetation types in the middle latitudes. In mountainous areas, diversity is further heightened by the variety of microclimates and **soil** conditions offered at different elevations.

The forests of East Asia are presumed to be both the oldest and were the least affected by climatic changes during the Quaternary Period. Today, they contain the greatest number of tree species and include representatives of very ancient genera, genera limited to Eastern Asia, genera shared only with western North America, and all genera found today in western Europe and eastern North America—the other two Northern Hemisphere expressions of the biome. Eastern North America ranks second in tree diversity, with European forests having the lowest number of species.

The soils developed under broadleaf deciduous forest tend to have relatively high natural fertility and were for thousands of years the best soils available to farmers. Only after the steel plow was invented in the eighteenth century could the thick sods of temperate grasslands be removed to expose an even richer soil. Agriculture is ancient and intensive in East Asia; as a result, very little of the original forest is intact today. The same can be said for European deciduous forests. The largest existing tracts of this biome are found in eastern North America, and almost all of this is second growth.

## CLIMATE

Temperate broadleaf deciduous forests are associated with humid subtropical (Koeppen's Cfa and Cfb **climate** types) and hot summer, humid continental (Koeppen's Dfa climate type) climates. Typically, precipitation is evenly distributed throughout the year in these climates, and average annual precipitation ranges between 750 and 1,250 mm (30–50 in). Winters are cold with at least some snow; summers are warm to hot. The growing season is six months or longer. Inhabitants of these regions experience four distinct seasons: spring, summer, autumn, and winter.

The influence of the Asian monsoons and the great continental mass of Eurasia are felt in eastern China. Summers are extremely humid and hot (several cities in the Yantgze drainage basin are called the "furnaces of China"), while winters are very dry and cold. Since the dry season coincides with the nongrowing season when most deciduous plants are dormant, the same types of trees found in other northern hemisphere regions of the biome thrive in China as well.

## VEGETATION

The temperate broadleaf deciduous forest has a multilayered structure (Photo 7.1). The tallest trees form a **closed canopy**. Beneath them is a small tree or sapling layer made up of the crowns of small trees, which even at maturity do not reach the forest canopy. A shrub layer of broad-leaved deciduous and, sometimes broad-leaved **evergreen** plants, follows with a rich **herb** layer of **perennial** forbs. Some scientists recognize a ground layer of **lichens**, clubmosses, and mosses distinct from the herb layer. The woody vine (**liana**), represented by such plants as wild grape (*Vitis* spp.) and poison ivy (*Toxicodendron radicans*), is another **growth-form** encountered in temperate broad-leaved deciduous forests.

Photo 7.1 Temperate broadleaf deciduous forest. As seen in this oak-chestnut second-grown forest in Virginia, foliage occurs at all levels in this forest type. (Photo by SLW.)

The seasonality of life processes is one of the key characteristics of this forest. A period of vegetative growth and activity and a distinct dormant period exist. Most obvious, perhaps, is the turning color of leaves from green to yellows, browns, reds, and oranges in the autumn. With shorter days and lower night temperatures, the green chlorophyll breaks down to unmask yellow (carotinoid) and red (anthocyanin) pigments in the leaves. The intensity of the color depends upon weather conditions and varies from year to year, but in good years, autumn foliage is a natural wonder that attracts tourists in large numbers.

Structural changes in the leaves precede the actual falling of the leaves in autumn. First, an abcission zone develops at the base of the leaf petiole, and then a protective material forms over the surface that will become exposed at leaf fall in order to prevent water loss. In the spring, activity begins before buds open as sap rises from the roots to the branches. Some trees, such as maples and hazels, flower before their leaves appear, but most flowering occurs during the initial leaf formation. The linden, limes, or basswoods (*Tilia* spp.) are among the few trees that flower in summer. Young leaves contain the highest amount of nitrogen, phosphorus, and potassium compounds, so many insects coordinate the hatching of their larvae to this springtime abundance of nutrients. Migratory birds time their arrival to the explosion of insects. As summer progresses, some of the plant nutrients are stored in the bark of twigs and stems, and some seeps out of the leaf cells and is washed away by the rains. However, calcium and magnesium compounds accumulate in the leaves and will be transported to the litter in the autumn. A thick **humus** develops under the deciduous forests to foster the nutrient-rich soils that broad-leaved trees tend to require (Röhrig 1991b).

Considerable energy is expended each year in the production of new foliage. Only where the growing season is sufficiently long can the annual renewal of leaves be possible. The overwhelming dominance of deciduous species in the humid climates of the middle latitudes demonstrates that this is a successful alternative to producing well-protected evergreen leaves.

Photo 7.2 The perennial forb trailing arbutus (*Epigaea repens*). It blooms early in the spring when there is a brief season of warm ground temperatures and bright sunshine on the forest floor. (Photo by SLW.)

Spring greenup begins at the forest floor with a flush of flowering perennial forbs (Photo 7.2). In North America, these include such wildflowers as spring beauty (*Claytonia virginica*), hepatica (*Hepatica americana*), and mayapple (*Podophyllum peltatum*). A few, such as skunk cabbage (*Symplocarpus foetidus*), can bloom

Photo 7.3 Early spring in a temperature broadleaf deciduous forest of the Appalachians. Small trees and shrubs flower and leaf out before the canopy trees. (Photo by SLW.)

while snow is still on the ground since the heat generated by the flower melts the overlying snow. Perennials complete their growth when they can take advantage of the short period when ground temperatures are moderate and canopy trees have no leaves. The shrub layer responds early in the spring, too (Photo 7.3), with spicebush (*Lindera benzoin*) taking the lead, followed shortly by blossoms on the redbud (*Cercis canadensis*), serviceberry or shadbush (*Amelanchier* spp.), and flowering dogwood (*Cornus florida*). Finally, the leaves on the canopy trees expand and shade the forest below, which slows the activity of smaller plants.

## SOILS

Two soil types are characteristic, both products of the podzolization soil-forming process. In cooler areas, alfisols (gray forest soils) dominate; more heavily leached, less fertile ultisols (red and yellow forest soils) are found in warmer climates. The organic layer contains nutrient-rich leaf litter that is deposited in autumn but begins to decay in spring, so that the nutrients within it are mineralized just as the plants start to need fertilizer for a rapid growth spurt. The leaf litter decomposes into a dark, slick layer of humus, which, as it washes through the **soil horizons**, stains them brown. Since this humus is much less acidic than the humus that forms under the **needle-leaved** trees of the boreal forest, **leaching** of the A horizon is less severe than in spodosols (or true podzols). Aluminum and iron remain with silica compounds as major mineral components of the A horizon. It is from the chemical symbols for the first two elements (Al and Fe, respectively) that the soil order name alfisol derives. Iron compounds give a reddish brown color to alfisols, and the red or yellow color to ultisols. Nutrients leached from the leaves or dropped to the ground in leaf fall mineralize and leach through the A-horizon to accumulate with humus in the B horizon. The B horizon is distinct, being a lighter brown than the A horizon. Tree roots (and farmers' plows) reach into the B horizon to retrieve mineral nutrients.

## ANIMAL LIFE

Temperate broadleaf deciduous forests offer a great variety of **habitats** and food resources for invertebrates and small vertebrates. Nutrient-rich foliage at various levels within the forest, sap, nuts, berries, and seeds, a rich decaying litter, and dead and dying wood all provide opportunities for different kinds of animals. The caterpillars of moths, butterflies, and other Lepidoptera such as bagworms, leaf rollers, and tentmakers are abundant foliage eaters. Some chew the leaves, others skeletonize them by eating all but the veins, still others mine the inner cells of leaves. Beetles (Order Coleoptera) and flies (Order Diptera) also are abundant.

Many kinds of amphibians (salamanders, frogs, toads, and tree frogs) occur in the moister parts of the biome. Reptiles (lizards, snakes, and turtles) are also diverse. The abundance of insects supports many insect-eating birds that exploit the resource in a variety of ways. Woodpeckers forage the bark of trees by chipping and drilling to find beetles and grubs. Members of the tit family (Paridae) actively glean insects from leaves, flycatchers (Tyrannidae) pursue insects moving through the canopy, and swallows (Hirundidae) patrol above the canopy. Thrushes (Turdidae) tend to forage on the ground. Among mammals, rodents (voles, mice, and squirrels), bats, and insectivores (e.g., shrews) are most diverse. Small predators such as foxes are typical. For the most part, large mammals have been eliminated by land use changes, overhunting, and deliberate campaigns of extermination. Wolves (*Canis lupus*) and cougars (*Felis concolor*) have been largely eliminated from the forests of eastern North America, as have elk (*Cervus elaphus*) and wood bison (*Bison bison*). In Europe, a remnant population of wisent (*Bison bonasus*), a close relative of the North American bison, survives in the Bialowieza Forest in Poland; but the forest-dwelling tarpan (*Equus caballus*), a close relative of the domestic horse, is extinct in the wild (Furley and Newey 1983).

The seasonal rhythm of plant production and insect activity is mirrored in the patterns of many vertebrates. The most common adaptation is timing the season of reproduction to coincide with the greatest abundance of food. Hibernation and migration are adaptations to cold and scarcity of food in winter. Frogs hibernate during the winter in the mud and debris at the bottom of ponds; toads and salamanders find shelter from cold and dessication in deserted rodent burrows, cracks and crevices in the soil, caves, or in rotting logs. Snakes also use animal burrows, rock crevices, caves, and even deserted anthills. In the North American forests, three-fourths of the birds migrate south in winter. Those that remain are omnivores like chickadees (*Poecile* spp.) or insectivorous birds like woodpeckers and nuthatches that can pry dormant insects from beneath the tree bark. Conditions are not severe enough for most mammals to migrate, but bats, particularly in the northern part of the biome, migrate south and hibernate in caves and other cavities. A few mammals such as woodchucks (*Marmota monax*) and eastern chipmunks (*Tamias striatus*) do hibernate during the coldest months (Kitchings and Walton 1991).

# MAJOR REGIONAL EXPRESSIONS

## North America

Temperate broadleaf deciduous forests are found in eastern North America east of the 95° W meridian and from 48° N to 35° N. On the Coastal Plain of the southeastern United States, with its long growing season and nutrient-poor soils, it gives way to needle-leaved and evergreen broad-leaved trees. Within the broad expanse of the biome is a patchwork of forest types, characterized by different assemblages of deciduous trees. This reflects the geological and climatological differences of the several **physiographic provinces** that make up the eastern part of the continent. Lucy Braun (1950), in the classic study of the deciduous forests of North America, identified nine different forest regions that make up the regional mosaic. Trees commonly encountered over much of the biome include oaks (*Quercus* spp.), hickories (*Carya* spp.), American beech (*Fagus grandifolia*), basswoods (*Tilia* spp.), and sugar maple (*Acer saccharum*). The American chestnut (*Castanea dentata*) used to join them, but this tree is now extinct as a member of the canopy layer. Several evergreen **conifers** are also part of the forest, most notably various pines (*Pinus* spp.) and eastern hemlock (*Tsuga canadensis*).

Among the forest regions identified by Braun is the Oak-Hickory-Pine Forest Region, which coincides fairly well with the Piedmont Plateau Physiographic Province and extends from Alabama to Pennsylvania east of the Appalachian Mountains. Elevations in the province run from 60 to 300 m (200–1,000 ft) above sea level. The climate is characterized by a growing season of 180 to 240 days and hot summers with high humidity. Annual precipitation varies from 1,090–1,345 mm (42–52 in). Soils tend to be deeply weathered ultisols with clay subsoils. Many kinds of oak occur in this forest region and many, adapted to fire, can resprout after a burn. Fourteen of the fifteen North American hickories are also associated with this forest type. While evergreen broad-leaved trees are essentially absent, evergreen needle-leaved trees are major components of the forest. In the more southerly areas, loblolly (*Pinus taeda*) and shortleaf (*P. echinata*) pines are common; in the more northerly areas, Virginia pine (*P. virginiana*) is prevalent. Typical understory trees are sweetgum (*Liquidambar styraciflua*), black tupelo (*Nyssa sylvatica*), sourwood (*Oxydendron arboreum*), flowering dogwood (*Cornus florida*), and red maple (*Acer rubrum*). The pine barrens of New Jersey, although on the Coastal Plain, are considered part of this forest region. The barrens are essentially a pygmy forest of stunted pitch pine (*Pinus rigida*); oaks such as blackjack oak (*Quercus marilandica*), bear oak (*Q. ilicifolia*), and dwarf chinkapin oak (*Q. prinoides*); and **heaths** such as mountain laurel (*Kalmia latifolia*) and *Gaylussacia frondosa*. Red maple, black tupelo and Atlantic white cedar (*Chamaecyprus thyoides*) occupy swampy terrain (Braun 1950; Vankat 1979; Barnes 1991).

The Oak-Chestnut Forest Region is found mainly in the Blue Ridge and Valley and Ridge physiographic provinces. (The name applied to this forest type is invalid today since chestnut is no longer a dominant, so sometimes the forest region is called the Appalachian Oak region instead.) These sections of the Appalachian Mountains have surface elevations ranging from 300–2,000 m (1,000–6,500 ft). Precipitation ranges from 760–1,780 mm (30–70 in), with a maximum of 2,300 mm (90 in) occurring in the Great Smoky Mountains in the Blue Ridge. The greatest number of forest communities occurs in the Southern Appalachians, where mixed stands are dominated by chestnut oak (*Quercus prinus*) and white oak (*Q. alba*) but also contain northern red oak (*Q. rubra*), black oak (*Q. velutina*), and scarlet oak (*Q. coccinea*). Tall, straight tulip polar (*Liriodendron tulipifera*) is common on moist lower slopes, and eastern hemlock occurs in cool ravines. In sheltered valleys, known locally as coves, a species-rich **community** of trees resembling the mixed mesophytic forest (discussed later) occurs. Cove forests can be distinguished by the added presence of sugar maple, American beech, eastern hemlock, tulip poplar, yellow birch (*Betula alleghaniensis*), yellow buckeye (*Aeschulus octandra*), white basswood (*Tilia heterophylla*), and silverbells (*Halesia carolina*). Above 1,400 m (4,500 ft), some of the less cold-tolerant species drop out, and the forest is one of northern hardwoods including sugar maple, yellow buckeye, yellow birch, and American beech. An understory of striped maple (*Acer pensylvanicum*), mountain maple (*A. spicatum*), and maple-leaved viburnum (*Viburnum alnifolium*) grows beneath the canopy. The ground layer contains many spring annual forbs. At the highest elevations (above 1,520 m or 5,000 ft), almost all the broad-leaved deciduous trees disappear, and forests of red spruce (*Picea rubens*) or Fraser fir (*Abies fraseri*) occur. Growing with these conifers may be mountain ash (*Sorbus americana*) and yellow birch (Braun 1950; Vankat 1979; Barnes 1991).

The Mixed Mesophytic Forest Region coincides with the unglaciated part of the Appalachian Plateaus Physiographic Province, namely the Cumberland Plateau that lies just west of the Southern Appalachians. The Cumberland Plateau has been dissected into a landscape of hills and hollows; most of the land is sloped. Precipitation is high, the growing season is long, and winters are relatively mild. The "mixed" in mixed mesophytic comes from the fact that this is an area of overlap for plants of generally more northerly distributions with those of generally more southerly distribution and those of more westerly distribution. In addition, some species are more or less restricted to the Southern Appalachians and Cumberland Plateau. As a consequence, this forest region has the greatest number of tree species in the American temperate broadleaf deciduous forest. Approximately 33 different trees reach the canopy layer, and several kinds share dominance at any given location. Nine deciduous trees tend to be dominant in all communities: sugar maple, northern red oak, tulip poplar, white basswood, cucumbertree (*Magnolia acuminata*), white ash (*Fraxinus americana*), black tupelo, black walnut (*Juglans nigra*), and—previously—American chestnut. A large number of smaller trees appear in the sapling layer, such as sourwood; striped maple; and deciduous magnolias including umbrella magnolia (*Magnolia tripetala*), bigleaf magnolia (*M. macrophylla*), and the Fraser magnolia (*M. fraseri*). Also occurring in the sapling layer may be flowering dogwood, redbud (*Cercis canadensis*), pawpaw (*Asimina triloba*), American hornbeam (*Carpinus caroliniana*), and eastern hophornbeam (*Ostrya virginiana*). The herb layer is equally diverse (Braun 1950; Vankat 1979; Barnes 1991).

The Southeastern Evergreen Forest, located on the Gulf-Atlantic Coastal Plain, from southern Texas to the mouth of the James River in Virginia, is one of the nine forest regions Braun included in the temperate broad-leaved deciduous forest, although within this forest type broad-leaved deciduous trees are largely restricted to bottomlands and swamps and areas such as ravines that are not susceptible to frequent wildfires. Trees such as sweetgum (*Liquidambar styraciflua*), post oak (*Quercus stellata*), turkey oak (*Q. laevis*), blackjack oak (*Q. marilandica*), and southern red oak

(*Q. falcata*) are found on better drained, nutrient-poor soils. In swamp forests, a deciduous conifer, the bald cypress (*Taxodium distichum*), grows with deciduous broad-leaved trees such as water tupelo (*Nyssa acquatica*), Ogeechee tupelo (*N. ogeche*), willows (*Salix* spp.), and sycamore (*Platanus occidentalis*). The land between river valleys was probably covered by fire-resistant longleaf pine (*Pinus palustris*) before European settlement. Today, as a consequence of both forest destruction and tree-farm management practices, much of that forest has been converted to slash pine (*P. elliottii*) and loblolly pine (*P. taeda*). Evergreen broad-leaved trees of the region include live oak (*Quercus virginiana*), southern magnolia (*Magnolia grandiflora*), and American holly (*Ilex opaca*). Upland shrub-bogs (pocosins) are dominated by evergreen shrubs, mostly heaths (Braun 1950; Vankat 1979; Barnes 1991).

West of the Appalachian provinces, in the drier and more continental regions of eastern North America, the forests become less diverse. Braun recognized a Western Mesophytic Forest region of oaks (in particular, northern red oak, white oak, and black oak) and several hickories. Important communities called cedar glades have developed on areas underlain by limestone and dolomite. The dominant species in cedar glades is red cedar (*Juniperus virginiana*). With it may grow post oak and chinkapin oak (*Quercus muehlenbergii*), shagbark hickory (*Carya ovata*), winged elm (*Ulmus alata*), and redbud. Another oak-hickory forest occurs in the westernmost parts of the biome, on the Interior Highlands and much of the Central Lowlands physiographic provinces. The Beech-Sugar Maple Forest Region occurs to its north on the glaciated plains of Indiana and Ohio, is, and the Sugar Maple-Basswood Forest Region is in the so-called Driftless Area of Wisconsin—an area bypassed by continental glaciers during the Pleistocene Epoch (Braun 1950; Vankat 1979; Barnes 1991).

The ninth and final forest region occurs on the glaciated areas of the northeastern United States, including parts of New York and New England. This is the Northern Hardwoods-Conifer Region in which white pine (*Pinus strobus*) and other pines, red spruce, balsam fir (*Abies balsamea*), and eastern hemlock mix with northern red oak, sugar maple, and American beech (Braun 1950; Vankat 1979; Barnes 1991).

Amphibians are well represented in the broad-leaved deciduous forests of eastern North America. Of particular note are the lungless salamanders (Plethodontidae), which reach their highest diversity in the southern Appalachians. The woodland group (**genus** *Plethodon*) has entirely dispensed with an aquatic larval stage and lives under logs and rocks away from the mountain streams in which they likely evolved. Some 30 kinds of lungless salamander are **endemic** to the Southern Appalachians. A few, such as the Peaks of Otter salamander (*Plethodon hubrichti*), have extremely small distribution areas, but others are widespread and are found throughout the region. The red-backed salamander (*P. cinereus*) is one of the most abundant vertebrates throughout much of the temperate broadleaf deciduous forest.

Among 11 kinds of lizard known from the deciduous forest region, the eastern fence lizard (*Sceloporus undulatas*) is most widespread. Many snakes occur, mostly members of the family Colubridae such as the common garter snake (*Thamnophis sirtalis*), rat snake (*Elaphe obsoleta*), common kingsnake (*Lampropeltis getula*), and racer (*Coluber constrictor*). However, venomous vipers (Family Viperidae), such as the timber rattlesnake (*Crotalus horridus*) and copperhead (*Agkistrodon contortix*), also occur (Kitchings and Walton 1991).

Birds are particularly diverse in the mature forests of this biome. A large proportion, as previously noted, are migratory insect eaters. As a group, most of these are what are known as Neotropical migrants because they winter south of the United States in Central and South America. One of the most common—and vocal—of these songbirds is the Red-eyed Vireo (*Vireo olivaceus*), a bird that gleans insects from leaf surfaces. A large number of migrants belong to the wood warbler group. They are most abundant where it is moist and there is dense understory of saplings and shrubs. In the northern part of the forest, common species include the Black-and-white Warbler (*Mniotilta varia*) and Chestnut-sided Warbler (*Dendroica pensylvanica*); in the southern forests, the Hooded Warbler (*Wilsonia citrinia*) and Yellow-throated Warbler (*Dendroica dominica*) are among the more common warblers. A large number of other songbirds come to the forest in springtime to breed, including Wood Thrush (*Hylocicla mustelina*), Hermit Thrush (*H. guttata*), Scarlet Tanager (*Piranga olivacea*), and Rose-breasted Grosbeak (*Pheucticus ludovcianus*). Most hawks, such as the Red-tailed (*Buteo jamaicensis*), Red-shouldered (*B. lineatus*) and Broad-winged (*B. platypterus*), migrate, at least from the northern half of the region. Among birds that stay year round in the forest are such common species as the Black-capped Chickadee (*Poecile atricapillus*), White-breasted Nuthatch (*Sitta carolinensis*), Blue Jay (*Cyanocitta cristata*) and Downy and Hairy woodpeckers (*Dendrocopus pubescens* and *D. villosus*, respectively). Larger, ground-dwelling birds such as Turkey (*Melagris gallopava*) and Ruffed Grouse (*Bonasa umbellus*) are resident in some areas, as are owls such as the screech Owl (*Otus asia*), Great-horned Owl (*Bubo virginianus*), and Barred Owl (*Strix varia*) (Kitchings and Walton 1991).

The short-tailed shrew (*Blarina brevicauda*) is perhaps the most abundant mammal over much of the region, although it is often overlooked because it feeds in the litter. Common bats are eastern pipistrelle (*Pipistrellus subflavus*) and big brown bat (*Eptesicus fuscus*). Characteristic rodents of the eastern forests of North America include gray squirrel (*Sciurus carolinensis*), fox squirrel (*S. niger*), and eastern chipmunk (*Tamias striatus*). The white-footed mouse (*Peromyscus leucopus*) and the deer mouse (*P. maniculatus*) are also abundant. Widespread omnivores such as striped skunk (*Mephitis mephitis*), raccoon (*Procyon lotor*), and black bear (*Ursus americanus*) inhabit these forests, and small carnivores such as long-tailed weasel (*Mustela frenata*), red fox (*Vulpes vulpes*), and bobcat (*Lynx rufus*) are quite abundant. The white-tailed deer (*Odocoileus virginianus*), an abundant animal of forest edge—and increasingly suburbia—has adapted well to human-dominated habitats. Less successful have been the cougar or mountain lion (*Felis concolor*); it was deliberately extirpated over much of the area, but it seems to be staging a comeback. The wolf (*Canis lupus*) has been now replaced by its smaller relative, the coyote (*Canis latrans*), a species of open habitats.

## East Asia

The East Asian temperate forests once extended unbroken from the tropical rainforest in the south to the boreal forest in the north. Much of the surface it covered has been geologically stable since the Paleozoic Era and, unlike Europe and eastern North America, remained largely untouched by glaciers during the Pleistocene. This stability has resulted in the preservation of several older lineages of plants not found in the other parts of the biome. Some genera are relicts (i.e., remain from former climatic conditions) of the Tertiary Period some 60 million years ago, such as two ancient deciduous gymnosperms: ginko (*Ginko)* and dawn redwood (*Metasequoia*). The age of the surface and the presence of several mountain ranges, and the environmental complexity they offer, account at least in part for the very high diversity that characterizes this part of the biome. Although the forests have been destroyed over most of their extent by millennia of intensive agriculture, large tracts still exist in mountains and in reserves around important temples that allow some reconstruction of what the original vegetation was like (Ching 1991).

Three forest types are identified in China. The richest part of the biome, the mixed mesophytic forest, is found in the southern Yangtze Valley, under a climate that is very hot and very humid in summer but cold and dry in the winter, Oaks predominate (e.g., *Quercus acutissima*, *Q. aliena*, *Q. chenii*, *Q. fabri*, and *Q. variabilis*) but are joined by hornbean (*Carpinus fargesii*), chestnut (*Castanea seguinii*), sweet gum (*Liquidamber formosana*), basswood (*Tilia henryana*), sassafras (*Sassafras tzuma*), and a relict broad-leaved tree, *Platycarya strobilacea.* Ginko and dawn redwood grow in this forest type. The mixed mesophytic forest borders a subtropical broad-leaved evergreen forest at its southern limit (Ching 1991).

Between the latitudes of 32° 30′ N and 42° 30′ N lies a warm temperate broad-leaved deciduous forest also dominated by oaks, most of which are different from those associated with the mixed mesophytic forest. Included are *Quercus dentata, Q. liaotungensis*, *Q. mongolica*, and *Q. serrata* (Ching 1991).

The most extensive forest type is found in the northern and northeastern provinces of China. This is the warm temperate mixed northern hardwood forest that contains more than 20 genera of trees. Common are maples (*Acer*), alder (*Alnus*) hornbeam (*Carpinus*), hackberry (*Celtis*), ash (*Fraxinus*), walnut (*Juglans*), poplar (*Populus*), oaks (*Quercus*), and basswoods (*Tilia*). The vegetation consists of several layers. Below the uppermost canopy layer is a second story of smaller trees such as maples, hornbeams, and mountain ash (*Sorbus*); a shrub layer containing plants such as dogwood (*Cornus* spp.), spicebush (*Lindera* spp.), and euonymus (*Euonymus* spp.); and a rich herb layer. This forest type extends into northern Korea and is also found on the Pacific side of Kyushu, Japan, between 37°30′ and 38° N (Ching 1991).

At higher elevations and latitudes in China, at the limits of broad-leaved deciduous forests, a cool temperate deciduous broadleaf forest occurs; it is dominated by birches (*Betula* spp.). A different type of cool temperate forest is found in northwestern Japan, a forest in which beech (*Fagus crenata* and *F. japonica*) and large oaks (*Quercus crispula*) grow above a thick undergrowth of bamboo (*Pleioblastus* spp. and *Sasa* spp.) (Ching 1991).

The native animal life of the remaining fragments of temperate broadleaf deciduous forests in eastern Asia is not well known. Birds and mammals similar to those of the European part of the biome were probably also members of these **ecosystems**. One unique animal is the endangered raccoon dog (*Nyctereuctes procyonoides*).

## Europe

European forests suffered major extinctions during the Pleistocene glacial periods and today are the least diverse of the three major northern hemisphere expressions of the biome. The dominant tree over a wide geographic range and a wide range of environmental conditions is the European beech, *Fagus sylvaticus*. Various subregions within the biome can be distinguished by the other trees that accompany beech, in particular the many oaks. Tree communities are distributed along environmental gradients created by the climatic trend from a maritime climate with relatively mild winters and cool summers in the west, to continental climates with cold winters and great annual temperature range in the east. Latitudinal position also plays a role in climatic differentiation, and broad north-south differences in forest composition are apparent. European plant geographers and botanists recognize two major regional forest types delineated according to latitude: the cool-temperate forest region of Middle Europe and the warm-temperate Submediterranean Forest Region just north of the Mediterranean zone. Within each, several subdivisions or forest provinces are demarcated longitudinally (Eyre 1968; Furley and Newey 1983; Jahn 1991).

Middle European forests are composed of beech with pedunculate oak (*Quercus robur*) and/or sessile oak (*Q. petraea*). The Atlantic province occurs in the cool, humid maritime climate regions of Great Britain and the coasts of the North Sea and Atlantic, where the pedunculate oak was once dominant. The Subatlantic Province is just inland on the continent; it extends east to the Elbe River and south across the Central Massif of France to the Pyrenees. Here sycamore (*Acer pseudoplatanus*), sessile oak, and lime (*Tilia* spp.) join beech as canopy trees. The Central European Province extends eastward from the Subatlantic to central Poland, where beech and sessile oak reach their easternmost limits. A deciduous forest dominated by pedunculate oak and lacking most tree species of the Atlantic and Subatlantic provinces composes the fourth and last province of Middle Europe, the Sarmatic Province, which reaches from Ukraine to the Ural Mountains. The largest remnant of these Middle European forests is the Bialowieza Forest, which covers a mere 1,250 $km^2$ (485 $mi^2$) straddling the Poland/Ukraine border. Bialowieza is dominated by hornbeam; associated trees are ash, alder, oak (*Quercus robur*), lime, maple (*Acer platanoides*), aspen (*Populus tremula*), elms (*Ulma scabra* and *U. campestris*), wild apple (*Malus silvestris*), Scots pine (*Pinus sylvestris*), and spruce (*Picea abies*) (Eyre 1968; Furley and Newey 1983; Jahn 1991).

The Submediterranean Forest Region was characterized by oaks such as *Quercus frainetto*, *Q. pubescens*, and *Q. pyrenaica*. Trees such as Spanish chestnut (*Castanea sativa*), elms (*Ulmus* spp.), and limes (*Tilia* spp.) were also important. The West Submediterranean Province, which occurs in northern Spain and southern France, was dominated by the oak *Q. pyrenaica* and box (*Buxus sempervirens*). The Middle Submediterraenan Province was distinguished by an absence of the key species of the West Province and the presence of eastern trees such as ash-elm (*Fraxinus ornus*), hophornbeam (*Ostrya carpinifolia*), and the oak *Quercus cerris*. The East Submediterranean Province contains maple (*Acer tataricum*), hornbeam (*Carpinus occidentalis*), and several oaks including *Quercus frainetto*, and *Q. peduncula* (Eyre 1968; Furley and Newey 1983; Jahn 1991).

Since the forests of Europe have been so decimated, most animals that survive are apt to be identified as creatures of gardens, hedgerow, and parks rather than deciduous forests per se. As far as birds are concerned, many of the same families found in American forests are represented in Europe: woodpeckers such as the White-backed (*Denrocopus leucotos*), Green (*Picus viridis*) and Gray-headed (*P. canus*);

thrushes such as Blackbird (*Turdus merula*), Song Thrush (*T. philomelos*), and Nightingale (*Luscinia megarhynchos*); tits such as the Great Tit (*Parus major*) and Blue Tit (*P. caeruleus*); and a nuthatch (*Sitta europaea*). The flycatchers, such as the Red-breasted (*Ficedula parva*) and Pied (*F. hypoleuca*), behave like the flycatchers of America but belong to an entirely different family (Muscicapidae versus Tyrannidae in the New World) (Shaeffer 1991). Certainly one of the most often heard birds across the forest region of Europe is the Old World Cuckoo (*Cuculus canorus*), which sounds more mellow than its clock namesakes.

Common small mammals in the European forests include rodents such as the garden dormouse (*Elionys quercinus*); the edible dormouse (*Glis glis*); yellow-necked mouse (*Apodemus flavicolis*); and red-backed mouse or bank vole (*Clethrionomys glareolus*); and insectivores such as European shrew (*Sorex araneus*) and long-tailed shrew (*S. mintus*). Roe deer (*Capreolos capreolus*) is the most common large mammal in the deciduous forest, but red deer (*Cervus elaphus*), fallow deer (*C. dama*), and wild boar (*Sus scrofa*) also inhabit wooded areas (Schaeffer 1991). In the eastern reaches of the biome, the forests were home to wisent (*Bison bonasus*) and tarpan (*Equus caballus*) well into the nineteenth century. Wisent have since been reintroduced; but the tarpan, a small wild horse, is extinct.

## South America

Temperate broadleaf deciduous forests **cover** a very small part of southern South America, but they are worthy of note because of their uniqueness. They are dominated by deciduous southern beeches (*Nothofagus* spp.), members of a genus restricted to lands of the former Gondwanan supercontinent (Donoso 1996).

In central Chile, southern beech forests (Photo 7.4) occur in a region of winter precipitation and summer drought and grow in the mountain ranges at elevations above the Mediterranean **scrub** vegetation that is characteristic of the region. They are composed mostly of roble (*N. obliqua*), actually a Spanish word for oak, which the southern beeches

Photo 7.4 Dense southern beech (*Nothofagus* sp.) forest in central Chile. (Photo by SLW.)

superficially resemble, and hualo (*N. glauca*) but with the locally important addition of the rare *N. alessandri*. These trees attain average heights of 20 to 25 m (65–80 ft) and are associated with numerous evergreen species related to the Mediterranean vegetation of Chile or to the evergreen laurel forests of southern Chile. The sparse shrub layer may contain bamboo (*Chusquea quila*). In the northernmost parts of the range in the Coastal Range (approximately 33°–34° S), small stands of this forest type occur due to the moisture from coastal fogs. Farther south in central Chile, roble-hualo forests are associated with the cooler and more humid south-facing slopes and ravines in the Coastal Range and with intermediate elevations of 650–2,500 m (2,000–8,000 ft) in the Andes (Schmaltz 1991; Donoso 1996).

In southern Chile and southwestern Argentina, deciduous *Nothofagus* forests and mixed *Nothofagus* forests occur in the Coastal range between 37° and 40° S, in the Central Valley between 38° and 41°30′ S, and in the Andes between 36° and 40° 30′ S. These areas have no summer drought, and total annual precipitation increases to the south. Due to the uplift of westerly flowing air masses off the Pacific by the coastal mountains and the Andes, the windward slopes receive 3,000–5,000 mm (120–200 in) of precipitation a year. Temperatures are moderated by the maritime influence but vary according to altitude. On well-drained, lowland sites the dominant species is roble (*N. obliqua*). Here it reaches heights of 40 m (130 ft) or more and diameters greater than 2 m (6 ft). The reddish-colored wood is extremely resistant to rot and has long been used for fencing, outdoor paneling, and beams—much like American chestnut or redwood in the United States. Also like American chestnut, roble is able to resprout after it has been cut. Roble often grows in association with another deciduous southern beech, raulí (*N. alpina*), also commercially important, and an evergreen southern beech (*N. dombeyi*). Beneath the tall southern beeches grow usually smaller evergreen trees such as laurel (*Laurelia sempervirens*); *Persea lingue*; ulmo (*Eucryphia cordifolia*); olivillo (*Aextoxicon punctatum*); avellano (*Gevuina avellano*); and *Drimys winteri*, the winter-blooming holy tree of native Araucanian people. Bamboos (*Chusquea* spp.) dominate the shrub layer. At the highest elevations of forest, the southern beeches mix with, and then give way to, other Gondwanan trees such as the evergreen conifer *Araucaria araucana* (Schmaltz 1991; Veblen et al. 1996).

Cool temperate *Nothofagus* forests occur from 37° 30′ to about 55° S on Tierra del Fuego. An evergreen southern beech (*N. dombeyi*) dominates lower elevations, but deciduous "lenga" (*N. pumilo*) and ñirre (*N. antarctica*) dominate at higher elevations. The latter occurs as a small tree 10–15 m (30–50 ft) tall until at treeline it forms a krummholz 2–3 m (6–10 ft) tall (Veblen et al. 1996).

## ORIGINS

Deciduous flowering plants such as the broad-leaved trees of the temperate broadleaf deciduous forest biome apparently descended from evergreen subtropical forms. Most groups that compose Northern Hemisphere forests today may have originated in subtropical China and differentiated into modern genera during the Cretaceous Period. The occurrence in China today of ancient relict species and the fact that China possesses the greatest variety of trees of any of the Northern Hemisphere parts of the biome support this contention. A land connection between North America and Eurasia was maintained until the mid-Tertiary Period and allowed **dispersal** of plants across the Northern Hemisphere that established a so-called Arcto-Tertiary Geoflora across both North America and Eurasia. During the late Tertiary (Miocene and Pliocene epochs), global cooling led to the disappearance of warmth-demanding plants from midlatitude forests. North America and Greenland separated from Eurasia, and the uplift of major mountain ranges created three isolated regions of Arcto-Tertiary vegetation. Independent evolution continued in each region, leading to innovation at the species level. As a result, the same genera are represented today by different species in North America, Europe, and East Asia. The close relationships of these species can be seen in appearance and in ecology. For example, the maple *Acer mono* in China is very similar to the sugar maple (*A. saccharum*) of North America. One can make similar comparisons among the basswood/limes, beeches, and elms of all three major expressions of the biome.

The global climate changes of the Pleistocene—the alternating cooling and warming that was associated with the advance and retreat of ice sheets particularly in North America and Europe—forced repeated north-south migrations of deciduous broad-leaved tree species. North America, where the major mountain ranges are aligned in an essentially north-south direction, had no barriers to migration, and most species survived the cold periods in pockets of favorable conditions in the southern parts of the continent. In China, trees were able to survive in Manchuria and along the southeast coast. However, in Europe, the mountain ranges run east-west and were barriers to warmth-loving species trying to move southward during glacial periods. Therefore, the Quaternary Period witnessed the extinction of whole families of plants in Europe but little plant extinction at any level in North America or China. For this reason, Europe today has the lowest plant diversity (Röhrig 1991c).

In the Southern Hemisphere, pollen records suggest that southern beech originated during the late Cretaceous Period in the high southern latitudes, in particular in what is today southern South America and the Antarctic Peninsula. Ancient forms may have been related to ancestors of the birches and beeches, but recent scientific work places the genus *Nothofagus* in a separate plant family (Nothofagaceae) (Hill and Dettmann 1996). The composition and geographic distribution of South American forests were affected by mountain-building episodes that occurred from the Paleocene Epoch until the late Miocene Epoch. The increase in ice on the Antarctic continent and in southern South America during the late Tertiary and Quaternary periods influenced climate and affected evolutionary trends. Plants from the **temperate zone** of the Northern Hemisphere were never able to disperse to southern South America, so after the separation of the South American continent from Antarctica, the trees of the *Nothofagus* forests evolved in isolation. During the Quaternary, altitudinal and latitudinal shifts in vegetation occurred in response to glacial and interglacial conditions, although not to as great a degree as in the Northern Hemisphere. The final deglaciation in southern South America began 13,000–10,000 years ago. The appearance of drought-adapted plants in the southern beech forests began about 5,000 years ago, and contemporary assemblages were fully developed by 3,000 years ago (Markgraf et al. 1996).

## HUMAN IMPACTS

Although the timing was different, all regions of temperate broadleaf deciduous forest became centers of agricultural development. In East Asia and Europe, widespread clearing of the forests for food production began a few thousand years ago; in North America some 400 years ago; while in South America, the process is under way right now. Even without forest clearing, the trees have long been exploited for construction, shipbuilding, potash production, leather tanning, and charcoal making. The lowland forests of Europe and China are essentially gone, but remnant forests are preserved in the mountains of China and, to a greater degree, in Japan and Korea. North America has large expanses of broad-leaved deciduous forest today, but almost all are second growth. Only the areas least suited for cultivation and grazing—areas too steep, too wet, or too nutrient poor—have old growth forest today.

In Europe, forest destruction began about 5,000 years ago with clearing and burning to create farmland. By the Bronze Age, heathlands and blanket bogs began to replace forests because of repeated burning and sheep grazing as permanent settlements developed across much of Europe. Increased clearing of forests occurred again in the Roman period of settlement expansion. Some regeneration took place when many villages were abandoned from the thirteenth through fifteenth centuries and during the Thirty Years' War (1618–48). Woodlands, however, were used as pasture for horses, cattle, sheep, and goats, and this practice undoubtedly changed the composition and structure of the understory. Oak trees were protected because people survived the winters on hogs fattened on acorns and they preferred oak timbers for house construction. Oak was also favored for charcoal making, and its bark was used in tanning. The leaves of broad-leaved trees were used as forage, so the trees were commonly coppiced by harvesting whole branches to be carried to livestock. Trees that resprout from their roots (e.g., ash, oak, lime, hornbeam, and hazel) were favored by this practice at the expense of the beeches that do not; therefore beeches began to disappear from the landscape. Princes of the state and church carefully controlled access to their woods and charged fees for peasants' use of the forest, whether it was gathering firewood, hunting game, or pasturing livestock. The open lands or commons came under extreme pressure from overuse and developed into treeless areas. The only area considered relatively unaltered forest today is in Bialowieza National Park in eastern Poland (Jahn 1991).

Reforestation occurred in various locations toward the end of the eighteenth century and into the nineteenth century. Germany, for example, planted oaks when war with Napolean prevented the shipment of tanning bark from The Netherlands. Agricultural trends at this time actually favored widespread reestablishment of wooded areas. The white potato, **introduced** from Andean South America, became used as winter food for livestock, which ended the need to use branches of broad-leaved deciduous trees. The sheep population declined in much of Europe as wool production became concentrated in Ireland and Australia. New crop varieties and the use of fertilizers meant that it was possible to achieve higher agricultural yields per **hectare**, so even though the human population continued to grow, they needed less arable land. Marginal tracts were abandoned. The industrial use of wood also declined when coal replaced charcoal as the main fuel, and stone and steel replaced timber as a construction material. Many of the trees that were planted at this time, however, were not broad-leaved but spruce, pine, and larch (Jahn 1991).

Native agricultural peoples occupied the deciduous forests of eastern North America by 3,000 years ago. They hunted and gathered in the varied foods of the forests as well as farmed. Fire was used to clear temporary plots for the cultivation of maize, beans, and squash, and also as a tool to drive animals toward groups of hunters. Fire was likely also used to clear brush around settlements and thereby prevent ambush by unfriendly tribes. Early English settlers reported 400 years ago that the forests were free enough of shrubs and saplings that a stagecoach could be driven through them (Williams 1989). Some 400 years ago, English and other European settlers introduced plow agriculture and new crops (in particular wheat) as well as domestic livestock (most significantly hogs, sheep, and cattle). They "Europeanized" the landscape and created a shifting mosaic of cultivated field, pasture, and woods. Timber was cut for fences to keep livestock out of the fields, to produce charcoal for smelting iron and potash for soap-making and wool cleaning, and to strip bark for tanning leather. Hogs ran in the woods to fatten on chestnuts, acorns, and hickory nuts. Cattle and sheep also grazed in the woods, initially in an open-range fashion. Pastureland and cultivated land were left fallow after a number of years and were allowed to revert to forest, but eventually that woodland would be cleared again. In the 1800s, other assaults on the forest began as the construction of railroads extended the reach of lumbermen to previously remote locations. The iron industry, at that time centered in the Valley and Ridge Physiographic Province, demanded wood for charcoal, and coal mines needed wood for mine supports. All but the most inaccessible and agriculturally marginal land was stripped of its forests at one time or another (Williams 1989).

Animal life also changed. Large predators such as wolves and cougars were extirpated, as were large herbivores like the elk and bison. Two enormously abundant birds, the Carolina Parakeet (*Conuropsis carolinensis*) and Passenger Pigeon (*Ectopistes migratorius*), both considered agricultural pests, became extinct in the nineteenth century.

The extensive forests one finds today are actually quite young (60–90 years) and certainly not exact replicas of the forests that existed in 1600. They result from major periods of farm abandonment associated with the opening of more productive lands in the West, with the Civil War, and with the Great Depression of the 1930s (Whitney 1994). The 1930s also saw the demise of the American chestnut, once the largest tree in the eastern forests and, in places, the overwhelming dominant. Chestnut blight, caused by a fungus (*Cryphonectria parasitica*), was first seen in New York City in 1906, probably accidentally introduced on Chinese chestnuts imported as ornamentals. The wind carried its spores and rapidly spread the fungus through the Appalachian areas south to North Carolina and west into Ohio, girdling and killing mature trees. Due to its ability to resprout from roots, chestnut may still be found in the shrub layer, but as trees reach a size and age at which they would reproduce, almost all are infected by the blight and killed (Vankat 1979). Watersheds stripped of protective forests led to destructive flooding and a recognition that forest restoration and management was necessary. One outcome of new attitudes was the Weeks Act of 1911, which established a system of National Forests, 10 of which were located in the Appalachian highlands (Loucks 1998).

The regenerated forests of eastern North America are by no means safe from human impact. Today they are under assault from air pollution and introduced insects and diseases. Damage is visible as crown and branch dieback, premature leaf drop, slowed growth, root decay, and increased rates of mortality. Die-offs began in the 1960s but accelerated in the 1980s, when hickories, walnuts, white oaks, beech, yellow buckeye, white basswood, tulip poplar, dogwood, sassafras, hemlock, and sugar maple all shown signs of decline. The eastern forests lie within and downwind of the old manufacturing belt of the United States. They also surround megalopolis, that densely populated and urbanized zone stretching along the east coast from Portland, Maine, to Norfolk, Virginia. Acid deposition in rain, fog, rime ice, and snow may directly damage leaves; the addition of excess nitrogen and sulfur changes the chemistry of the soil and increases its acidity, thereby slowing the bacterial decay processes vital to nutrient cycling and increasing available aluminum to levels that may become harmful to plants. Trees are also being subjected to high doses of ground-level ozone generated from vehicular exhaust. The combination of effects weakens the trees and makes them more susceptible to insects and diseases that otherwise they would be able to fight off. Dying trees open the forest canopy and allow more sunlight to penetrate, which warms and dries the forest floor and creates conditions ill-suited to frogs and salamanders and the regeneration of canopy trees. Slower tree growth can translate into less flowering and fruiting and can affect the populations of animals that depend upon acorns and nuts. Among the many problems affecting the eastern forests are sapstreak disease in sugar maples, caused by the fungus *Ceratocystis coerulescens;* the hemlock woolly adelgid (*Adelges tsugae*) that defoliates and kills hemlocks; beech bark disease that occurs when the bark of beech is damaged by the beech scale (*Cryptococcus fagisuga*) and thereby opened to invasion by fungi (*Nectria coccinea or N. galligena*); and anthracnose (*Discula destructiva*), a fungus that is killing dogwoods. The caterpillar of the introduced gypsy moth (*Lymantria dispar*) defoliates oaks, hickories, and other trees and contributes to the weakening and dying of trees in areas it has infested (Loucks 1998). The forests are also threatened by urban sprawl, chip mills that clear-cut forests to produce pulpwood, and climate change.

Clearing of the deciduous *Nothofagus* forests of South America began only a little more than 100 years ago. This was done mostly to open land for agriculture, but the roble and raulí were also cut for shipbuilding and charcoal production. Today, much of the original forest is gone, and reforestation has largely involved plantations of nonnative Monterey pine (*Pinus radiata*) (Schmaltz 1991). The replanting of commercially important native trees is gaining popularity in some places, however.

## HISTORY OF SCIENTIFIC EXPLORATION

By the time the scientific age began, the forests of Europe and East Asia were pretty much gone. It was a search for alternative resources, including trees and other plants, which led some western European countries to explore and colonize the New World. Thomas Hariot, a member of England's first colonization effort at Roanoke Island in 1585, studied the natural history of the site and was the first to describe plants and animals from the temperate deciduous forest of America. Other expeditions followed, first at the coastal colonies, and later inland, pushing farther and farther west to the Blue Ridge and beyond. Reports always included references to the natural bounty of the region. One of the earliest examples, Captain John Smith's 1612 *A map of Virginia. With a description of the country, commodities, people, government and religion*, demonstrates the combination of inventory, description, and promotion that characterized much of the natural history writing of the Colonial Period. John Lawson surveyed the Carolina colony to evaluate its potential for European settlement and in the process inventoried native plants and animals. He published his results in 1714 as *The History of Carolina*. Between 1712 and 1726, the famous British naturalist and painter Mark Catesby traveled through Virginia and colonies to its south collecting and illustrating plants and animals. Based on his study of New World natural history, Catesby developed a classification system for organisms that predated that of Linnaeus. For the most part, Linnaeus either adopted or slightly modified the names Catesby gave newly discovered forms, so that the scientific names of many species first found in the southeastern part of the United States are attributed to Catesby. The published results of his work, *Natural History of Carolina, Florida, and the Bahamas, Containing the Figure of Birds, Beasts, Fishes, Serpents, Insects, and Plants* was richly illustrated and represents one of the first comprehensive surveys of plants and animals in the British American colonies. Other general works were by Americans. Thomas Jefferson's *Notes on the state of Virginia* (1787), for example, was a significant scientific contribution in its day (Linzey 1998; American Philosophical Society n.d.; Dickinson n.d.).

A tradition of searching for new plants was stimulated by the desire in Europe to find exotic species for formal gardens and new botanical gardens. Sponsored by wealthy European patrons, botanists traveled the wilds of eastern North America looking for new plants. The Bartrams, John and later his son William, were Americans hired for this purpose in the mid-1700s. The French botanist Francois Andre Michaux was another. In 1818 and 1819, Michaux published the three-volume *The North American Sylva, or a Description of Forest Types of the United States, Canada and Nova Scotia* (Paris and Philadelphia: C. D. Hautel) (American Philosophical Society n.d.; Wieboldt 1989).

In more modern times, the classic work on the eastern deciduous forests of North America was produced by E. Lucy Braun. In her 1950 book, *Deciduous Forests of Eastern North America*, she classified the forest into the nine types still recognized today and provided detailed descriptions of the composition of each. Her goal was to give an accurate picture of the presettlement forest. The book is still the standard reference on the biome in America. Since the 1950s, ecological studies have become more important than descriptive work or natural history. Many of the pioneering studies on ecological **succession**, energy flow, and nutrient cycles were conducted in the forests of the eastern United States, and this work continues today.

## REFERENCES

American Philosophical Society. n.d. "Mark Catesby." http://www.amphilsoc.org/library/exhibits/nature/carolina.htm.

Barnes, Burton V. 1991. "Deciduous Forests of North America." Pp. 219–344 in E. Röhrig and B. Ulrich, eds., *Temperate Deciduous Forests*. Ecosystems of the World, 7. Amsterdam: Elsevier.

Braun, E. Lucy. 1950. *Deciduous Forests of Eastern North America*. Philadelphia: Blakiston.

Ching, Kim K. 1991. "Temperate Deciduous Forests in East Asia." Pp. 539–555 in E. Röhrig and B. Ulrich, eds., *Temperate Deciduous Forests*. Ecosystems of the World, 7. Amsterdam: Elsevier.

Dickenson, Ann. n.d. "Curator's notes." For Mark Catesby, The Natural History of Carolina, Florida, and the Bahama Islands: A Hypertext Exhibit. http://www.people.virginia.edu/~jad7e/catesby/contents.html.

Donoso, Claudio. 1996. "Ecology of *Nothofagus* forests in Central Chile." Pp. 271–292 in Thomas T. Veblen, Robert S. Hill, and Jennifer Reed, eds., *The Ecology and Biogeography of* Nothofagus *Forests.* New Haven: Yale University Press.

Eyre, S. R. 1968. *Vegetation and Soils.* 2nd edition. Chicago: Aldine Publishing Company.

Furley, Peter A. and Walter W. Newey. 1983. *Geography of the Biosphere: An Introduction to the Nature, Distribution and Evolution of the World's Life Zones.* London: Buttersworth.

Hill, Robert S. and Mary E. Dettmann. 1996. "Origins and diversification of the genus *Nothofagus.*" Pp. 11–24 in Thomas T. Veblen, Robert S. Hill, and Jennifer Reed, eds., *The Ecology and Biogeography of* Nothofagus *Forests.* New Haven: Yale University Press.

Jahn, Gisela. 1991. "Temperate Deciduous Forests of Europe." Pp. 377–502 in E. Röhrig and B. Ulrich, eds., *Temperate Deciduous Forests.* Ecosystems of the World, 7. Amsterdam: Elsevier.

Kitchings, J. Thomas and Barbara T. Walton. 1991. "Fauna of the North American Temperate Deciduous Forest." Pp. 345–370 in E. Röhrig and B. Ulrich, eds., *Temperate Deciduous Forests.* Ecosystems of the World, 7. Amsterdam: Elsevier.

Linzey, Donald W. 1998. *The Mammals of Virginia.* Blacksburg, VA: The McDonald & Woodward Publishing Company.

Loucks, Orie. 1998. "In changing forests, a search for answers." Pp. 85–97 in Harvard Ayers, Jenny Hager, and Charles Little, eds., *An Appalachian Tragedy: Air Pollution and Tree Death in the Eastern Forests of North America.* San Francisco: Sierra Club.

Markgraf, Vera, Edgardo Romero, and Carolina Villagian. 1996. "History and paleogeography of South American *Nothofagus* Forests." Pp. 354–386 in Thomas T. Veblen, Robert S. Hill, and Jennifer Reed, eds., *The Ecology and Biogeography of* Nothofagus *Forests.* New Haven: Yale University Press.

Röhrig, Ernst. 1991a. "Introduction." Pp. 1–5 in E. Röhrig and B. Ulrich, eds., *Temperate Deciduous Forests.* Ecosystems of the World, 7. Amsterdam: Elsevier.

Röhrig, Ernst. 1991b. "Seasonality." Pp. 25–33 in E. Röhrig and B. Ulrich, eds., *Temperate Deciduous Forests.* Ecosystems of the World, 7. Amsterdam: Elsevier.

Röhrig, Ernst. 1991c. "Floral composition and its evolutionary development." Pp. 17–23 in E. Röhrig and B. Ulrich, eds., *Temperate Deciduous Forests.* Ecosystems of the World, 7. Amsterdam: Elsevier.

Schmaltz, Jürgen. 1991. "Deciduous forests of southern South America." Pp. 557–578 in E. Röhrig and B. Ulrich, eds., *Temperate Deciduous Forests.* Ecosystems of the World, 7. Amsterdam: Elsevier.

Shaeffer, M. 1991. "Fauna of the European Temperate Deciduous Forest." Pp. 503–525 in E. Röhrig and B. Ulrich, eds., *Temperate Deciduous Forests.* Ecosystems of the World, 7. Amsterdam: Elsevier.

Vankat, John L. 1979. *The Natural Vegetation of North America, An Introduction.* New York: John Wiley & Sons.

Veblen, Thomas T., Claudio Donoso, Thomas Kitberger, and Alan J. Rebertus. 1996. "Ecology of southern Chilean and Argentinean *Nothofagus* forests. Pp. 293–353 in Thomas T. Veblen, Robert S. Hill, and Jennifer Reed, eds., *The Ecology and Biogeography of* Nothofagus *Forests.* New Haven: Yale University Press.

Whitney, Gordon G. 1994. *From Coastal Wilderness to Fruited Plain. A History of Environmental Change in Temperate North America 1500 to the Present.* New York: Cambridge University Press.

Wieboldt, Thomas F. 1989 "Early botanical exploration and plant notes from the New River, Virginia section." Pp. 149–159 in *Proceedings of the Eighth Annual New River Symposium*, Oak Hill, West Virginia. National Park Service.

Williams, Michael. 1989. *Americans and Their Forests: A Historical Geography.* New York: Cambridge University Press.

# 8 BOREAL FORESTS

## OVERVIEW

A nearly continuous belt of cone-bearing **needle-leaved** trees stretches across northern North America and Eurasia. People refer to this forest and **biome** by various names: Boreal (meaning "northern") Forest, Taiga (originally a Russian term), **Evergreen** Needleaf Forest, and Northern Coniferous Forest. The term Boreal Forest will be used here to emphasize the geographic fact that this forest is limited to the high **latitudes** of the Northern Hemisphere. This biome, which appears monotonously the same over hundreds of square miles, contains 29 percent of earth's total forest area.

The Boreal Forest is characterized by a small number of tree **species** in a mosaic of plant communities determined by local **soil**, drainage conditions, and fire history. Bogs and other wetlands are common. Four genera of **conifers** dominate on both continents: spruces (*Picea*), firs (*Abies*), pines (*Pinus*), and larches (*Larix*). Two genera of **broad-leaved** trees are common in areas of recent or frequent **disturbance**: birches (*Betula*) and aspens or poplars (*Populus*). Different species in each of these genera occur in different parts of the biome.

The northern limit of the Boreal Forest corresponds to the Arctic tree line, where the Boreal Forest meets the tundra biome. In North America, the tree line angles across the continent from about 68° N in Alaska to 58° N in eastern Canada, while in Eurasia it largely parallels the Arctic coastlines a few hundred kilometers inland, dipping southward to any major extent only in central Siberia (Larsen 1980). The establishment of trees poleward of tree line is apparently prevented by a continuous layer of permafrost—permanently frozen ground—close to the surface and not by present climatic conditions.

The forest forms a belt some 1,000 km (600 mi) wide across both continents. In North America, major southward extensions occur along the Coastal Ranges, the Cascade range, Sierra Nevada, and the Rocky Mountains in the west and the Appalachian Mountains in the east. A large proportion of the Boreal Forest region

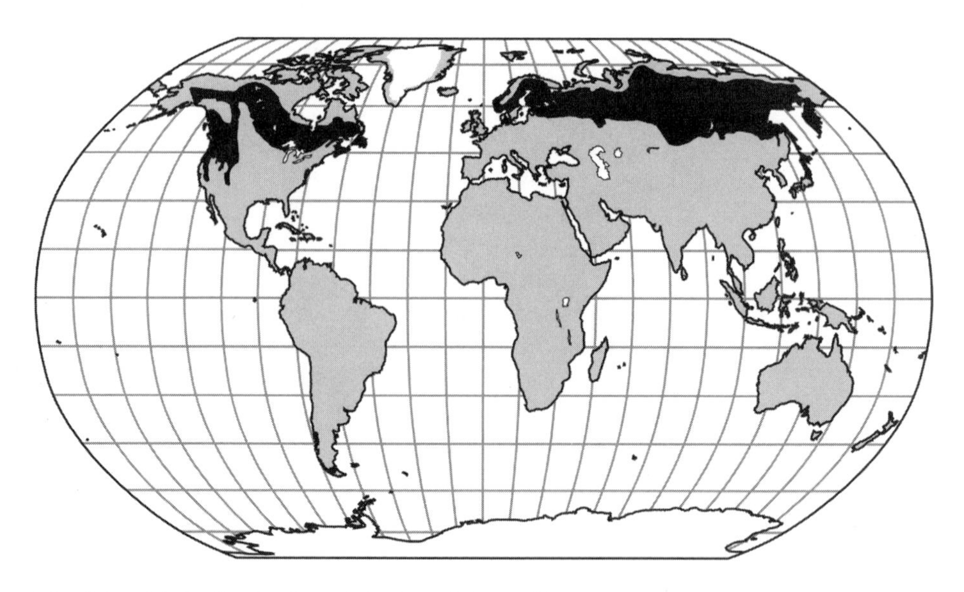

Map 8.1 World distribution of the Boreal Forest Biome.

is covered by lakes and bogs, consequences of Pleistocene glaciation and Recent **vegetation** cycles, which will be described later. The advancing and retreating ice sheets of the Pleistocene shaped the earth by removing preexisting soils; gouging depressions; and/or depositing vast, irregular hills and sheets of sands and gravels. Once the ice retreated, the depressions filled with rainwater to form myriad lakes (some, such as the Great Slave and Great Bear lakes in Canada and the Great Lakes, are enormous). Ridges of glacial deposits blocked streams to cause additional ponding. Before plants could invade the ice-free ground, winds redistributed sands to create sand dunes, especially along the beaches of the larger lakes. The surface plants of the Boreal Forest grow on therefore varies greatly in its material composition and drainage. This variability is reflected in the distribution of plant communities, with some favoring dry sites and others wet and some able to tolerate low nutrient conditions while others cannot.

## CLIMATE

The boreal forest biome corresponds with subarctic and cold continental **climate** types (Dfc, Dfd, and Dwd climate types in the Koeppen system). Winters are typically long and severely cold, with up to six months having an average temperature at or below freezing. Summers are short and cool to warm. Only 50 to 100 days may be frost free. Except where moderated by closeness to the sea, a distinguishing characteristic of the climate is the great range in temperatures experienced in a year. The greatest range in mean monthly temperatures (56° C or 130° F) occurs in eastern Siberia; in North America, the maximum range of mean monthly temperatures (44° C or 110° F) occurs in interior Alaska (Bonan and Shugart 1989). Verkhoyansk, Russia, holds the record for temperature extremes; it has reported a minimum daily low of –68° C (*minus* 90° F) and a maximum daily high of 32° C (*plus* 90° F). Mean annual precipitation is 15–20 inches, but this is sufficient moisture to support forests because of the generally low temperatures and hence low evaporation rates. Winters are typically dry (winter drought is extreme in eastern Siberia); more than half the yearly precipitation occurs in summer (Bonan and Shugart 1989).

In North America, the northern limits of the biome coincide with the summer position of the Arctic Front, a meeting place of different air masses. To the north of the front cold, dry Arctic and continental polar air dominates (MacDonald et al. 1998). The southern border of the biome corresponds with the winter position of the front and the line where the annual minimum temperature averages –40° C (–40° F). This relationship between temperature minimum and the southern limit does not hold in Eurasia (Gates 1993).

## VEGETATION

The Boreal Forest consists of a single **canopy** or tree layer, which at any given location may be composed of only one or two different kinds of tree. Few large shrubs grow there. The ground layer may contain some low shrubs, often members of the **heath** family (Ericaceae), and scattered **herbs**, such as wood sorrel (*Oxalis* spp.). Often a continuous ground **cover** of mosses and/or **lichens** occurs. The coniferous trees are typically evergreen, although the larches (sometimes called tamarack in North America) are a conspicuous exception, especially in the autumn when their needles turn golden before they drop. The retention of needles year round allows plants to photosynthesize whenever temperatures permit and is viewed as an adaptation to a short growing season. The needles of evergreen conifers are very dark green, which lets the foliage absorb maximum solar radiation to increase leaf temperatures and promote **photosynthesis** earlier in the spring than air temperature alone might allow (Photo 8.1). The long winters mean low

Photo 8.1 Boreal Forest near Denali National Park, Alaska. The shape and dark tone of the evergreen conifers is characteristic. An understory of heath shrubs is conspicuous here because a road has permitted sunlight to enter the forest edge. (Photo by SLW.)

temperatures and the unavailability of water while it is frozen in the soil. Leaves exposed to the winter wind are susceptible to damage from water loss from evaporation, and several properties of the needles of boreal species act to minimize such loss. First, the narrow shape of the leaf reduces the surface area through which water is transpired. Furthermore, the pores (**stomata**) through which plants give off water vapor are sunken and thereby somewhat shielded from drying winds. The needles also have thick waxy coatings or cuticles, which further waterproofs them. Trees are essentially dormant during the dark, cold winter. The foliage undergoes a process of hardening in late autumn that increases its resistance to frost and prevents damage to the needles at temperatures as low as of –60° C (–76° F) in some species. Without hardening, spruce needles are killed at –7° C (19° F) (Walter 1985). In areas where winters are exceptionally cold and dry, such as in the interior regions of eastern Siberia, even these properties are not enough to permit survival, and the **deciduous habit** becomes necessary to prevent destruction of the life-giving photosynthetic surfaces. Under such conditions, larches and broad-leaved deciduous birches and aspens prevail.

Boreal conifers are usually conical or spire shaped, which may help shed snow and lessen the breakage of branches. Across most of the forest belt, the stature of the trees is not great (15–24 m or 50–80 ft), and their height decreases as one approaches the arctic tree line, where dwarfed conifers intermingle with tundra plants. There are some spectacular exceptions to this generality. In the southward extensions of the biome, on the wetter slopes of mountains in the Pacific Northwest of Canada and the United States, giants grow; western hemlock (*Tsuga heterophylla*), Douglas fir (*Pseudotsuga menziesii*), and—in California—the giant Sequoia (*Sequoia gigantea*) and the coastal redwoods (*Sequoia sempervirens*) attain heights well over 60 m (200 ft).

Latitudinal **zonation** within the Boreal Forest is evident. At its northern limits, the forest penetrates into a mixing zone or ecotone with tundra. The conifers are scattered and dwarfed since only the parts covered by snow escape winter damage. Reproduction is by layering, a slow process in which branches touching the ground will sprout roots and send up a leader branch to form a new individual. Layering is a type of cloning or **asexual reproduction** that occurs when environmental conditions prevent seed germination. Dwarfed deciduous broad-leaved trees may also survive at or near tree line. Representative trees are identified in the following regional descriptions.

Southward of the forest-tundra ecotone lies an open-canopied coniferous forest with a continuous ground cover of dwarfed shrubs, mosses, and lichens. According to some scientists, it is this zone that is most properly referred to as taiga. Moving still farther south, one encounters the wide swath of closed-canopy, needle-leaved evergreen forest characteristic of the boreal forest biome. At the southern borders of the biome, another ecotone or mixing zone is found, or rather several since vegetation types south of the Boreal Forest tend to vary from west to east. Inland from the western mountains of North America, the Boreal Forest mixes with the shortgrass and tallgrass prairies of the temperate grassland biome as an aspen parkland. From the Great Lakes eastward, the Boreal Forest ecotone with the temperate broadleaf deciduous forest biome is a mixed forest of needle-leaved evergreen trees such as eastern white pine (*Pinus strobus*) and eastern hemlock (*Tsuga canadensis*), and broad-leaved deciduous trees such as northern red oak (*Quercus rubra*), sugar maple (*Acer saccharum*), and American beech (*Fagus grandifolias*). (Note that in some schemes, this mixed forest is considered part of the temperate broadleaf deciduous forest biome.) In Eurasia, the southern ecotone is primarily a forest-steppe (grassland) transition.

## The Vegetation Mosaic

Different plant communities are found in the Boreal Forest according to drainage conditions, nutrient levels, frequency of fire, and **successional** stage. In mature stands, some of the conifers prefer dry, well-drained sites; others concentrate in moister, poorly drained areas. (See the regional descriptions for the trees found in each situation.) A **landscape**-scale pattern of forest and wetland also repeats (Photo 8.2). The most distinctive boreal wetland is the northern bog or muskeg (an Algonquian Indian term meaning "trembling earth" relating to the fact that when one walks on these, the surface ripples as though walking across a trampoline). Bogs are composed of thick mats of peat moss (*Sphagnum* spp.) (Photo 8.3) overlying and floating upon a body of standing water, at last during the spring (Larsen 1982). Small shrubs, usually heaths such as Labrador tea (*Ledum groenlandicum*) or bilberries and cranberries (*Vaccinium* spp.), and herbaceous plants—often the same found in the Arctic tundra—may grow on the mat. The sedges (*Eriophorum* spp.) commonly called cottongrasses are indicator species. Terrestrial orchids such as bog rose (*Arethusa bulbosa*) are also associated with bogs. Larches may encroach from the margins. The bog environment is highly acidic (**pH** = 4.0) and low in nutrients. Among the unusual plants adapted to survive these conditions are the carnivorous plants that trap and

Photo 8.2 Muskeg circled by black spruce in northern British Columbia, Canada. (Photo © Gunter Marx/CORBIS.)

Photo 8.3 Sphagnum moss. It holds moisture like a sponge and perpetuates the acidic, boggy conditions of muskeg. (Photo by SLW.)

digest insects as a source of nitrogen, such as pitcher plants (*Sarrracena purpurea*) and sundews (*Drosera* spp.). Bogs may develop as steps in the succession from glacial ponds to dry land, or they may be created by **paludification** of spruce forests—a process described in the next section. It has been estimated that almost one-third of western Siberia is covered with bogs and other types of wetland, perpetuated in part by the spring flood of north-flowing rivers still blocked by ice in their downstream reaches when the southern upper watershed has thawed.

Fens (marshes usually occupied by sedges) and swamps (wooded wetlands) occur where flowing water is found. Swamps created by beavers damming streams are short lived but important **habitats**. Alder (*Alnus* spp.), willow (*Salix* spp.), and poplar (*Populus* spp.) thrive in such waterlogged situations and provide food and cover for a number of mammals and birds.

## Vegetation Cycles

Classic textbook ecological succession probably doesn't occur in the Boreal Forest. Rather an ever-shifting mosaic of plant communities exists in response to fire, soil moisture, depth to permafrost, depth of the organic layer on the forest floor, and nutrient depletion—often in combination with each other. The complex processes involved and their interconnections are not fully understood, but some of the probable ways one plant **community** is replaced by another in cycles of forest growth and decline are identified here.

Fire is the factor most scientists feel dominates vegetation cycles in the Boreal Forest (Photo 8.4). Fires are common and recur at intervals of 50–200 years in much of North America, although up to 500 years in the wetter, eastern parts of the biome (Bonan and Shugart 1989). Many fires range over small areas (4–5 ha or 10–12 ac), but some of the largest forest fires in the world, which have blackened well over 100,000 ha (250,000 ac), are known from this biome (Johnson 1992). Jack pine (*Pinus banksiana*) and Scots pine (*P. sylvestris*) are among the few tree species with bark thick enough as adults to tolerate average temperature fires. However, forest fires in this biome typically are very hot **crown** fires or hot surface fires that kill adult trees and thereby foster regeneration of whole stands. The tree species that invade the burned site are determined by the source of available seeds. Two

Photo 8.4 The patchy nature of severe burn. Taken in Yellowstone National Park about six months after the fires of 1988, this photo shows pioneer fireweed (*Epilobium* sp.) already in bloom in a patch of forest that had been reduced to ash. Fire plays a major role in the Boreal Forest biome. (Photo by SLW.)

conifers in the Boreal Forest—jack pine and black spruce (*Picea mariana*)—maintain closed (serotinous) cones on their branches and depend on fire to open the cones and let the seeds fall to the ground. These pines will replace themselves on the burned area. Common pioneers that invade from beyond the burn area are birches and aspens (or poplars), broad-leaved trees with light, wind-dispersed seeds that readily get to newly opened sites. These trees also sprout from roots or stumps, so they may regenerate themselves if fires occur frequently enough (i.e., before the broad-leaved species have been replaced by conifers) (Bonan and Shugart 1989). The usual consequence of fire, thus, is an even-aged stand of a single species. Since fires are often spotty in their distribution, a patchwork of stands of different trees and/or different ages develops over time.

In the absence of fire, the more slowly dispersing white spruce (*Picea glauca*) will invade aspen and birch stands on well-drained sites in North America (Photo 8.5). After some 150 years, the original, short-lived aspen or birch will die off and their seedlings will be unable to grow in the shade of the maturing spruce, so white spruce will come to dominate the site. As the spruce mature and take over, however, they reduce the amount of nitrogen in the soil, which is essential to plant growth. The spruce trees draw nutrients out of the soil and incorporate them into their trunks, branches, and needles. But since the spruce needles decay very slowly when shed, fewer nutrients are returned to the soil than are removed in a given unit of time. The gradual depletion of nitrogen in the soil will slow the growth and impair the health of the spruce, which makes them vulnerable to attack by insects and diseases. Over the next 125–200 years, the spruce die back and open the forest canopy.

Photo 8.5 An even-aged stand of aspens, conspicuous with their white trunks. They will eventually give way to the white spruce now growing in their shade. (Photo by SLW.)

As sunlight reaches the forest floor, it warms the layer of undecayed organic matter on the forest floor and decomposition rates increase, which releases nitrogen in forms usable by plants (Pastor et al. 1987). The sunlight penetrating to the forest floor also can dry out the ground-layer mosses and organic matter and make the site prone to fire and then invasion by birch and aspen. Even in the absence of fire, the **open canopy** allows the seeds of sun-loving birch and aspen that are blown onto the site to germinate. The seedlings thrive because low-nitrogen soils are not a problem to birch and aspen since both have nitrogen-fixing bacteria in nodules on their roots. The annual decay of the deciduous, nutrient-rich leaves of the broad-leaved trees builds up the nitrogen levels in the soil to make the site suitable once again for invasion by white spruce. And so the cycle is repeated.

Nitrogen is a major limiting factor for the growth of conifers in the Boreal Forest. In addition to the impact of the spruce, mosses and lichens have been implicated in the depletion of soil nitrogen and the subsequent die-off of needle-leaved trees. On drier sites, lichens form a ground cover beneath an open canopy of pine or spruce. The lichen mat reflects sunlight and contributes to a lowering of soil temperature. At lower temperatures, decomposition slows, which may decrease nutrients (especially nitrogen) to levels below that necessary to maintain tree growth. In closed forests, the moister conditions favor mosses over lichens. Mosses, which lack roots and a true **vascular** system, absorb precipitation through their foliage and hence trap any nutrients dissolved in it before the moisture reaches the ground. This may accelerate nutrient depletion in the soil. Mosses retain so much moisture that they can create waterlogged, **anaerobic** conditions at the ground surface in which decomposition is inhibited. In addition, mosses may insulate the soil and cause lower temperatures, further slowing rates of decomposition and recycling of nutrients between vegetation and soil. As the dead needles and dead mosses fail to decompose, the depth of the organic layer on the forest floor becomes greater [averaging 20–30 cm (8–12 in) under black spruce stands in interior Alaska]; this may retard the regeneration of spruce by preventing the roots of seedlings from reaching the soil beneath (Bonan and Shugart 1989). Slowly, the spruce forest will be replaced by an open, treeless community of mosses, low shrubs, and herbs in a process known as paludification (Larsen 1980). This may set the stage for encroachment by the *Sphagnum* mosses, which perpetuate a cool, wet, acidic environment and create a bog.

Some scientists have implicated mosses in the destruction of spruce forests in other ways. For example, mosses and the highly acidic soil environment they promote may kill the fungi that partner with tree roots to extract nutrients from the soil (see "Soils" next) or create a soil water acidic enough to release aluminum from the mineral portion of the soil (Klinger 1991). Aluminum is toxic to most plants. In areas of permafrost, the insulating layer of mosses can reduce temperatures to the point that permafrost rises close enough to the surface to destroy all but the most shallow-rooted trees.

Photo 8.6 The profile of a spodosol with the ashy layer that gives the soil type its name. In this instance, the layer is about 3.5 cm (1.5 in) deep. (Photo by SLW.)

## SOILS

Soils of the Boreal Forest are young and shallow. The acidic nature of litter composed of conifer needles is a major influence in the formation of diagnostic horizons. In the soil-forming process known as podzolization, the acidic water that filters through the dead needles on the forest floor and into the ground removes two of the three major mineral components of soil from the top inch or two of the soil. Compounds of iron and of aluminum are washed downward to leave behind a thin but quite visible powdery, white layer of silicon oxide as the soil's A horizon (Photo 8.6). Early Russian soil scientists described this layer as ash (*podzol* in Russian) and gave the soil type its common name of podzol. In the U.S. system of soil classification, this soil is known as a spodosol, a name that preserves the original Russian root.

The iron and aluminum compounds collect a few inches below ground in the B horizon, as do other compounds that are washed out of the A horizon, including **humus**—black, partially decayed organic material. The iron and the little amount of humus present give the B horizon its characteristic reddish-brown color. The B horizon may be a foot or two deep and lies upon the weathered glacial deposits or bedrock of the C horizon.

Spodosols (podzols) are low in natural fertility for at least two reasons. Many of the important plant nutrients are bases, and the acidic soil water is able to leach them out of the topsoil. Also, because of prolonged periods of low temperatures, the rate at which dead leaves and needles decompose is very slow. This means that at any given time the soils contain very little humus, which ordinarily holds nutrients in the soil long enough for them to be exchanged with plant roots. Nutrients, especially nitrogen but also phosphorus, are significant limiting factors to plant growth in the Boreal Forest as previously described. Mycorrhizal fungi that grow on the outside of plant roots are vital components of soil life and are associated with many species of tree. Their threadlike hyphae are able to scavenge phosphorus from the soil and process nitrogen compounds into forms that plants can use, and they exchange these at the plants' roots for energy-rich sugars produced by the photosynthesizing trees. Mycorrhizae may bind the forest together below the soil surface in a nutrient distribution network. The mushrooms often abundant in the Boreal Forest are the above-ground fruiting bodies of some of the mycorrhizal fungi; many are edible and provide food for wildlife and humans.

## ANIMAL LIFE

The needles of boreal conifers have little nutrient value for herbivorous animals, so animals depend largely on the plants of the forest floor or the broad-leaved trees and shrubs in bogs or burned areas for browse. Some mammals and birds are able to open pine and spruce cones to extract the seeds. However, cone production is variable from year to year, and the animals depending on them may be forced to leave the Boreal Forest during winters of seed shortages.

Among mammals and birds, a number of species occur in the Boreal Forests of both North America and Eurasia. They are called panboreal in terms of their distribution patterns. Some will be discussed here. Species limited to one continent will be identified in the regional descriptions later.

Moose (*Alces alces*), called elk in Europe, is the largest member of the deer family and the largest plant-eating mammal of the Boreal Forest. Moose browse on trees, shrubs, and wetland and pond plants; they prefer the foliage of willows and poplars along the shores of streams and ponds. Another large deer, the elk or wapiti (*Cervus elaphus*), called red deer in Europe, is associated with open sites in the Boreal Forest, where it grazes on grasses when the ground is free of snow. During the winter, it will browse the twigs and foliage of shrubs and trees that extend above the snow. The Boreal Forest is the winter home of another deer, the barren ground caribou (*Rangifer tarandus*) called wild reindeer in Eurasia. The springtime migration of huge herds of these animals onto the Arctic tundra for calving is one of nature's spectacles. However, a woodland form of the caribou also lives in the Boreal Forest.

Caribou feed on lichens, especially the so-called reindeer mosses (*Cladonia* spp.), as well as the leaves of broad-leaved plants in summer. In winter, they eat what lichens and twigs they can find by digging through the snow but largely depend on fat stores accumulated during summer months.

Large carnivores of the Boreal Forest include wolf (*Canis lupus*), lynx (*Felis lynx*), and brown or grizzly bear (*Ursus arctos*). The brown bear is actually omnivorous and subsists on grasses, sedges, roots, bulbs, and berries, as well as small mammals and fish; occasionally, however, it will take the very young or the infirm among large mammals. Red fox (*Vulpes vulpes*) and members of the weasel family, such as wolverine (*Gulo gulo*), ermine (*Mustela erminea*), and least weasel (*M. nivalis*)—the world's smallest carnivore—prey on smaller mammals. The animals just named are all furbearers; they and others like them have long been important in the economies of peoples inhabiting the forest.

Small mammalian herbivores such as mice, voles, squirrels, and hares are ecologically important in that they support the populations of carnivores. One of the notable features of animal life in the Boreal Forest is the periodic population increases and crashes many species undergo. The textbook example is the approximately 10-year cycle of snowshoe hare (*Lepus americanus*) and lynx in North America. By feeding on woody browse, populations of hare build up after fire or other disturbance creates shrub communities. With an increase in their chief prey species—the snowshoe hare—lynx populations also grow. As the hare populations peak, browse decreases. When the hares have depleted their food supply, their death rates increase and hare numbers decline. Faced with low numbers of their primary prey species, lynx may direct their hunting to other species, such as grouse or voles, and cause periodic declines in those species; eventually, however, the lynx populations plummet. With fewer predators on the scene and recovery of the vegetation, the hare population is able to increase once again. Lynx revert to hunting their preferred prey, and grouse populations rebound. With abundant hares, the lynx population again increases and the cycle repeats itself (Larsen 1980).

Among resident bird species in the Boreal Forest are grouse, such as the panboreal Willow Ptarmigan (*Lagopus lagopus*); Common Raven (*Corvus corax*); woodpeckers; and a number of owls, of which the Great Gray Owl (*Strix nebulosa*) is panboreal. Most songbirds are migratory and are abundant only during the brief summers when they come to the Boreal Forest to breed, but some tits or chickadees are resident. Birds that depend on the seeds of the conifers, such as the Red and White-winged crossbills (*Loxia curvirostra* and *L. leucopatera*, respectively) and Redpoll (*Acanthis falmea*) are wanderers that continually search for areas with good cone production. They will remain in the Boreal Forest throughout the winter during good cone production years but migrate south during poor ones. Their erratic mass migrations southward are known as irruptions. In the United States, crossbills can be found in Arizona and Virginia when irruptions occur.

Flying insects have a significant presence in the Boreal Forest. Mosquitoes and blackflies are particularly aggravating to people and other animals, but insects are also the main food of many of the songbirds that come to the northern forest to breed each year. Other native insects are considered major pests by foresters because they can defoliate and kill whole stands of needle-leaved trees. In North America, the main culprit is the spruce budworm (*Choristoneura fumiferana*); in Eurasia the Siberian silkworm (*Dendrolimus superans sibiricus*) and the big coniferous long-horned beetle (*Monochamus urossivi*) are particularly damaging (Bonan and Shugart 1989). These insect pests undergo periodic outbreaks similar to the population explosions of snowshoe hare or lynx previously mentioned. The population cycles of these insects may be related to the age and condition of the forest since they tend to attack only weakened or dying trees.

## MAJOR REGIONAL EXPRESSIONS

### North American Boreal Forest

In North America, a nearly unbroken belt of coniferous forest stretches across the continent. The main variant is found in the Pacific Northwest, where maritime influence moderates temperature extremes and coastal mountain ranges intercept moist air blowing off the Pacific; as a consequence, 125–175 cm (50–70 in) of precipitation a year is usual. The montane conifer forests southward and farther inland in the Cascades, Sierra Nevada and the Rocky Mountains are also considered extensions of the Boreal Forest, as are high-elevation forests in the Appalachians. Brief descriptions of these are given after the characterization of the main portion of the North American Boreal Forest.

The North American Boreal Forest has nine common, dominant kinds of trees. Six are conifers, and three are broad-leaved (Woodward 1995). While the structure and appearance of the forest are quite uniform across the entire belt of Boreal Forest, the dominant trees in a given **genus** tend to replace themselves from east to west, with white spruce (*Picea glauca*) being most widely distributed. As for the spruces, the highest diversity is found in Canada's Maritime Provinces and northern Maine, where white, red (*P. rubens*), and black spruce (*P. mariana*) occur. White spruce prefers better-drained slopes; black spruce grows in the moist, poorly drained, fireproof lowlands, where it may be joined by tamarack (*Larix larina*) at the edges of bogs. Westward across Canada to the Rocky Mountains, red spruce disappears and only white and black spruce occur, but they are separated by different ecological preferences. In the Rockies, white spruce is joined by Englemann spruce (*P. englemanni*). White spruce is absent in the westernmost parts of the biome. In the Pacific coastal forests from Alaska south into Washington state, the only spruce found is Sitka spruce (*Picea sitchensis*). As for the true firs, balsam fir (*Abies balsamea*) occurs east of the Rocky Mountains; alpine fir (*A. lasiocarpa*) occurs in the Rockies and westward. The closely related two-needled jack pine (*Pinus banksiana*) and lodgepole pine (*P. contorta*) also replace each other at the Rocky Mountains, with jack pine occurring to the east and lodgepole pine in the mountains and westward (Larson 1980; Pielou 1988).

Broad-leaved trees are short-lived members of early successional communities that invade open territory after flood, fire, or other disturbance has removed the conifers and allowed sunlight to reach the forest floor. The common species are paper birch (*Betula papyrifera*), quakening aspen (*Populus tremuloides*), and balsam poplar (*Populus balsamifera*) (Larson 1980; Pielou 1988).

The large herbivores of the Boreal Forest in North America include moose, woodland and barren ground caribou, and elk. Among smaller mammals, red squirrels (*Tamiasciurus hudsonicus*) and flying squirrels (*Glaucomys sabrinus*) are herbivores of the treetops, while snowshoe or varying hare (*Lepus americanus*), voles, and mice forage for seeds, twigs, or fungi on the forest floor. These small animals support an array of carnivores, such as fisher (*Martes pennanti*), pine marten (*M. americana*), and long-tailed weasel (*Mustela frenata*), which move adeptly through the tree canopy in pursuit of prey. Lynx, wolves, foxes, and coyotes (*Canis latrans*) prey on hare and small ground-dwelling rodents and birds. Mink (*Mustela vison*) are usually found near open water where they hunt small mammals, fish, and crayfish. Playful river otters (*Lutra canadensis*) seek invertebrates, fish, and frogs in the streams and ponds. The powerful wolverine is an omnivore in summer and a carnivore and scavenger during the winter.

Beaver (*Castor canadensis*) and muskrat (*Ondatra* zibethica) burrow in stream banks and feed in ponds. The beaver also makes its own ponds by damming streams. Beaver was certainly the most important furbearer in the history of settlement in North America. The desire for its pelt in Europe to make fashionable men's hats supported the French and English fur trades and drew European trappers and explorers far into the interior of the American continent in the 1600s and 1700s. Another rodent of the Boreal Forest and one unique to North America is the porcupine (*Erethizon dorsatum*), which foresters consider a pest because it eats tree bark in winter.

Conifer forests and muskeg are home to the spruce grouse (*Canachites canadensis*), while the Sharp-tailed Grouse (*Pedioecetes phasianellus*) prefers open burn sites. Boreal Owl (*Aegolus funerus*) and little Saw-whet Owl (*A. acadius*) silently hunt in the woods at night, as the Barred Owl (*Strix varia*) does east of the Rockies. The raven, a year-round resident and object of much Native American lore, is an omnivore and feeds on carrion to survive the long winters. Its relative the Gray Jay or Whiskyjack (*Perisoreus canadensis*) is a resident species; it is relatively tame and frequently visits campsites. Other resident species are woodpeckers such as the Three-toed (*Picoides tridactyla*), Black-backed (*P. arcticus*), and large, red-crested Pileated Woodpecker (*Dryocopus pileatus*), which feed on insects but can extract dormant insects or their larvae from beneath tree bark in the winter; this gives them a year-round food supply, unlike species that depend upon flying insects only (Larson 1980; Pielou 1988).

Songbirds are primarily migratory, with only 10 percent of them year-round residents (Larsen 1980). In summer, the Boreal Forest is home to such insect-eaters as Swainson's Thrush (*Catharus ustulatus*) and the Hermit Thrush (*C. guttatus*), whose flute-like calls fill the spring woods. More than a dozen wood warblers add color and song to the forest, and each has its own preferred place to nest and forage. For example, the Cape May Warbler (*Dendroica tigrinia*) prefers stands of black spruce; the Bay-breasted Warbler (*D. castanea*) prefers open coniferous forest; and the Palm Warbler (*D. palmarum*) selects sites at the edges of bogs. Other songsters who come to the northern forest to breed include the Purple Finch (*Carpodacus purpureus*) and the White-throated Sparrow (*Zonotrichia albicollis*), whose high whistle "Canada, Canada, Canada" is a sure sign summer is coming. The few resident songbird species include the Black-capped and Boreal chickadees (*Poecile atricapillus* and *P. hudsonicus*, respectively) and the Red-breasted Nuthatch (*Sitta canadensis*). As mentioned above, some birds (e.g., Red Crossbills, White-winged Crossbills, and Redpolls) depend on spruce, fir, or pine seeds; they are residents in some years but in other years, when cone production has been poor, they migrate to more southern latitudes for the winter (Larson 1980; Pielou 1988).

Of insects, the biggest nuisances are mosquitoes and blackflies. The spruce budworm is most prominent for its effects on trees.

**Forests of the Pacific Northwest**. South from the Kenai Peninsula of Alaska along the fjorded coast of British Columbia, across the Olympic Peninsula of Washington state and down to the Coastal Range of northern California is a unique outlier of the Boreal Forest that harbors giant trees, often festooned with lichens, that thrive in the abundant rainfall and frequent fogs of the windward slopes of the mountains. This forest is sometimes referred to as a temperate rainforest because so much rainfall is received. Conifers dominate, but the trees differ from those of the Boreal Forest proper. Most common is the great Douglas fir (*Pseudotsuga menziesii*), which is not a true fir despite its common name. It is joined by towering western hemlock (*Tsuga heterophylla*), Sitka spruce, and western arbor vitae (*Thuja plicata*). In the coastal valleys and river bottoms of northern California, groves are dominated by the redwood (*Sequoia sempervirens*), which can attain heights greater than 90 m (300 ft) (Vankat 1979). The one claimed to be the world's tallest tree is 112 m (367.8 ft) tall (Van Gelder 1982).

The mammals of the Pacific Northwest are generally the same as or related to those found across the continental Boreal Forest, although there are some important differences. Caribou are missing, but another member of the deer family, the black-tailed variety of mule deer (*Odocoileus hemionus*), is present from Washington into California. Bobcats (*Felis rufus*) and mountain lion (*Felis concolor*) represent the cat family. Spotted and striped skunks (*Spigale putorius* and *Mephitis mephitis*, respectively) are found from Washington southward. Red squirrel can be found in the northern parts of the region; the chickaree (*Tamiasciurus douglasi*) replaces them to the south. Townsend's chipmunk (*Eutamias townsendi*) can be found from California to southern British Columbia. The coastal areas of Washington and Oregon are home to the so-called mountain beaver or sewellel (*Aplodontia rufa*), the single member of a family of rodents **endemic** to North America (Van Gelder 1982).

Perhaps the most famous bird of the Pacific Northwest is the Spotted Owl (*Strix occidentalis*), whose declining numbers have become a rallying point for the conservation of old-growth forests. Most other birds have wider distributions in the Boreal Forest, but a few are restricted to this region. Examples are the Chestnut-backed Chickadee (*Poecile rufescens*) and the Varied Thrush (*Ixoreus naevius*). Stellar's Jay (*Cyanocitta stelleri*), a common thief at picnic tables, is found in conifer forests through the western mountains (Van Gelder 1982).

**Cascades and Sierra Nevada Conifer Forests**. The mediterranean climate patterns influence the windward (western) slopes of the interior ranges of the westernmost tier of states in the United States—the volcanic Cascades and the granitic Sierra Nevada. They receive more than half of their precipitation as snow during the winter months. Total annual amounts of precipitation vary from 25–35 cm (10–14 in) on the lower slopes to 125 cm (50 in) on the upper slopes. Many peaks extend above tree line, which ranges from 2,000 m (6,500 ft) in the north to about 3,500 m (11,500 ft) in the south. Conifers dominate three altitudinal zones of vegetation: the subalpine zone just below tree line, the upper montane zone, and the lower montane zone. Mountain hemlock (*Tsuga mertensia*) and lodgepole pine dominate the subalpine zone, where whitebark pine (*Pinus albicaulis*) and sometimes red fir (*Abies magnifica*) also may be found. In the upper montane zone, Douglas fir and white fir (*Abies concolor*) dominate; sugar pine (*Pinus lambertiana*) and incense cedar (*Calocedrus decurrens*) may also occur (Vankat 1979). Scattered groves in the central and southern Sierra Nevada are home to giant sequoia (*Sequoia gigantea*) (Photo 8.7), among them the singular specimen named "General Sherman," which has been called the world's largest living organism by virtue of its great volume. General Sherman is 84 m (275 ft) tall, 31 m (103 ft) in circumference, and has a trunk volume of 1,485 m$^3$ (52,500 ft$^3$). It is somewhere between 2,500 and 3,000 years old. Lower montane conifer forests are characterized by yellow pines: Jeffrey pine (*Pinus jeffreyi*) at higher elevations and ponderosa pine (*P. ponderosa*) at lower elevations (Van Gelder 1982).

Photo 8.7 Giant sequoia (*Sequoia gigantea*) on the cloudy windward slopes of the Sierra Nevada. (Photo by SLW.)

Large mammals of the conifer forests of the Cascades and Sierra Nevada include bighorn sheep (*Ovis canadensis*), mule deer, black bear, coyotes, and mountain lion. Smaller mammals are abundant and include spotted skunk, lodgepole chipmunk (*Eutamias speciosus*), chickaree (*Tamiasciurus douglasi*), golden-mantled ground squirrels (*Citellus lateralis*), and bushy-tailed woodrats (*Neotoma cinerea*). Several weasels associated with the Boreal Forest live in these mountains, including ermine, long-tailed weasel, fisher, and marten (Van Gelder 1982).

Many birds typical of the greater Boreal Forest also live in the coniferous forests of these mountains. The White-headed Woodpecker, (*Picoides albolarvatus*) can be found in sugar pine and ponderosa pine stands at elevations above 1,200 m (4,000 ft) in the Cascades and Sierra Nevada. The birds one is most apt to see include Stellar's Jay, Clark's Nutcracker (*Nucifraga columbiana*), and Mountain Chickadee (*Poecile gambeli*), all also found in the Rocky Mountain conifer forests (Van Gelder 1982).

**Rocky Mountain Conifer Forests**. The Rocky Mountains, from western Alberta to Colorado and discontinuously into Arizona, also have three altitudinal vegetation zones dominated by coniferous trees. The climate at these high elevations is more continental than farther west, with greater temperature ranges between summer and winter. Heavy snowfall is typical; total annual precipitation varies from 40 to 150 cm (15–60 in). Typical conifers of the subalpine zone are Englemann spruce (*Picea englemani*) and subalpine fir (*Abies lasiocarpa*), with bristlecone pine (*Pinus aristata*)—some of them reputedly earth's oldest living organisms—at tree line. Spruce, fir, and pine acquire a krummholz **growth-form** (a stunted tree with most of its branches close to the ground and spindly growth, if any, above the average depth of winters' snow) near tree line (Photo 8.8). In many places where fire has destroyed the spruce and fir forest, extensive, even-aged stands of lodgepole pine and quakening aspen are common in the subalpine zone. Douglas fir dominates the upper montane zone in the Rocky Mountains, and ponderosa pine dominates the lower montane regions.

Bighorn sheep (*Ovis canadensis canadensis*), elk, mule deer, and moose are the large herbivores of the Rockies. Black bear and mountain lion are other large mammals occurring there. Many squirrels are associated with conifer forests, among them red squirrel, the tassel-eared Abert's squirrel (*Sciurus aberti*), golden-mantled ground squirrel (*Citellus lateralis*), and least chipmunk (*Eutamias minimus*). The last two are frequently seen begging handouts from visitors to the region's National Parks. Snowshoe hare, beaver, muskrat, and porcupine are other occupants of the Rocky Mountain forests and wetlands. Birds of the Rockies are similar to those of the Cascades and Sierra Nevada (Van Gelder 1982).

**Appalachian Mountain Conifer Forests.** The elevation of the coniferous zone in the Appalachians increases southward. In New Hampshire and Vermont, the zone occurs between 800 and 1,200 m (2,600–4,000 ft); it occurs only above

Photo 8.8 Krummholz on the San Francisco Peaks near Flagstaff, Arizona. (Photo by SLW.)

1,100 m (3,600) in New York, and above 1,600 m (5,250 ft) in the Great Smoky Mountains of Tennessee and North Carolina. Precipitation increases from north to south, but there is less snow to the south.

The northern Appalachian Mountains are essentially a continuation of the Boreal Forest of eastern Canada. Red spruce and balsam fir are the common needle-leaved evergreen trees and paper birch the most common broadleaf species in the northern section of the Appalachians. Both balsam fir and paper birch reach their southern limits in Shenandoah National Park in the Blue Ridge Mountains of northern Virginia. The distribution of fir along the Blue Ridge has a gap of a few hundred miles, and this separation has apparently been great enough and long enough for a different species to evolve in the Southern Appalachians. Fraser fir (*Abies fraseri*) crowns Mt. Rogers—at 5,597 feet Virginia's highest peak—near the Virginia-North Carolina state line and is the only fir in the even higher Great Smoky Mountains. Yellow birch (*Betula alleghaniensis*) replaces the paper birch south of Shenandoah. Red spruce occurs at lower elevations than fir and is thus more widespread in the Middle and Southern Appalachians. An understory of heaths (e.g., rhododendrons, azaleas, huckleberries, and the like) is common in red spruce stands. Mosses, clubmosses, and lichens form a ground layer beneath both spruce and spruce-fir forests (Vankat 1979).

The eastern mountains of North America have fewer mammals than the western mountains, in part due to overhunting in the past and in part because of habitat change. The Appalachians are still home to black bear, mountain lion, and moose. White-tailed deer (*Odocoileus virginianus*) replaces mule deer in the east and is very abundant throughout the Appalachian chain. Small carnivores such as pine marten and fisher can be found in the northern areas, as can the snowshoe hare and flying squirrel. Long-tailed weasel and red squirrels, not so closely tied to conifers as the hare and flying squirrel, range throughout the mountain chain.

Bird life differs from the western mountains, but many of the migratory species that breed in the Boreal Forest proper also breed in the high-elevation conifer forests of the Appalachians. Examples are Golden-crowned kinglet (*Regulus satrapa*), Ruby-crowned Kinglet (*R. calendula*), Dark-eyed Junco (*Junco hyemalis*), Purple Finch, White-throated Sparrow, and some of the thrushes and wood warblers. Among resident species also found in the great stretch of northern forest are Common Ravens, Black-capped Chickadees, and Red-breasted Nuthatches.

## Eurasian Boreal Forest

Only 14 kinds of trees dominate the Boreal Forest across the entire Eurasian continent; 10 of these are conifers (Woodward 1995). None extends the full breadth of the continent; instead they are all spatially limited and help define subregions within the biome, as follows.

**Fennoscandanavian Subregion.** Two species, Norway spruce (*Picea abies*) and Scots Pine (*Pinus sylvestris*), dominate drier sites of the Boreal Forest in Finland, Norway, and Sweden. Norway spruce generally is associated with regions where proximity to the sea moderates the annual temperature range and where fire frequency is low. The Scots pine, on the other hand, occurs on more continental and more fire-prone sites (Larsen 1980). Few shrubs are associated with the forest.

Other than tree saplings, rowan (*Sorbus aucuparia*) and goat willow (*Salix caprea*) shrubs may grow in the spruce forest; moose and mountain hare (*Lepus timidus*) both heavily browse them. Dwarf shrubs such as cranberry (*Vaccinium vitis-idaea*) and bilberry (*V. myrtillus*) may also occur at moister locations. Associated with drier sites are heather (*Calluna vulgaris*) and *Empetrum hermaphroditum*. Lichens (especially *Cladonia* spp.) cover the ground of pine forests and provide winter forage for domestic reindeer herds. Moose and roe deer (*Capreolus capreolus*) occur in low numbers in the Fennoscandanavian forests, largely because of hunting pressures. Members of the grouse family found in these forests include the large capercaille (*Tetrao urogallus*), Siberian spruce grouse (*Falcipennis falcipennis*), and hazel hen (*Tetrastes bonasia*). Mountain hare and red fox (*Vulpes vulpes*) undergo the population fluctuations characteristic of Boreal Forest animals (Larsen 1980).

**European Russia Subregion.** Scots Pine dominates much of European Russia's Boreal Forest, but spruce (*Picea excelsa*) and birch (*Betula pendula*) are also prevalent. Siberian spruce (*Picea obovata*) and Siberian larch (*Larix sibirica*) are present in the northwestern regions; Siberian fir (*Abies sibirica*) and Siberian cedar pine (*Pinus sibirica*) become common in the northeast. The typical forest is referred to as a green-moss community since it has a ground cover of feather mosses (*Pleurozium shreberi* and *Hylocomium proliferum*) (Larsen 1980).

**Western Siberian Subregion.** Western Siberia is underlain with patches of permafrost and therefore is poorly drained; vast areas flood in the spring when ice blocks the north-flowing Ob River near its mouth. Large bogs cover about a third of the area. The forest is composed primarily of Siberian larch, Sukachev's larch (*Larix sukaczevii*), Siberian cedar pine, and Siberian spruce (Larsen 1980). Single-species stands of larch characterize sandy soils in the north where lichens (*Cetrana colluta* and *C. nivalis*) form a continuous ground cover (Forestry Resource Assessment 2000).

**Central and Eastern Siberia Subregions.** Central and Eastern Siberia are better drained but have a drier and colder climate than Western Siberia; they support Siberian spruce and Siberian cedar pine (Larsen 1980). East of the Yenesei River, spruce disappear and over 2.5 million km$^2$ (950,000 mi$^2$) of forest is dominated by open stands of Dahurian larch (*Larix dahurica*), which often reaches only a meter or so (a few feet) high. This larch is associated with permafrost; its shallow root system lets it survive in the thin active layer of ground that thaws each summer. It is frequently accompanied by Japanese stone pine (*Pinus pumila*) (Eyre 1968, Larsen 1980; Walter 1985). Boreal Forest continues eastward into northeastern China and northern Japan. At its easternmost limits on the Eurasian continent, the diversity of conifers increases, and the composition of the forest more resembles that of the west coast of North America (Walter 1985).

Mammals of the Eurasian Boreal Forest are very similar to those of North America. In addition to the panboreal forms mentioned in the overview, several furbearer species are found only in Eurasia, but they have close relatives in North America. These include European beaver (*Castor fiber*), European mink (*Mustela lutreola*), European pine marten (*Martes martes*), and the sable (*Martes zibellina*)—the luxuriously fine furs of which were worn as a sign of royalty. Eurasia also has its own species of river otter (*Lutra lutra*) and flying squirrel (*Pteromys volans*).

Birdlife, too, is similar to that of North America, although it lacks the colorful wood warblers and many other summer migrants that breed in the northern forests of the United States and Canada. Two very large grouse—Capercaillie (*Tetrao urogallus*) and Siberian Spruce Grouse (*Falcipennis falcipennis*)—and the Hazel Hen (*Tetrastes bonasia*) are limited to Eurasia. Tits are represented by the Willow Tit (*Parus montanus*) and the Siberian Tit (*Parus cintus*), the range of which actually extends across the Bering Strait into Alaska.

## ORIGINS

The rise of conifers took place from the late Cretaceous Period some 70 million years ago and continued through the Tertiary Period. During much of the Tertiary, when global temperatures were more uniformly mild than today, coniferous trees apparently grew mixed with broad-leaved deciduous trees in the Arcto-Tertiary forests that blanketed the middle and high latitudes of the northern continents. A general cooling trend started toward the end of the Tertiary (the Miocene Epoch) in response to the changing latitudinal positions of the continents. As a result, a separation occurred between coniferous and broad-leaved trees, the former becoming a boreal-type forest and the latter a temperate forest type. Rapidly decreasing temperatures in the higher latitudes during the Pliocene Epoch set the stage for the ice ages characterizing the Pleistocene (Larsen 1980).

While modern trees probably evolved in the Late Tertiary Period, the plant assemblages that characterize the boreal forest biome today have only recently come together in these northern latitudes. Large continental ice sheets covered most of the North American distribution area, as well as the Scandinavian part of the Eurasia continent, during the Pleistocene Epoch. The plants and animals of the northern forests were pushed southward into ice-free, midlatitude refuges. Much of what is today the United States was covered by an open spruce forest unlike any of the modern regional boreal communities in terms of species composition. Across much of Siberia, it had been too dry during the glacial periods of the Pleistocene for snow to accumulate into vast ice sheets; however, the ground was frozen and trees were unable to survive, so vast expanses were covered by a cold grassland, known as steppe-tundra. The end of the last ice age and the establishment of modern environmental conditions came late to the higher latitudes of both continents. The land was open to invasion by trees only about 5,000 years ago. Dominant trees may have completed the migration northward only in the last 2,000 years (Pielou 1991; Woodward 1995).

Large mammals, most of which became extinct around 11,000 years ago, dominated the animal life of the Pleistocene forest of North America. The largest mammals included the elephant-like mastodon (*Mammut americanum*), woolly mammoth (*Mammuthus* spp.), giant ground sloths (*Megalonyx* spp.), horses (*Equus* spp.), and woodland muskoxen (*Bootherium bombifrons*). Mastodons browsed spruce and were particularly abundant in the eastern part of the continent, where they may have helped to maintain an open canopy to the forest. Large carnivores, including the dire wolf (*Canis dirus*), saber-toothed cat (*Smilodon fatalis*), and short-faced bear (*Arctodus simus*) also inhabited the ice age forest. These giants shared the forest with many smaller species—and even some large ones that survived the wave of extinction at the end of the Pleistocene and were able to reinvade northern lands once the trees established habitats for them. Such is the case for moose, caribou, fisher, martin, ermine, red squirrel, and beaver.

## HUMAN IMPACTS

Although large areas of Boreal Forest remain intact, human activities have caused significant impacts, and will cause further affects. Hunting, trapping, and logging have been major agents of change on both continents. Even today, livelihoods in the Boreal Forest may depend on trapping furbearers. Logging has gone through several phases. The first target was the pines and taller trees for sawtimber, but increasingly spruce are in demand for pulpwood in the paper industry. The large stands made up of only one or two species are most economically clear-cut. The cut-over areas may then be replanted with conifer species, which eliminates the broad-leaved stage of forest renewal. The landscapes are increasingly dominated by carefully managed, even-aged, single-species stands of industrially important conifers.

The removal of old-growth forest has led to a decrease in plants and animals specialized for that habitat type and an increase in generalists that prefer disturbed sites. Studies of forestry practices in northern Finland show that among the threatened old-growth specialists are lichens, orchids, marten, Capercaillie, and Siberian Tit (Esseen et al. 1997). In the Pacific Northwest, the spotted owl is only one of several species threatened by elimination of old-growth forests (Ricketts et al. 1999).

In the Fennoscandanavian region, wolves, wolverine, brown bear, and lynx are rare, and their populations are deliberately maintained at low levels by hunting. Moose and roe deer are on the increase, partly as a consequence of a low numbers of predators. The reduction of cattle grazing in the forest in recent years, an increase in forage as a result of clear-cutting, and regulated hunting are other factors favoring growing populations of these large herbivores. Nonnative mammals have been introduced and may affect native plants and animals. The North American white-tailed deer (*Odocoileus virginianus*) was established in Finland by 1934. American mink and muskrat have also been introduced. The raccoon dog (*Nyctereutes procyonoides*), native to southeastern Siberia, has invaded most of Finland (Esseen et al. 1997). Raccoon dogs were initially released in European Russia between 1927 and 1957 in hopes of increasing fur production; they then extended their range westward to reach Sweden in 1945. In the milder climates west of Russia, their fur fails to develop the long hairs valued by the fur industry, and they have become a nuisance species that feeds on native small game animals (Nowak 1991).

In North America, predator control has reduced or eliminated the wolf and the grizzly bear from much of the Boreal Forest. Logging and dam building contribute to the fragmentation of the forest and disruption of migration routes of larger mammals. Especially in the Rocky Mountain area, destruction of forest habitats has occurred as a result of copper and iron mining, domestic livestock grazing, and gas and oil development. Increasingly, recreational use is destroying forested areas; ski resorts and associated development are major culprits. Where human populations are dense, urbanization and air pollution are affecting conifer forests. It is estimated than less than 4 percent of the original redwood forest remains because of the combined losses from logging and urban development. The ozone generated when vehicular exhaust reacts with sunlight is particularly damaging to evergreen needle-leaved trees because they are exposed year round. In the Sierra Nevada, the effects of ozone are evident in white fir, Jeffrey and ponderosa pine, and lichens (Ricketts et al. 1999).

The greatest human impact on the Boreal Forest may be yet to come. We are living in a time of global climatic change, the consequences of which are unknown. Most scientific models predict that global warming will be greatest in the latitudes that the Boreal Forest occupies, and no one agrees what the effects of this warming might be. Slight increases in temperature might increase the rate of nutrient cycling, and thereby increase the production of timber. Greater increases, however, could lead to the destruction of the forest, which is a major source of timber for the world and a major part of the economies of Canada and Russia. With so much at stake, the Boreal Forest is a major focus of researchers studying global change (Kasischke and Stocks 2000).

## HISTORY OF SCIENTIFIC EXPLORATION

Early investigation of the Boreal Forest had an economic focus, as both continents exploited this great forest belt for its furs, timber, and—later—pulpwood. In Canada in the 1700s and 1800s, naturalist-explorers for Hudson's Bay Company paid close attention to the location of trees along the Arctic tree line since they represented shelter for trappers in winter as well as firewood and the raw material for snowshoes and sleds (MacDonald et al. 1998). More modern work traces to scientists such as Hare (1950), who tried to correlate Boreal Forest types with regional climate; Rowe (1972), who recognized the complexity of communities within the biome and identified 35 Boreal Forest regions across Canada; and Larsen (1980), who produced a more generalized categorization with seven Boreal Forest regions in North America and called attention to the significant role of fire in this vegetation (Elliott-Fisk 2000).

Long-term studies on Boreal Forest ecology in the Soviet Union began in the Serebryanyi Bor Experimental Forest in the Moscow Region in 1945; by then, however, cutting and grazing had greatly altered the forest. Studies expanded in 1957 to include other forest types, including the spruce forests at Podushkinskoe forest (Rysin and Nadezhdina 1968).

The coniferous forest biome was one of the Integrated Research Programs in the Ecosystem Analysis Section of the U.S. International Biological Programme funded by the National Science Foundation from 1970–74. Studies focused on the H. J. Andrews Experimental Forest in Oregon, a forest dominated by Douglas fir and western hemlock.

Despite these efforts, the Boreal Forest biome remains relatively unexplored by science (Elliott-Fisk 2000). Modeling the ecosystem to better understand it has been undertaken by G. B. Bonan and H. H. Shugart and their colleagues (e.g., Sirois et al. 1994). The response of Arctic tree line to climate change continues to be a focus of research for others (e.g., Elliott-Fisk 1983; MacDonald et al. 1998). Yet another body of research concentrates on understanding forest dynamics to develop sustainable forestry practices.

## REFERENCES

Bonan, G. B. and H. H. Shugart. 1989. "Environmental factors and ecological processes in boreal forests." *Annual Review in Ecology and Systematics* 20: 128.

Elliott-Fisk, D. L. 1983. "The stability of the northern Canadian tree limit." *Annals of the Association of American Geographers* 73: 560–576.

Elliott-Fisk, Deborah L. 2000. "The taiga and boreal forest." Pp. 41–73 in Michael G. Barbour and William Dwight Billings, eds., *North American Terrestrial Vegetation.* 2nd edition. Cambridge: Cambridge University Press.

Esseen, Per-Anders, Bengt Ehnstrom, Lars Ericson, and Kjell Sjoberg. 1997. "Boreal forest." Pp. 16–47 in Lennart Hannson, ed., *Boreal Ecosystems and Landscapes: Structures, Processes, and Conservation of Biodiversity.* Copenhagen: Munksgaard International Publishers.

Eyre, S. E. 1968. *Vegetation and Soils.* Chicago: Aldine Publishing Company.

Forestry Resource Assessment. 2000. "AN Boreal coniferous forest (Ba)." http://www.fao.org/forestry/fo/fra/index.html. Food and Agriculture Organization of the United Nations.

Gates, David M. 1993. *Climate Change and Its Biological Consequences.* Sunderland, MA: Sinauer Associates, Inc.

Hare, F. K. 1950. "Climate and zonal divisions of the boreal forest formation in eastern Canada." *Geographical Review* 40: 615–635.

Johnson, Edward A. 1992. *Fire and Vegetation Dynamics. Studies from the North American Boreal Forest.* Cambridge: Cambridge University Press.

Kasischke, Eric S. and Brian J. Stocks. 2000. "Introduction." Pp. 1–6 in Eric S Kasischke and Brian J. Stocks, eds., *Fire, Climate Change and Carbon Cycling in the Boreal Forest.* Ecological Studies 138. New York: Springer.

Klinger, Lee F. 1991. "Peatland formation and ice ages: A possible Gaian mechanism related to community succession." Pp. 247–255, in S. H. Schneider and P. J. Boston, eds., *Scientists on Gaia.* Cambridge: M. I. T. Press.

Larsen, James A. 1980. *The Boreal Ecosystem.* Physiological ecology series. New York: Academic Press.

Larsen, James A. 1982. *Ecology of Northern Lowland Bogs and Coniferous Forests.* New York: Academic Press.

MacDonald, Glen M., Julian M. Szeicz, Jane Claricoates, and Kursti A. Dale. 1998. "Response of the central Canadian treeline to Recent climatic changes." *Annals of the Association of American Geographers* 88: 183–208.

Nowak, Ronald M. 1991. *Walker's Mammals of the W*orld, Fifth Edition, Volume II. Baltimore: The Johns Hopkins University Press.

Pastor, J., R. H. Gardner, V. H. Dale, and W. M. Post. 1987. "Successional changes in nitrogen availability as a potential factor contributing to spruce decline in boreal North America." *Canadian Journal of Forest Research* 17 (11): 1394–1400.

Pielou. E. C. 1988. *The World of Northern Evergreens.* Ithaca: Comstock Publishing Associates.

Pielou, E. C. 1991. *After the Ice Age, the Return of Life to Glaciated North America.* Chicago: University of Chicago Press.

Ricketts, Taylor H., Eric Dinerstein, David M. Olson, Colby J. Loucks, et al. 1999. *Terrestrial Ecoregions of North America, A Conservation Assessment.* Washington, DC: Island Press

Rowe, J. S. 1972. *Forest Regions of Canada.* Publication No. 1300. Ottawa: Canadian Forestry Service, Department of the Environment.

Rysin, L. P. and M. V. Nadezhdina, eds., 1968. *Long-term Biogeocenotic Investigations in the Southern Taiga Subzone.* Jerusalem: Israel Program for Scientific Translations.

Sirois, L., G. B. Bonan, and H. H. Shugart. 1994. "Development of a simulation model of the forest-tundra transition zone of northeastern Canada." *Canadian Journal of Forest Research* 24: 697–706.

Van Gelder, Richard G. 1982. *Mammals of the National Parks.* Baltimore: The Johns Hopkins University Press.

Vankat, John L. 1979. *The Natural Vegetation of North America.* New York: John Wiley & Sons.

Walter, Heinrich. 1985. *Vegetation of the Earth and Ecological Systems of the Geo-biosphere.* 3rd edition. New York: Springer-Verlag.

Woodward, F. I. 1995. "Ecophysiological controls of conifer distributions." Pp. 79–94 in William K. Smith and Thomas M. Hinckley, eds., *Ecophysiology of Coniferous Forests.* San Diego: Academic Press.

# 9 TUNDRA

## OVERVIEW

The tundra **biome** primarily consists of the treeless **landscapes** in the Northern Hemisphere that encircle the Arctic Ocean poleward of the Boreal Forest. Similar **vegetation** also can be found in high-altitude situations in the middle **latitudes**. Tropical mountains may also extend above treeline and have treeless **habitats**; however, the structure of the vegetation and the composition of the plant and animal life are quite different than what one finds outside the **tropics**. While true tundra does not occur in Antarctica, a few organisms do inhabit ice-free areas. The word tundra comes from the Finnish word *tunturi*, which originally referred to the barren lands of northern Finland and was later applied to similar treeless areas throughout northern Eurasia and North America (Bliss 1997).

An extremely short growing season, continuous permafrost close to the surface, very shallow **soils**, low precipitation, drying winds, and nutrient-poor conditions are major limiting factors to plant growth in this biome. Arctic Tundra has relatively few **species**, and many of these are circumpolar in distribution, which means they are found on both the North American and Eurasian continents. Mosses, **lichens**, sedges, **perennial forbs**, and dwarfed shrubs compose the vegetation **cover**. Mirroring latitudinal trends, high-elevation, treeless areas similarly have fewer species than lower-elevation zones. An extreme impoverishment of terrestrial plant and animal life occurs in the Antarctic region, where only two species of **vascular plants** have been discovered.

The scientists of each country possessing tundra have devised their own schemes for classifying latitudinal zones that occur within the biome. In North America, the biome is often divided into High Arctic, Middle Arctic, and Low Arctic zones. High Arctic refers to the habitat encountered on the northern islands of Canada's Arctic Archipelago, such as Ellesmere, Baffin, and the Queen Elizabeth islands. Middle Arctic is the type of tundra found primarily on the level Arctic Coastal Plain of continental North America, and Low Arctic—the greatest proportion of

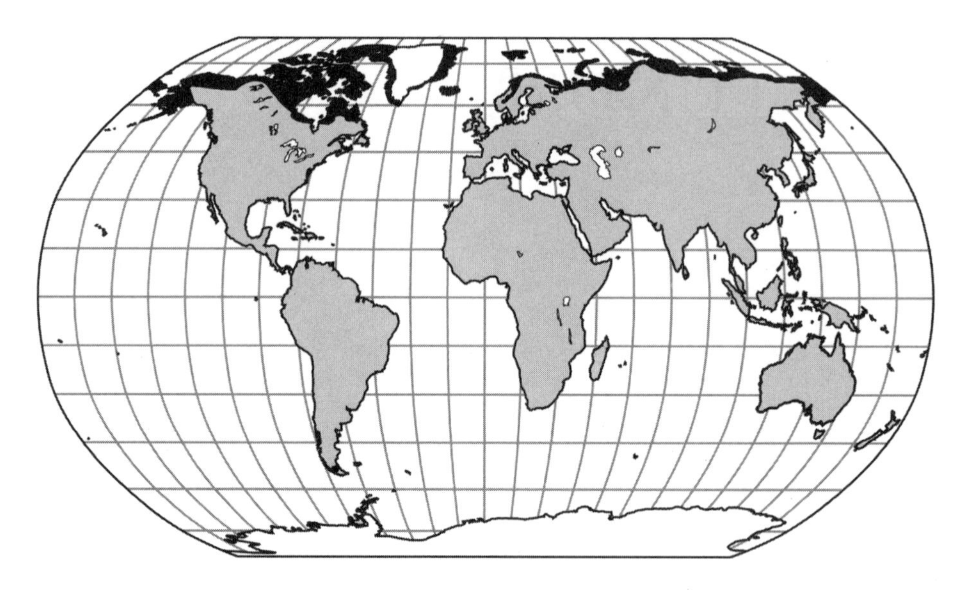

Map 9.1 World distribution of the Tundra Biome.

tundra biome—occurs in upland areas where slopes provide better drainage and depth to permafrost is greater than in the other two zones. In Russia, the latitudinal pattern is similar, but four zones are identified. From north to south, they are polar desert, northern or arctic tundra, typical tundra, and southern tundra.

Tundra vegetation, regardless of its location, is very easily damaged and very slow to recover when plants are trampled, killed, or removed. Today, much of the biome remains largely intact, but it is threatened by mineral exploration and development, air pollution, **climate** change, and tourism.

## CLIMATE

The tundra climate (ET in the Koeppen system) has a mean annual temperature below 0° C (32° F) and a growing season of less than 50 days. The southern limits of arctic tundra roughly coincide with areas where mean July temperatures are between 4° and 5° C (approx. 40° F). Winters are very cold—mean monthly temperatures in winter may be –30°C (–22° F) and also dark up to 24 hours a day for weeks at latitudes above the Arctic Circle. Annual precipitation averages 200–500 mm (8–20 inches) and usually occurs as snow.

Similar conditions of cold can be found at high elevations, but other aspects of the climate differ significantly between arctic and alpine tundra situations. The midlatitudes and tropical regions of earth have different light regimes compared to those of polar regions. Seasonal difference in day length may be pronounced in the midlatitudes (though never as extreme as poleward of the circles) and nearly nonexistent in the tropics. Furthermore, sunlight is more intense, and atmospheric pressure is lower at high elevations. It is typical in tropical alpine situations that temperature differences are greater between day and night than between summer and winter, while in the arctic, temperature in the daily cycle has relatively little variation. On high mountains, a large temperature difference can occur between sunny and shaded areas as well, and each passing cloud can cause rapid temperature changes. Alpine areas tend to experience higher maximum wind speeds than polar areas. The highest windspeed ever measured on Earth—371 km/h (231 mi/hr) occurred on the summit of Mt. Washington, New Hampshire, in April 1934. Precipitation on mountains varies according to slope and prevailing wind direction. Generally, maximum precipitation occurs at intermediate elevations, so above tree line, there may be little precipitation but perhaps frequent fog. The actual elevation at which tree line occurs depends upon latitude; in polar latitudes, tree line occurs at sea level and in equatorial regions at approximately 4,000 m (13,000 ft). A great deal of variation, however, is the result of regional and local climatic patterns (Sarmiento 1986).

## VEGETATION

Common **growth-forms** of arctic and alpine tundra plants include mosses and lichens (nonvascular plants or cryptogams), graminoids (sedges and grasses), perennial forbs, and dwarfed shrubs. Most are small in stature and hug the ground for warmth and protection from drying winds. Vascular plants with renewal buds at or just above the ground surface (chamaephytes and hemicryptophytes are the most usual Raunkiaer **life-forms** in the tundra) often grow clumped as **tussocks** (bunched groups of sedges or grasses), mats (many individuals clustered together) (Photo 9.1), or **cushions** (a single plant with multiple stems). The outermost individuals, branches, or leaves may be killed by winter cold and drought, but the accumulation of dead material helps insulate plants or stems at the center of the clump, thereby protecting the renewal buds and allowing the species to persist. The **rosette** growth-form, in which a whorl of leaves surrounds a central stalk, is another common adaptation of tundra plants that protects a central, above-ground renewal bud.

Photo 9.1 The dwarf shrub alpine azalea (*Loiseleuria procumbens*) in mats only 2.5 cm (1 in) high. (Photo by SLW.)

The extremely short growing season means that many plants require more than one season to complete their life cycle. Growth rates in tundra plants are often very slow, and many plants are very long lived. On average, the lifespan of a tundra plant is 20 years, but the root systems of some sedges may be a few hundred years old, and some **crustose** lichens may be more than 1,000 years old (Furley and Newey 1983). The flower buds of many forbs develop over two to three growing seasons; it may take even longer for the buds of grasses and sedges to develop. The catkins of willows that trail along the ground need four seasons to develop (Bliss 1997; Wielgolaski 1997b).

Dwarf shrubs must be small enough to be covered by winter's snow, which is often only a few inches deep. Many trail along the ground surface where they can absorb maximum heat from the sun-warmed soil in spring and summer. Some of these shrubs are **evergreen**; others are **deciduous**. The evergreens are slow growers. Some, like arctic bell-heather or moss plant (*Cassiope* spp.), withdraw chlorophyll from the leaves at the end of the growing season, so that their winter foliage is red. They then green up again in spring. Other plants have leaves that remain green and functional for two or more years. In spring, nutrients are transferred from old leaves to new buds, which open much later than the leaf buds of deciduous species. Maintaining leaves for several growing seasons means evergreen species have lower nutrient requirements than those species that must replace their foliage each year. Evergreen species, therefore, tend to do well on acidic, low-nutrient soils. Deciduous shrubs such as willows (*Salix* spp.) and birches (*Betula* spp.), on the other hand, grow quicker but demand more nutrients. Toward the end of the growing season, nutrients that were manufactured that year are moved into the roots and woody stems for storage. Come spring, these nutrients are rapidly returned to growing leaf buds, and within two to three weeks the shrubs are in full leaf (Bliss 1997; Wielgolaski 1997b).

Perennial forbs, sedges, and grasses have underground structures (e.g., roots, **rhizomes**, and bulbs) for the storage of energy and nutrients during the nongrowing season. Some maintain a few green leaves above ground for storage—an advantage where very shallow soils limit the development of underground storage tissues. They may begin to **photosynthesize** even before the snow melts in early spring because sunlight can penetrate through snow to a depth of about 50 cm (20 in). For those plants that do not maintain green leaves through the winter, the development of the next year's leaves and flowers begins later in the growing season. They remain protected in buds through the winter and then—drawing upon the stored reserves of the previous growing season—rapidly explode into growth as soon as there are a few hours with temperatures above freezing early in the spring. Flowering and seed production are very variable from year to year and depend upon weather conditions, so **vegetative reproduction** is common (Bliss 1997).

Hairs are common on leaves, stems, and especially the sepals of flower buds of many tundra plants. The hairs help protect against drying winds and also work to warm the sensitive, growing plant tissues. Dark hairs, such as found on some fleabanes (*Erigeron* spp.) and arctic poppies (*Papaver* spp.), can absorb heat directly from the sun. Light-colored hairs, such as those on the catkins of willows and on cotton-grasses (*Eriophorum* spp.), act as the roof of a greenhouse, to let in sunlight but trap the heat radiated from the plant (Wielgolaski 1997b).

The arctic tundra has distinct latitudinal zones with differences in growth-form, percent of ground cover, and kinds of plants present. The zone just north of the

Photo 9.2 Tussocks of the sedge *Eriophorum*, popularly called cottongrass. They cover slopes in low arctic tundra near Unalakleet, Alaska. (Photo by SLW.)

Boreal Forest (Low Arctic Tundra in North American terminology) is most diverse in both species and growth-forms and encompasses the greatest extent of the tundra biome. This area is mostly a rolling upland surface where, because of the depth of permafrost, soils are well drained to the depths that roots can penetrate. The vegetation is characterized by dwarf shrubs and tussock-forming cotton-grasses (*Eriophorum* spp.—despite the common name, a sedge) (Photo 9.2). An open shrub layer composed of willows (*Salix* spp.) and birches (*Betula* spp.) ranges in height from 0.5–2.0 m (1.6–6.5 ft). **Subshrubs** 5–20 cm (2–8 in) high are chiefly members of the **heath** family, such as alpine bearberry (*Arctostaphylos alpina*), bog bilberry (*Vaccinium uglinosum*), arctic bell-heather (*Cassiope* spp.), and Labrador tea (*Ledum* spp.). Some slopes are primarily tussock grasslands dominated by cottongrass. A ground cover 5–10 cm (2–5 in) thick is composed of **fruticose** lichens, such as the so-called reindeer moss (*Cladonia rangifera*) and snow lichen (*Cetraria nivalis*) in drier areas and hairy cap (*Polytrichium* spp.) and other mosses in wetter areas. Total vegetation cover is 80–100 percent (Bliss 1997; Chernov and Mataveyeva 1997).

Farther north, on the Arctic Coastal Plain and some of the arctic islands, the Middle Arctic Tundra occurs in areas of low relief and poor drainage, where only a thin layer of earth thaws each year—the so-called active layer. Typically, a mosaic of low geometric earthen features known as patterned ground occurs. Patterned ground often appears as a set of polygons outlined by rocks or plants on the surface, but stripes and netlike patterns are also frequent. It apparently develops for a number of reasons related to freeze-thaw action and ecological **succession**. In waterlogged soils, freezing water raises the surface in frost boils and can push larger particles aside, thereby sorting particles so that finer ones remain at the center of a polygon and larger ones form the perimeter. The finer particles hold water when the active layer thaws, so the process is repeated in successive years. However, summer drought also can initiate the formation of patterned ground by creating cracks in the surface that collect water. When the water freezes, ice wedges form and push soil particles upward and aside. As time goes on, the cracks widen and provide a waterlogged situation conducive to the growth of mosses. Because the site stays waterlogged, dead mosses do not decay but form a deposit of peat upon which future generations of moss grow raised above the general surface layer. Interlaced cracks result in a netlike pattern of mosses across the land (Zoltai and Pollett 1983; Chernov and Mataveyeva 1997).

Three habitat types are repeated across an area of patterned ground: patches, rims, and troughs. The patch is the saturated ground on the inside of a polygon that is composed of fine-grain soil materials. The patch is surrounded by a raised rim of better-drained conditions created by the growth of vegetation and/or the accumulation of coarser soil particles. Between the rims of adjacent polygons are troughs—waterlogged depressions that began as dessication cracks. The size of these features varies; polygons may be 7–15 m (20–50 m) across and the rims up to 1 m (3 ft) high. Where a slight slope permits the active layer to flow, the polygons become elongated, eventually to the point where the rims become rock stripes running down the length of the slope. In these areas of patterned ground, the vegetation consists essentially of mosses and sedges growing in areas of saturated soil. The center of a patch may have no plants at all. Foxtail (*Alopecurus* spp.) and other grasses sometimes accompany the sedges (*Carex* spp. and *Eriophorum* spp.). A number of genera of mosses are represented, including *Aulocomnim*, *Calliergoni*, *Drepanocladus*, *Sphagnum*, and *Tomenthyphum*. Liverworts (*Ptilidium* spp.) may also occur with the mosses. The few lichens, dwarfed shrubs, and forbs prefer better-drained sites. If they occur, they will be on the rims of polygons (Bliss 1997; Chernov and Matveyeva 1997).

On the Arctic coastal plain and more southerly islands of the Arctic Ocean, where the land is drier and better drained, no patterned ground is found, and the vegetation is sometimes referred to as a polar semidesert. Cushion plants, usually mountain avens (*Dryas* spp.), and lichens dominate these areas. The cushions range 0.2–1.0 m (0.6–3 ft) in diameter and 2–5 cm (0.8–2 in) in height. Rosette plants—such as saxifrages and arctic poppies, as well as sedges (*Carex* spp.) and woodrushes (*Luzula* spp.) are sparse. Lichens and mosses cover 30 to 60 percent of the surface. The surface may have a black crust formed by crustose lichen and **cyanobacteria** (Bliss 1997; Chernov and Metveyeva 1997).

On the northernmost islands of the Arctic Ocean, High Arctic tundra plants exist in a landscape that is primarily bare rock surfaces. The rock often has been split into sharp, angular fragments by freeze-thaw action. The very few species of vascular plants that grow here survive only in the most sheltered nooks and crannies. No woody shrubs occur; forbs are rosette plants, cushion plants, or mat-forming species and are found as scattered, isolated individuals. Saxifrages are most numerous. Crustose lichens and mosses grow on some rock surfaces (Bliss 1997; Chernov and Matveyeva 1997).

## SOILS

In general, soils of the tundra are thin, young, and nutrient poor. These conditions are related to the underlying permafrost and the fact that much of the area was covered by continental ice sheets until just a few thousand (in some areas only a few hundred) years ago. Permafrost is close enough to the surface that the active layer (the layer that thaws in spring and summer and in which plants are rooted) ranges from between 20 cm (7.5 in) and 300 cm (120 in). The direction a slope is facing, the particle sizes in the soil, and the type of vegetation on the surface all influence the depth of permafrost.

Mature soils with distinct horizons are rare because of their young age and slow rates of soil-forming processes and also to the constant churning of the active layer from the expansion and contraction of soil water as it repeatedly freezes and thaws. Three main soil-forming processes are at play in the tundra: gleization, **paludification**, and brunification. Gleization is most widespread in poorly drained soils on slopes and on level uplands underlying tussock sedges with abundant shrubs or wetland grasses and sedges (Bliss 1997), where most organic material remains at the surface. The usually waterlogged condition of the soil means a lack of oxygen slows bacterial decay. Since downward movement of water is generally lacking, the A horizon is not leached but becomes acidic as a result of overlying vegetation and peat. The B horizon is characteristically mottled, bluish- or greenish-gray, the result of chemical reactions with iron under **anaerobic** conditions. In the U.S. soil taxonomy, soils developed from this process are called inceptisols, in Canada they are known as gleysols, and in Russia they are classified as tundra or meadow soils (Bliss 1997).

Paludification refers to the accumulation of peat from a plant cover of sedges and mosses. Resulting soils are primarily organic in composition and acidic (**pH** = 4.5 to 6.0). In the United States, these are classified as histosols, whereas in Russia they are called bog or half-bog soils.

Brunification is a soil-forming process in drier or better-drained areas where iron binds to clay-**humus** particles and thus cannot be removed from the A horizon, as happens in podzolization (see Chapter 8). The result is a reddish-brown A horizon. Such soils in the Arctic are variously categorized in the U.S. system as inceptisols, entisols, or mollisols. In Canada, they are identified as cryosols or brunisols; in Russia they are termed Arctic Brown Soils (Bliss 1997).

## ANIMAL LIFE

Arctic tree line represents the northern limit of many groups of organisms, including amphibians and terrestrial mollusks. (Reptiles scarcely make it into the Boreal Forest.) The number of orders, families, genera, and species continues to decline the closer to the pole one goes. However, the typical pattern is that the few animals, especially if herbivores, occur in large numbers, and each prey species supports a number of different predatory animals. Typically, food chains are short and simple. This is one fact that makes the tundra so vulnerable to destruction. The loss of a single species can negatively affect a number of animals that depend on it for food.

The few animals that reside year round in the tundra biome must be adapted to a very short season of plant production and to a very long season of cold when plants are dormant. In general, the winter is too long for hibernation to be an option. The arctic fox (*Alopex lagopus*), for example, is able to withstand temperatures of –50° C (–58° F). The ability to accumulate a thick layer of fat, a trait possessed by musk oxen (*Ovibos moschatus*) and brown bear (*Ursos arctos*) among others, provides mammals with insulation and a store of energy for use when food is in short supply. The typical body shape of high-latitude, warm-blooded animals is adapted to conserving body heat. The trunk of the body is large to reduce the ratio of surface area to volume and minimize heat loss to the environment. Limbs and ears tend to be short, compared with relatives in warmer climes, because these appendages are also pathways for the loss of body heat. Among mammals, a double coat of fur, the undercoat of which is dense and fine, provides insulation. Among birds, thick plumage allows the ptarmigan (*Lagopus* spp.) to withstand bitter cold (Bliss 1997).

Some resident mammal and bird species may undergo seasonal color changes, and the browns of summer become white in the winter as camouflage against the snow-covered terrain. Arctic fox, the arctic hare (*Lepus arcticus*) of North America, ermine (*Mustela erminea*), and ptarmigan are prime examples. Others, such as lemmings (*Lemmus* spp. and *Dictrostonyx* spp.), avoid being seen by carrying on their winter activities beneath the snow, where they feed on the roots, stems, and buds of sedges and other plants.

Lemmings play a dominant role in another characteristic of life in the arctic—the repeating cycle of population explosions and crashes that affects both plants and animals. When their food supply is abundant, the lemming population increases rapidly. Birth rates and the survival rates of the young of their predators [arctic fox, short-tailed weasel (*Mustela nivalis*), ermine, Snowy Owl (*Nyctea scandiaca*), Pomarine Jaegers (*Stercocarius pomarinus*) and others] then increase since lemmings are their main food. However, the lemmings soon deplete the tundra plants; a shortage of forage results in declines in their reproduction rates and increases in their mortality rates—in other words the population crashes. The brown lemmings or true lemmings (*Lemmus* spp.) undergo mass migrations at times of food shortage, and many starve. With fewer lemmings, the predators' birth rates and survival rates also decline. Weasels and foxes tend to move toward the coast during low lemming years, and arctic foxes spend much of the winter on the sea ice scavenging polar bear kills. Pomarine Jaegers breed only during peak lemming years. Snowy owls have smaller clutches in low lemming years, and many move south into the midlatitudes in winter, where apparently most starve to death. With fewer lemmings to graze the tundra plants, the vegetation recovers—having been well aerated by lemming burrowing and well fertilized by their droppings during the population peak. Once again, this will mean abundant food for lemmings, and while their predators' populations are still low, the lemming populations rebound to repeat the cycle. On average, a lemming population peak occurs every four years (Chernov and Matveyeva 1997).

Seasonal migration is a common response to the rigors of the arctic winter. Huge herds of barren-ground caribou (*Rangifer tarandus*) come from wintering grounds at the edge of the Boreal Forest to calve on the Arctic Coastal Plain. Millions of birds come to the tundra to nest. Shorebirds such as plovers (*Pluvialis* spp.) and, especially, sandpipers of the **genus** *Calidris* come north in summer to feast on insects and raise their young along the shores of the numerous ponds and lakes. Thousands of White-fronted Geese (*Anser albifrons*), Brant (*Brenta bernicla*), and Snow Geese (*Chen caerulescens*) come to the tundra to nest, as do many ducks. The dabblers, such as Pintail (*Anas acuta*), Northern Shoveler (*A. clypeata*), Green-winged Teal (*A. crecca*), and Wigeon (*A. penelope*), are most common in inland sites;

sea ducks such as Oldsquaw (*Clangula hyemalis*) and eiders (*Somateria* spp.) are prevalent along the arctic coast and shores of islands. The world record holder for distance in migration, the Arctic Tern (*Sterna paradisea*), breeds near tundra lakes and then flies to subantarctic seas to catch the Southern Hemisphere summer. Eggs and chicks produced in such great numbers by the ground-nesting shorebirds and waterfowl attract other birds that are predators and scavengers, such as gulls (*Larus* spp.) and jaegers (*Stercorarius* spp.). Passerine or songbird species are relatively few in number; Lapland Longspur (*Calcarius lapponicus*), Snow Bunting (*Plectrophenax nivalis*), wagtails (*Motacilla* spp.), and pipits (*Anthus* spp.) are some that breed on the tundra of both Eurasia and North America. They attract the Peregrine Falcon (*Falco peregrinus*) and other hawks to breed in the arctic in summer (Bliss 1997).

Many of the birds come north to feed on the insects so abundant in the summer tundra. Adults and larvae of mosquitoes and craneflies (order Diptera), springtails (order Collembola), and stoneflies (order Plecoptera) are important seasonal food resources in the tundra **ecosystem**. As is true for all classes of organisms, the diversity of insects decreases poleward, but the abundance of those that do occur makes up for lack of variety in terms of energy and nutrients available to the organisms that feed on them.

## MAJOR REGIONAL EXPRESSIONS

### Arctic

The descriptions of plants, soils, and animal life previously given hold true throughout the Arctic regions of both North America and Eurasia. The fact that many plants and animals are circumpolar makes it unnecessary to go into separate details here for each. However, it should be realized that species composition have regional and local variations. Many species are restricted in their distributions, although almost always they are closely related (i.e., congeners) to a species that occurs in another part of the tundra. For example, the brown lemming (*Lemmus sibirucus*) is circumpolar in that it is found on both continents, but it is replaced by congeners in two locations: *L. lemmus* in Scandinavia and *L. amurensis* in parts of eastern Siberia. The Arctic tundra has five kinds of collared lemming (*Dicrostonyx* spp.). *D. torquatus* occurs across much of Eurasia, while *D. groenlandicus* is widespread across North America. These two are so similar that in the past they were classified as the same species. The other three lemmings in the genus are very restricted in their distributions, one to Wrangell Island off northeastern Siberia, another to St. Lawrence Island in the Bering Sea, and the third to northern Quebec and Newfoundland, Canada (Nowak 1991). Comparable patterns can be found among other groups of plants and animals.

A final example of a difference between North American tundra and Eurasia pertains to the species *Rangifer tarandus*, known as caribou in America and reindeer in Eurasia. Caribou congregate into vast herds, each numbering in the tens of thousands—and in some cases, hundreds of thousands—to migrate to calving grounds on the Arctic Coastal Plain of Alaska and Canada. They produce one of nature's great spectacles, and their presence is a major factor in modern land-use issues regarding development of the Arctic. The endangered subspecies, the white Peary's caribou, is found on northern islands of the Canadian Arctic Archipelago; it crosses the sea ice to move from island to island. Wild reindeer have nearly disappeared from the Eurasian continent, where they were domesticated perhaps 3,000 years ago. Domestic reindeer herds have generally replaced their wild ancestors across the tundra. A few small herds of wild reindeer remain in Scandanavia and Russia (Nowak 1991). The population on the remote Norwegian territory of Svalbard, as is often the case on islands, is unique. The animals have shorter legs, a shorter face, and longer winter hair than other reindeer. They do not migrate but accumulate thick layers of fat during the summer from which they derive energy during the winter (Elvebakk 1997).

## Antarctic

The antarctic region consists of the Antarctic continent, offshore islands, and islands of the Scotia Arc—the group of islands that stretches from the tip of South America to the Antarctic Peninsula. (The islands of Kerguelen and South Georgia are usually placed in a subantarctic zone that is considered transitional between the **temperate** regions of South America and the cold, dry areas of the Antarctic proper.) The antarctic, where mean annual temperatures are below 0° C (32° F), is subdivided into northern (sometimes called coastal or maritime) and southern (or continental) zones. The Northern Antarctic consists of the northwestern Antarctic Peninsula and adjacent islands, the South Orkney and South Shetland islands, and the island of Bouvetoya, where one month of the year has an average temperature above freezing. They receive at least 500 mm (20 in) of precipitation a year. Most of the land is covered with ice, but on sheltered north-facing slopes at elevations below 100 m (350 ft) and some flat areas that are protected by snow in the winter, ice-free areas support grass-moss or grass-lichen-moss communities. The hairgrass *Deschampia antarctica* is one of two flowering plant species occurring in the antarctic. The other is a cushion plant, *Colobanthus quitensis*, which, like the hairgrass, is limited to the northern zone (Kanda and Komárková 1997).

The Southern Antarctic encompasses most of Antarctica. Ice-free areas make up less than 5 percent of the surface and are found primarily along the coast, in dry valleys, and on nunataks—mountain peaks that stick up through the great ice sheet that covers most of the continent. The warmest month of the year never averages above freezing and may be as low as –15° C. Precipitation occurs mostly as snow and averages 145–190 mm (5.5–7.5 in) a year. Because it is so dry, no patterned ground occurs like that found on the Arctic Coastal Plain. Under the extreme conditions of the Southern Antarctic, few organisms are able to survive. No flowering plants grow in the southern zone, but bacteria, algae, lichens, and a few mosses are found (Kanda and Komárková 1997). Lichens are the most visible component of the terrestrial vegetation of continental Antarctica, where some 90 species have been identified. Only 30 percent are **endemic**; most of the rest are either found also in the Arctic or are cosmopolitan in their distribution (Castillo and Nimis 1997).

An interesting group of microorganisms and crustose lichens dwell in the pores in rocks. Invisible from the surface, they are called cryptoendolithic life forms. The only truly terrestrial animals in the Antarctic are invertebrates such as rotifers (phylum Rotifera), nematodes (phylum Nematoda), water bears (order Tardigrada), mites (order Acarina), and springtails (order Collembola) (Kanda and Komárková

1997). Mites (especially Cryptostigmata and Prostigmata) and the flightless springtails are the most successful groups on the islands of maritime Antarctica, where they must contend with soil temperatures frequently below freezing [the average is –16° C (3° F) where snow cover is less than 10 cm (2.5 in) deep and –8° C (18° F) where snow depth is greater than 70 cm (27 in)]; where freeze-thaw action is frequent; and where drought and wet and dry cycles are common. Surface dwellers must also tolerate seasonal differences in light intensity and, in some areas, increased ultraviolet (UV-B) radiation. The mite *Alakozetes antarcticus* is common in nutrient-rich areas where seabirds and marine mammals have congregated. The springtail *Cryptopygus antarcticus* occurs where there is vegetation in maritime and continental Antarctica (Block 1997).

No land mammal or land bird breeds south of South Georgia, although millions of seabirds visit the region to breed, as do two seals. The Adelie Penguin (*Pygoscelis adeliae*) is the most common bird in the Antarctic; among the approximately 50 other birds that breed there are Emperor Penguin (*Aptenodytes forsteri*), Gentoo Penguin (*Pygoscelis papua*), Chinstrap Penguin (*P. antarctica*), Black-browed Albatross (*Diomedea melanophus*), Antarctic Petrel (*Thalassoica antarctica*), and Wilson's Storm Petrel (*Oceanites oceanicus*). These birds feed at sea, while the Antarctic Skua (*Catharacta antarctica*) takes eggs and chicks from the often-huge colonies of nesting seabirds. Fur seals (*Arctocephalus gazella*) and southern elephant seals (*Mirunga leonina*) rest and breed on land, particularly on the islands of the Scotia Arc, but feed at sea (Kanda and Komárková 1997).

Cyanobacteria are the major nitrogen-fixing organisms in terrestrial habitats of the Antarctic. Most nutrients, however, come to the land from the sea—transferred by seabirds and sea mammals. The dominant food chain is detritus based with nutrients coming largely from the waste products, feathers, fur, and carcasses of animals that feed elsewhere. In fact, the nutrient levels in some colonies become high enough to be toxic to plants, and these sites become bare of vegetation (Kanda and Komárková 1997).

## Arctic-Alpine Tundra

Mountain ranges in North America and Eurasia that have peaks above tree line harbor vegetation resembling the arctic tundra; indeed, many of the plants found at higher elevation are the same ones found in the Arctic (Photo 9.4). The similarities between arctic-alpine tundra and arctic tundra diminish with latitude until by 30° N use of the prefix "arctic" is no longer appropriate. Unlike plant life, however, animal life in the mountains is quite distinct from that of the arctic tundra. Climatic differences are also great, as previously noted in the section on climate. Plants must be adapted to withstand intense solar radiation, greater exposure to ultraviolet wavelengths, and rapid and extreme temperature change, in addition to a very short growing season. Many high-elevation cushion plants orient their leaves vertically or have leaves and buds covered with hairs to help deflect the sun's rays. Some have a natural antifreeze that keeps tissues from freezing when air temperatures are below 0° C (32° F). Animals must be able to tolerate reduced amounts of oxygen in the thinner air and strong winds. Some alpine mammals, such as guanaco of South America, have higher amounts of hemoglobin in the blood than lowland species to bind more oxygen. Common adaptations among arthropods involve reduced body

Photo 9.3 Arctic alpine tundra in the Beartooth Mountains of Montana. The vegetation consists of lichens, mosses, and dwarfed shrubs closely related to those at the arctic. Rock polygons or patterned ground are visible in the foreground. (Photo by SLW.)

size, presumably because it is easier for smaller individuals to find shelter among loose rock particles; dark color, which helps to convert more sunlight to heat; reduced wing size and even flightlessness, perhaps to prevent being blown off mountain sides; and the ability either to withstand the formation of ice in body tissues or to undergo supercooling so that their tissues do not freeze until well below the freezing point of water (Somme 1997).

In western **North America**, two great chains of high mountains extend south from northern Alaska and Canada—the Cascades-Sierra Nevada range and the Rocky Mountain cordillera. In the east, the older Appalachian Mountains are much lower in elevation but, because of the climate, tree line is reached on several peaks, including Mt. Katahdin (Maine), Mt. Washington (New Hampshire), and Mt. Marcy (New York). In eastern Canada, arctic-alpine communities are found on the higher peaks of the Gaspe Peninsula (Quebec) and the highlands of southern Labrador. The alpine (i.e, above tree line) communities are composed of the same growth-forms that characterize the Arctic. Dwarfed shrubs—many creeping along the ground—are deciduous willows or evergreen heaths (e.g., *Cassiope* spp., *Empetrum* spp., *Ledum* spp., and *Vaccinium* spp.). **Herbs** are almost all perennials and include cushion plants such as moss campion (*Silene acaulis*), sandwort (*Arenaria* spp.) and whitlow-wort (*Paronychia* spp.); and rosette-forming plants such as the saxifrages, sedges (*Carex* spp. and *Krobesia* spp.), and grasses. Iceland purslane (*Koeniga islandica*) is among the very few **annual** forbs. The perennial plants usually reproduce vegetatively by tillers and rhizomes but occasionally by seed. Many of these plants have thin roots near the surface, where the soil first warms up, in addition to a large taproot where they store energy and nutrients. Most also have an overwintering above-ground stem and preformed leaf and flower buds that let them grow rapidly once the snow melts, bloom within a few weeks and—if the growing season extends long enough—set seed within six to eight weeks (Campbell 1997).

The plants found on a given mountain chain depend on the unique history and situation of the range. During the Pleistocene, the high peaks were glaciated and plants have since had to return from refuges in the lowlands. Therefore, while some have the arctic **affinity** previously mentioned, others derive from the plants found on the nearby lowlands. The general trend is that the farther south the alpine environment, the fewer arctic elements in its plant life. For example, on Wrangell Mountain in Alaska, over 70 percent of the alpine plants are arctic species. In the southern Rockies on Wheeler Peak, in the Sangre de Cristo Mountains of New Mexico, only 32 percent of the species can be categorized as arctic-alpine, while 28 percent are broadly western United States species, and 40 percent are found exclusively in the Rocky Mountains (Campbell 1997).

Vertebrates of arctic-alpine areas are characterized by a near absence of reptiles and amphibians. Birds and mammals are mostly herbivores. One of very few birds that can be found in the arctic-alpine zone in winter is the White-tailed Ptarmigan (*Lagopus leucurus*), a true resident species in the western mountains. Birds that breed in the high-elevation habitats but winter elsewhere include Water Pipit (*Anthus spinoletta*), rosy finches (*Leucostricta* spp.), and Rock Wren (*Salpinctes obsoletus*). Golden Eagle (*Aquila chrysaetos*), Prairie Falcon (*Falco mexicano*), and Red-tailed Hawk (*Buteo jamaicensis*) are some of the predatory birds that hunt in the alpine tundra.

Resident small mammals may be abundant where they have adequate cover, such as in sedge meadows and rock piles. Pikas (*Ochotona princeps*), deer mice (*Peromyscus maniculatus*), and voles (*Microtus oeconomus*) remain active all year, storing hay or seeds below ground. Mice and voles may become social and share communal nests in winter for greater warmth (Campbell 1997). Ground squirrels (*Spermophilus* spp.) and marmots (*Marmota* flaviventris and *M. caligata*), on the other hand, are true hibernators. With some rodents available throughout the winter, small carnivores such as weasels and martens can hunt above tree line all year.

One of the differences between arctic and arctic-alpine communities is the absence of large herds of hoofed animals in the latter. Large mammalian herbivores do graze in small groups on high mountain pastures in summer, however. These include mountain goat (*Oreamnos americana*) (native only to the mountains from southeast Alaska to western Montana) and mountain sheep, the thin-horned Dall's sheep (*Ovis dalli*) in Alaska south to northern British Columbia and the mountain bighorn sheep (*Ovis canadensis canadensis*) in the Rocky Mountains and Sierra Nevada. They may be joined by grizzly bear (*Ursos arctos*) and coyote (*Canis latrans*).

Among invertebrates, springtails (order Collembola) are the most abundant in the soil. Arthropods such as bumblebees, flies, and butterflies are common pollinators of alpine plants. Insects blown onto snowfields from lower elevations provide food for rosy finches.

In **Eurasia**, the major mountain chains such as the Alps, Tien Shan, and Himalaya run parallel to and separated from the Arctic Tundra proper rather than being directly connected to it as the north-south running ranges of North America are. For that reason, fewer arctic species are found in the arctic-alpine zone, but the same growth-forms are characteristic. The **zonation** is similar to what is found on mountains of western North America, and both plant and animal species are often close relatives of American forms. In the Alps, tree line varies from about 1,800 m (5,900 ft) in the outer ranges to 2,400 m (7,800 ft) in the interior mountains. A lower alpine zone of dwarf shrubs—primarily heaths—and an upper alpine zone of sedge (*Carex* spp.) grasslands are recognized. About half of the plants derive from southern and central Europe; the other half represent species that have dispersed to the Alps from the Arctic or from the highlands of central Asia (Grabherr 1997). Botanists have studied these mountains extensively, but because people have heavily affected the mountains for thousands of years it is difficult to determine what is natural.

In the Tien Shan—a complex of several ranges that stretches from Kirgistan into Xinkiang, China—tree line occurs at roughly 3,000 m (10,000 ft) and snow line somewhere above 4,000 m (13,000 ft). Due to its interior position in the large Eurasian landmass, very little precipitation [160–450 mm (6–18 in)] is received annually. Most of it falls in summer, but even then it may come as snow. In this dry climate, the alpine vegetation is largely a high-elevation grassland (steppe) dominated by sedges (*Kobresia* spp. and *Carex* spp.) but also containing perennial forbs [for example, asters (*Aster* spp.), gentians (*Gentian* spp.), cinquefoils (*Potentilla* spp.), and violets (*Viola* spp.)], grasses, mosses, and lichens. Botanists recognize several altitudinal zones, ranging from cushion plants (*Dryadanthe tetrandra*) on the exposed rock surfaces in the uppermost alpine zone, through a variety of grassland communities (moist meadows, steppe meadows, cold steppe, dry steppe) to semi-desert and desert zones 3,300–3,000 m (11,000–10,000 ft). *Artemesia rhodantha* dominates the vegetation of the last two. Moist meadows harbor the greatest diversity of vertebrates. Breeding songbirds include Horned Lark (*Eremophila alpestris*), Brandt's Rosy Finch (*Leucostice brandti*), White-winged Snow Finch (*Montifringilla nivalis*), and Alpine Chough (*Pyrrhocorax gracilia*). Rodents are represented by a vole (*Stenocranius gregalis*), a marmot (*Marmota baibacina*), and a pika (*Odotona macrotis*). Argali (*Ovis ammon*), wild sheep, graze in the alpine meadows. Among mammalian carnivores are wolf (*Canis lupus*), snow leopard (*Panthera uncia*), ermine (*Mustela erminea*), brown bear (*Ursos arctos*), and red fox (*Vulpes vulpes*) (Zlotin 1997).

Alpine plant communities in the Himalayan Mountains exist as a string of high-elevation islands. They are the southernmost Northern Hemisphere tundra ecosystems, the highest, and the most isolated. They are also the most diverse in terms of species composition. Treeline occurs between 3,500 and 4,000 m (11,500–13,000 ft); the highest plant growth is reported at 6,400 m (21,000 ft). There are no woody plants that are also found in the arctic, but about 20 herbs are common, among them alpine bistort (*Bistorta vivipara = Polygonum vivipara*), a sedge (*Carex atrata*), fireweed (*Epilobium angustifolium*), Iceland purslane (*Koeniga islandica*), mountain sorrel (*Oxyria digynia*), and a meadow rue (*Thalictum alpinum*). Iceland lichen (*Cetraria islandica*) and reindeer lichen (*Cladonia rangifera*) are other arctic species that occur in the Himalayas, although they do not form carpets there as they do in the north. Dwarf shrub thickets of *Rhododendron* near tree line give way in the lower alpine zone to cushion-forming *R. anthopogen* and *R. setosum*, which shelter liverworts, mosses, and fruticose and **foliose** lichens. Between 4,500 and 4,800 m

(14,700–15,700 ft), a mosaic of rhododendron and sedge (*Krobesia* spp.) communities occurs. Above that, the upper alpine zone is characterized by mats of *Krobesia pygmaea* less than 2 cm (1 in) high and dwarf *Rhododenron nivale* that get no higher than 5 cm (2 in). Along streams where the sedges form a closed cover, rosette-forming louseworts (*Pedicularis* spp.) or primroses (*Primula* spp.) can be found (Miehe 1997). The arctic-alpine zone in the western Himalayas is grazed by markhor (*Capra falconeri*) and in the eastern ranges by blue sheep or bharal (*Pseudois nayaur*); it is also the domain of the snow leopard (*Panthera uncia*) (Miehe 1997).

## High-Altitude Tropical Environments

Even the tropics have peaks that extend high above tree line and bear alpine ecosystems. The tropical systems differ in appearance and species composition from the arctic-alpine communities of the midlatitudes; they bear names unique to their geographic location. The vegetation of the cold, humid high elevations of the northern Andes (11° N to 8° S; from Venezuela to northern Peru) is called *paramo*. Central America has a few outliers, such as on Cerro Chirripio, Costa Rica's highest peak. The uppermost life zone on East African mountains is referred to as the afro-alpine zone, while it is known as tropical-alpine on the mountains of Indonesia and New Guinea (Monasterio and Vuilleumier 1986). The largest area of tropical alpine habitat occurs in the Andean Cordillera, so most of the following discussion focuses on the South American paramo but includes some comparisons to the afro-alpine zone.

Between the Tropic of Cancer and the Tropic of Capricorn day length and temperature patterns vary little seasonally, so alpine life does not have to contend with long periods of cold and dormancy. The other attributes of the high mountain environment described for the midlatitudes still pertain, however, and are sometimes aggravated by the more direct rays of the tropical sun. Daily and even hourly temperature changes are pronounced. In the Andean paramo, morning temperatures can be at or even below freezing and midday temperatures can rise to 23° C (73° F) (Luteyn 1992). The air is thin and deficient in oxygen. An upward flow of air usually bathes the mountaintops in clouds by early afternoon and rapidly chills the air. Such sudden changes in temperature do not allow plants time to adjust; they must be ready to tolerate or avoid freezing all of the time. Some of the growth-form adaptations are the familiar traits of tundra species: ground-hugging forms that help plants absorb heat from the soil; rosettes, cushions, and tussocks that conserve heat and protect new growth; and fine hairs on leaves, stems, and buds to trap heat. In addition, tropical mountain plants commonly have a silvery color to reflect excessive radiation, especially among large plants. Leaves on dwarfed shrubs are often small and thickened (i.e., sclerophyllous or hard-leaved) so less surface is exposed to cold. Giant rosette plants, a growth-form absent from arctic tundra and arctic-alpine tundra, are the most conspicuous element of the vegetation, although they are not present everywhere. A giant rosette plant may stand from 1 m (3 ft) to almost 10 m (30 ft) above the ground, its stem protected from freezing by a thick sheath of dead leaves and its flower bud protected by hairs or thick, hairy leaves that fold over the bud when the sun is not shining (Körner 1999).

Tropical alpine vegetation is composed of many more species than counterparts in the midlatitudes and high latitudes of the Northern Hemisphere. The plants

Photo 9.4 Paramo in Ecuador near its border with Colombia. Giant rosettes of *Espeletia* and tussock sedges are typical of this high-elevation wet grassland. (Photo © Jeremy Horner/CORBIS.)

tend to be derived from the tropical or temperate lowlands plants nearby, and many are unique to a given peak or cluster of peaks. Between 200 and 400 species of vascular plants have been described from a number of paramo sites. Luteyn (1992) estimated that 60 percent of paramo plants are found only above tree line in the northern Andes. Very few long-distance colonizers from the arctic (or subantarctic) have become established on tropical mountains; Iceland purslane (*Koeniga islandica*) is one arctic species that has successfully dispersed along the Rockies and Andes all the way to Tierra del Fuego (Hedberg 1992).

The paramo is largely a wet grassland dominated by tall bunch or tussock grasses (*Calagrostis* spp., *Festuca* spp. and others) and rosette plants (Photo 9.4). They may be accompanied by dwarfed bamboo (*Chusquea* spp.), shrubs from several plant families (Asteraceae, Ericaceae, Melastomaceae, and Hypericaceae), and sedges. A dense mat of mosses, lichens, and low cushion plants usually grows on the ground. *Espeletia* spp. and *Coespeletia* spp. show one of the typical arborescent growth-forms that distinguish tropical from temperate alpine vegetation. The stem is a thick column 1–3 m (3–10 ft) high and covered in dead leaves that droop down the sides. At the top is a rosette of living leaves, long elliptical, silver in color, and covered with hairs. The dead leaves collect dust and debris and serve both to protect the stem from freezing temperatures and to provide shelter for invertebrates, mice, and small birds (Diaz et al. 1997). The other form that giant rosette plants assume is that of a large flower stalk a meter (3 ft) or more tall emanating from a basal rosette of leaves. This is the form plants in the genus *Puya*, members of the bromeliad or pineapple family, take. Puyas occur at lower elevations in the Andes, but those of the paramo are distinguished by a thick coat of hairs on the flower stalk that almost completely covers the blossoms.

Cushion plants usually dominate swampy or boggy areas in the paramo, and scattered forbs from several plant families, even orchids, dominate open rocky areas. In an upper alpine zone [approximately 4,100–4,800 m (13,5000–15,750 ft), where permanent ice and snow begin] fragmented rock (scree) called *superparamo* is characterized by scattered rosette plants (Luteyn 1992).

Vertebrates are poorly represented at these high elevations and vary according where you are in the Andean Cordillera. A few endemic frogs and very few reptiles (one or two species of lizard, maybe a snake) are represented. Birds are the most abundant and diverse group. Many resident species are small and eat insects either from the ground or from bushes. Hummingbirds, which are able to go into torpor when temperatures drop, are important pollinators. Of the few mammals associated with the paramo, most are herbivores. Rodents are most diverse; shrews (*Cryptotis thomasi*) most abundant. Paramo rabbits (*Sylvilagus* spp.) and opossums (*Didelphis* spp.) occur, and the long-tailed weasel (*Mustela frenata*) is a common predator (Luteyn 1992).

The same five growth-forms dominate the afro-alpine vegetation belt and the paramo: giant rosette plants, tussock grasses (e.g., *Festuca* spp.), rosette plants, cushion plants, and hard-leaved shrubs. Giant lobelia (e.g., *Lobelia telekii*) is in some ways similar to puya, although it also can secrete a watery liquid (pectin) into the rosette to a depth of 10 cm (4 in). At night, only the top 1 cm (0.5 in) freezes, while the remaining water insulates the plant's leaves (Hedberg 1992). The exceptionally long bill of a large sunbird (*Nectarinia johnstoni*) seems adapted to collecting nectar from deep within the lobelia blossom (Moreau 1966). Giant groundsels (*Dendrosenecio* spp.) (Photo 9.5) take the place of the arborescent *Espletia* of South America and grow even taller and often form branches. Some mosses take a

Photo 9.5 Giant groundsels (*Dendrosenecio* sp.) of the afro-alpine zone on Mt. Kenya and on other high peaks in East Africa. (Photo © Galen Rowell/CORBIS.)

peculiar form in areas of wet soil. The cushions are detached from the ground and rolled into balls, presumably turned over by the nightly freezing and heaving of the ground (Hedberg 1992). The species diversity of the afro-alpine belt is markedly less than that of the Andean paramo. The high peaks of East Africa (the Virunga volcanoes, the Ruwenzori, Mt. Meru, Mt. Kenya, and Mt. Kilimanjaro) are isolated from one another and are not part of a long, continuous *cordillera* like the Andes. Movement of plant species among the mountains is difficult and rare, so each tends to harbor species unique to it. The **dispersal** of northern species to East Africa is even more problematic, since the great mountain ranges of Eurasia run east-west, rather than north-south as with North American mountains. Species have no corridors through which they can readily move, and only a few northern groups, such as saxifrages (*Saxifraga* spp.) and primroses (*Primula* spp.), have reached tropical Africa (Hedberg 1992).

## ORIGINS

The origins of arctic and arctic-alpine tundra communities are still poorly known. It is generally assumed that arctic and alpine tundra species evolved from forest and **scrub** plants and animals of lower latitudes and elevations. The outlines of their history are based on the geologic history of continental drift and associated mountain building, with some collaborating evidence from the fossil record—especially the fossil record of mammals. The general picture suggests a possible center of origin for arctic plants and animals in the high plateaus of central Asia, the Tibetan Plateau in particular, during the Tertiary Period. The Early Tertiary was wetter and warmer than today and without high mountains in North America. Tropical and subtropical species inhabited most of North America. However, in the Late Tertiary, global climate cooled as a consequence of the poleward movement of the northern continents, and the Asian plateau species were able to move westward through Eurasia and northward and eastward into eastern Siberia and—whenever a land bridge (Beringia) existed—into Alaska. At the same time, the Rocky Mountains formed and reached their present elevation late in the Tertiary. The young Rocky Mountains opened a new environment for colonization by plants from the surrounding lowlands that could adapt to the cold and other aspects of high elevations (Campbell 1997). Plants and animals rapidly evolved to adapt to arctic or alpine conditions beginning in the late Miocene and continuing through the Pliocene into the early Pleistocene. Before the ice ages of the Pleistocene, a set of about 1,500 circumpolar plant species existed (Axelrod and Raven 1985; Löve and Löve 1974). With the Rocky Mountains as one source area and the Arctic as another, the advances and retreats of continental ice sheets during the Pleistocene made possible a mixing of cold-adapted species moving north and south along a Rocky Mountain corridor. When ice accumulated on the northern regions of North America and mountain ice caps extended to lower elevations, arctic species moved southward. As the ice receded, arctic species returned to the north, and some of their alpine associates accompanied them. Many species, however, did not survive the rigors of the Pleistocene; the contemporary set of circumpolar arctic plants numbers about 900 (Bliss 1997).

Not all of the Arctic was covered with ice during the Pleistocene glacial periods. Much of northeastern Siberia, western Alaska, the Canadian Arctic islands, and—

most importantly—Beringia—the land bridge that formed between Asia and North America whenever the world's water became stored on land as ice and sea levels dropped 100–200 m (300–600 ft)—were ice free. These areas were colder and drier than today and supported a vegetation of grasses and sedges with *Artemisia* forbs and small shrubs. This vegetation is sometimes referred to as steppe-tundra or mammoth-steppe, although the plants it contained are quite different from those of contemporary steppes (temperate grasslands) or tundra; it was, however, home to mammoths—and a host of other large mammals now extinct. Beringia was important because it allowed the dispersal of cold-adapted plants and animals from one continent to the other, with the main flow being from the Old World to the New (perhaps because Eurasia was essentially ice free while much of North America was often blocked by ice). Those entering North America from Eurasia included the ancestors of such modern tundra animals as caribou (*Rangifer taradanus*), musk ox (*Ovibos moschatus*), fox (*Alopex* spp.), hare (*Lepus* spp.), ground squirrel (*Spermophilus* spp.), and lemming (*Lemmus* spp.) (Bliss 1997).

The origins of the plant life of Antarctica have recently been reinterpreted in light of a revised **taxonomy** of lichens. The lichens of the continent are believed to be the products of long-distance dispersal events that occurred no later than the Quaternary Period. The evidence supporting this contention is the large proportion of wide-ranging or strictly polar species present, the low number of endemics, the absence of species common in the sub-Antarctic, and the relatively large number of genera compared to species in the contemporary lichen **community** (Castillo and Nimis 1997).

The Andean Cordillera rose as a consequence of the separation of South America from Africa and reached elevations high enough to have alpine zones in the tropics by the late Pliocene or early Pleistocene epochs. These zones represented a new, unoccupied habitat in South America. They were colonized by plants that migrated upward from the montane zone and adapted to the cold, that dispersed overland from temperate regions to the south, or that dispersed by long-distance dispersal from alpine habitats in temperate North America. Isolation on the various high peaks allowed for speciation so that today the paramo has some 30 endemic genera and hundreds of endemic species. About one-half of the genera are tropical-Andean in origin, and the other half are temperate in origin. Of the latter group, 10 percent are Holarctic (found in both North America and Eurasia). These could have invaded only along the Rocky Mountain-Andean corridor after the formation of the Isthmus of Panama toward the end of the Pliocene Epoch. They mostly occur in the superparamo and include certain species of sphagnum moss (e.g., *Sphagnum compactum*). An austral-antarctic element comprises another 10 percent of the plants of temperate origin and are mainly associated with high-latitude lakes and bogs. The cushion plant *Azorella aretioides* and mat-forming *A. pedunculata* of Ecuador are examples. The rest of the temperate species most likely came in from the south along a more or less continuous dispersal route along the southern Andes (van der Hammen and Cleef 1986).

The origin of East Africa's high peaks is associated with the Western and Eastern rift valleys, which probably date from the late Miocene. Many of the mountains along the Western Rift, such as the Virunga volcanoes, are volcanic, but the Ruwenzori are a block of ancient crystalline rock thrust up between two branches of the rift. The Eastern Rift or Great Rift Valley also has high volcanic mountains associated with it, including Mount Kenya, Mount Elgon, and Mount Kilimanjaro, last active about 36,000 years ago (Taylor 1996). These peaks create an archipelagolike arrangement of high-elevation sites, each isolated from the others and all isolated from arctic, antarctic, and arctic-alpine sources of cold-adapted species. The species that live on them today, therefore, derive primarily from afro-montane plants at lower elevations. At times during the Pleistocene when the global climate was cooler, the afro-alpine zone probably expanded at lower elevations and the degree of isolation was reduced so some exchange of plants and animals between mountains occurred. However, the large proportion of species unique to each peak suggests isolation was the norm (Hedberg 1986; Hedberg 1992).

## HUMAN IMPACTS

The arctic tundra of North America is largely intact because it was sparsely populated by essentially hunting-fishing-and-gathering peoples (Inuit) well into the twentieth century. Significant human impact is limited to the close vicinity of scattered settlements, abandoned military sites, mining operations, and transportation networks. Tundra vegetation, however, is easily damaged or destroyed and very slow to recover, so expanding human activity in the Arctic threatens its integrity. The main threats come from oil and mineral exploration and development, and the necessary construction of base camps, roads, pipelines, and ports. These can interfere with the migration routes of the great caribou herds and the movements and foraging of wolf, bear, and musk ox. Devastating oil and chemical spills and toxic runoff from mine tailings are other potential threats associated with this development (Ricketts et al. 1999).

Ecotourism in the Arctic is increasingly popular as people come to witness some of North America's last wilderness, the caribou migrations, the autumn congregation of polar bears at the shores of Hudson Bay, and other wildlife wonders. Without proper management, visitors can disturb nesting colonies of waterfowl and shorebirds and caribou on their calving grounds and destroy the very spectacles they come to see. With an increased human presence—whether of workers or tourists—one can expect the call for removal of nuisance or dangerous wolves, polar bears, grizzly bears, or wolverines (Ricketts et al. 1999).

The Eurasian tundra suffers from similar impacts and threats with the important addition of overgrazing by domestic reindeer. Traditional nomadic and seminomadic societies (e.g., Lapps and Samoyedes) maintain large herds, but political and economic considerations are increasingly confining their movements, which prevents the recuperation of overgrazed pastures.

In Antarctica, a significant proportion of ice-free land, especially on the Antarctic Peninsula, has been destroyed by research stations as country after country established territorial claims to the continent. This is an unfortunate consequence of the international Antarctic Treaty, written in part to conserve the natural environment. The Treaty requires nations to establish bases to become eligible for participation in mineral extraction, commercial harvest of the seas, and research. The bases result in buildings, ground and air traffic and pollution, garbage storage facilities, airfields, and trampling. Noise and the human presence are stressful to animals, as shown by a decline in breeding success in some penguin colonies. Increasingly popular land-based ecotourism also takes a toll by disrupting vegetation and animal life (Kanda and Komárková 1997), although fluctuations in Adelie Penguin populations on islands west of the Antarctic Peninsula may be the result of natural environmental change (Fraser and Patterson 1997). **Introduced** plants, insects, and livestock affect ecosystems of the subantarctic (Gremmen 1997; Chevrier et al. 1997). The impacts of commercial overexploitation of krill—the base of oceanic food chains—on penguin populations and terrestrial nutrient cycles and of global climate change are yet to be seen.

Alpine ecosystems worldwide are affected by fire and the grazing of domestic livestock, in particular sheep and cattle. In the Alps, this has been going on for thousands of years, and it is now difficult to determine what natural vegetation might be like. Even in the sparsely populated Himalayas. the impact of grazing animals is noticeable up to the highest elevations plants grow. Plant species richness has decreased, and unpalatable plants have become more abundant. Herding peoples have increased and improved pasturage below tree line by cutting and burning woody plants in the upper montane zone. On Annapurna South in Nepal, the prostrate junipers (*Juniperus* spp.) and dwarf deciduous rhododendrons (*Rhododendron* spp.) of south-facing slopes are regularly burned. Elsewhere in Nepal, on north-facing slopes burning has eliminated evergreen rhododendron scrub and converted the vegetation to one of willow (*Salix* spp.) scrub and tall herbs. A distinct plant community has been reported from alpine areas in the Himalayas of Assam, where yaks, cattle, sheep, and goats are regularly grazed. It is composed of cosmopolitan weeds such as docks (*Rumex* spp.), nettles (*Urtica* spp.), and knotweeds (*Polygonum* spp.), although native primroses (*Primula roylei* and *P. strumosa*) maintain dominance. Where yaks are concentrated, *Scopalita lurida*, normally a rare plant, is locally abundant (Schweinfurth 1983).

Agricultural practices are ancient in the Andes, but pre-Columbian peoples had their permanent settlements below the paramo zone. They used the paramo for hunting or gathering wild plants. The giant rosette plants were (and continue to be) useful for a variety of purposes. The leaves were used for roofing and insulation in the walls, to fill mattresses, and to manufacture clothing. The fleshy pith of some species is edible, raw or cooked; roots could be processed into a butterlike product; and the aromatic resins in the stem could be burned as incense. In Colombia and probably elsewhere, the whole plant was burned and the remaining ash was applied to the potato fields as fertilizer and to reduce the acidity of the soil (Peres 1998).

When Spanish domination began, native peoples sought refuge in the bleak wilderness, but in the early 1600s, the Indians were forced into settlements in the lower paramo. The Spanish introduced European cattle and sheep—the first domestic livestock and large grazing animals—to the high Andes. To expand pastureland, timberline forests were cut and burned, which extended the coverage of paramo to elevations below natural tree line. In Venezuela, giant rosette plants can be seen in farmland at 2,700 m (8,900 ft), while remnant patches of forest occur up to 3,060 m (10,000 ft) (Peres 1998). Similarly, in the vicinity of Quito, Ecuador, it is difficult to delineate tree line because of the occurrence of paramo grasses below the timberline *Polylepis* forest. Indeed, there is considerable debate today about just how natural the paramo actually is and to what degree it is a product of human land use.

Cattle eat the soft, fleshy leaves of cushion plants, which may negatively affect the germination of the giant rosette species that use the cushion plants as stable environments—protected from the churning of daily freeze-thaw cycles—in which to grow. Cattle (and mules) also consume the leaves of the giant rosette plants and may strip the sheath of dead leaves from the stems by scratching against the few tall rubbing posts in the pasture. In their search for forage in a habitat marginal for cattle, the domestic animals also break off chunks of turf and expose the slopes to greater erosion (Peres 1998).

The construction of the Pan American Highway provided easy access to the lower reaches of the paramo in several locations. This allowed an increase in settlement and recreational use of the high-elevation zones. Grazing and agricultural pressures have increased markedly with the rapid population growth experienced in the second half of the twentieth century. In some areas. native paramo species are disappearing and being replaced by **exotic** weeds. Off-road vehicles are an increasing threat; they destroy the crust of mosses and lichens that add nutrients to the soil and shield it from wind erosion (Peres 1998). Yet away from major population centers, large tracts of paramo are still intact (Luteyn 1992).

## HISTORY OF SCIENTIFIC EXPLORATION

Russian scientists led biological research on the arctic tundra beginning in the early nineteenth century. The most important and influential work of the period was Middendorf's (1869) *A Journey in the North and East of Siberia, II.* Toward the end of that century and continuing into the early twentieth century, Russian interest in the North increased, and a number of botanical and zoological expeditions, such as the Russian Polar Expedition of 1900–1903, yielded rich collections of tundra plants and animals. In the 1920s, Norwegian and Russian studies focused on the islands of Novaya Zemlya, which became the model for arctic nature. Several groups of organisms became well known, including vascular plants, mosses, lichens, birds and mammals. Scientists interested in understanding the processes by which species adapt to extreme environments used arctic birds and mammals as their objects of study. Lemmings became the classic case study of population cycles. Investigation into the interrelationships between plants and animals of the tundra started in the 1930s to form a scientific basis for reindeer husbandry. The Soviet Union, by virtue of its northern location, was interested in settling and exploiting other natural resources in the Arctic and in the 1930s developed classifications and maps of tundra vegetation types to aid in planning for economic development. Research intensified in the 1960s and 1970s with Soviet scientists' participation in the International Biological Programme (IBP) (Chernov and Matveyeva 1997). American scientists also got deeply involved in tundra studies as a consequence of the IBP. Detailed studies of terrestrial ecosystems were conducted at Barrow, Alaska, and Truelove Lowland, Devon Island, Northwest Territory, Canada (Bliss 1997). The descriptive phase of tundra exploration is pretty much complete; ecologists today are more interested in how the arctic ecosystems function and study energy flows, nutrient cycling, community stability, and cyclical phenomena related to soil-plant-animal interactions (Chernov and Matveyeva 1997). Others are using the tundra and the forest/tundra ecotone as a laboratory for studying the affects of global climate change.

Paramo, bleak wilderness, received its name from Spanish explorers in the mid-1500s but was not scientifically described until Alexander von Humboldt traveled in the Andes between 1799 and 1804. Based on his observations, Humboldt published the first profiles of altitudinal zonation of vegetation and put forth the idea that climate determines vegetation. For this early work, he is today regarded as the father of plant geography. Botanists from several nations collected plants high in tropical Andes over the next couple of centuries, but it was in the early twentieth century that the real study of South American tropical alpine environments got under way, led by such people as Carl Troll of Germany and J. Cuatrecasas (1968) of Colombia. Research continues today with major programs at the Universidad de los Andes (Bogota, Colombia), Herbario Nacional Colombino (Bogota, Colombia), and Pontificia Universidad Catolica (Quito, Ecuador) in conjunction with Aarhus University (Denmark). CIELAT (Centro de Investigaciones Ecologicas de los Andes Tropicales), established by M. Monasterio in Colombia, has produced important work in paleoecology and community structure. The ECOANDES project (Investigaciones de Ecosistemas Tropandinos), a cooperative effort between Colombia and the Netherlands, is involved in studies of vegetation composition, function and dynamics as the history of the paramos (Luteyn 1992).

The snow-capped mountains of Africa were discovered by Europeans only in the mid-1800s. In 1889, Hans Meyer succeeded in reaching the summit of Kibo on Kilimanjaro, Africa's highest peak (Taylor 1996). The afro-alpine vegetation belt was first described and illustrated by the Swedish team of Fries and Fries in 1922, and the giant rosettes plants they presented have fascinated people ever since (Hedberg 1992). The most complete and influential ecological investigations of the zone were conducted by Olov Hedberg (1964), who studied how the high-altitude environment selected for the growth-forms characteristic of the plants growing there and who proposed a scheme of altitudinal zonation on African mountains that incorporated the notion of an afro-alpine zone. He and a number of younger researchers continue to study the unique ecosystems on Africa's highest mountains.

Much current research in arctic-alpine and tropical alpine habitats, as in the arctic tundra itself, is related to climate-change problems and human impacts.

## REFERENCES


Axelrod, D. I. and P. Raven. 1985. "Origins of the Cordilleran flora." *Journal of Biogeography* 12: 21–47.

Bliss, L. C. 1997. "Arctic ecosystems of North America." Pp. 551–683 in Wielgolaski, F. E., ed., *Polar and Alpine Tundra.* Ecosystems of the World, 3. Amsterdam: Elsevier.

Block, William. 1997. "Ecophysiological strategies of terrestrial arthropods in the maritime Antarctic." Pp. 316–320 21 in B. Battaglia, J. Valencia, and D. W. H. Walton, eds., *Antarctic Communities: Species, Structure and Survival.* Cambridge: Cambridge University Press.

Campbell, J. S. 1997. "North American alpine ecosystems." Pp. 211–261 in Wielgolaski, F. E., ed., *Polar and Alpine Tundra.* Ecosystems of the World, 3. Amsterdam: Elsevier.

Castillo, Miris and Pier Luigi Nimis. 1997. "Diversity of lichens in Antarctica." Pp. 15–21 in B. Battaglia, J. Valencia, and D. W. H. Walton, eds., *Antarctic Communities: Species, Structure and Survival.* Cambridge: Cambridge University Press.

Chernov, Y. I. and N. V. Mataveyeva. 1997. "Arctic ecosystems in Russia." Pp. 361–507 in Wielgolaski, F. E., ed., *Polar and Alpine Tundra.* Ecosystems of the World, 3. Amsterdam: Elsevier.

Chevrier, M., P. Vernon and Y. Frenot. 1997. "Potential effects of two alien insects on a sub-Antarctic wingless fly in the Kerguelen islands." Pp. 424–432 in B. Battaglia, J. Valencia, and D. W. H. Walton, eds., *Antarctic Communities: Species, Structure and Survival.* Cambridge: Cambridge University Press.

Cuatrecasas, J. 1968. "Paramo vegetation and its life forms." *Colloquium geographicum* 9: 163–186.

Diaz, A., J. E. Péfaur, and P. Durant. 1997. "Ecology of South American páramos with emphasis on the fauna of the Venezuelan páramos." Pp. 263–310 in Wielgolaski, F. E., ed., *Polar and Alpine Tundra.* Ecosystems of the World, 3. Amsterdam: Elsevier.

Elvebakk, Arve. 1997. "Tundra diversity and ecological characteristics of Svalbard." Pp. 347–359 in Wielgolaski, F. E., ed., *Polar and Alpine Tundra.* Ecosystems of the World, 3. Amsterdam: Elsevier.

Fraser, William R. and Donna L. Patterson. 1997. "Human disturbance and long-term changes in Adélie penguin populations: a natural experiment at Palmer Station, Antarctic Peninsula." Pp. 445–452 in B. Battaglia, J. Valencia, and D. W. H. Walton, eds., *Antarctic Communities: Species, Structure and Survival.* Cambridge: Cambridge University Press.

Furley, P. A. and W. W. Newey. 1983. *Geography of the Biosphere, An Introduction to the Nature, Distribution and Evolution of the World's Life Zones.* London: Butterworths.

Grabherr, Georg. 1997. "The high-mountain ecosystems of the Alps." Pp. 97–121 in Wielgolaski, F. E., ed., *Polar and Alpine Tundra.* Ecosystems of the World, 3. Amsterdam: Elsevier.

Gremmen, Niek J. M. 1997. "Changes in the vegetation of sub-Antarctic Marion island resulting from introduced vascular plants." Pp. 417–423 in B. Battaglia, J. Valencia, and D. W. H. Walton, eds., *Antarctic Communities: Species, Structure and Survival.* Cambridge: Cambridge University Press.

Hedberg, O. 1964. "Features of afroalpine plant ecology." *Acta Phytogeographica Suecia* 49: 1–144.

Hedberg, O. 1986 "Origins of the Afroalpine flora." Pp. 443–468 in Francois Vuilleumier and Maximina Monasterio, eds., *High Altitude Tropical Biogeography.* New York: Oxford University Press.

Hedberg, O. 1992. "Afroalpine vegetation compared to paramo: Convergent adaptations and divergent differentiation." Pp. 15–29 in H. Balslev and J. L. Luteyn, eds., *Paramo, An Andean Ecosystem under Human Influence.* New York: Academic Press.

Kanda, H. and V. Komárková. 1997. "Antarctic terrestrial ecosystems." Pp. 721–761, in Wielgolaski, F. E., ed., *Polar and Alpine Tundra.* Ecosystems of the World, 3. Amsterdam: Elsevier.

Körner, Christian. 1999. *Alpine Plant Life. Functional Plant Ecology of High Mountain Ecosystems.* Berlin: Springer.

Löve, A. and D. Löve. 1974. "Origin and evolution of the arctic and alpine floras." Pp. 571–603 in J. D. Ives and R. G. Barry, eds., *Arctic and Alpine Environments.* London: Methuen.

Luteyn, J. L. 1992. "Paramos: Why study them?" Pp. 1–14 in H. Balslev and J. L. Luteyn, eds., *Paramo, An Andean Ecosystem under Human Influence.* New York: Academic Press.

Miehe, Georg. 1997. "Alpine vegetation types in the central Himalaya." Pp. 161–183 in Wielgolaski, F. E., ed., *Polar and Alpine Tundra.* Ecosystems of the World, 3. Amsterdam: Elsevier.

Monasterio, Maximina and Francois Vuilleumier. 1986. "Introduction: High tropical mountain biota of the world." Pp. 3–7 in Francois Vuilleumier and Maximina Monasterio, eds., *High Altitude Tropical Biogeography.* New York: Oxford University Press.

Moreau, R. E. 1966. *The Bird Faunas of Africa and Its Islands.* New York: Academic Press.

Nowak, Ronald M. 1991. *Walker's Mammals of the World.* 5th edition, Vol II Baltimore: The Johns Hopkins University Press.

Peres, Francisco L. 1998. "Human impact on the high paramo Landscape of the Venezuelan Andes." Pp. 147–183 in Zimmerer, K.S. and K. R. Young, eds., *Nature's Geography,*

*New Lessons for Conservation in Developing Countries.* Madison: The University of Wisconsin Press.

Ricketts, Taylor H., Eric Dinerstein, David M. Olson, Colby J. Loucks et al. 1999. *Terrestrial Ecoregions of North America, A Conservation Assessment.* Washington, DC: Island Press.

Sarmiento, Guillermo. 1986. "Ecological features of climate in high tropical mountains." Pp. 11–45 in Francois Vuilleumier and Maximina Monasterio, eds., *High Altitude Tropical Biogeography.* New York: Oxford University Press.

Schweinfurth, U. 1983. "Man's impact on vegetation and landscape in the Himalayas." Pp. 297–309 in W. Holzner, M. J. A. Werger, and I. Ikusima, eds., *Man's Impact of Vegetation.* The Hague: Dr. W. Junk Publishers.

Somme, L. 1997. "Adaptation to the alpine environment in insects and other terrestrial arthropods." Pp. 11–26 in Wielgolaski, F. E., ed., *Polar and Alpine Tundra.* Ecosystems of the World, 3. Amsterdam: Elsevier.

Taylor, David. 1996. "Mountains." Pp. 287–306 in Adams, W. M., A. S. Goudie, and A. R. Orme, 1996. *The Physical Geography of Africa.* Oxford: Oxford University Press.

van der Hammen, Thomas and Antoine M. Cleef. 1986. Development of high Andean páramo flora and vegetation." Pp. 153–201 in Francois Vuilleumier and Maximina Monasterio, eds., *High Altitude Tropical Biogeography.* New York: Oxford University Press.

Wielgolaski, F. E. 1997a. "Introduction." Pp. 1–5, in Wielgolaski, F. E., ed., *Polar and Alpine Tundra.* Ecosystems of the World, 3. Amsterdam: Elsevier.

Wielgolaski, F. E. 1997b. "Adaptation in plants." Pp. 7–10, in Wielgolaski, F. E., ed., *Polar and Alpine Tundra.* Ecosystems of the World, 3. Amsterdam: Elsevier.

Zlotin, R. J. 1997. "Geography and organization of high-mountain ecosystems in the former U.S.S.R." Pp. 133–159 in Wielgolaski, F. E., ed., *Polar and Alpine Tundra.* Ecosystems of the World, 3. Amsterdam: Elsevier.

Zoltai, S. C. and F. C. Pollett. 1983. "Wetlands in Canada: their classification, distribution, and use." Pp. 245–268 in A. J. P. Gore, ed., *Mires—Swamp, Bog, Fen, and Mire.* Ecosystems of the World, 4B. Amsterdam: Elsevier Scientific Publishing Company.

# II FRESHWATER BIOMES

## INTRODUCTION

Freshwater **ecosystems** are a problem to define as **biomes** because **vegetation** corresponding to regional **climates** is lacking. Elements of the physical environment—such as the nature of the substrate, depth of water, velocity of flow, and water chemistry—determine the patterns of life in these systems, in which large plants are relatively minor components. Many scientists would argue that if streams, lakes, marshes, and the like are to be considered in the framework of biomes at all, they constitute a single entity, as they are usually connected to each other not only physically but functionally. In this scenario, streams would represent the flowing water expression of the biome, and lakes would be the still water aspect. Yet freshwater ecologists usually specialize in one or the other: they are either stream ecologists or limnologists. The types of **habitats** and the organisms inhabiting streams and lakes are sufficiently distinct to require different expertise among scientists and usually are discussed in separate textbooks. Therefore, for the purposes of this book, it was decided to divide the freshwater realm into two biomes: a stream biome and a lake biome.

This two-part scheme still leaves some problems, namely what to do about communities developed in waterlogged **soils** on frequently or permanently flooded sites (wetlands such as freshwater marshes, swamps, mires, bogs, and fens) and how to treat lakes that do not contain freshwater but have brackish or even highly saline waters. It was decided to group freshwater wetlands in the stream biome, for many are directly associated with the floodplains of rivers. However, another author might with equal justification consider them a separate biome. Brazilians, for example, refer to the Pantanal (see Chapter 10) as one of their country's major biomes. Lakes, as a matter of physical geography, are defined by the fact that they are cut off from the great world ocean and not by virtue of their water chemistry. The Caspian Sea is the world's largest lake in terms of surface area, but its waters are salty. The Great Salt Lake and the Dead Sea are even saltier than the ocean. In keeping with the basic definition of a lake, considering that few organisms inhabiting saline lakes are derived from the sea, and that many salt lakes originated from the evaporation of former freshwater lakes, these saline bodies of water are treated as just one type of lake and are included in the lake biome.

The distribution of life in freshwater biomes is controlled in part by the substrate. Different forms tend to inhabit rocky bottoms than sandy and muddy bottoms. Water chemistry and water temperature are also important factors. Especially significant are the amounts of dissolved oxygen and carbon dioxide and the amount and kind of dissolved salts. Amounts of oxygen and carbon dioxide vary with water temperature; the amounts and kinds of dissolved salts vary according to local geology as well as the seasonal precipitation patterns and the vegetative cover of surrounding lands that influence the volume of water and sediment flowing into a lake or stream. Turbulence in the water determines where nutrients may settle to the bottom and when and where mixing occurs. Turbulent flow is a major habitat factor in streams, and plants and animals must adapt by avoiding it or by fastening themselves to rocks or other solid substrates in the channel. Water turbulence in lakes is generated by the wind and mixes the upper parts of water column to help keep nutrient particles and plankton suspended near the surface where sunlight can reach them.

Seasonal change is a fact of life in all aquatic environments. At higher **latitudes**, freezing temperatures in winter impose one set of challenges and warmer water in summer another. Most **temperate zone** lakes experience vertical mixing of the whole water column in spring and autumn because water density varies with temperature. As water cools, its density increases until it reaches maximum density at 3.94° C (39° F). Cool water, therefore, sinks to the bottom of a still body of water. However, it is a peculiarity of water that continued cooling toward the freezing point results in a lessening of density, so the coldest water rises to the surface and carries with it any nutrients that had settled to the bottom. Ice floats on top of the lake: with the change of phase at the freezing point, solid ice is about 8.5 percent lighter than liquid water at 0° C (32° F).

In the **tropics**, seasonal change is more closely related to dry seasons and wet seasons than to seasons based on different temperatures. Water levels rise and fall in lakes, and streams may flood their banks at one time of year and dry up to become a series of isolated pools at other times. Life in these waters must adapt to such extreme fluctuations.

Much of the nutrients in stream ecosystems derive from land plants whose overhanging leaves and dead material fall into the water. This detritus is the foundation of food chains in flowing water, where phytoplankton is necessarily insignificant. In the still water of lakes, however, the phytoplankton are the important primary producers for grazing food chains. Larger plants in both freshwater biomes are restricted to shallow, slow-moving or still water, where they provide shelter for numerous types of animals. The common **growth-forms** are emergents such as cattails (*Typha* spp.) and reeds (*Juncus* spp.) that root in the bottom but send stiff erect stems that bear their flowers above the water level; hydrophytes such as pond lilies (*Nympaea* spp.) and pondweeds (*Elodea* spp.) that root in the bottom but have leaves submerged or floating in the water and renewal buds below water; or truly floating plants such as duckweed (*Lemna* spp.) or water hyacinth (*Eichhornia crassipes*) that are not anchored to the channel bottom at all.

The abundance of insects in freshwater biomes is a major way that freshwater and marine biomes differ. Insect nymphs are essential components of benthic communities and food for other invertebrates as well as vertebrates. Such groups as caddisflies (order Trichoptera), stoneflies (order Plectora), flies and midges (order Diptera), and damselflies and dragonflies (order Odonata) are abundantly represented. Other common invertebrates include snails, freshwater mussels, and crayfish.

Fish are the most important vertebrates in these aquatic biomes, and this is especially true of streams. Lakes have a greater abundance and variety of amphibians, particularly frogs and salamanders. Various turtles, water snakes, and crocodilians represent the reptiles in freshwaters. A few birds are closely associated with fast-moving streams, and many more diving birds and shorebirds are associated with lakes. Mammals that inhabit freshwater full time are rare and occur mostly in tropical rivers [e.g., river dolphin (*Inia geoffrensis*) and Amazon manatee (*Trichechus inunguis*)], but others, such as capybara (*Hydrochoreus hydrochaeris*), hippopotamus (*Hippopotamus amphibius*), otter (*Lutra canadensis*), and moose (*Alces alces*) regularly enter streams and lakes.

Since freshwater aquatic biomes lack the prerequisite plant growth-forms to allow for distinct formations on the different continents, the format of the following chapters departs from that used for terrestrial biomes and does not identify major expressions of the biomes in the same way. Much of the research on both streams and lakes has been conducted in the **temperate** regions of Europe and North America, and information from these areas of the world is used in most textbooks available in English. For that reason, the general information presented on stream and lake environments and the organisms inhabiting them apply to most of the surface waters in those parts of the world. It seemed unnecessary to repeat the information for specific water bodies on those two continents but important to indicate areas where these conditions do not prevail. In the stream biome, therefore, tropical streams receive separate treatment, with examples coming from some of the world's largest rivers. The other distinct form of riverine habitat that receives special attention is that of the freshwater wetland. Again, some of the world's largest wetlands are highlighted. With lakes, separate treatment is given to some of the world's largest and deepest, not only because of their superlative characteristics but because their very size gives them different attributes than the thousands of smaller lakes dotting the Earth. In this regard, the Great Lakes of North America, Lake Baikal in Russia, and the Great Lakes of tropical Africa are featured. Lake Titicaca on the Andean Altiplano in Peru and Bolivia is unique as the world's highest navigable lake and presents some of the limitations of high-altitude lakes. A third category of lakes singled out for separate description is that of saline lakes. Discussions of the Caspian Sea, Great Salt Lake, and Lake Nakura acknowledge the considerable variation exhibited in lakes of this type. Lastly, the recently mapped subglacial lakes of Antarctica are described.

Freshwater aquatic biomes in general are young in geologic terms, especially when compared to marine ecosystems. Most of today's lakes date only to the Pleistocene epoch, the main exceptions being Lake Baikal and the Great Lakes in Africa's Rift Valley. Most rivers in the Northern Hemisphere either came into being after the continental ice sheets of the Pleistocene receded for the last time (about 10,000 years ago) or were greatly modified by environmental changes of ice ages. Lakes tend to be short lived; they fill in and become overgrown by terrestrial vegetation in a few thousand years. As a consequence, freshwater organisms show much lower overall diversity than marine life does and have relatively few **endemic** forms. The exceptions in the latter case are found mostly in those very old lakes just mentioned.

Human impacts on the freshwater biomes come directly from overfishing, dredging, or oil spills and indirectly from poor land-use practices and consequences of industrialization. Common to both biomes are problems such as nutrient enrichment (eutrophication), acidification and toxic chemical pollution largely derived from airborne particles, siltation resulting from the removal of natural vegetation on the watershed, and the introduction of **exotic species**. These negative changes threaten a large number of invertebrates, some unique vertebrates, and some commercial fisheries. Much of the recent interest and research in aquatic ecosystems is directed at restoration of natural habitats and learning how best to conserve what still remains essentially intact.

# 10 RIVERS AND STREAMS

## OVERVIEW

Flowing waters, or the lotic environment, make up only 0.0001 percent of the world's water (Allan 1995). They present a unique set of conditions for life. The channelized water of a stream moves in one direction, downslope, and yet plants and animals do not all get washed out of a river system because they have evolved adaptations that allow them to maintain a particular horizontal position in a stream. This has not been the case for floating algae (phytoplankton), which therefore normally play a minor role in stream ecosystems. Life in streams depends mostly upon sun energy, which is converted to food by land plants growing along the banks and overhanging the river and enters the water as dead leaves and fallen branches and stems. Nutrients and the very water in a stream also have their sources on the surrounding catchment area or watershed, and both are constantly flushed from the system so that continual input is required for a stream ecosystem to maintain itself. Rainwater may run off the land directly into a stream, or it may first percolate into the soil and groundwater and enter a stream through the sides or bottom of the channel. Either way, as the water passes through the soil and litter, it picks up dissolved salts—including important nutrients like potassium, calcium, and silica—and acids that, along with bedrock geology, determine water chemistry (Giller and Malmqvist 1998). Water hardness is a measure of the calcium and magnesium occurring in water as salts of bicarbonate, sulfate, or chloride. In general, the greater the hardness, the more varied life is in a stream (Allan 1995). Overland flow from the catchment area also carries silts and other particles into the stream to contribute to a stream's sediment load.

Streams are influenced by the regional climate, most obviously by the amount and seasonality of precipitation. Most streams will exhibit variation in discharge—the volume of water passing a particular point as it flows downstream—and water levels from season to season. As the flow changes, the amount of sediments carried or deposited by a stream varies, and thus the composition and depth of channel sands and gravels are never constant.

Different types of streams are recognized according to their annual pattern of water flow. Permanent streams maintain water in their channels all year. In arid regions, permanent rivers typically receive almost all of their water in humid areas near their source; they are known as exotic rivers where they flow through the world's drylands. Intermittent streams are those that carry water only for three or more months; much of the year their channels are dry. Intermittent streams are characteristic of arid and semiarid climate regions, where they flow during the rainy seasons. They are also found in highly seasonally tropical climate regions and—as a consequence of spring snowmelt—in humid, temperate regions, where they are generally limited to small tributary streams. Ephemeral streams occur in deserts; water flows in their channels only for a few hours or days after significant rainfall, an event that is rare and unpredictable from year to year (Giller and Malmqvist 1998).

Since streams are essentially linear habitats with unidirectional flows, the lower portions are influenced by what happens in smaller drainages and watersheds upstream. There appears to be a hierarchical or nested organization of stream processes and stream life, from tiny headwater streams to the large river flowing across the floodplain it has created downstream. A common way scientists talk about this organization is by referring to stream order. A first-order stream is an initial, tiny rivulet in the uplands, the first channelized flow at a river's source. When two first-order streams join, they create a second-order stream; two second-order streams combine to form a third-order stream, and so forth. The Mississippi and Nile rivers are tenth-order streams, and the Amazon is a twelfth-order stream. Lower-order streams tend to have fewer species than higher-order ones, but the organisms are often more specialized and their occurrence is more apt to be restricted to a particular river system (Giller and Malmqvist 1998).

No two streams are alike. Differences are generated by bedrock, latitude, shapes and sizes of watersheds, climate, vegetation on the watershed, and, especially today, surrounding land use. The generalized descriptions that follow, therefore, will not accurately reflect all streams and pertain most closely to permanent streams in temperate zones. Some major exceptions will be described separately.

Life in streams can be classified according to where it occurs. The pleuston, often composed of insects, lives on the water surface and uses the high surface tension of water for support. Epilithic bacteria, fungi, and algae are trapped in a gelatinous slime along with detritus particles and silts as a living film on rocks and wood fragments in the stream channel. Algal mats on stones and rocks are referred to as the periphyton; but algae may also grow on larger plants and then are classified as **epiphytes**. Larger plants, macrophytes, are generally confined to shallow, slower waters near the edge of a stream. Most animals are members of the benthos, the **community** living on the channel bottom. They are largely invertebrates—mollusks, crustaceans, and, most especially, the larvae of insects. Some invertebrates live deeper within the channel sediments as members of the hyporheos (Giller and Malmqvist 1998).

Insects are certainly the dominant group in stream ecosystems and are represented by a number of forms that invaded streams from the land. Several insect groups are believed to have evolved in cool, running water, including midges (Family Chironomidae), damselflies and dragonflies (Order Odonota), mayflies (Order

Ephemeroptera), and stoneflies (Order Plectoptera). These freshwater insects are widely distributed around the world (Giller and Malmqvist 1998).

Fish are obviously the most important and ubiquitous of vertebrates in streams; different families are dominant on the different continents. In North America, darters (Family Percidae), minnows (Family Cyprinidae), and suckers (Family Catostomidae) are most diverse; in South America, characins (Family Characidae) and catfishes (Family Siluridae) dominate. The fish of the rivers of Asia consist mainly of minnows, loaches (Family Cobitadae), and hillstream loaches (Family Homalopteridae); while in Europe the ice ages eliminated most freshwater fishes and left brown trout (*Salmo trutta*) most characteristic (Giller and Malmqvist 1998).

Vertebrates other than fish are relatively uncommon in streams. A few amphibians, such as the hellbender (*Cryptobranchus alleghaniensis*), are restricted to flowing water habitats. Among reptiles, alligators, crocodiles, and caiman can be found in slow-moving streams in the tropics and subtropics and the dwarf caiman in fast-flowing rivers in South America. Birds are perhaps best represented by dippers (*Cinclus* spp.), which live by mountain streams and swim under water to probe for the insect larvae they eat. They have a dense, waterproof plumage and nasal flaps that close to prevent their breathing in water. Different dippers occur in the Rocky Mountains of North America, the Andes of South America, and in mountains in Europe and eastern Asia. Some ducks are associated with fast mountain streams. Examples include the Harlequin Duck (*Histrionicus histrionicus*) found in the Pacific Northwest of North America, Labrador, Iceland, Greenland, and Siberia; the Torrent duck (*Merganetta armata*) of the Andes; and the Brazilian Merganser (*Mergus octosetaceous*) (Giller and Malmqvist 1998).

Finally, several mammals are adapted to stream living. Some, though good swimmers, do spend part of their time out of water. Examples include beaver (*Castor canadensis*) and river otter (*Lutra canadensis*) in North America, capybara (*Hydrochaeris hydrochaeris*) in South America, duck-billed platypus (*Ornithorhynchus anatinus*) in Australia, and hippopotamus (*Hippopotamus amphibius*) in Africa. River dolphins are completely confined to the water environment. South America has two kinds, known as boutos or botos: one (*Inia geoffrensis*) in the Amazon and Orinoco river systems and the other (*I. boliviensis*) in the upper Madeira River system in Bolivia. South Asia has the susus (*Platanista minor*) of the Indus River system and the Ganges susus (*P. gangetica*), which is found in the Ganges but also occurs in the Brahmaputra and other river systems. The whitefin dolphin or baiji (*Lipotes vexillifer*) is found in the Chang Jiang (Yangtze River) of China. All of these river dolphins have poor eyesight and depend upon echo location to find the mollusks, crustaceans, and fish they eat (Giller and Malmqvist 1998; Nowak 1991).

Humans have had major impacts on streams through land-use practices, air pollution, flood control and irrigation projects, and introductions of nonnative organisms. Water chemistry, sediment load, nutrient levels, acidity, flow patterns, stream bank vegetation, and the aquatic community have been altered in most streams.

## THE STREAM ENVIRONMENT

The idealized stream begins at a spring or seep at high elevations and becomes a torrent in the mountains, where the course is steep and full of waterfalls and the flow is shallow, fast, and turbulent. The stream bottom is solid rock, and the chan-

Photo 10.1 The principal habitats in lower-order streams: turbulent riffles; smooth-flowing runs; and calm, deep pools. (Photo © Gary W. Carter/CORBIS.)

nel contains large boulders. Downstream, the character of the stream changes as the watershed increases in size and more water flows in the channel. The gradient become less steep, the water is less turbulent (whitewater), and the stream course begins to meander as the river starts to cut laterally into the sides of its valley. In the meanders, the outer banks are undercut, but deposition of sands and gravels occurs on the inside of each bend. This creates a regular pattern of shallow riffles, intermediate runs, and deep pools—each a distinctly different habitat (Photo 10.1). Riffles consist of small piles of gravel and cobbles and tend to be distributed at intervals five to seven times the width of the channel on alternating sides of a stream. Water velocity is higher through the riffles than elsewhere. Runs are areas of smooth flow at moderate to swift velocities and shallow to moderate depths. This fast-moving water is able to carry sediments and therefore often looks muddy. Pools, in contrast, are moderate to deep depressions in which water velocity is slow and fine particles settle out to cover the bottom (Allan 1995; Jenkins and Burkhead 1994).

As a river continues downstream, it typically gets larger and larger because more tributaries (lower-order streams) discharge their water into it, and water seeps in from the groundwater. The river becomes wider, deeper, and faster, but its slope or gradient flattens. In lower reaches, the river bottom is usually sandy (Allan 1995). As a river widens its valley, it builds a floodplain—a flat valley bottom of alluvium deposited each time the river overflows its banks. Receding waters may leave depressions on the floodplain filled with water—that is, floodplain lakes—and low-lying land close to the channel waterlogged and covered with freshwater marshes or swamps. The inhabitants of floodplain rivers must be able to tolerate changes in water volume and the effects of flooding. Animals may undergo migrations between wet-season and dry-season habitats that involve movements either upstream and downstream or laterally between the channel and the floodplain (Giller and Malmqvist 1998).

The water environment varies between upstream and downstream reaches of a river and within the water column and across the stream at any point. Many differences result from the friction between moving water and the sides and bottom of the channel that cause flow to be slower but more turbulent in those areas. The atmosphere also exerts a drag on moving water at the surface. Flow is smooth (laminar) and swiftest in the center of the water column. Stones and boulders in the channel introduce small but important differences because their lee or downstream sides offer still water and shelter from the fast currents near the center of a stream. Finally, the shade of overhanging plants near the banks can cause differences in light across the width of a channel (Moss 1998).

## MICROORGANISMS

Bacteria, fungi, algae, and protozoans in streams are mostly epilithic. The slimy outer layers of their cell walls act as glues to fasten them to permanently wet rocks. Some lie flat against the rock surface, but others have attachment pads from which they extend out into the water. Still others are able to move within the thin, slimy film on the rock. The physics of flow is such that a so-called boundary layer less than a millimeter (0.004 in) thick exists between rock and moving water where velocity approaches zero and prevents these organisms from being scoured from the rock surfaces. The composition of the epilithic community varies according to the availability of light. In sunlit streams, fewer bacteria but more algae, primarily diatoms, grow than in shaded streams (Moss 1998).

## PLANTS

Macrophytes in fast-flowing upland streams include red algae (*Lemanea, Hildenbrandia, Bartrochospermum*), mosses (e.g., *Fontinalis*), and liverworts, and—in the tropics and subtropics—unusual flowering plants of the family Podostemonaceae. These last have no roots but are anchored to rocks with holdfasts—flattened disks similar to those of some seaweeds. The entire plant is only a few centimeters (an inch or so) tall. Most macrophytes are found on the lee side of rocks where they form crusts or small tufts (Moss 1998).

In the lower reaches of rivers, water depth and high sediment load may prevent submerged plants from growing on the stream bottom. Nearer the banks, however, where water is more shallow, rooted emergents and submerged plants may find suitable habitat. These are mostly flowering plants that have evolved from terrestrial ancestors; because they still have flowers pollinated by wind or insects, they bear them above the water. A few are ferns. Submerged plants usually have little woody tissue; air-filled spaces give them support instead. Their leaves are often finely dissected, a trait that may increase their efficiency in absorbing the carbon dioxide dissolved in the water. They tend to spread by **stolons** or **rhizomes**. Small plants such as shoreweed (*Littorella uniflora*) hug the bottom in tight mats, a growth-form that may reduce the chance of being pulled out by water turbulence

Photo 10.2 A floating mat of water hyacinth (*Eichhornia* sp.) on a river in Vera Cruz, Mexico. (Photo by SLW.)

(Moss 1998). Pondweeds (*Potamogeton* and *Elodea*) and plants with floating leaves are typical in slower, nutrient-rich waters in the midlatitudes. In the lowland parts of large tropical rivers, mats of floating plants such as water hyacinth (*Eichhornia crassipes*) (Photo 10.2) commonly move downstream along the sides of the river beyond the main channel (Allan 1995).

## ANIMALS

Since most animals in streams move freely, they have evolved a number of adaptations to prevent their being swept downstream. Permanent attachment would likely be nonadaptive since the animals could easily be stranded and exposed to the air when water levels dropped. (Among the few attached organisms are freshwater sponges, but they stay on the undersides of submerged rocks.) Common adaptations include a streamlined body shape in which the greatest width of the body is found about 36 percent of the way along the length of the body. This shape presents the least resistance to flowing water and greatly reduces the chances that an animal will be dislodged. Nymphs of the mayflies *Rithrogena* and *Epeorus* exemplify the streamlined body. In addition, their bodies are flattened so they can crawl under stones and keep from being washed away in that manner also. A number of other modifications to the animal body are found among stream invertebrates. Suckers or suction pad feet help leeches, limpets, and snails adhere to surfaces, as do the friction pads made of small moveable spines on the undersides of water-penny beetles (Family Psephidae). Hooks, grapples, and clawlike legs form the same function for other insect larvae. The larvae of caddisflies construct cases of heavy particles that act as ballast, to keep them from floating up into the currents. Blackfly larvae (*Simulium* spp.) attach themselves to rocks by hooking onto pads of silk they create, and the larvae of butterflies and moths temporarily fasten themselves in cocoons. Every now and then—sometimes seasonally, sometimes daily, depending upon the animal—these invertebrates let go and drift downstream short distances. This drift may help them find new food sources or avoid predators. Flying adults make the return trips upstream (Moss 1998).

The small bottom-dwelling fishes of fast, upland streams have also evolved a variety of ways to remain on the bottom. They generally have rounded bodies with a flattened underside. The back in cross section is arched, and the mouth is often turned downward. The pectoral or side fins are muscular and spiny and are used to wedge themselves into crevices and between rocks. Some have fins modified as suckers or friction pads (Moss 1998).

The energy source for animals in a stream is largely dead leaves falling into the water. Aquatic fungi are important in breaking this detritus down. Among the invertebrates are coarse-particle feeders, shredders such as insect larvae and crustaceans (for example, the amphipod *Gammarus* spp.) that consume the softer parts between the veins of dead leaves. Others, such as midge, beetle, caddisfly and cranefly larvae, concentrate on rotting wood—although their real food may be the microorganisms decomposing the wood. Fine-particle feeders are filter feeders, deposit feeders, and scrapers. Among filter feeders are blackfly larvae that have fine hairs fringing their mouths and some mayfly nymphs that have fine hairs on their legs for the purpose of filtering food out of the water. Some caddisfly larvae build fine nets to collect small particles from the water. The previously mentioned organisms provide food for a variety of carnivores, including, among invertebrates, leeches, water mites (Order Hydracarina), and still other insect larvae. Many small bottom-dwelling fish depend upon the invertebrates of the benthos for food. Larger fish then eat them. At the end of the food chain are fish-eating birds such as mergansers and fish-eating mammals such as otters (Moss 1998).

The composition of the animal community tends to change in the progression from first-order streams to floodplain streams, as described in what is known as the river continuum concept (Vannote et al. 1980). Lower-order streams are inhabited by animals such as trout that require cool temperatures and high concentrations of dissolved oxygen in the water. Organisms associated with hard surfaces are also more prevalent in these upland streams. Higher-order streams have many more animals that feed in silt deposits and can tolerate waters that are warmer and more nutrient rich but are lower in oxygen content. Insect larvae and crustaceans are less varied than upstream; the benthos of lower parts of a river tends to be dominated by oligochaete worms, midge larvae, and bivalve mollusks (Moss 1998). However, the river continuum is not the only, nor perhaps even the dominant, distribution pattern. Fishes sort themselves within the same stretch of water. Different minnows, for example, occur according to differences in water flow, depth in the water column, type of stream bottom, and available food. Closely related darters rarely occur in the same pool, although they may occupy the same reaches of a river. The composition of the benthos in a given reach of a river varies more according to **microhabitat** than to distance from the source (Giller and Malmqvist 1998).

## TROPICAL RIVERS

### Amazon

The Amazon is the world's largest river by discharge [average = 175,000 $m^3$/sec (6.18 million $ft^3$/sec)] and by size of drainage basin [5.8 million $km^2$ (2,239,500 $mi^2$)]. Only the Nile is longer [6,600 km (4,100 mi) for the Nile versus 6,200 km (3,850 mi) for the Amazon]. The amount of freshwater discharged from the Amazon is more than four times that of the next largest river, the Congo/Zaire. The water level in the Amazon and its major tributaries fluctuates greatly during the year; average differences between low water and high water run between 7 m (23 ft) and 13 m (43 ft). Parts of the floodplain may be under water 8–10 months a year. The seasonal changes in water levels are so regular that some Amazonian fishes depend upon diets of seeds and fruits from the flooded forests (Goulding 1980).

The rivers of the Amazon drainage system are traditionally classified into three types according to water chemistry and sediment load. Whitewater streams have their headwaters in the geologically young Andes Mountains to the west. They are full of sediments that give them a muddy appearance and make them actually more of a light yellowish-brown than white. The Rio Napo in Ecuador is a good example. Such streams are transparent to depths of less than 0.5 m (1.5 ft) but contain dissolved nutrients, so they often have mats of floating plants. The greatest production of phytoplankton in the Amazon system occurs where and when the floodplains are inundated. The phytoplankters provide food for zooplankters, which in turn are important food for juvenile fishes. The other two types of streams, blackwater and clearwater, have few sediments and few nutrients and hence few aquatic plants. Blackwater streams originate on the ancient Guiana Plateau north of the Amazon basin and contain humic acids that discolor them and make them quite acidic. Many look like strong tea; others are truly black. Transparency ranges from 0.5 to 4 m (1.5–13 ft). The Rio Negro is a prime example of a blackwater stream (Photo 10.3). The clearwater streams enter the Amazon from the south; they are also nutrient poor but lack the humic acids of blackwater streams and are transparent to depths of 4–10 m (13–33 ft) (see also Chapter 1) (Goulding 1980).

Work by Michael Goulding (1980) on the Rio Machado, a clearwater tributary of the whitewater Rio Madeira, demonstrated that fish life in the Amazon is a link among all three river types as well as between river and forest. Some of the characins are migratory. From the beginning to the peak of the high-water period, they leave the nutrient-poor clearwater stream to swim downstream to spawn in the whitewater river. Others move from the floodplain to the main river channel to spawn. After spawning, the fishes return to the tributaries and spread through the flooded forest, where they fatten for several months on the seeds and fruits that fall into the water from the trees above. Tambaqui (*Colossomoa macropomum*), the largest scaled fish in the Amazon with a length up to 90 cm (35 in) and weight up to 30 kg (65 lb), has specialized teeth forward in the mouth that look like molars and are able to crush the hard seeds of barriguda rubber trees (*Hevea spruceana*) and fruits of jauari palms (*Astrocaryum jauary*). The much smaller pacu characins (*Mylossoma* spp.) also eat fruits and may play a role in dispersing the tiny seeds of *Cecropia* and figs (*Ficus* spp.) that pass through their guts undigested (Goulding 1980).

A second type of migration occurs as water levels drop. Some adult fishes leave the nutrient-poor clearwater tributaries then and enter the whitewater main

Photo 10.3 Junction between the blackwater Rio Negro and whitewater Rio Solimoes near Manuaus, Brazil. (Photo © Arvind Garg/CORBIS.)

streams, where they swim upstream, a migration known by local fishermen as the *piracema.* Fishermen eagerly haul in the *peixe gordo* (fat fish) such as the jaraqui (*Semaprochilodus* spp.) that come seasonally from the flooded forest. The survivors later enter upstream tributaries, where they replenish the fish populations of rivers too nutrient poor to support locally reproducing populations. As a result, a spatial pattern develops among age-classes. Upper tributaries contain older (2- to 4-year-old) fishes; lower tributaries have younger fishes. Catfishes such as dourada (*Brachyplatystoma falvicens*) also migrate, but since they travel near the bottom, their movements have been harder to document. It may be that they are pursuing the migrating characins (Goulding 1980).

Fish diversity in the rivers of the Amazon is enormous. Goulding (1980) estimates some 2,500–3,000 species will eventually be described. More than 80 percent of the known fishes come from two groups: the characins and the catfishes (siluroids). Each group has 11 families with representatives in the Amazon drainage system. The characins include the small, 1–2 cm (0.5–0.75 in) long tetras (*Hemigrammus, Hyphessobrycon, Cheirodon,* and *Paracheirodon*) so well known among aquarium owners, the notorious piranhas (e.g., *Serrasalmus* spp.), and some large and important food fishes such as the tambaqui. Catfishes range from small, blood-sucking forms 2–3 cm (0.75–1.25 in) in length to giants 2 m (6 ft) long and weighing 150 kg (330 lb) (Goulding 1980). Other fishes found in the Amazon include gymnotids that are able to generate electric discharges; most are too weak to be detected without special instruments, although that of the electric eel is powerful. Members of three ancient fish families live in these waters. The bony-tongues (Osteoglossidae) are represented by the giant pirarucu (*Arapaima gigas*) and by the aruanã (*Osteoglossum bicirrhosum*), a fish of the aquarium trade. The lungfishes

(Order Dipnoi) are represented by *Lepidosiren paradoxa*, a fish of the lower reaches of whitewater streams that survives the dry season buried in the mud. The third family, Nanidae, is represented by the leaf-fish (*Monocirrhus polycanthus*) that in appearance, as its common name suggests, mimics leaves (Goulding 1980).

Several groups of primarily marine fishes are also represented in the waters of the Amazon. These include cartilaginous fishes such as stingrays and bull sharks (*Caracharinus leucas*) and bony fishes such as herrings (Pellona, Family Clupeidae), croakers or drums (Family Sciaenidae), puffers (Family Tetraodontidae), and anchovies (Family Engraulidae) (Goulding 1980).

Among the fishes of the Amazon is the most diverse group of predatory freshwater fishes in the world. Best known are the piranhas, although they are rarely the vicious killers popularly portrayed. The Amazon has some 20 different piranhas; not all are carnivores, and even the carnivores eat seeds and fruits. The largest is the black piranha (*Serrasalmus rhombeus*), an adult of which may be 40 cm (16 in) long and weigh 2 kg (4.5 lb). With strong jaws and razor-sharp, triangular teeth, this fish-eater ambushes its prey and bites out chunks of flesh. It congregates in small groups. It is the smaller *piranha caju* (*S. hattereri*) that forms large schools, and its feeding frenzies are those most photographed. Piranhas inhabit quiet waters and are attracted by the noise and splashing of injured animals (Goulding 1980). They are generally not interested in humans!

Other interesting animals in the Amazon are the sidenecked turtles such as the yellow-spotted Amazon turtle (*Podocnemis unifilis*) and the giant river turtle (*P. expansa*). Instead of retracting the head toward the shell by bending the neck vertically—as Northern Hemisphere turtles do, these Southern Hemisphere forms wrap the neck along their sides beneath the shell. The giant river turtle, weighing up to 50 kg (110 lb), was once herded by the thousands into pens by indigenous peoples to be fattened like river cattle (Smith 1999).

There are several mammals of interest. The Amazon manatee (*Trichechus inunguis*) is the only exclusively freshwater sirenian and the smallest of the manatee species. Found throughout the Amazon drainage system, it inhabits calm, whitewater rivers and floodplain lakes with deep channels connecting them to large rivers. It can be found at the mouth of the river and along the Atlantic coast, where the enormous discharge of the Amazon keeps the surface waters fresh. Manatee feed on aquatic plants such as grasses, bladderworts, and waterlilies (Smith 1999; Sirenian International 2000–2002; Nowak 1991). The pink Amazon river dolphin (*Inia geoffrensis*), a cetacean restricted to freshwater, also prefers slow-moving whitewater streams. It eats fish exclusively and feeds in the flooded forest. The river dolphin sometimes associates with the giant otter (*Pteronura brasiliensis*), which, though a terrestrial animal, spends much time in the rivers hunting fish and crabs (Nowak 1991).

## Major African Rivers

The rivers of tropical Africa flow across plateau surfaces at elevations higher than the Amazon Basin. In any given drainage, life may be restricted to various reaches of the river by rapids and inland deltas. The Niger and Nile rivers in the northern hemisphere and the Zambezi in the southern hemisphere have savanna-river lifeforms, while the Congo/Zaire in equatorial Africa is characterized by forest-river organisms. The Congo/Zaire experiences a relatively small annual fluctuation in water level of about 3 meters (10 ft), since lakes and swamps in the upper reaches absorb much of the rainwater and moderate flows downstream. As a forest river, the waters are acidic and often low in oxygen. An estimated 1,000 different fishes inhabit the river system—less than half the number found in the Amazon. Compared to the lakes of Africa (see Chapter 11), the Congo/Zaire has very few cichlids (32), but among them are three endemic genera (*Teleogramma*, *Steatocranus*, and *Heterochromis*) known from the rapids of the Lower Zaire (Lowe-McConnell 1991).

In the savanna-rivers such as the Nile, Niger, and Zambezi rivers, the water is open to the sun, and water levels have a greater seasonal variation. The annual floods inundate huge floodplains and connect floodplain ponds and swamps. The flooding waters are enriched by decayed plant matter; algae, bacteria, and zooplankter populations respond with rapid proliferation. The algal blooms support a large variety of aquatic insects and other invertebrates. Aquatic vegetation also grows rapidly, and as the waters recede and the plants die down, they provide cover for juvenile fish and food for adults. As the dry season progresses, fishes are confined to the river channel or trapped in floodplain ponds that become deoxygenated before the rains begin again. The river itself may cease to flow and may become a series of stagnant pools that also become deoxygenated. Organisms that survive in the savanna rivers must be able to withstand these poor conditions and respond rapidly and prolifically to the wet season time of plenty (Lowe-McConnell 1991).

## MAJOR FRESHWATER WETLANDS

Rivers and their floodplain or delta wetlands are parts of the same ecosystem, although they may be separated from each other seasonally. Wetlands are often very important conservation areas and thus worthy of mention here. One of the largest floodplain wetlands on Earth is the Pantanal in Brazil (Photo 10.4), where it is recognized as a biome in its own right. The upper Paraguay River and its tributaries inundate over 100,000 km$^2$ (38,600 mi$^2$) for up to seven months (November to May) each year. This watery region contains thousands of small lakes renewed each year by the flood, marshes and swamps, and vast areas covered with floating plants. Scattered islands of higher ground stand above the flood and serve as refuges for land plants and animals. About 40 different plants make up the floating vegetation—the highest diversity in the world. Among them are water hyacinth (*Eichhornia* spp.), water cabbage (*Pistia* spp.), and floating fern (*Salvinia auriculata*). The mats of floating plants rise and fall with the changing water levels and are thick enough to support the weight of the jacana (*Jacana jacana*), which with its long, widely spread toes can feed and even nest on the mats. The extremely abundant small crocodilian, the jacaré (*Caiman jacaré*), also nests on the mats, where its eggs are vulnerable to coatimundi (*Nasua nasua*) and to forest dogs. The dead plant matter that accumulates and decays beneath the floating mats creates conditions of low oxygen in the water. The world's most diverse community of air-breathing (pulmonate) snails occurs in the Pantanal, and a number of air-breathing fishes also live there (Moss 1998).

Animal life occurs in splendid variety and abundance in the Pantanal, although the diversity isn't nearly as great as that of the Amazon. Some 405 fishes have been described, most from groups that are dominant throughout South America: tooth carps or cyprinodonts, spiny cichlids, and catfish or siluroids. Many are tolerant of

Photo 10.4 The watery world of the species-rich Pantanal, a wetland formed along the upper Paraguay River system in Brazil. (Photo © Stephanie Maze/CORBIS.)

low oxygen levels. Some river fishes, however, require better oxygenated water and undergo migrations upstream and downstream and across the rivers and floodplains. During the migrations of fishes moving upstream to feed on detritus or to spawn, fish such as dourados and pacus (see description of Amazon) can turn the waters silver. The rivers of the Pantanal are also home to large populations of piranhas and a few fishes of marine origin, including stingrays and a flounder (Moss 1998, Dugan 1993; Garda 1996).

In addition to jacaré, both spectacled (*Caiman crocodilus*) and broad-nosed caiman (*Caiman latirostris*) are found. These two reptiles, however, are near extinction due to poaching for their skins. A number of turtles and snakes live there, but only the anaconda (*Eunectes murinus*) is abundant. A total of 657 birds have been recorded from the Pantanal. Water birds are numerous and include the large Jabiru stork (*Jabiru mycteria*) with its striking black and red head and neck, Wood Stork (*Mycteria americana*), 17 herons, 6 ibises, kingfishers, ducks, grebes, rails, and so forth. Herbivorous mammals include the capybara (*Hydrochoreus hydrochaeris*) that graze in small herds in the floating vegetation; the tapir (*Tapirus terrestris*); and the marsh deer (*Blastocerus dichotomus*), which with its wide, spreading feet is well adapted to life in the wetlands. Predators include jaguar (*Panthera onca*), four wild dogs, and several small cats (Moss 1998).

In northern South America, along the Orinoco River, there is another large area of floodplain wetlands, the llanos (see Chapter 3). This is a patchwork of seasonally and permanently flooded marshes, slow-moving rivers, oxbow lakes, and seasonally flooded palm (*Mauritia* spp. and *Coperinicaia tectorum*) savannas and grasslands. The waters are home to stingray (*Dasyatis americana*), electric eel (*Electrophorus electricus*), red piranha (*Serrasalmus natterei*), and spectacled caiman (*Caiman crocodilus*).

Birds are abundant and diverse and include many of the same wading birds found in the Pantanal. The amphibious capybara is at home here, as are white-tailed deer (*Odocoileus virginianus*) and jaguar (Dugan 1993).

On the African continent are a number of large inland deltas and floodplains with vast wetlands developed upon them. Among them is the Inner Niger Delta in arid Mali. At full flood, an area of some 320,000 km$^2$ (123,500 mi$^2$) is inundated, which provides an environment vital to the survival of the region's human inhabitants. A floating grass (*Echinoclea stagnina*) is characteristic, but flooded acacia (*Acacia kirkii*) forest plays a most important role in the wetland ecosystem. The flooded forest is the site of large nesting colonies of herons and cormorants. The droppings from these birds fertilize the water and support an important **fishery** (Dugan 1993).

North and east of the Niger River, in southern Sudan, is The Sudd, the 500-km (300 mi) long floodplain of the White Nile. Thick with emergent plants such as papyrus and bulrushes and interlaced with channels, it received its name from an Arabic word for obstacle, since it was a barrier to north-south transport on the Nile (Dugan 1993).

The largest inland delta in the world occurs in southern Africa at the northern rim of the Kalahari Desert. This is the Okavango Delta in Botswana. About half the delta region is a permanent swamp; the rest is seasonally flooded. Floating water lilies, emergent papyrus, and palms (*Hyphanae* spp.) characterize the vegetation. Animals include 15 species of antelope, including those adapted to life in swamps and floodplains such as the sitatunga (*Tragelephas spekei*), water buck (*Kobus ellisiprymnus*), puku (*K. vardoni*), and lechwe (*K. lechwe*). The antelopes plus hippopotamuses and birds such as the Little Bee-eater (*Merops pusillus*), Jacana (*Actophilonus africana*), Malachite Kingfisher (*Alcedo cristata*), and African Fish Eagle (*Haliaeetus vocifer*) attract many ecotourists to the Moremia Game Reserve on the Delta (Dugan 1993).

Several significant large wetlands are found in the midlatitudes of Eurasia. Among the better known is the Camargue on the delta of the Rhone River in France. Known for its feral white horses, the Camargue has long been the site of salt production. Shallow, saline pools in which seawater is evaporated have become home to a number of shorebirds and waders, including Avocet (*Recurvirostra avosetta*), Kentish Plover (*Charadrius alexandrius*), Gull-billed Tern (*Gelochelidon nilotica*), Slender-billed Gull (*Larus genei*), Little Tern (*Sterna albifrons*), and Sandwich Tern (*S. sandvicensis*). Perhaps most famous, though, is the Greater Flamingo (*Phoenicopterus ruber roseus*), which has become a symbol of conservation success in France. In Siberia, the floodplain of the River Ob, more than 50,000 km$^2$ (20,000 mi$^2$) in area, is the largest waterfowl breeding and molting area in Eurasia. Among rare species concentrated there are the Siberian Crane (*Grus leucogeranus*) and White-tailed Eagle (*Haliaeetus albicilla*) (Dugan 1993).

Among the many significant freshwater wetland areas in North America are the prairie potholes region that lies between the Missouri and Mississippi rivers, the bottomlands along the lower Mississippi River, the Great Dismal Swamp in Virginia and North Carolina, the Okefenokee Swamp in Georgia, Big Thicket in Texas, and the unique **evergreen** shrub bogs known as pocosins on the coastal plain of the Carolinas. However, the Everglades—the river of grass—and neighboring Big Cypress Swamp are perhaps the best known. The Everglades (Photo

Photo 10.5 The Everglades, a "river of grass" with tree-covered hummocks. (Photo by SLW.)

10.5) is one of the largest freshwater marshes in the world. Fed by waters from Lake Okeechobee, the marsh once covered 10,000 km$^2$ (3,800 mi$^2$) between the lake and the mangrove swamps fringing southern Florida. It is dominated by sawgrasses (*Cladium* spp.) that grow to 4 m (13 ft) tall. Aquatic plants such as waterlilies (*Nymphaea* spp. and *Nuphar* spp.) are also found. Hardwood trees grow on hummocks raised above the saturated zone and are home to tree snails (*Liguus* spp.) otherwise known only from Cuba and the island of Hispaniola. Although famous for alligators (*Alligator mississippiensis*), the Everglades also support a variety of water birds, including herons, Wood Stork, Roseate Spoonbill (*Ajaia ajaja*), ibises, and Anhinga. Part of the same freshwater system, Big Cypress Swamp is a forested wetland dominated by bald cypress (*Taxodium distichum*) but including other trees such as pond cypress (*T. ascendens*), slash pine (*Pinus elliotti*), and pond-apple (*Annona glabra*) (Dugan 1993).

## HUMAN IMPACTS

It may be impossible to find a large or floodplain river that human activities have not altered. Freshwater is a vital resource for human populations everywhere, and rivers have been dammed or their waters otherwise diverted to supply water to fields or settlements for thousands of years. Wherever human use is significant, the physical structure of the stream environment has been altered, as has water chemistry and the composition of river life. Dams have effects both upstream and downstream. Above a dam, a river is transformed into a lake, and river organisms are replaced by those of still waters. Sediments, including suspended organic matter, are trapped above the dam which depletes the downstream reaches of a major source of nutrients. A dam and the impoundment it creates are barriers to the movement of river organisms. Downstream, the seasonal cycle of low and high water is altered as water is released to serve human purposes such as electric power generation, flood control, or irrigation. The flow may become more regular than under natural conditions or it may fluctuate more extremely depending upon the primary purpose of the dam. If the outflow from the dam is derived from the reservoir surface, the waters may be warmer than predam temperatures downstream; if the release is from the bottom of the dam, water temperatures can be much colder (Moss 1998; Giller and Malmqvist 1998). In the Zambezi River, the regularized flow has had negative effects in a variety of ways. Lack of a low-water period deprived baboons of the mud flats from which they had extracted mollusks, insects, amphibians, and reptiles. Without the rainy season high-water period, floodplain ponds were not flushed clean, and they became choked with aquatic plants such as water fern and water lettuce, which degraded the habitat of amphibians, reptiles, and birds. The abnormal releases of water during the dry season destroyed the nests of bank-dwelling birds and reptiles (Jeffries and Mills 1990).

The composition of river life is often changed deliberately in efforts to increase production of game fishes. For example, American brook trout (*Salmo frontinalis*) and rainbow trout (*Oncorhynchus mykiss*) were introduced into British streams in the late nineteenth century, and European brown trout (*Salmo trutta*) has been introduced into California for sportfishing. Since such fishes are often fish eaters themselves, they can wipe out native forms. Exotic species are also introduced to streams for other purposes or are introduced accidentally. In the United States, for example, mosquitofish (*Gambusia affinis*) has been introduced as a biological control agent. Tilapia and carp have been introduced to control aquatic weeds, one of which, water hyacinth, is itself a widespread exotic native to South America (Giller and Malmqvist 1998). Bait fish and other organisms are also frequently transferred from one drainage to another, so sometimes it is difficult to tell where they are truly native. Crayfish are an example in the United States. The rusty crayfish is native to streams in Ohio, Kentucky, and Indiana but has been spread elsewhere and is wiping out local crayfish through hybridizing with them. North America has three-fourths of the world crayfish species and the highest rate of extinction (Perry et al. 2001). The zebra mussel (*Dreissena polymorpha*), introduced into the St. Clair river between Lake Huron and Lake Erie, is spreading through the rivers of the central United States. It is considered a major threat to all freshwater mussels.

Water pollution constitutes another major human impact on stream environments. It can take several forms. Acidification of water has resulted from mine drainage in coal-mining areas and the planting of **needle-leaved** trees in the watershed, but the effects of air pollution are more widespread. Sulfur dioxide and nitrogen oxides released into the atmosphere become acids when they settle into a stream. Low **pH** can inhibit the reproduction of fishes and amphibians. Another type of water pollution involves adding nutrients (eutrophication) in the form of organic wastes (sewage) or fertilizer into groundwater or runoff from lawns and farmland. In the temperate zone, this can lead to phytoplankton blooms and the growth of large filamentous algae (e.g., *Cladophora*) in streams; in the tropics it can cause massive growths of water hyacinths (Giller and Malmqvist 1998). The subsequent decay of this plant material reduces the dissolved oxygen content of the water and triggers the demise of the native river community. In lowland rivers, a pollution community of **anaerobic** bacteria including *Beggiatoa*, flagellates, and filamentous bacteria such as the sewage bacterium *Sphaerotilus natans* can result (Moss 1998).

A third type of pollution is that by toxic compounds such as mercury, organic pesticides, and polychlorinated biphenyls (PCBs). Those toxic chemicals that do not break down in nature but persist and accumulate in organisms are most significant. They are biologically magnified as they pass up the food chain and can have both lethal and sublethal consequences to organisms such as predatory fishes and dippers at or near the top of the food chain. Among the effects are mutations, physical abnormalities, and reproductive impairment. A recently discovered and controversial example is the feminization of male fishes due to estrogens in sewage entering streams (Giller and Malmqvist 1998).

Land use in the catchment area and along the stream banks exerts changes on stream habitats. Clear-cutting of forests can increase siltation in streams and result in the suffocation of mussels and the eggs of amphibians and fishes. Removal of riverside plants reduces shade and cover for life in the stream. Birdwatching, picnicking, and fishing from the bank can trample plants and disturb nesting animals, while boating can erode river banks, destroy aquatic plants, and disturb wildlife (Giller and Malmqvist 1998).

## ORIGINS

While river systems are more long lived than lakes, they are considerably younger habitats than most marine environments. Rivers are easily altered by tectonic changes (mountain building, rifting, volcanism) and by climate change. The oldest—such as the Nile River in Africa and the New River in North Carolina, Virginia, and West Virginia—are likely no more than 300 million years old. Most of the major midcontinent rivers of North America, on the other hand, originated from or were greatly modified by the meltwaters of receding ice sheets during the late Pleistocene Epoch (Allan 1995).

Most freshwater organisms apparently evolved relatively recently in geologic time from marine ancestors. **Vascular plants** and insects had terrestrial origins. Bony fishes probably first arose in freshwater during the Middle or Late Mesozoic Era. Compared with either marine life or life on land, freshwater forms exhibit low diversity. This may be due to a combination of factors, including limited periods of environmental stability and confinement to discrete drainage systems that prevented dispersal and escape from catastrophic environmental changes (Moss 1998). Even so, some fishes and other freshwater life-forms have been able to disperse widely. The darters, for example, originated in Europe but crossed a North Atlantic land bridge into North America before the Eocene Epoch. Ancestral minnows came to North America by way of a land connection to Asia sometime before the Oligocene Epoch. Modern genera have developed on the North American continent since the Miocene Epoch (Jenkins and Burkhead 1994).

Freshwater life-forms are able to disperse from drainage system to drainage system in a variety of ways. The headward erosion of streams means that first-order streams of one system can eventually intercept and divert headwater streams of another and pirate both the channel and the life in it. In other situations, the headwaters of different drainage systems may emanate from the same swampy area.

Such is the case in South America, where the headwaters of the Amazon and Paraguay rivers begin in the same wetland, thereby allowing the exchange of at least smaller organisms. To the north, the Amazon and Orinoco river systems are connected by what is called the Casiquiare Canal, a natural stream resulting from the headward erosion of the Rio Negro.

In the lower reaches of rivers, exchanges can become possible during periods of flood. And with lowering sea level, streams once separated by bodies of saltwater may join into a single system. During the glacial episodes of the Pleistocene epoch, the Chesapeake Bay was the valley of the Susquehanna River, into which the Potomac, James, and other rivers flowed as tributary streams. With the later rise of sea level, ocean waters flooded the valley, and each river became an isolated drainage system. Today, these river systems have many fishes in common as a result of their past connections.

## HISTORY OF SCIENTIFIC EXPLORATION

Rivers were frequently the route to the interior of newly discovered continents and explored by Europeans from the 1500s onward. The search for the source of the Nile was one of the epic adventures of the nineteenth century. The source of the Amazon was pinpointed only with the use of a global positioning system (GPS) in 2001! Much of our information on river life is recently acquired, and much is yet to be learned.

Before the sixteenth century, it was believed that rivers formed as a result of the infiltration of ocean water into continental rocks, where it reemerged cleansed and in the form of springs and seeps. Only in 1647 it was proposed that rainwater stored in the ground was the source of freshwater streams (Allan 1995). Knowledge of rivers and the life within them expanded as European empires expanded. In what is today the United States, the descriptions of the lower Roanoke River by Thomas Harriot (1588) were among the first studies of rivers. He mentioned the presence of gar (*Lepisosteus osseus*). John Banister (1650–1692) and John Lawson (1709) contributed considerably more information on fishes of the mid-Atlantic region, but the first truly scientific surveys, descriptions, and analyses awaited Edward Drinker Cope (1840–97), America's "Master Naturalist." At the same time Cope was working, the leading American ichthyologist, David Starr Jordan (1851–1931), initiated a series of river expeditions that culminated in *The Fishes of North America*, written by Jordan and Barton W. Evermann and published between 1896 and 1900 (Jenkins and Burkhead 1994).

In South America, Alfred Russel Wallace (1853) first described the three types of rivers (whitewater, blackwater, and clearwater) in the Amazon system (Goulding 1980). The collectors and **taxonomists** of the late nineteenth and early twentieth centuries greatly increased knowledge on the classification and distribution of freshwater life-forms, so that aquatic organisms came to play an important role in the development of the theory of **plate tectonics**. For example, insects demonstrated that South Africa had separated from Antarctica very early in the fragmentation of Gondwana (Giller and Malmqvist 1998).

Current research continues on the taxonomy, distribution patterns, and **ecology** of stream life. There tends to be strong emphases on stream restoration ecology and management for the conservation of **biodiversity**.

## REFERENCES

Allan, J. David. 1995. *Stream Ecology: Structure and Function of Running Waters.* London: Chapman & Hall.

Dugan, Patrick. 1993. *Wetlands in Danger, A World Conservation Atlas.* New York: Oxford University Press.

Garda, Eduardo Carlos, ed., 1996. *Atlas do Meio Ambiente do Bra*sil. Brasilia: EMBRAPA-SPI.

Giller, Paul S. and Björn Malmqvist. 1998. *The Biology of Streams and Rivers.* Oxford: Oxford University Press.

Goulding, Michael. 1980. *The Fishes and the Forest: Explorations in Amazonian Natural History.* Berkeley: University of California Press.

Harriot, Thomas. 1588. *Briefe and true report of the new found land of Virginia.* 1972 Reprint of 1590 version by Dover Publications, Inc., New York.

Jeffries, Michael and Derek Mills. 1990. *Freshwater Ecology: Principals and Applications.* London: Belhaven Press.

Jenkins, Robert E. and Noel M. Burkhead. 1994. *Freshwater Fishes of Virginia.* Bethesda, MD: American Fisheries Society.

Lowe-McConnell, Rosemary. 1991. "Ecology of cichlids in South American and African waters, excluding the African Great Lakes." Pp. 60–85 in Miles H. A. Keenleyside, *Cichlid Fishes: Behaviour, Ecology, and Evolution.* London: Chapman & Hall.

Moss, Brian. 1998. *Ecology of Fresh Waters; Man and Medium, Past to Future.* 3rd edition Oxford: Blackwell Scientific.

Nowak, Ronald M. 1991. *Walker's Mammals of the World.* 5th edition, Vol. II. Baltimore: The Johns Hopkins University Press.

Perry, William L., Jeffrey L. Feder, and David M. Lodge. 2001. "Implications of hybridization between introduced and resident *Orconecte*s crayfishes." *Conservation Biology* 15: 1656–1666.

Sirenian International. 2000–2002. "Amazonian manatee." (http://sirenian.org/amazonian.htm).

Smith, Nigel J. H. 1999. *The Amazon River Forest, A Natural History of Plants, Animals, and People.* New York: Oxford University Press.

Vannote, R. L., G. W. Minshall, K. W. Cummings, J. R. Sedell, and C. E. Cushing. 1980. "The river continuum concept." *Canadian Journal of Fisheries and Aquatic Science* 37: 120–37.

Wallace, Alfred Russel. 1853. *A Narrative of Travels on the Amazon and Rio Negro.* London: Reeve (cited in Goulding 1980).

# 11 LAKES AND PONDS

## OVERVIEW

A lake is any water-filled depression on land that is cut off from an inflow of ocean water. Any such body of water—from a puddle to one of the Great Lakes—can be considered a lake, although many people call smaller, shallower bodies of water ponds. Brönmark and Hansson (1998) define a lake as a body of water in which wind primarily mixes the water column (see "The Lake Environment" later) and define a pond as one in which temperature changes induce mixing.

Lakes account for only 0.009 percent of the water on Earth (Allan 1995), but they occur in great variety. Lakes are found in a variety of **climate** regions. They differ in how they were initially formed, and they vary greatly in size and depth. No two lakes present identical environmental conditions nor have the same organisms inhabiting them. Indeed, lakes are probably the most varied aquatic environment on Earth (Payne 1986). Nonetheless, some generalizations are possible. One is a matter of age. Lakes are temporary features on the landscape and most are quite young, geologically speaking (Photo 11.1). Most are only 10,000–20,000 years old (Moss 1998). The very few geologically old lakes—such as Lake Baikal, estimated to be between 50 million and 75 million years old, and Lake Tanganyika, estimated to be between 1.5 million and 6 million years old—are associated with rift valleys (Burgis and Morris 1987). Eventually, lakes fill with sediments and turn into marshes and swamps; later ecological **succession** changes them into whatever the surrounding terrestrial **biome** is. Age affects the numbers and uniqueness of **life-forms** inhabiting a lake, but the overall nature of a particular lake is strongly controlled by its water chemistry—a product of geology and climate.

Since lakes serve as catchments for streams and overland flow off the surrounding basin, they are readily affected by land use in the drainage basin and are easily polluted. Yet many of the world's people depend on lakes for their drinking water and for other domestic, agricultural, and industrial uses, so many deliberate efforts

Photo 11.1 Glacial lakes. Most of the world's lakes are small, shallow, and—in a geologic time frame—only in existence for a short while. Like these in Maine, most were formed as a result of glacial activity in the Pleistocene Epoch. (Photo by SLW.)

around the world strive to reduce the negative impacts of human populations on bodies of freshwater. People actually have been responsible for the creation of a vast number of lakes for use as stock ponds, fish ponds, reservoirs, hydroelectric power, and recreation sites.

**Ecology** had its birth as a science in the study of lakes. Much as islands have, lakes have became outdoor laboratories. Classic studies of nutrient cycles and energy were conducted in lakes. More recently, organisms inhabiting lakes have stimulated scientists to reexamine mechanisms for the origin of new **species** and the very definition of a species.

## THE LAKE ENVIRONMENT

Much of the research on lakes has been conducted until recently in the **temperate zone** of the Northern Hemisphere, so temperate lakes have become the standard with which others are compared. The water in a typical temperate zone lake is deep enough to become stratified in summer, when the upper layer becomes warmed. This upper layer, the epilimnion, is mixed regularly by the wind. A bottom layer out of reach of the effects of the wind, the hypolimnion, consists of cooler, denser water. Separating the two is a narrow zone of rapid temperature change known as the thermocline. In the open water of a lake, phytoplankton live in the epilimnion, the depth of which corresponds to the depth at which most light penetrates, and so it is also called the euphotic zone.

Although nutrients are carried into lakes by streams, usually these are a minor contribution to the overall nutrient content, and most nutrients need to be recycled to sustain life in the open water. Since all particles tend to sink and be lost to the surface waters where the phytoplankters are active, they must be retrieved through surface mixing (which wind accomplishes) or upwelling of water from below. The latter is achieved by seasonal temperature changes. Freshwater reaches its maximum density at 3.94° C (39° F). As the water at depth cools below that temperature in the autumn, it becomes less dense and rises to the surface to initiate a turnover of the lake water.

Since the density of water decreases as its temperature approaches 0° C (32° F), during winter, lakes freeze from the top down. A complete cover of ice acts as an insulator so only the shallowest of ponds freeze solid. The lake may become stratified again in winter, with the coldest water, at temperatures of approximately 1° C (34° F), forming the epilimnion and water at temperatures close to 4° C (39° F) lying below the thermocline. This phenomenon allows organisms to survive the winter in or near the lake bottom without themselves becoming frozen solid. In spring, as the surface waters are warmed to 4° C (39° F), they become denser and sink back toward the bottom, again mixing the lake water (Brönmark and Hansson 1998; Moss 1998). By summer, surface waters in midcontinent lakes in humid, temperate climates may reach 20° C (68° F), while bottom layers are at about 4° C (39° F). In more maritime climates, the temperature range is narrower; surface water may be 18° C (64° F) and deep waters 10° C (50° F). Tropical lakes may have surface waters with temperatures as high as 29° C (84° F) and bottom waters at 25° C (77° F) (Moss 1998).

Not all lakes mix twice a year. Shallow lakes may mix completely whenever the wind blows. Lake George in Uganda, less than 2.5 m deep (8 ft), becomes stratified by day but mixes again at night. On the other hand, the cool bottom waters in the world's deepest lakes, such as Baikal and Tanganyika, never mix with surface water. Not only temperature and pressure, but also an accumulation of salts in the deep waters creates a density greater than water higher in the water column. With no addition of oxygenated water from the surface, the deep waters of such lakes are lifeless (Moss 1998). (This is a significant contrast to the deep sea, which receives currents of cold oxygenated water from the surface and contains a diverse group of animals.)

Some lakes mix only very rarely. The Dead Sea, the saltiest lake on Earth, is fed from below by freshwater springs bubbling through a deep, ancient layer of precipitated salts. In 1979, enough freshwater came into the bottom waters to significantly lower density and cause the whole lake to mix for the first time in 300 years (Moss 1998).

In the **tropics** where there are no significant temperature changes through the year, mixing is infrequent. The density of the deep water can be increased due to the dissolved organic products of decay. The bacteria that decompose dead plant and animal matter settling down from above in these oxygen-depleted depths produce deadly hydrogen sulfide gas and methane. These gases are held at the bottom under pressure until a rare turnover event occurs. When that happens, the gas escapes, the oxygenless water rises to the surface, and fish kills and human deaths result. In August 1986, Lake Nios in Cameroon turned over and released an odorless gas (in this case, perhaps carbon dioxide from volcanic emissions trapped in the lake bottom) that came to the surface and killed some 1,500 people living nearby (Moss 1998).

Temperature is important in forcing the vertical circulation of midlatitude lakes and also because the amount of oxygen and carbon dioxide that can be dissolved in water varies with temperature. Both oxygen and carbon dioxide, gases essential to most life-forms, are more soluble in cold water. These gases are dissolved into the water at the surface where the lake water is in contact with the atmosphere. They may be depleted from deeper water during the summer by biological processes but are replenished with the autumn turnover.

Lakes change over time as nutrients and sediments increase. Nutrient content provides a way of classifying lakes. Oligotrophic lakes are those with few nutrients, clear water, few phytoplankters, and a low diversity of animal life. These are usually deep and/or very young lakes and have small numbers of those organisms that inhabit them. At the opposite end of the spectrum are eutrophic lakes, which are nutrient rich. Their water may be greenish and clouded by the abundance of phytoplankters. They are shallow due to the accumulation of sediments on the lake bottom; therefore, most of the water column is warmed during summer months. Warming, along with the decay of dead organisms, reduces the amount of oxygen dissolved in their waters and influences the types of organisms that can live in them. Eutrophic lakes support large numbers of individuals in a relatively few species. Between the extremes of oligotrophic and eutrophic categories are mesotrophic lakes with enough nutrients, dissolved gases, and depth to support a diverse **community** of animals. A fourth category in this classification scheme is that of dystrophic lakes, those that have black or brown water from a high content of humic or tannic acids derived from peat or the vegetation on the lake's watershed. The water is very acidic and very low in carbonates and hosts a distinctive community of acid-tolerant aquatic plants (Riemer 1993).

Lakes can also be described according to stages in a theoretical succession marked by changes in the plant life. Succession begins with a deep, young oligotrophic lake. Phytoplankters colonize the new lake. When enough sediments have accumulated along the edge of the lake to provide both water shallow enough for light to penetrate to the bottom and a substrate for anchoring plants with roots, a littoral zone has been created. In this zone, **vascular plants** gain a foothold. Some, such as pondweeds, grow totally submerged though rooted in the lake bottom. Other plants, such as water lilies, will root in the bottom but have leaves floating on the surface of the water. Where very shallow water occurs close to shore, emergents colonize. These are plants such as cattails and reeds that have upright stems and leaves extending above the water surface. The emergents trap sediments washing in from the surrounding land area and slowly build the land outward into the lake. At the same time, sediments accumulate on the lake bottom and cause the lake to become shallower and shallower. The surface area of open water continually shrinks; eventually the lake fills in and becomes a wetland. Woody shrubs such as alders (*Alnus* spp.), willows (*Salix* spp.), and buttonbush (*Cephalanthus occidentalis*) invade at the edges. Given enough time, the plants of the surrounding landscape take over, and the lake is obliterated. The whole process may take a few hundred years in small lakes (Jeffries and Mills 1990; Riemer 1993). In contrast, Lake Baikal, after millions of years, is still oligotrophic.

## PHYTOPLANKTON

**Cyanobacteria** and different kinds of algae make up the phytoplankton in lakes. The actual kinds and amounts present vary according to climate, season, water chemistry, nutrient levels, and the composition of the animal life that feeds upon them. These tiny organisms absorb nutrients through their cell walls and are the major primary producers of food for the rest of the lake community. Diatoms, their cell walls strengthened by silica in intricate designs, form the spring bloom in most lakes. They take advantage of the upwelling of nutrients that occurs as a consequence of spring mixing. Once silica and certain other nutrients are exhausted, the diatoms populations crash. An increase in green algae follows them. Lakes and ponds have several types of green algae. Among them are single-celled forms with flagella such as *Chlamydomonas* and colonial types such as *Volvox.* Mainly filamentous forms, such as *Spirogyra,* live on the lake bottom. Cyanobacteria, or blue-green algae, are more tolerant of low oxygen and are able to use nitrogen dissolved in the water; their abundance increases later in the summer when conditions have become less advantageous for the green algae. Most cyanobacteria in lakes are filamentous forms; they are dominant in eutrophic lakes. Other members of the phytoplankton are golden-brown algae and dinoflagellates. The golden-brown algae are mostly single celled; some are covered with siliceous or calcareous plates. Euglenoids, algae that have chlorophyll and **photosynthesize** but also engulf food particles, also occur; *Euglena* species are often abundant in polluted water (Brönmark and Hansson 1998; Jeffries and Mills 1990).

## PLANT LIFE

The larger plants in a lake are primarily associated with the littoral zone of shallow water close to shore (Photo 11.2). Free-floating plants such as duckweed (*Lemna* spp.), floating fern (*Salvinia* spp.), water lettuce (*Pistia stratiotes*), and water hyacinth (*Eichhornia crassipes*) have leaves and flowers exposed to the air. Other free floaters, such as water milfoils (*Myriophyllum* spp.) and bladderworts (*Utricularia* spp.), have most of the plant submerged and only the flower above water. Most floating plants are attached to the bottom. Water lily (*Nymphaea* spp.), spatterdock (*Nuphar* spp.), and floating hearts (*Nymphoides* spp.) are prime examples from temperate lakes. Submerged plants rooted in the lake bottom such as quillworts (*Isoëtes* spp.), pondweeds (*Elodea* spp. and *Potamogeton* spp.), and coontail (*Ceratophyllum*) are necessarily restricted to shallow water where sunlight can reach them. A final group of vascular plants are the emergents confined to the outer fringe of the lake (Riemer 1993). These include cattails (*Typha* spp.), reeds (*Phragmites* spp.), rushes (*Juncus* spp.), sedges (*Carex* spp.), golden club (*Orontium aquaticum*), pickerelweed (*Pontederia cordata*), and arrowweed (*Sagittaria* spp.). On saturated beach sands, they may grow with plants such as bulrush (*Scirpus* spp.), rose pogonia (*Pogonia ophioglossoides*)—a terrestrial orchid—and insectivorous sundews (*Drosera* spp.).

Photo 11.2 The water lilies (*Nymphaea* sp.) and emergent pickerel weed (*Pontederia cordata*), some of the larger aquatic plants that grow in the shallow littoral zone of lakes. They are an early stage in the ecological succession toward swamp and finally forest. (Photo by SLW.)

## ANIMAL LIFE

Among the zooplankton in lakes are rotifers, tiny filter feeders 0.1–1.0 mm (0.003–0.04 in) long, and crustaceans of the orders Cladocera and Copepoda. The cladocerans are represented by phytoplankter eaters such as water fleas (for example, *Daphnia* spp. and *Moina* spp.) and carnivores such as *Leptodora* and *Polyphemus.* The copepods among the zooplankton include small particle feeders such as *Diaptomus* and those that seize their prey, such as *Cyclops.* Other crustaceans in lakes include opossum shrimp (*Mysis relicta*), apparently restricted to temperate lakes (as are cladocerans); more widespread isopods (for example, the water louse *Asellus aquaticus*); amphipods such as freshwater shrimps (*Gammarus* spp.); and crayfish—generalists and the largest of freshwater crustaceans (Brönmark and Hansson 1998; Moss 1998).

Freshwater sponges such as the pond sponge (*Spongilla lacustris*) may be attached to solid surfaces. They are greenish or yellowish in color and may contain symbiotic algae (zoochlorellae).The sponges, which are filter feeders, are supported by embedded silica spicules and are toxic. Also occurring are hydroids with symbiotic algae; being cnidarians, they use stinging cells to capture prey. Flatworms such as *Planaria* are abundant, as are various annelid worms, including leeches, and oligochaetes. Leeches have suckers at either end on their ventral surface to attach to prey. Some feed on blood, but others consume prey. Oligochaete worms are deposit feeders. They build tubelike burrows in soft sediments on a lake bottom and let their tails, which act as gills, protrude into the water column (Brönmark and Hansson 1998; Moss 1998).

Both gastropod and bivalve mollusks can be found in lakes. Snails scrape algae and detritus from rock and other surfaces. Both those that pass water over a gill (prosobranchs) and those that draw air into a lunglike cavity (pulmonates) occur. Bivalves include large unionid mussels 10–15 cm (4–6 in) long, such as *Anodonta, Unio* and *Elliptio,* and small sphaeriid clams only 2–20 mm (0.08–0.8 in) long, such as the pea clam (*Pisidium* spp.). Both filter particles out of the water. The bivalves live in soft sediments, while the gastropods are associated with hard surfaces (Brönmark and Hansson 1998, Payne 1986).

Most insects occur as larvae and are important components of the benthic community because they provide food for many fish. Most common are caddisflies (Order Trichoptera), flies (Order Diptera), and beetles (Order Coleoptera). The herbivorous caddisfly larvae build cases over their backs of leaves, twigs, pebbles, or shell fragments; each kind constructs its own distinctive type of case. Particularly abundant among fly larvae are craneflies (*Dicranota* spp.), mosquitoes (*Culex* spp. and others), and midges (*Chironomus* spp.). Indeed, midge larvae are among the most widespread insects in lakes. Other insect larvae are able to suspend themselves in the water column by adjusting gas-filled cavities to maintain buoyancy. An example is the phantom midge (Family Chaoboridae); its name derives in part from its being transparent and hence nearly invisible (Brönmark and Hansson 1998; Jeffries and Mills 1990; Payne 1986).

Adult insects are important predators on the surface of lakes and ponds. These include damselflies (Suborder Zygoptera), dragonflies (suborder Anisoptera), and true bugs (Order Hemiptera) such as pondskaters (*Gerris* spp.). Pondskaters use hairs on their feet and the tension of the water surface to walk on water and feed on terrestrial insects trapped on top of the water. The water boatman (*Notonecta* sp.), another bug, hangs under the water surface with air trapped in its underwings and body hairs and swims after aquatic invertebrate prey (Brönmark and Hansson 1998; Jeffries and Mills 1990).

Many different fishes inhabit lakes. Those streamlined for pursuit, such as trout (*Salmo* spp.), swim in search of prey. Others, such as pike (*Esox lucius*), wait in ambush. They are built for fast acceleration with long, slender bodies and large tail fins. Those living among submerged and floating plants, such as bluegill (*Lepomis macrochirus*), are built for maneuverability and are deep bodied with strong pectoral fins. Fishes consume a variety of food items; many are generalists or change the preferred elements of their diet as they grow. Pumpkinseed sunfish (*L. gibbosus*), for example, eat soft invertebrates when they are juveniles but focus on snails as adults. Toothlike structures (pharyngeal teeth) develop in their throats and allow them to crush snail shells. Bluegills, on the other hand, feed on the zooplankton even as adults, while walleye go from consuming zooplankters as juveniles to eating other fish as adults (Brönmark and Hansson 1998).

Other important vertebrate predators in lakes include amphibians, the tadpoles of frogs and toads, and adult frogs. Reptiles are represented by snakes, turtles, alligators, and caiman. A number of birds feed in lakes. So-called dabblers such as the Mallard (*Anas platyrhynchos*), wigeons (*Mareca* spp.), and Green-winged Teal (*Anas carolinensis*) are surface feeders and collect what they can by tipping tail up, head down into the water. They eat aquatic plants, invertebrates, and occasionally small fish. Grebes (*Podiceps* spp. and others), Common Merganser (*Mergus merganser*), Common Loon (*Gavia immer*), and Anhinga (*Anhinga anhinga*) dive and swim underwater in pursuit of fish and crustaceans. Wading birds such as herons, egrets, and ibises stalk fish from shore (Burgis and Morris 1987).

Finally, mammals can also be part of the lake community. Plant eaters include beaver (*Castor canadensis*), muskrat (*Ondatra zibethica*), and moose (*Alces alces*). Carnivores may include otter (*Lutra canadensis*), mink (*Mustela vison*), raccoon (*Procyon lotor*), and fish-eating bats.

## MAJOR LAKES

### Large Temperate Freshwater Lakes

**Great Lakes of North America.** Together, the five Great Lakes contain 20 percent of the total freshwater on the surface of Earth and, with the waterways that connect them, form the largest freshwater system on the planet. Lake Superior (the name comes from the French for uppermost lake) is the largest, coldest, and least affected by human activities. With a surface area of 80,000 km$^2$ (31,000 mi$^2$) and a maximum depth of 406 m (1,333 ft), Superior contains the second-greatest volume of freshwater of any lake, surpassed only by Russia's much deeper Lake Baikal. Lake Superior's waters flow into Lake Huron through the 100-km-long (60 mi) St. Mary's River, the rapids of which have been bypassed by the canal and locks at Sault Ste. Marie since 1855. Lake Huron, the second largest of the Great Lakes, is the fifth-largest freshwater lake in the world. Georgian Bay at its northeastern edge is large enough that it has been dubbed the Sixth Great Lake. Lake Huron has a maximum depth of 228 m (750 ft). The Straits of Mackinaw connect it to Lake Michigan, the third largest of the Great Lakes, and equalizes the water levels of the two lakes. Lake Michigan reaches a maximum depth of 281 m (923 ft). Some of its waters flow out through the narrow straits leading into Huron, and some flows into the Mississippi River system by way of the Illinois River and Chicago Sanitary and Ship Canal (University of Wisconsin Sea Grant Institute 2001a).

Lake Huron connects to Lake Erie via a 145-km (90 mi) link composed of the St. Clair River, Lake St. Clair, and the Detroit River. Lake Erie, the fourth-largest Great Lake, has an average depth of only 19 m (62 ft) [maximum depth = 64 m (210 ft)]. At its eastern end, it empties into the 56-km-long (35 mi) Niagara River, which plunges almost 61 m (200 ft) over Niagara Falls into Lake Ontario. Lake Ontario is the smallest of the Great Lakes but third deepest, with a maximum depth of 244 m (802 ft). It is also probably the most polluted. Lake Ontario's waters flow out past Thousand Islands into the St. Lawrence River and then some 1,600 km (1,000 mi) to the Atlantic Ocean. The St. Lawrence Seaway, which opened in 1958, is a series of locks in the Thousand Islands reach of the river that lifts ships 225 feet and provides access for ocean-going vessels to the ports of the Great Lakes (University of Wisconsin Sea Grant Institute 2001a).

Early European explorers described these huge bodies of water as inland seas remarkable for the fact that their water was not even slightly saline. Indeed, the salt content is measured in parts per million (ppm) rather than in parts per thousand, as is the case in the sea or saline lakes (see following). Each of the five lakes is different and reflects the type of bedrock over which streams entering a respective lake flow. The drainage basins of Superior and much of Lake Huron lie on slowly **weathering** igneous rock of the Canadian Shield, so streams contribute very little dissolved mineral matter. The salinity of Lake Superior, the most oligotrophic of the lakes, is 35 ppm. Drainage into the other lakes flows over glacial **drift** that overlies sedimentary rock. Lake Ontario's offshore waters have a salinity of 170 ppm. The other lakes lie between the extremes marked by Superior and Ontario (Mortimer 1979).

These are temperate lakes and, with the exception of shallow Lake Erie, experience seasonal patterns of mixing and phytoplankton blooms. Lake Erie freezes over in winter. The others maintain open waters with temperatures of about 2.2° C (36° F). Thermal stratification develops by early June except on Superior, where cool pools of water persist offshore, and western Erie, which is so shallow winds mix the entire water column. Before extensive pollution, a spring bloom of diatoms fed the zooplankton, mostly vertically migrating opossum shrimp (*Mysis relicta*). By August, the surface waters begin to cool, become denser, and lower the thermocline. Turnover occurs by early December and brings nutrients that sank during the summer back toward the surface and carries oxygen-rich waters to depth (Mortimer 1979).

A thermal bar or strong contrast in temperatures between shallow inshore waters and cooler waters offshore acts as a barrier to the transport of materials that streams feed into the lakes. Most nutrients, and hence production of new life, were—before the lakes became nutrient-enriched through human activities—limited to shallow embayments such as western Lake Erie, Saginaw Bay, and Green Bay (Stoermer 1979). Large populations of cold-water fishes such as Atlantic salmon (*Salmo salar*), lake trout (*Salvelinus namaycush*), lake whitefish (*Coregonus clupeaformis*), lake herring (*C. artedii*), deepwater cisco (*C. johannae*), and deepwater sculpin (*Moxocephalis thomponsi*) existed when European peoples first settled on the shores of the Great Lakes. Subsequent land use, the introduction of **exotic** species, and cultural eutrophication (the increase of nutrient content at rates well above the natural processes of lake aging) have by now greatly altered the aquatic communities of the lakes (Smith 1979).

Several ecological changes occurred at about the same time. The original lake system had been deficient in phosphorus; this had favored diatoms, which build silica skeletons, over other forms of algae. However, increased phosphorus levels from polluted runoff into the lakes stimulated population explosions among the diatoms to the point they depleted dissolved silica in the water, and their populations crashed. Green algae and cyanobacteria were able to thrive in the absence of the diatoms and began to produce summer blooms. Further increase in phosphorus led to a decline in nitrogen compounds below the levels green algae could tolerate. Cyanobacteria, however, can fix nitrogen directly from the water, and they came to dominate the phyoplankton in the summer. Cyanobacteria are relatively poor food for zooplankton, so this change affected the entire food chain. As increased numbers of algae decreased the transparency of the waters, plankton growth became more restricted to waters closer to the surface and favored surface-feeding fishes such as **introduced** alewife (*Alosa pseudoharengus*) over native cisco and lake herring that generally feed at greater depth. The smaller lakes of Erie and Ontario felt these affects the most, but cyanobacteria blooms were also occurring in the Duluth embayment of Lake Superior by the 1970s (Stoermer 1979).

As the lakes changed from cool, clear oxygen-rich bodies of water to eutrophic lakes, the fish community changed. Lake trout, the top predator in deep-water lakes, was markedly reduced in numbers by the mid-nineteenth century, and the Atlantic salmon, only known from Lake Ontario since they could not pass by Niagara Falls, was extinct in the lake by 1890. Deforestation of the watershed and damming of tributary streams to provide power for mills have raised the temperatures of stream waters flowing into the lakes. All the characteristic fishes were near their natural southern limits, and warming water temperatures stressed them. Together with overharvesting by commercial fisherman, changing water chemistry and temperature caused most fishes to become rare. Deepwater cisco, once common in Huron and Michigan, is now extirpated from both lakes. Naturalized exotic species also contributed to the change in Great Lakes fish communities. The first alewives were established in the lakes by 1873. These are anadromous fish that spawn in freshwater streams and usually spend the adult phase of their life cycles at sea. However, they are also able to survive as adults in lakes. It is believed that they entered the Great Lakes from the Atlantic Ocean by way of the Hudson River and Erie Canal. The presence of predatory Atlantic salmon in Lake Ontario may have prevented an earlier arrival via the St. Lawrence River. Once established, alewife populations consumed the larger plankton. This depletion of an important food source hurt native fishes, such as lake herring and emerald shiners that fed in shallow water, and deepwater cisco and deepwater sculpin, which were food for lake trout (Smith 1979). Alewife populations crashed in the 1980s, and as a consequence recovered lake herring populations in western Lake Superior (University of Wisconsin Sea Grant Institute 1998).

Another important invader of the Great Lakes was the sea lamprey (*Petromyzon marinus*), a parasite on large fish such as the lake trout first recorded in Lake Ontario in the 1880s. It first appeared in Lake Erie in 1921 after entering via the Welland Canal. From Erie, it quickly invaded all the upper Lakes and was particularly aggressive in lakes Huron and Michigan. By 1958, however, lamprey populations were being controlled by regular application of a chemical larvicide to the streams where the adults spawned (Smith 1979; University of Wisconsin Sea Grant Institute 2001b). Warming waters favored lamprey reproduction but acted against lake trout. Lake trout reproduction in Lake Ontario ended in 1950, and the fish was essentially extinct as a wild-breeding fish by 1960. Lake trout have since been restored through stocking programs, but significant natural reproduction is occurring only in Lake Superior (Hansen and Peck n.d.).

**Lake Baikal.** Lake Baikal in Siberian Russia is the world's deepest lake and one of only two lakes on the planet deeper than 1,000 m (3,280 ft). (The other is Lake Tanganyika, which is described later.) Baikal lies in a steep-sided rift valley some 650 km (400 mi) long and has a maximum depth of 1,620 m (5,315 ft). Although winter temperatures get very low in Siberia and ice covers the lake for four to five months, the winters are relatively dry and snow cover is limited. Sunlight is able to penetrate the ice and warm the water. Bottom waters stay at 3–3.6° C (37–38° F) all year. In summer, the temperature of surface waters rises to close to 20° C (68° F). Autumn cooling ensures a turnover of even the deepest waters, so the lake is well mixed, and bottom waters contain dissolved oxygen.

Lake Baikal is the world's oldest lake and as such contains many **endemic** groups. Thirty-five percent of plants and 65 percent of animals are found nowhere else. Several of the unique species of Baikal are closely related to marine animals. The small Baikal seal (*Phoca sibirica*), for example, is closely related to the Caspian seal. The omul (*Coregonus autumnalis migratorius*), a fish that spawns in rivers, appears to have entered from the Arctic during a glacial period. At higher **taxonomic** levels, 87 genera of animals and 11 families are endemic to the lake—indications of long isolation from similar **habitats**. Thirty-four of the 35 genera of freshwater fishes are found only in Baikal. An endemic family of fish, Comephoridae, has two species. The total number of fishes, about 50, is quite low for a lake this size, but that is not surprising for such an oligotrophic body of water. It has no amphibians (Burgis and Morris 1987).

## Tropical Lakes

The tropics have relatively few lakes. Several conditions distinguish tropical lakes from those of the temperate zone. At low and moderate elevations, freezing temperatures are never experienced, but strong seasonality may be experienced as annual fluctuations in water level related to distinct rainy and dry seasons. The fishes of the tropics are often also distinctly different from those found in the middle **latitudes**, and more species overall are found. The tropics are home to living fossils such as lungfishes (Dipnoi) and bony-tongued fishes (Osteoglossidae), the latter among the oldest of bony fishes. Lungfishes date back 300 million years to the Devonian and breathe air in modified swim bladders. Their modern descendants are found in shallow lakes, where they are able to survive the dry season by burrowing into the mud and forming a water-tight cocoon in which they enter a state of torpor. Different types of lungfish exist on each continent. An exclusively tropical suborder of fishes, Characoidea, includes the tetras (*Hemigrammus* spp.) common in the aquarium trade (Burgis and Morris 1987; Payne 1986).

The fish families that dominate each tropical landmass vary. In southern Asia, the most abundant and diverse groups are carps (Cyprinidae) and catfishes (Siluroidea). Africa has carp, catfish, and characoids, but cichlids (Cichlidae), including *Tilapia* spp., dominate in the large lakes (see following). Cichlids typically bestow parental care on their young either through guarding their nests or brooding them in their mouths. The many, many species of cichlids are differentiated by behavioral characteristics and subtle differences in mouth parts that enable them to pursue different food-gathering strategies. Large numbers of very closely related cichlids can thus occupy the same lake. In tropical South America, native carp are entirely absent; they are replaced by more than 1,000 species of characoids. Cichlids and catfish are also diverse (Payne 1986).

**Great Lakes of East Africa.** The Rift Valley of East Africa is actually a series of troughs that branch into a western and eastern rift on either side of a plateau. Long, deep lakes have formed in the valley bottoms; the largest are Lake Tanganyika and Lake Malawi. On the plateau between the rifts, Lake Victoria fills a shallow basin. Lakes Tanganyika, Malawi, and Victoria are each known for their great **variety** of cichlid fishes. The tremendous diversity within this family has raised important scientific questions about how so many closely related forms came into being and how they have been able to survive together in what appears to be a continuous habitat.

Lake Tanganyika is the second-deepest and probably second-oldest lake in the world (after Lake Baikal). It fills its valley completely so that most of the shoreline exhibits a steep dropoff. The only shallow water is found at the northern and southern ends, more than 600 km (370 mi) apart. The submerged valley floor lies 700 m (2,300 ft) below sea level, and the greatest depth of the lake is 1,470 m (4,800 ft). This lake is never fully mixed, and deoxygenated water lies 100–200 m (320–650 ft) below the surface. The deep waters have a high concentration of salts, due in part to water entering the lake from Lake Kivu to the north by way of the Ruzizi River. With no oxygen and high salt content, nine-tenths of the lake is without life. Interestingly, the Earth's interior heat and compression due to the weight of overlying water raise water temperature near the bottom of the lake. The coldest water is found at a depth of about 800 m (2,600 ft) (Burgis and Morris 1987).

The present outflow from Lake Tanganyika is via the Lukaga River, a tributary of the Congo River. Directly across the lake from the Lukaga River is the Malagavasi River, which contains many fishes native to the Congo and strongly suggests that, before rifting, the two rivers were parts of the same stream. Since its formation, Lake Tanganyika has accumulated 126 species of cichlid fishes, all endemic to the lake. Of the other 67 species of fish from 13 different families, about 70 percent are restricted to this body of water. The most abundant open-water fishes are herringlike fishes known as Tanganyika sardines or *dagaa.* These small 10-cm-long (4 in) fish are mostly adult *Stolothrissa tanganicae* and *Limnothrissa miodon* and the juveniles of four kinds of *Late.* They feed on copepods that may be part of a food chain that begins with methane-using bacteria in the boundary zone between the upper oxygenated and lower deoxygenated waters. The sardines constitute an important **fishery** for the human inhabitants of the lake's shores (Burgis and Morris 1987; Moss 1998; Payne 1986).

Among other interesting life-forms in Lake Tanganyika are a freshwater jellyfish (*Limnocnida tanganikae*—not endemic to the lake, however), a flightless caddisfly, and two aquatic snakes—one a pit viper that swims in the open water and feeds on sardines and the other a cobra that lives on shore and fishes at night. The lake also has some 60 different gastropods, 50 percent of which are found nowhere else (Burgis and Morris 1987).

Lake Malawi lies in its own rift valley in East Africa. It is the fourth-deepest lake in the world, and, like Lake Tanganyika, its deepest waters are without oxygen. Younger than Tanganyika, it nevertheless has been in existence long enough to acquire a rich and unique group of fishes and one that is completely different from Lake Tanganyika's. It has no sardines, yet Lake Malawi has more kinds of fishes—245—than any other lake. Most of them (200 species) are cichlids, 95 percent of which are found nowhere else. Many are restricted to highly localized areas in the lake and divide the lake's resources among themselves by feeding in different ways in different **microhabitats**. Some, for example, are algae eaters; others eat invertebrates; some consume other fish. At least one sucks the fry from the mouth of brooding parents, while still others eat the scales and fins of their own kind. The same food types are harvested by different species in sandy shoreline habitats, on submerged rocks, and in the open water. The cichlids of the open water, together with one type of carp, compose an important fishery. The unique cichlids also have commercial value in the aquarium trade (Burgis and Morris 1987; Payne 1986).

Lake Victoria has a large surface area [69,000 km$^2$ (26,000 mi$^2$)] but, with a maximum depth of 80 m (260 ft), is relatively shallow. Much of the shoreline is marshy. Like the Rift Valley lakes, Victoria also has a diverse group of fishes (208 species) dominated by endemic cichlids (170 species, 99 percent of which are found nowhere else). They originated as inhabitants of rivers. Thirteen nonendemic cichlids still migrate up streams to spawn (Burgis and Morris 1987).

## High-Altitude Lakes

**Lake Titicaca.** Lake Titicaca—on the northern part of the high Andean plateau called the Altiplano—is billed as the world's highest navigable lake. The surface of the lake lies 3,810 m (12,500 ft) above sea level and has an area of 8,560 km$^2$ (3,300 mi$^2$). The lake is composed of two basins, Lago Pequeño or the Huiñaimarca with a maximum depth of 40 m (130 ft) and Lago Grande with a maximum depth of 285 m (935 ft). They are connected by the Tiquina Strait, some 800 m (2,600 ft) wide. Titicaca serves as the collection basin for rivers in a closed drainage system and is slightly brackish (salinity = 1 ppt). Most other Andean freshwater lakes have considerably lower salinities (Dejoux and Iltis 1992).

Low diversity generally marks the life in Lake Titicaca. The most conspicuous vegetation of the lake is the band of the tall emergent reeds known as totoras (*Schoenoplectus totora*), the raw material for the famous reed boats or balsas of Lake Titicaca. They grow rooted in the bottom in water 2–4.5 m (6–15 ft) deep. Successive bands of submerged aquatic plants grow shoreward of the totoras in the littoral fringe. *Ranunculus trichophyllus* grows in the shallowest water closest to shore, then in increasingly deeper water, zones of the pondweeds *Elodea potamogeton*, *Potamogeton* spp., and *Myrophyllum elatinoides* occur. A dense growth of shorter-rooted, submerged plants dominated by species in the **genus** *Chara* occurs on the open-water side of the totoras to depths of 10 m (30 ft). Each of these zones forms a distinct habitat for animal life in the lake. In all, only about a dozen aquatic macrophytes are known from the lake. The totoras, *Elodea potamogeton*, and *Myriophyllum elatinoides*, are endemic (Raynal-Roques 1992; Iltis and Mourguiart 1992).

The mollusks of Lake Titicaca are numerous (approximately 30 species) and make up the dominant animals of the benthos. Unlike other tropical lakes, aquatic insects are of only secondary importance, and only the midges (chironomids, order Diptera) are well represented (Dejoux 1992a, 1992b). The fish **fauna** consists of a diversified group of small fishes from the genus *Oreistias*, endemic to the waters of the Andean Altiplano. These are classified in the Cyprinodontidae but, unlike others in that family, lack ventral fins and possess only one gonad. They are very variable in phenotype and have been grouped into four complexes of closely related species and subspecies. Most are confined to the zones of macrophytic vegetation. Only *O. ispi* is known from deep open water; it feeds on zooplankters. Among the others are *O. mulleri*, a bottom feeder found near the outer edge of the *Chara* zone; *O. agassii*, an omnivore found in all zones; and *O. luteus* and *O. olivaceous*—consumers of macrophytes, zooplankters, amphipods, and mollusks—which are abundant in the *Chara* and totora zones. A second endemic genus, *Trichomycterus*, is probably represented by only two species. Two introduced fishes are the main predators: rainbow trout (*Salmo gairdneri*) and pejerrey (*Basilichthys bonaniensis*) (Lauzanne 1992). Rainbow trout was introduced in 1941–42 to develop a commercial fishery and inhabits all of Lago Grande and part of Lago Pequeño (Loubens 1992). The pejerrey is native to rivers and estuaries in subtropical South America (Uruguay, Argentina, and southern Brazil), where it is important commercially. It was released into Lake Poopó in 1946 and invaded Lake Titicaca via the Río Desaguadera in 1955 or 1956. Today, pejerrey inhabits all of the Lake and its inflow rivers. It reaches a maximum length of 56 cm (22 in) and weight of 2.5 kg (5.5 lb) and is now, economically, the most important fish in Lake Titicaca (Loubens and Osorio 1992).

The Altiplano presents a challenging environment for amphibians, so it is not surprising that only a few are known from Lake Titicaca. The toad *Bufo spinulosus* is a typical Andean species. Two smaller toads, *Pleuroderma marmorata* and *P. cinerea*, come to the lake to breed. Also known is a tree frog, *Gastrotheca boliviana*, one of the marsupial frogs in which females brood small numbers of eggs and tadpoles in a pouch on their backs. The true frog, *Telmatobius mamoratus culeus*, is the most characteristic amphibian. It has a stout body with numerous highly vascularized skin folds that aid in respiration through the skin. *Telmatobius* inhabits a variety of lake and wetland habitats and displays a wide range of morphological variation that has made its taxonomy complex. The local Aymara and Quechua peoples regard the large forms as sacred animals that call the rains (Vellard 1992).

Among the birds that permanently inhabit Lake Titicaca are Short-winged Grebe (*Rollandia micropterum*), Puna Ibis (*Plegadis ridgwayi*), and five species of ducks that feed among the totoras and submerged aquatic vegetation: Yellow-bellied Pintail (*Anas georgica*), Speckled Teal (*A. flavirostris*), Puna Teal (*A. versicolor*), Cinnamon Teal (*A. cyanoptera*), and the Andean Ruddyduck (*Oxyura jamaicensis*). American Coot (*Fulica americana*) is the most abundant of the rails that nest and feed among totoras; the Giant Coot (*Fulicula gigantea*) is the rarest. The Andean Gull (*Larus serranus*) is also a permanent resident (Dejoux 1992c).

## Saline Lakes

Saline lakes hold 0.008 percent of Earth's water (Mandaville 2001). Seventy percent of that is in the Caspian Sea, the world's largest lake in terms of surface area. Most such lakes derive from interior drainage: rivers flow into them, but there is no outflow. As a consequence, most water is lost through evaporation, and this tends to concentrate salts. A chemical salt is a compound that, when dissolved in water, separates into positively charged and negatively charged particles. The most common salts in lakes are chlorides, sulfates, and carbonates that derive from sedimentary rock of marine origin or volcanic rock. Truly saline, or endorheic, lakes are widely distributed. Salinity varies from 3 ppt to more than 300 ppt in some hypersaline bodies of water such as Great Salt Lake and the Dead Sea. (Sea water has an average salinity of 35 ppt.) The Dead Sea, the saltiest lake on Earth, has 10 times the salt concentration of seawater (Mandaville 2001; U.S. Geological Survey 2001).

Lakes with high concentrations of chlorides support only very simple communities. The **Great Salt Lake**, for example, has just a few types of planktonic algae and diatoms that brine shrimp (*Artemia franciscana*) and the larvae of the brine fly (*Ephydra* spp.) eat. It has no other aquatic life-forms, but the brine shrimp support millions of breeding and migrating shorebirds and waterfowl. These include the world's largest breeding colonies of White-faced Ibis (*Plegadis chihi*) and California

Gulls (*Larus californicus*), as well as a significant population of White Pelicans (*Pelecanus erythrorhyncos*). Among migrating birds that find vital habitat on the shores of the lake are Eared Grebe (*Podiceps caspicus*), Wilson's Phalarope (*Steganopis tricolor*), American Avocet (*Recurvirotra americana*), and Snowy Plover (*Charadrius alexandrinus*) (U.S. Geological Survey 2001).

In contrast to the hypersaline Great Salt Lake, the **Caspian Sea** has an average salinity of only 10–12 ppt. Fed primarily by the Volga River, it covers about 400,000 $km^2$ (154,000 $mi^2$) and stretches from the cold, continental desert region of Kazakhstan in the north 1,030 km (640 mi) to the subtropical regions of mediterranean Iran in the south. Its surface is 28 m (92 ft) below sea level. The Caspian reaches a maximum depth of 1,025 m (3,400 ft) in the southern part of the lake, while the northern end lies on a shallow shelf about 4.4 m (14 ft) deep. Physical variation results in a variety of habitats. The northern end, which receives most of the freshwater flowing into the lake, is the richest biological region (Bolshov 1999).

The phytoplankton in the Caspian Sea is composed of about 450 species and subspecies—mostly diatoms, algae, and cynobacteria. Some 100 species make up the zooplankton; about one-third are rotifers. Benthic animals are largely endemic and are believed to be relicts from the Tertiary Period, when the Caspian was part of the globe-spanning Tethys Sea. The Caspian has about 125 fishes from 17 families. The families with the greatest numbers of species are carp (Cyprindae), gobies (Gobiiadae), and herring—both shads (*Alosa* spp.) and sprats (*Clupeonella* spp.). The most commercially important fishes, however, are the several species of sturgeons that provide 90 percent of the world's supply of caviar. These include the Russian sturgeon (*Acipenser gueldenstadt*), the stellate or starry sturgeon (*A. stellatus*), and the beluga sturgeon (*Huso huso*). Fishes such as sprats spend their entire lives in the sea, but the anadramous sturgeons, among others, migrate upstream into rivers when they are mature adults. The Caspian is an extremely important wintering, resting, and staging area for migratory birds from all over the lands of the former Soviet Union. The only mammal in the Caspian Sea is a seal (*Phoca caspica*). It spends the winter in the ice-covered lake regions of the north, where the females give birth. In summer, the seals migrate to the deeper, colder waters at the southern end of the Sea (Bolshov 1999; Tirdád n.d.; Nowak 1991).

Soda lakes are found in the craters of a string of volcanoes in the Eastern Rift Valley of East Africa. They occur at elevations of 1,870–2,000 m (6,000–6,500 ft), and the salts held in high concentrations are carbonates and bicarbonates. These lakes are often very productive. **Lake Nakuru** in Kenya contains a coiled blue-green algae (*Spirulina platensis*). More than a million lesser flamingos (*Phoeniconaias minor*) breed at the lake and filter the algae from the water in their strangely shaped beaks. (Flamingos are the only birds restricted to saline or soda lakes. Different species breed in saline lakes high in the Andes of southern Peru, Bolivia, and northern Chile.) Other consumers of the blue-green algae are a copepod (*Lovenula africana*) and a small cichlid fish (*Oreochromis alcalicus grahami*). The rest of the lake's animal life consists of five rotifers, four waterboatmen, and the larvae of two midges. Over 400 birds have been recorded from Lake Nakuru National Park. In addition to lesser flamingos, the lake is home to greater flamingoes (*Phoenicopterus ruber*), which feed upon copepods and midge larvae, and cormorants (*Phalacrocorax* spp.), Anhinga (*Anhinga anhinga*), and white pelicans (*Pelecanus* sp.) that are fish eaters (Burgis and Morris 1987; Moss 1998).

Many saline lakes are shrinking and turning into salt pans or playas as their water is withdrawn from inflow streams for irrigation. The Aral Sea is the prime example (see Human Impacts) but by no means the only one. Others, however, undergo periodic expansion in response to cyclical climatic changes. The Caspian Sea and the Great Salt Lake in Utah are two good examples. As the lakes expand, the salt content becomes diluted, which creates adverse conditions for the few species that are specialists in highly saline lakes.

Not all lakes in arid lands are salty. Lake Chad in West Africa remains relatively fresh because it shrinks during dry periods and the salts are precipitated on the abandoned lake bed. (Wet and dry cycles run over several years in West Africa, and Lake Chad has three recognizable stages: Greater Chad, Normal Chad and Lesser Chad.) The waters of Lake Turkana in East Africa taste salty, but freshwater plants and animals live in them (Burgis and Morris 1987; Payne 1986).

## Subglacial Lakes

The most recently discovered lakes lie under the Antarctic ice sheet. More than 70 such subglacial lakes have been found; the largest, **Lake Vostok**, bears the name of the Russian research station located on the ice above it in one of the coldest places on Earth. Lake Vostok's existence some 1,000 km (600 miles) north of the South Pole was unknowingly revealed in 1961 when the Russian pilot R. V. Robinson remarked upon the unusually smooth surface of the ice near the station. In the 1960s and continuing into the 1990s, seismic surveys and remote sensing by airborne and satellite radar confirmed the presence of liquid water 3,700–4,200 m (12,150–13,800 ft) beneath the ice and allowed the mapping of the lake. The discovery of Lake Vostok was announced to the public in 1996. The lake, probably filling a rift valley, is roughly the size of Lake Ontario. It is 224 km (140 mi) long, 48 km (30 mi) wide, and has a maximum depth of about 500 m (1,600 ft) deep near its southern end. The Antarctic ice sheet floats across the water and, in the absence of the frictional drag generated when ice passes over a rock base, remains flat and smooth (Siegert 1999).

No one knows why water is in a liquid state in Lake Vostok and the other subglacial lakes of Antarctica. Geothermal heating from Earth's interior, as happens in the Great Lakes of Africa, is a possibility. Also, no one knows yet if there is life in these lakes, but scientists expect to find unique forms of bacteria when they finally devise a way to sample the waters without contaminating them with microorganisms from the surface or from the ice mass itself. Bacteria, fungi, diatoms and other tiny life-forms blown to Antarctica have been found at various depths in the ice near Vostok Station. The greatest depths sampled, 1,500–2,405 m (4,900–7,890 ft), contained viable spore-forming bacteria (*Bacillus* spp.), albeit in very low numbers. The ice at these depths is approximately 200,000 years old (Abyzov 1993). It is possible that some bacteria, in a dormant state, were carried in the bottom ice and released into the lake when basal ice came into contact with the lake waters and melted. Some 50 m (165 ft) of sediments have accumulated on the lake bottom, presumably from rock debris so carried across the lake (Siegert 1999). Another possibility is that bacteria were in the lake before it was sealed off from the rest of the world by the ice that began to accumulate in the Miocene epoch. Bacteria are the form of life most apt to have survived and evolved to withstand the extreme conditions of Lake Vostok and other subglacial Antarctic lakes. Continuing research will someday reveal the natural history of these bodies of freshwater.

## HUMAN IMPACTS

Lakes have been vital sources of drinking water and food for human populations for millennia. They are also traps for whatever flows off a watershed, so they are affected by land use practices in their basins. Human activities have manipulated, disturbed, and altered lakes. The floodplain lakes of the Yangtze River (Chang Jiang), for example, have long been used for rearing various native and introduced species and today are increasingly part of an integrated system of lake farming (Jeffries and Mills 1990; Chang 1987). Other lakes were disturbed as water entering them was rerouted onto agricultural fields. The Aral Sea is the most notorious example. During the Soviet era, its water level dropped 15 m (50 ft), its surface area was halved, and its salinity tripled as inflow waters were diverted to irrigate cotton. Salt-encrusted ground surrounds the dying lake. Before irrigation projects robbed it of its freshwater intake, the Aral Sea had 20 kinds of fish, 12 aquatic plants, and 200 invertebrates. Today there are no fish. Only one higher plant, *Zostera nana*, remains. The most common organism is an introduced bivalve, *Abra ovata*. The people who once lived on its shores now find themselves 120 km (75 mi) away from the water, and many suffer lung diseases from blowing salt dust (Moss 1998).

Introduced organisms are major problems in lakes around the world. Predatory fish can change the composition of native species. The Nile perch (*Lates nilotica*) has been introduced to Lake Victoria and several other East African lakes, where it feeds on native cichlids. The introduction of *Cichla ocelaris* into Lake Gatun, Panama, in 1967, was followed by the loss of six of the most common fishes. Tarpon and terns that fed upon these missing fishes declined in numbers, but mosquitoes—the larvae of which had been eaten by the lost fish—increased (Moss 1998). In the Great Lakes of North America, the sea lamprey (*Petromyzon marinus*) entered the lakes and led to a sharp decline in native whitefish (*Coregonus* sp.) (Payne 1986). Another exotic having a significant impact in the Great Lakes and elsewhere is the zebra mussel (*Dreissana polymorpha*), native to the Black Sea. In 1985–86 it was accidentally introduced into the river that connects St. Clair Lake with Lake Erie. It had moved into the western end of Lake Erie by 1988 and a few years later into Lake Huron. By 1995, the zebra mussel was reported in 18 American states and two Canadian provinces; halting its spread appears to be impossible. The small [2–2.5 cm (0.75–1.0 in)] zebra mussels attach to hard surfaces including rocks and the shells of native mussels and severely overgraze the phytoplankton upon which the natural lake system depends (Abell et al. 2000; Moss 1998). They pose a major threat to the survival of native mussels.

In addition to aquaculture, water withdrawal, and the introduction of exotic species, humans also negatively affect lakes through overfishing, nutrient enhancement (eutrophication) from agricultural runoff and acid precipitation, and pollution by pesticides and industrial wastes.

## ORIGINS

Lakes originate in many ways and tend to occur in clusters in different parts of the world. Most lakes are the result of glacial activity—their basins were gouged out of bedrock as ice sheets advanced or they formed when streams from melting glaciers were blocked by the deposits of stones, gravels, and sands left at the edge of the ice as the glaciers retreated. The Great Lakes of North America are prime examples of the former type of glacial lake, while most of the 10,000 lakes of Minnesota are examples of the latter. Since most lakes were formed by glacial activity, the greatest number of lakes occur in the areas of North America and Eurasia once covered by the Pleistocene ice sheets.

The world's deepest lakes are associated with locations where movements of the Earth's crust resulted in the creation of long, deep troughs or rift valleys. Lake Baikal in Russia, the world's deepest lake, was formed in this manner, as were Lake Tanganyika, Lake Malawi, and the other large elongated lakes of East Africa. In South America, Lake Titicaca lies in a trough between the two ranges of the Andes that developed during the early Pleistocene epoch and filled with meltwaters from mountain glaciers during the successive glacial periods in the Quaternary Period (Lavenu 1992). In North America, Lake Tahoe, 501 m (1,640 ft) deep, was also formed by rifting; and it is believed that Lake Vostok in Antarctica is in a rift valley that has been covered with ice for perhaps 15 million years (Siegert 1999).

Still other lakes were formed as a result of volcanic activity. Crater Lake in Oregon, Lago Atitlan in Guatemala, Barombi Mbo in Cameroon, and Lake Bosumtwi in Ghana are examples of the nearly circular lakes that fill depressions created by the collapse of volcanic cones. Lake Kivu in western Congo was created when a lava flow blocked a river.

Natural lakes can be found in areas never covered by ice or directly affected by rifting or volcanoes. In limestone areas, such as in Florida, they fill sinkholes. In desert basins where there are no outlets, they form from the collection of water running off surrounding uplands during infrequent rains. Along the lowland stretches of rivers, the changing course of the stream may leave segments of the former river isolated from the mainstream to create what are known as oxbow lakes in the United States and billabongs in Australia. Lakes may form in depressions on the floodplain; they fill (or are replenished) with water during the yearly flood but remain separate from the river most of the year, as is the case along the Amazon River. Some 1,760 floodplain lakes occur beside the Yangtse River downstream from Yichang, China. Most are only about 10 m (30 ft) deep; they vary in surface area from 300 to 3,000 ha (740–7,400 ac) (Burgis and Morris 1987; Payne 1986).

Nonetheless, most lakes occur in regions that were covered by ice sheets beginning about 100,000 years ago and continuing until about 10,000–12,000 years ago. Any preexisting lakes in these areas would have been destroyed. Only those species that were able to escape the advancing ice and survive in lower latitudes would have been available to colonize the new lakes formed at the end of the ice age. There has been little time for speciation to occur, so endemism and specialization is very low in these lakes and those formed from volcanic activity or wandering stream courses. However, a very few rift valley lakes such as Baikal, Tanganyika, and Malawi are very old, and these have many unique forms and many specialists (Moss 1998). Lake fish seem descended from river species. Only in Baikal, among the freshwater lakes, is there a conspicuous marine origin for some organisms.

## HISTORY OF SCIENTIFIC EXPLORATION

Alexander the Great, a student of Aristotle, discovered the Caspian Sea in 330 B.C. He found that it contained no marine fishes and thus decided it was not an arm of the great world Ocean that he sought and that he believed encircled all the lands of Earth. One of his commanders, however, saw the Caspian seal (*Phoca caspica*), and Alexander took this to mean that the Caspian had once been linked to the Ocean (Keay 1991).

The lakes of Africa became known to the European world in the nineteenth century as the race began to discover the resources of this vast continent. Heinrich Barth discovered Lake Chad in the early 1850s. In their quest to discover the source of the Nile, Richard Burton found Lake Tanganyika, and John Hanning Speke found Lake Victoria (Keay 1991).

The beginnings of modern ecology can be traced to studies of freshwater lakes. In the 1930s, G. Evelyn Hutchison of Yale University studied tiny Linsley Pond outside of New Haven, Connecticut, and formulated ideas about productivity and self-regulating systems in nature. Inspired by Hutchinson, Raymond Lindeman (1942) studied Cedar Lake Bog in Minnesota and described the link between nutrient cycles and energy flow. Howard T. Odum (1957), who worked at Silver Springs, Florida, in the 1950s, further refined the concepts, which became the foundation for systems ecology (Hagen 1992; Moss 1998).

Collections of tropical aquatic plants and animals began in the nineteenth century by mostly European explorers and naturalists. In the early twentieth century, more scientific expeditions were conducted, such as the German Sunda expedition to lakes in Indonesia in 1928 and the Cambridge University expedition to East African lakes in 1930–31. The Smithsonian Institution established a research station on Lake Gatun, Panama, after World War II. As for so many biomes, the International Biological Programme (1966–72) created international teams to study lakes as **ecosystems**. While much of the research was conducted in North America and Eurasia, tropical lakes were included.

Much of the early work focused on the role of the physical factors of the environment in determining the composition of life in lakes. In the 1980s, however, evolutionary biology began to influence researchers so that the origins and complexities of life in lakes became a dominant theme. The study of the cichlid species flocks in East African lakes has rekindled interest in how new species of animals came into being and even in what constitutes a true species. Research on Lake Vostok is aimed, in part, at understanding life in extreme environments, including possibilities for life on other planets in our universe.

## REFERENCES

Abell, Robin A. et al. 2000. *Freshwater Ecoregions of North America: A Conservation Assessment.* Washington, DC: Island Press.

Abyzov, Sabit S. 1993. "Microorganisms in the Antarctic ice." Pp. 265–295 in E. Imre Friedmann, ed., *Antarctic Microbiology*. New York: Wiley-Liss, Inc.

Allan, J. David. 1995. *Stream Ecology: Structure and Function of Running Waters.* London: Chapman & Hall.

Bolshov, Alexandr A. 1999. "Main priorities for conservation of biodiversity of Caspian region." Annex 6 in Caspian Environmental Programme, Biodiversity Caspian

Regional Thematic Center, First Regional Workshop in Kokshetau, Kazakhstan. http://www.caspianenvironment.org/biodiversity/first/annex6.htm.

Brönmark, Christer and Lars-Anders Hansson. 1998. *The Biology of Lakes and Ponds.* New York: Oxford University Press.

Burgis, Mary and Pat Morris. 1987. *The Natural History of Lakes.* Cambridge: Cambridge University Press.

Chang, W. Y. B. 1987. "Fish Culture in China." *Fisheries* 12(3): 11–15.

Dejoux, Claude. 1992a. "The Mollusca." Pp. 311–336 in C. Dejoux and A. Iltis, eds., *Lake Titicaca, A Synthesis of Limnological Knowledge.* Dordrecht: Kluwer Academic Publishers.

Dejoux, Claude. 1992b. "The insects." Pp. 365–382 in C. Dejoux and A. Iltis, eds., *Lake Titicaca, A Synthesis of Limnological Knowledge.* Dordrecht: Kluwer Academic Publishers.

Dejoux, Claude. 1992c. "The avifauna." Pp. 460–469 in C. Dejoux and A. Iltis, eds., *Lake Titicaca, A Synthesis of Limnological Knowledge.* Dordrecht: Kluwer Academic Publishers.

Dejoux, Claude and André Iltis. 1992. "Introduction." Pp. xv–xx in C. Dejoux and A. Iltis, eds., *Lake Titicaca, A Synthesis of Limnological Knowledge.* Dordrecht: Kluwer Academic Publishers.

Hagen, Joel. 1992. *An Entangled Bank: The Origins of Ecosystem Ecology.* New Brunswick, NJ: Rutgers University Press.

Hansen, Michael J. and James W. Peck. n.d. "Lake Trout in the Great Lakes." http://biology.usgs.gov/s+t/m2130.htm.

Iltis, André and Phillippe Mourguiart. 1992. "Higher plants: distribution and biomass." Pp. 241–253 in C. Dejoux and A. Iltis, eds., *Lake Titicaca, A Synthesis of Limnological Knowledge.* Dordrecht: Kluwer Academic Publishers.

Jeffries, Michael and Derek Mills. 1990. *Freshwater Ecology: Principals and Applications.* London: Belhaven Press.

Keay, John, ed., 1991. *History of World Exploration.* New York: Mallard Press.

Lauzanne, Laurent. 1992. "Fish fauna: Native species." Pp. 405–419 in C. Dejoux and A. Iltis, eds., *Lake Titicaca, A Synthesis of Limnological Knowledge.* Dordrecht: Kluwer Academic Publishers.

Lavenu, Alain. 1992. "Origins." Pp. 3–15 in C. Dejoux and A. Iltis, eds., *Lake Titicaca, A Synthesis of Limnological Knowledge.* Dordrecht: Kluwer Academic Publishers.

Lindeman, Raymond. 1942. "The trophic dynamic aspect of ecology." *Ecology* 23: 399–418.

Loubens, Gérard. 1992. "Introduced species: *Salmo gairdneri* (Rainbow trout)." Pp. 420–426 in C. Dejoux and A. Iltis, eds., *Lake Titicaca, A Synthesis of Limnological Knowledge.* Dordrecht: Kluwer Academic Publishers.

Loubens, Gérard and Francisco Osorio. 1992. "Introduced species: *Basilichthys bonariensis* (the 'Pejerrey')." Pp. 427–448 in C. Dejoux and A. Iltis, eds., *Lake Titicaca, A Synthesis of Limnological Knowledge.* Dordrecht: Kluwer Academic Publishers.

Mandaville, S. M. 2001. "Saline lakes (The largest, highest and lowest lakes of the world!)." http://www.chebucto.ns.ca/Sciences/SWCS/saline1.html.

Mortimer, Clifford. 1979. "Props and actors on a massive stage." Pp. 1–12 in John Rousmaniere, ed., *The Enduring Great Lakes.* New York: W. W. Norton & Company.

Moss, Brian. 1998. *Ecology of Fresh Waters, Man and Medium, Past to Future.* 3rd edition Oxford: Blackwell Science.

Nowak, Ronald M. 1991. *Walker's Mammals of the World.* 5th edition, Vol. II. Baltimore: The Johns Hopkins University Press.

Odum, Howard T. 1957. "Trophic structure and productivity of Silver Springs." *Ecological Monographs* 27: 55–112.

Payne, A. I. 1986. *The Ecology of Tropical Lake and Rivers.* New York: Wiley & Sons Ltd.

Raynal-Roques, Aline. 1992. "Macrophytes: The higher plants." Pp. 223–231 in C. Dejoux and A. Iltis, eds., *Lake Titicaca, A Synthesis of Limnological Knowledge.* Dordrecht: Kluwer Academic Publishers.

Riemer, Donald N. 1993. *An Introduction to Freshwater Vegetation.* Malabar, FL: Krieger.

Siegert, Martin J. 1999. "Antarctica's Lake Vostok." *American Scientist* 87: 511–517.

Smith, Stanford. 1979. "Pushed toward extinction: The salmon and the trout." Pp. 33–46 in John Rousmaniere, ed., *The Enduring Great Lakes.* New York: W. W. Norton & Company.

Stoermer, Eugene F. 1979. "Bloom and crash: Algae in the Lakes." Pp. 13–20 in John Rousmaniere, ed., *The Enduring Great Lakes.* New York: W. W. Norton & Company.

Tirdád. n.d. "Fishes of northern Iran, Southern Caspian sea basin." Gorgán homepage: http://medlem.spray.se/davidgorgan/fishes.html.

University of Wisconsin Sea Grant Institute. 1998. "Lake Herring." http://www.seagrant.wisc.edu/greatlakesfish/lakeherring.html.

University of Wisconsin Sea Grant Institute. 2001a. "Gifts of the glaciers." http://www.seagrant.wisc.edu/communications/GlacialGift/.

University of Wisconsin Sea Grant Institute. 2001b. "Sea Lamprey." http://www.seagrant.wisc.edu/greatlakesfish/sealamprey.html.

U.S. Geological Survey. 2001. "Great Salt Lake, Utah." USGS: http://ut.water.usgs.gov/greatsaltlake/index.html.

Vellard, Jean. 1992. "Associated animal communities: the Amphibia." Pp. 449–457 in C. Dejoux and A. Iltis, eds., *Lake Titicaca, A Synthesis of Limnological Knowledge.* Dordrecht: Kluwer Academic Publishers.

# III MARINE BIOMES

## INTRODUCTION

Oceans cover 71 percent of Earth's surface and contain 97.2 percent of the planet's water. Considering that life occurs throughout the seas but survives in only a thin film on, just below and just above the continental surfaces, it can be said that oceans represent 99 percent of the habitable space on Earth. "To a first approximation, therefore, the **ecology** of earth is the ecology of its seas" (Barnes and Hughes 1999, p. 1). Four oceans are usually recognized (a fifth, the Antarctic or Great Southern Ocean, is sometimes viewed as a separate entity). These are, in order of diminishing size, the Pacific, which itself covers nearly half (46 percent) of the surface area of Earth, the Atlantic, the Indian, and the Arctic oceans. Numerous seas, gulfs, and bays—parts of the ocean nearly enclosed by land—bear separate names from the ocean they connect to. Some, such as the Mediterranean Sea, the Persian Gulf, and Hudson's Bay, have restricted exchange of waters with the open ocean due to narrow entrances and the configuration of their floors. Others, such as the South China Sea and the Arabian Sea, are more open. (Note that the term "sea" is used in several ways. One is as previously defined, but totally landlocked bodies of saltwater also are called seas—for example, the Dead Sea and the Caspian Sea. And the word is used as a synonym for ocean as in the expression "on the high seas." The "Seven Seas" of the ancient world were the Adriatic Sea, Black Sea, Caspian Sea, Mediterranean Sea, Persian Gulf, Red Sea, and the Indian Ocean.)

Attempts to apply the **biome** concept to this vast realm are complicated by the fact that the feature that distinguishes terrestrial biomes is the structure of the natural **vegetation**, dominated by conspicuous flowering plants and shaped by adaptation to regional **climate**. In the open ocean, microscopic single-celled algae (Protista) and some photosynthesizing bacteria (Monera) fulfill the function of plants on the land as the major producers of food energy. The distribution of these organisms is very variable in both time and space and reflects seasonal differences in light and wind patterns; the depth, temperature, salinity and nutrient content of the water; the movement of water in waves and tides; and the great circulation systems of surface and subsurface ocean currents.

Some proposed regionalizations come from the distribution of animal life (zoogeography) in the seas (Ekman 1953; Briggs 1974). Temperature, and hence **latitude**, is an important determinant of patterns of life in the sea. Briggs, for example, divided the seas into seven latitudinal groups and then delineated one or more biogeographic regions within each one. His latitudinal groups are Arctic, Cold-Temperate Northern Hemisphere, Warm-Temperate Northern Hemisphere, Tropical, Warm-**Temperate** Southern Hemisphere, Cold-Temperate Southern Hemisphere, and Antarctic. The various regions to be found within each latitudinal group are a consequence of the circulation patterns of the oceans—themselves a product of global wind patterns and the barriers created by landmasses. The Atlantic and Pacific oceans, confined by the Americas on one side and the Old World Eurasian-African landmass on the other, flow around great central gyres. The western boundaries are characterized by warm ocean currents such as the Gulf Stream, while eastern boundary currents such as the California Current are cool. The warm currents on the western sides of the oceans draw tropical waters into the midlatitudes and even high latitudes, and the cool currents bring cold polar waters equatorward well into the midlatitudes, which prevents a purely latitidunal **zonation** of surface seawater temperatures.

To a degree, large algae, such as kelps, and seagrasses fit such a latitudinal regional scheme. Kelps are limited to the Temperate and Arctic/Antarctic waters. The kinds of seagrasses found in tropical waters differ from those found in Temperate and Arctic/Antarctic regions. And coral **reefs** and mangroves are strictly tropical in distribution (Lüning and Asmus 1991). However, seagrasses, coral reefs, and mangroves introduce another complication of regionalizing the marine environment—the striking difference between coastal environments, shallow water environments of the continental shelf, and conditions of the open sea. If biomes were based on the dominant, small floating plant life of the seas, light and nutrient availability would have to be factored in as well. Water depth, upwelling, evaporation rates, sediment load, depth of mixing layer and all sorts of other variables would have to be—and have been—considered. A review of proposed classification schemes for marine biomes can be found in Longhurst (1998), who offers a system based on phytoplankton of the open sea. No one classification system has become widely used, although the Briggs's latitudinal groups are often used in a descriptive sense.

This section offers a simplified scheme in which three major marine biomes are identified: coastal biome, continental shelf biome, and open sea biome. For each of these, major variations that occur with respect to latitude, boundary currents, or geological conditions are treated as major regional expressions. This reflects a traditional division of marine **ecosystems** according to the nature of the **habitats** rather than the composition and structure of the living communities within them (Smetacek 1988). The content of the following chapters, therefore, is not organized in exactly the same way as content was handled in the chapters on terrestrial biomes; and, although some of the same terms may be used, it is not intended to suggest that terrestrial and marine biomes are directly comparable.

The rest of this chapter introduces the marine environment and its many inhabitants. For many of us land dwellers, this is an alien realm filled with unfamiliar creatures, so particular attention is given to those microscopic plants and animals that float passively in the water (phytoplankton and zooplankton, respectively) and to the general types of invertebrates that dominate the food chains. These are forms of life either absent from the land or usually neglected in coverage of terrestrial biomes, where large life forms are both conspicuous and distinguishing characteristics.

## MARINE ENVIRONMENTS

Standing onshore or flying over the sea, the ocean appears to be a uniform, continuous habitat that stretches beyond the horizon for hundreds or thousands of miles and assumes different moods only according to the weather above it. This is a misleading view, because the environmental conditions are complex and ever changing in both horizontal and vertical dimensions. Important physical and chemical aspects of the ocean include light; temperature; salinity; waves; currents; tides; and the inputs, transport, and sedimentation of inorganic and organic matter.

Sunlight provides the energy for much of the life in the seas (at hydrothermal vents in the oceanic abyss, chemosynthetic bacteria use chemical energy from inorganic compounds). Although water is transparent, light energy can penetrate only so far. Dissolved and particulate matter in the water absorbs and scatters incoming radiation from the sun. The longer wavelengths of visible light (e.g., red) are absorbed close to the surface, and only the shortest—those in the blue and green part of the spectrum—continue to any depth. Thus, on sunlit days, the ocean appears to be blue or blue-green. The actual depth to which sunlight penetrates depends on the clarity of the water. In clear particulate-free (and hence nutrient-poor) oceanic water, the amount of light at 50 m (160 ft) depth is only 10 percent of what strikes the surface. The light reaching 100 m (350 ft) down in the water column is only 1 percent of that at the surface. As a rule of thumb, the depth at which 1 percent of the surface light penetrates is the boundary between the euphotic (lighted) zone and the aphotic (unlighted) zone beneath it. In more opaque, nutrient-rich areas with lots of plankton and particles that absorb light, the euphotic zone is shallow; it may be only 10 m (30 ft) deep or less. It is also shallow when much light is reflected at the surface, such as when there are waves or when the sun is at a low angle. The depth of the lighted zone determines that restricted region where photosynthesizing organisms can live and produce food and oxygen. At lower intensities of light, most algae use more energy in life processes (respiration) than they can produce through **photosynthesis**. At 1 percent light levels, energy production exactly compensates for energy use, so the maximum depth of the euphotic zone is also known as the compensation point. Some shortwave visible light does penetrate to greater depths, and a few, slow-growing algae do use this energy. By 150 m (500 ft), in very clear water, only 0.1 percent of light received at the surface is left, and at 200 m (650 ft) only 0.01 percent (Angel and Harris 1977; Barnes and Hughes 1999; Smetacek 1988; Lüning and Asmus 1991). A diver looking skyward can perceive daylight to 800 m (2,600 ft). In depths less than 250 m (800 ft), a bright circle or ellipse of light, called Snell's circle, is evident because of the refraction of light at the surface. Marine animals and humans use this to determine the position of the sun. Below 250 m, light is diffuse and the position of the sun cannot be known. Aided by specially adapted eyes, deep sea fish can perceive light at greater depths than humans—in very clear water as deep as 1,300 m (4,000 ft) (Angel and Harris 1977). Still, considering the total volume of ocean waters, 97.5 percent of the sea environment is without light (Barnes and Hughes 1999).

Water temperature varies vertically with depth and horizontally with latitude. Incoming solar energy consists of the visible light just discussed and of infrared or heat energy. The infrared is absorbed in the top 1 m (3 ft) of the ocean to warm the uppermost water according to season and latitude. However, wind-generated waves mix the warmed layer with the waters just below to depths of 10 m (30 ft) or more. The heat is distributed evenly through this mixed layer, so temperature is uniform in this so-called surface zone. Below the surface zone is a transitional layer, the thermocline, in which temperature rapidly decreases with depth. Below the thermocline is the deep zone, in which further drop in temperature is extremely gradual. No heat is transported downward from the surface layer; in most circumstances, temperature in the deep zone is a nearly constant 3° C (37° F) all year. The coldest waters, near the sea floor, range between 0.5° and 2.0° C (33–35.5° F). (Unlike freshwater, saltwater undergoes a continual increase in density as temperature decreases to the freezing point at –1.9° C or 28.5° F.) (Barnes and Hughes 1999).

Surface water temperatures vary from –1.9° C (28.5° F) in polar regions to 26–30° C (79°–86° F) in tropical seas. [In the shallow waters of the Persian Gulf, temperatures can reach 35° C (95° F).] Since water can absorb much heat energy without a change in temperature and also conserves heat energy well, water temperature changes only slightly between day and night, and only within a few meters at the top of the water column. In the open ocean, the daily temperature change in surface waters is less than 0.3° C (0.5° F), in shallow coastal waters less than 3.0° C (5.5° F) (Barnes and Hughes 1999). As expected, equatorial waters gain more heat year round from solar radiation than polar areas, and tropical waters, in general, are warmer than elsewhere. However, due to the rotation of Earth and to gravity, the waters are in motion, much as the atmosphere is, and excess heat from the **tropics** flows poleward in warm surface currents. The strong trade winds of the tropics blow the surface waters westward across each ocean basin to pile up the warm waters off the eastern shores of the continents. The waters are then diverted by the land barrier into the midlatitudes and form warm ocean currents that move clockwise along the western edge of the sea in the Northern Hemisphere and counterclockwise in the Southern Hemisphere. In the Northern Hemisphere, they continue across the northern margins of the oceans as the North Atlantic Drift, which warms northwestern Europe, and the North Pacific Drift, which carries warm waters eastward to British Columbia, Washington state, and Oregon (the Pacific Northwest). As the trade winds move warm surface waters away from the eastern side of the oceans, they are replaced by cold water upwelling from 100–200 m (300–650 ft) below. Hence, cool currents occur in the tropics and subtropics off the western coasts of continents and create very different conditions for sea life than what occurs at the same latitude off the eastern coasts. The upwelling zone in the Benguela Current that extends from 29° to 15° S off the coast of Namibia has water temperatures of 12–14° C (54°–57° F) in contrast to the 20° C (68° F) more typical of water masses at those latitudes. The Benguela Current, as well as the Humboldt and Peru currents off western South America, are associated with some of the world's largest fisheries (anchovies, sardines, and horse mackerel, in particular). Other cold currents, such as the Labrador Current, are associated with waters moving equatorward from polar regions. Where these cold currents meet warm currents, they tend to concentrate nutrients in the contact zone, and these regions, too, are associated with major fisheries. Examples include the cod **fishery** once found on the Grand Banks off Newfoundland and the important fishery off Japan where the Kurishio and Oyoshio currents converge.

Salinity is a measure of dissolved particles in seawater. Most of these electrically charged particles or ions are sodium and chloride, but magnesium, calcium, sulfate, and bicarbonate ions are also abundant. [Some algae (coccolithophores) and many animals combine the calcium and bicarbonate to form skeletons of insoluble calcium carbonate.] All other elements are present in at least trace amounts. Average salinity is 35 parts per thousand (ppt), but salinity varies regionally in surface waters due to evaporation (which increases salinity), precipitation (which decreases salinity), and the discharge of freshwater from the land by rivers. In polar areas, the formation of ice will increase salinity. Usually salinity is lower and more variable in the surface layer and higher and constant in the deep zone. The transition zone between surface waters and the deep—when discussing salinity—is called the halocline.

Temperature and especially salinity affect the density of water. Warm water is less dense, or lighter, than cold water; freshwater is less dense than saltwater. Differences in density thus develop between the surface zone and the deep zone; these differences prevent the mixing of the two, since the lighter water floats on top of the denser water below. Again there is a transition zone in which density changes rapidly (usually increases) with increasing depth. This is called the pycnocline and is a very important barrier to the exchange or recycling of nutrients between the euphotic zone and deeper waters, as discussed later.

Particles in water have a tendency to sink. When they settle out of the euphotic zone, they represent a loss of nutrients for the algae (and animals) living there. Nutrients are not readily recycled as in **soil**-based ecosystems, and the supply must be constantly replenished if the phytoplankters are to continue photosynthesizing, growing, and reproducing. Under warm, calm conditions, a situation is set up that resists the mixing of upper nutrient-impoverished water with lower, nutrient-enriched waters. Year round in the tropics and during the summer in the midlatitudes, the surface water is warmed and becomes less dense than the water below it. The warmer water floats as a distinct, separate layer on top of the colder water, and a stable system is set up; the water column is said to be stratified. The only way to bring the denser water to the surface is by some mechanical means of mixing the water. This does occur when large waves are generated by major storms or by upwelling since the waters rising from depth carry nutrients that had settled out back toward the surface. In autumn and winter in the midlatitudes, as the surface waters chill, stratification breaks down and some of the nutrients recirculate from lower levels. But warm tropical seas may stay stratified and nutrient poor to become essentially oceanic wastelands (Smetacek 1988).

While prevailing winds drive surface currents, water density drives vertically circulating waters in the world ocean. Dense waters sink off Antarctica and flow equatorward along the sea floor. During summer, the high density of the water is a consequence of cold water from melting ice; in winter it is a consequence of freezing: only the water in sea-water freezes to leave a more highly saline solution as seawater. This also happens in the Arctic Ocean off Greenland, where a cold bottom current can be traced as far as 40° S. These two currents act as huge conveyor belts to slowly move the ocean waters from the surface to great depth and rise again toward the surface in zones of upwelling caused by obstructions in the sea floor or the margins of continental masses. A complete circuit may take 2,000 years. Vertical circulation is extremely important to life in the deep sea. While at the surface, in contact with the atmosphere in polar regions, the cold water is able to dissolve significant amounts of oxygen. This it carries to depth, so, unlike the situation in lakes, most bottom waters of the sea are well oxygenated. The region of minimum oxygen actually occurs between 100 m (300 ft) and 1,000 m (3,200 ft), a zone that receives relatively little oxygen from the surface waters or from deep waters (Barnes and Hughes 1999).

Other important water movements include waves and tides. These are most significant in shallow waters and will be discussed in the chapters on coastal and continental shelf biomes.

## MARINE HABITATS AND COMMUNITIES

Life in the oceans occurs throughout the water column as well as on and in the bottom materials (the substrate). Commonly, distinctions are made between the open water habitat or pelagic zone and the bottom or benthic zone. Environmental conditions—and quite often plant and animal communities also—vary with distance from the shoreline to give rise to another set of terms to distinguish among them. The littoral zone is the nearshore or intertidal zone, where organisms may be alternately exposed to the air and submerged by the sea with the ebb and flow of the tides. The neritic zone is that of shallow waters [from the low tide mark to a depth of approximately 200 m (650 ft)] overlying the continental shelves; out beyond the shelf in deep waters is the oceanic zone. This zonation forms the basis of the marine biomes presented in the following chapters.

## LIFE IN THE OCEAN

Life in the ocean differs significantly from that on land. Certain fungi, flowering plants, insects, and four-limbed vertebrates so dominant on land are poorly represented in the seas. As previously mentioned, monerans and protists are the major producers in marine communities. The sea is rich with respect to animal life: of the 34 animal phyla scientists recognize, 29 have representatives in the sea and 14 of these are restricted to the oceans (Ormond et al. 1997). This increased variety at the level of major groupings, however, is not found at the **species** level. Of an estimated 20 million different kinds of organisms on Earth, fewer than 250,000 are marine, and most of these live in or on the bottom. The ocean is still a largely unknown realm, however, and scientists continue to discover new organisms. Since 1980, two new phyla (Loricifera and Cycliophora) have been described, as have two new classes in well-known phyla (Class Remipeda, a group of crustaceans, and a group of concentricycloid echinoderms) (Barnes and Hughes 1999). On December 21, 2001, a new deep-sea squid with 6-m-long (20 ft) tentacles was announced. The creature is known only from photographs and may be a member of a new family.

To bring some general order to the diversity of life in the seas, organisms are often classified according to size, mobility, and location in the water column. A few, by virtue of their buoyancy, are permanently at the surface, partly in the water and partly out. They are blown by the wind from place to place and are collectively known as the pleuston. The Portuguese man-of-war (*Physalia* spp.) and the by-the-wind sailor (*Velella* spp.), both cnidarians with gas-filled sacs, are prime examples. Another group of organisms is found on or under the surface film of the water and is called the nueston. The sea strider (*Halobates* spp.), an insect and relative of the pond strider of freshwater lakes, is the only marine organism that spends its life above the water. Other animals use the surface tension to hang under the surface film. The stalked barnacle (*Lepas fasciculatus*) is a crustacean member of the nueston that attaches itself to floating material. Two gastropod mollusks are other examples: *Ianthina* secretes a raft of froth it hangs on, while *Glaucus* produces bubbles of air in the gut to keep it afloat. Neuston animals are all carnivores and are largely limited to tropical waters (Angel and Harris 1977; Barnes and Hughes 1999).

Plankton is the general term given to all organisms that are suspended in the ocean, unable to swim laterally against the currents and tides. They vary in size from bacteria less than 0.1 μm in diameter to jellyfish more than a meter (3 ft) across the umbrella. The smaller forms are the basis of food chains in the oceans and are described in more detail later. Active swimmers are referred to as the nekton. They are large enough [20 mm–20 m (0.7 in–65 ft)] to move against the force of currents and tides and include members of many phyla and classes, including crustaceans, sharks, fishes, and mammals. Those fishes that live close to the bottom are called demersal and are distinguished from pelagic nekton that live higher in the water column.

Both macroalgae such as kelp and flowering plants such as the seagrasses attach to the substrate and are necessarily limited to the shallow waters where sunlight can penetrate to the sea floor. Some multicelled animals are confined to the benthic zone and live either on or in the bottom materials. Some are immobile or sessile and, at least as adults, are attached to hard surfaces. Examples include sponges, coral polyps, and barnacles. Other benthic animals are mobile and crawl or otherwise propel themselves over or through the substrate. Worms, seastars, anemones, mussels, and crabs are examples. The group of animals, and sometimes animals and plants together, that inhabits the bottom of the sea are collectively referred to as the benthos. Nektonic members of both the benthos and the pelagic **community** may be part of the plankton during their larval stages.

## A PLANKTON PRIMER

Various kinds of plankton are recognized. A particular organism in this group of floaters is called a plankter. Plankters may be classified according to size, **taxonomy**, or ecological role (Table 12.1).

In taxonomic terms, there are bacterioplankton, phytoplankton, protozooplankton, and metazooplankton. The last two together are known as zooplankton, the planktonic animals. Photosynthetic members of the plankton, the phytoplankton, are one-celled algae with chlorophyll and other light-sensitive pigments and **cyanobacteria**; these tiny organisms 0.2 to 200 μm in size are the chief producers of open-sea communities. The smallest predominate in tropical waters, the larger

**Table 12.1**
**Classification of Phytoplankters by Size**

| Size class | Length | Taxonomic composition |
|---|---|---|
| Ultraplankton | 0.02 μm to < 2 μm | chiefly monerans |
| Nanoplankton | 2 μm to < 20 μm | protists |
| Microplanton | 20 μm to < 200 μm | protists |
| Macroplankton | 200 μm to < 2000 μm | animals |
| Megaplankton | 2000 μm + | animals |

in temperate waters. They are restricted to the euphotic zone and provide food for other **life-forms**, either when consumed directly or when their cell contents leak out and create a pool of dissolved organic matter. (Some 40 percent of the cell contents may leak out of an algal cell. Organic matter is also leached out of dead cells. Together, the resulting dissolved organic matter in seawater is a significant food source for bacteria and some animals.) Consumers are made up of bacteria, fungi, protozoans (one-celled animals) and metazoans (multicelled animals) that range in size from foraminiferans to whales (Barnes and Hughes 1999; Smetacek 1988).

Being tiny has several advantages in the marine environment. A small, light organism sinks more slowly than a larger, heavier one. Small size thus helps photosynthesizing plankton stay in the euphotic zone and helps tiny consumers stay where the producers are. Small size correlates with rapid rates of reproduction and allows for a quick response to the highly variable and changeable conditions of light and nutrient load in the surface waters, especially when conditions become favorable. (An organism 10 μm in size has a generation time of 1 hour.) Energy is channeled into the formation of new individuals rather than the continued growth of a single cell or organism. Small size, usually combined with a nonspherical shape, maximizes surface area relative to volume and permits rapid, efficient uptake of nutrients from the water (Barnes and Hughes 1999).

Bacterioplankton consist of two basic types. Smaller (<1 μm), free-living forms consume dissolved organic matter. Larger bacteria are attached in clumps to particulate organic matter, debris that results from the death of phytoplankton and zooplankton, the leftovers from the meals of herbivores, feces, and discarded tissue such as molted exoskeltons. Bacteria are decomposers and mark the beginning of detritus food chains since they are eaten by zooplankton or are the recyclers that return mineral nutrients to the producers (Barnes and Hughes 1999; Smetacek 1988).

Phytoplankters range in size from 2–200 μm. Diatoms are the main group of nonmotile types. They are enclosed in rigid exoskeletons of glassy silica minerals that give them distinct shapes and make it possible to identify them rather easily to **genus** level. Diatoms provide food for most herbivorous zooplankters that have some type of sievelike feeding apparatus to filter particles from the water. They are particularly abundant and diverse in nutrient-rich places and seasons. A few diatoms, though, are adapted to nutrient-poor conditions; *Planktoniella sol*, for example, is an indicator of warm, stratified water. Most genera are found throughout the oceans of the world, but some species are restricted in distribution, as is the small-celled *Thalassiosira parthenia*, which is limited to the upwelling zone off southwestern Africa (Smetacek 1988).

Dinoflagellates are larger, motile phytoplankters able to move up the water column against gravity. They engage in vertical migrations of up to 10 m (30 ft) daily to rise to the euphotic zone for light and sink to lower depths to obtain nutrients. Two strategies have evolved to maximize absorption of nutrients. One is to be a small sphere, thereby maximizing surface area relative to volume. The other is to have a shape as far from spherical as possible. Larger forms often have all sorts of projections or strange shapes that increase surface area; these are common features among forms living in the low-nutrient waters of the tropics. Two whiplike flagella more or less perpendicular to each other provide for a spiraling swimming motion in some dinoflagellates, which allows them to rise through the water.

Phytoplankters have complicated life cycles that involve periods of rapid cell division or blooms and resting periods when the cells are encysted spores. Diatoms bloom in the spring in the midlatitudes. The "red tides" of spring, however, are dinoflagellate blooms. (These cells release toxic by-products of their metabolism; red tides are often associated with the poisoning of shellfish and sometimes the death of people who eat them.) So are the autumn blooms characteristic of temperate waters and that involve the genus *Ceratium* (Smetacek 1988). In the midlatitudes, succession of dominant phytoplankters is seasonal and depends upon the availability of nutrients, especially organic vitamins that these cells cannot synthesize themselves. One type of phytoplankter can bloom only after another type has conditioned the water with the needed vitamin. Barnes and Hughes (1999) describe a simple sequence in the Sargasso Sea, the center of the North Atlantic gyre, in which a diatom that requires an external source of Vitamin $B_{12}$ dominates during early spring. As this diatom multiplies and its numbers build up, it depletes the $B_{12}$ in the water, which causes its reproduction rate to decline. However, this diatom releases Vitamin $B_1$, a vitamin needed by the cocclithophore *Coccolithus*, which then blooms after the diatoms. The coccolithophore releases $B_{12}$, which allows for a second bloom of diatoms.

Protozooplankters, organisms such as ciliates like *Paramecium* and the amoeboid foraminiferans and radiolarians, may take up dissolved organic compounds directly or may consume bacteria as part of a detritus food web. It has been estimated that 60 percent of the energy flow in marine ecosystems passes along a so-called microbial loop: bacteria take up dissolved organic matter and are then consumed by flagellates and ciliates that release dissolved organic matter into the water that is taken up by bacteria and so on (Barnes and Hughes 1999). Other protozooplankters are herbivores that feed on the cells of phytoplankters. These have rapid reproduction rates and become very abundant during or after the spring blooms of diatoms, after upwelling events, and during the red tides and autumn blooms of dinoflagellates—in other words whenever their food is abundant. Some protozooplankters have symbiotic relationships with algae, such as the large foraminiferans that carry so-called gardens of dinoflagellates (Smetacek 1988).

The metazooplankton is a diverse group that can be divided into those that feed on suspended particles through sievelike apparatuses and those that selectively trap large particles with some type of raptorial or clawlike device. Most metazooplankters are suspension feeders. They include copepods; euphausids [large shrimplike crustaceans such as krill (*Euphausia superba*)]—the main food of penguins, seals, and whales in the Antarctic; thaliceans (large, gelatinous organisms); and the larvae of many polycheate worms, mollusks, echinoderms, decapod crabs, and barnacles that, as adults, are part of bottom communities. The raptorial metazooplankters include different copepods, the mostly tropical chaetognaths or arrow worms, cnidarians [the largest planktonic creatures, including jellyfish that in Arctic waters may grow up 2 m (6 ft) in diameter], ctenophores or comb-jellies, the larvae of fishes, and the larvae of such mollusks as nudibranches or sea slugs (Smetacek 1988).

Zooplankters often undergo daily vertical migrations. By means of cilia and flagella, smaller organisms move on average less than 400 m (1,300 ft), while larger forms may move 600 to 1,000 m (2,000–3,300 ft) up and down the water column. The typical pattern is for zooplankters to spend the daylight hours at depth and rise to the surface at dusk. By the middle of the night, they are dispersed throughout the surface waters but regroup at the surface before dawn. By sunrise, they have descended to depth to wait out the day. The reasons for this migration are not well understood, but it does provide a mechanism for organisms that cannot move horizontally to move to enriched phytoplankton patches. The surface few meters of water under the influence of winds may be moving at a different speed and direction than deeper waters. When zooplankters deplete the food in the surface layer in an evening's foraging, they can sink during the day to rise in a new part of the surface waters where the phytoplankton may yet be ungrazed. Phytoplankton is patchily distributed at the surface for reasons other than grazing pressure. The cells become concentrated due to what is known as Langmuir circulation. Persistent gentle winds cause the surface waters of the sea to roll as a series of long, parallel cylinders. Adjacent cylinders move in opposite directions to create streaks of phytoplankton and other particles where the cells become concentrated in alternating zones of upwelling or downwelling. Other phenomena that concentrate phytoplankton include the eddies that spin off the large boundary currents on the eastern edges of the continents. Eddies from the Gulf Stream may have a diameter up to 250 km (150 mi) and depths of 1,000 m (3,000 ft) and persist as a water mass as long as three years. Another place phytoplankters concentrate is at shelf fronts, the transition zone between shallow, well-mixed coastal waters and the stable, stratified waters of the deep sea (Barnes and Hughes 1999).

## NONPLANKTON ANIMAL LIFE

Nektonic and benthic organisms encompass a huge range of animal phyla. The vast majority are creatures of the benthos. These include sedentary poriferans such as sponges; cnidarians such as sea anemones, corals, sea pens, and hydroids; bivalve mollusks such as mussels and oysters; and crustaceans such as barnacles. More mobile benthic forms include echinoderms such as sea stars, brittle stars, and sea cucumbers; gastropod mollusks such as snails; cephalopod mollusks such as the octopus and giant squid; and crustaceans such as crabs and lobsters The pelagic zone is home to squid and shrimp and a host of vertebrates, particularly the cartilaginous fishes such as sharks; and the teleost or bony fishes, a huge array of which are distributed in and adapted to vertical zones from surface to bottom (see Chapter 14 for more details). Reptiles are represented by 4 wide-ranging sea turtles, 50 sea snakes, and the saltwater crocodile (*Crocodylus porosus*) that inhabits coastal waters of Southeast Asia and northern Australia. Although all birds nest on land, several spend much of their life at sea and land on its surface to rest and feed. Some are great divers and swimmers. The flightless penguins, a strictly Southern Hemisphere group, are especially well adapted to swimming after oceanic prey including squid, fish, and krill. Their counterpart in the Northern Hemisphere, the Great Auk (*Pinguinus impennis*), is now extinct, but smaller relatives such as puffins, murres, and auklets–though able to fly and come to land to roost and breed–depend upon the offshore reaches of the sea for life. Gannets, boobies, and tropicbirds are found in both hemispheres. They fly well out to sea from their rookeries on rocky coasts and plunge into the water from high above to snare their prey. Other birds—gulls, terns, oystercatchers, cormorants, and brown pelicans—are dependent on nearshore waters and coasts. In contrast to these land-based foragers of the sea, truly oceanic birds such as shearwaters, storm petrels and albatrosses, spend much of their lives on the wing catching updrafts from waves far out to sea.

Among marine mammals are the sirenians, the herbivorous dugongs and manatee, and the now extinct Stellar's sea cow. These are totally aquatic animals found in coastal waters, estuaries, and sometimes rivers. Dugongs and manatee are essentially tropical in distribution; the sea cow was the only member of the order to live in cold waters (Bering Sea). Pinnipeds (14 sea-lions, fur seals, and eared seals; walrus; and 19 true seals) must haul out on land or ice for breeding. All are carnivores that consume fish, squid, and crustaceans from the pelagic zone or animals of the benthos. The mammals most completely adapted to life in the sea are the cetaceans or whales, dolphins, and porpoises. Two groups are distinguished. The whale-bone whales have plates of baleen through which they sieve planktonic crustaceans and fishes from sea. The world's largest animal, the blue whale (*Balaenoptera musculus*), is a member of this group. The other whales, the toothed whales, include the sperm whale (*Physeter catodon*) that feeds on fish and squid, the narwhal (*Monodon monoceros*)—the unicorn of the sea—and the killer whale or orca (*Orcinus orca*) that eats seals, porpoises, and penguins but seems to prefer fish and squid. Porpoises and dolphins are also placed in the toothed whale category (Angel and Harris 1977; Nowak 1991).

## MARINE BIOMES

The ocean environment is divided into three major biomes in this book. As mentioned earlier in this chapter, the biome concept doesn't exactly fit the ocean realm. It is a classification system designed for terrestrial environments dominated by **vascular plants**. However, most textbooks on marine ecology and other works discuss the oceans in terms of three broad habitat types, even though they are all interconnected. These three habitats—the coastal or intertidal zone, the subtidal habitats of the continental shelf, and the deeps of the open sea—are here considered biomes. The coastal biome takes on different aspects depending upon the nature of the coastal substrate (i.e., rocky shores versus sandy shores) and latitude (salt marshes at the upper limits in temperate regions. mangrove forests in tropical areas, kelp forests in temperate waters, and coral reefs in tropical seas). These different ecosystems will be considered major regional expressions of the coastal biome. Similarly for the continental shelf biome, regional expressions will be determined according to substrate (soft-sediment benthic communities versus rock or otherwise hard-surface benthic communities) and temperature and nutrient loads in the pelagic environment (warm versus cold waters and upwelling zones). For the deep sea biome, the unique communities assembled at hydrothermal vents will be treated as a major regional expression.

## ORIGINS

It is generally believed that life on Earth first arose in the sea, so it is not surprising that many genera not known on land survive in the oceans and so-called living fossils are still occasionally discovered. Since the late Permian, tropical surface water apparently has not exceeded 33° C (91° F), and tropical conditions have persisted somewhere on Earth. Today, the upper temperature limit of tropical marine organisms remains approximately 35° C (95° F), which reflects their adaptation to a stable temperature regime. Although glacial episodes occurred 300 million years ago (near the Carboniferous/Permian boundary), the most significant cooling began in the Early Tertiary. With this began the seasonal changes that still affect the midlatitudes and high latitudes and the confinement of warm waters (>20° C or >68° F) to the tropical region. Early in the Tertiary, evidence indicates the range of temperatures between the poles and the equator was less than 5° C (9° F). Fifty million years ago, during the Eocene Epoch, waters in the North Pacific were still near 18° C (64° F). Global temperatures declined 38 million years ago (Eocene/Oligocene boundary) while Antarctic was still ice-free. A second drop in global temperatures occurred near the middle of the Miocene epoch, and by 14 million years ago, glaciers covered Antarctica. Cold-adapted organisms may have evolved then. Two million years ago, at the beginning of the Pleistocene Epoch, periods of glaciation began again, and ice accumulated, especially on the northern landmasses. The permanent ice cover of the Arctic Ocean, thought to be about 700,000 years old, formed during these Pleistocene ice ages. In terms of the history of life in the seas, the recent—in geological terms—accumulation of permanent ice at both poles suggests that cold-adapted marine organisms are much younger than tropical forms (Lüning and Asmus 1991).

Water temperatures are not the only factor that has not remained constant through geologic time. Circulation patterns in the world's oceans have changed as the continental masses moved in response to **plate tectonics**. Sea level has lowered and risen in response to the fluctuating climates and glacial and interglacial periods of the Pleistocene Epoch—sometimes exposing and sometimes flooding continental shelves. The patterns and consequences of these events are described in the chapters on marine biomes that follow.

## REFERENCES

Angel, Martin and Tegwyn Harris. 1977. *Animals of the Oceans, the ecology of marine life.* New York: Two Continents Publishing Group.

Barnes, R. S. K. and R. N. Hughes. 1999. *An Introduction to Marine Ecology.* 3rd edition. London: Blackwell Scientific.

Briggs, J. C. 1974. *Marine Zoogeography.* New York: McGraw-Hill.

Ekman, S. 1953. *Zoogeography of the Sea.* London: Sidgwick and Jackson.

Longhurst, Alan. 1998. *Ecological Geography of the Sea.* New York: Academic Press.

Lüning, K. and R. Asmus. 1991. "Physical characteristics of littoral ecosystems, with special reference to marine plants." Pp. 7–26 in A. C. Mathieson and P. H. Nienhuis, eds., *Intertidal and Littoral Ecosystems.* Ecosystems of the World, 24. Amsterdam: Elsevier.

Nowak, Ronald M. 1991. *Walker's Mammals of the World.* 5th edition, Vol. II. Baltimore: The Johns Hopkins University Press.

Ormond, Rupert F. G., John D. Gage, and Martin V. Angel. 1997. "Foreword, The value of biodiversity." Pp. xiii–xxii in Rupert F. G. Ormond, John D. Gage, and Martin V. Angel, eds., *Marine Biodiversity, Patterns and Processes.* Cambridge: Cambridge University Press.

Smetacek, Victor. 1988. "Planktonic characteristics." Pp. 93–130 in H. Postma and J. J. Zijlstra, eds., *Continental Shelves.* Ecosystems of the World, 27. Amsterdam: Elsevier.

# 12 COASTAL BIOME

## OVERVIEW

The coastal biome occurs in a zone of transition between land and sea. Also known as the littoral zone (from the Latin *litus* meaning shore), it extends from the highest reach of sea spray on the shore out to the depth where storm waves cannot disturb the sediments on the sea bottom (usually 60 m or 200 ft). Three life zones are evident within the coastal biome. The supralittoral zone is that land between the upper limit of sea spray and the upper limit of high tide—a zone physically affected by its proximity to the ocean but never submerged in it. The eulittoral zone, also known as the intertidal zone, is that region of the coast periodically exposed to the atmosphere and then submerged beneath the sea with the ebb and flow of the tides. Finally, the sublittoral zone or subtidal zone is always submerged and extends from the low-tide mark out to the 60 m (200 ft) depth line. (In geomorphology, this zone is referred to as the nearshore zone to distinguish it from the offshore part of the continental shelf where bottom sediments are not moved by wave action.) (Christopherson 1998; Nienhuis and Mathieson 1991).

In this book, subtidal communities are placed in the continental shelf biome (Chapter 13), but they could also be correctly considered part of the coastal biome. Here we limit the coastal biome to those communities directly affected every day by both the sea and the air, namely the sea spray and intertidal zones.

## THE COASTAL ENVIRONMENT

The shape and composition of coasts vary tremendously around the world, but certain aspects of the physical environment are common to nearly all; these include the mechanical force of waves and longshore currents and the fluctuations in water level caused by the tides. Organisms that live in the coastal biome must tolerate submergence in saltwater for varying lengths of time and exposure to the air, which means the possibility of drying out and the need to tolerate often rapid temperature

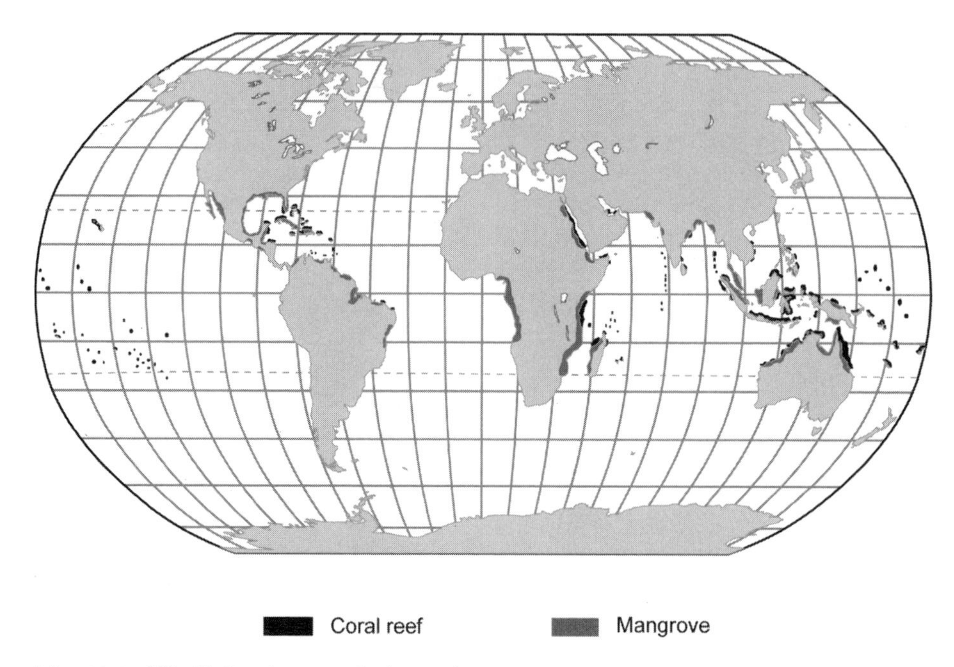


Map 12.1 World distribution of salt marsh, mangrove, and coral reefs.

change. If in the surf zone, they must also be able to bear (or avoid) the brunt of crashing waves. Organisms in the sea-spray zone are typically terrestrial forms tolerant of high concentrations of salt and a wet habitat.

The wind generates waves. The water in a wave actually moves in vertical circles or orbits rather than forward. Only the wave shape moves toward the coast to carry energy to the land. Water particles in circular motion at the sea's surface transfer energy downward; this causes deeper water particles to also orbit but in circles of decreasing diameter until most energy is dissipated at depths equal to about one-half the wavelength. When the sea bottom is the same depth as the lowest orbit, its sediments are stirred by the vertically circling water particles, and the boundary between nearshore and offshore is made. Sediments will be resuspended and nutrient levels often increased in the nearshore area due to the wave action (Shramm 1991). The sea bottom disrupts the circular orbits of the water and causes them to become elliptical and unstable. The result is that the wave gets steeper, becomes unstable, and eventually breaks apart—in other words, it becomes a breaker that crashes and rushes onto the shore. As the water moves to shore, it will carry sands and silts with it. Organisms living in the surf zone and on the coast have to contend with the force of moving water and often abrasion by the sediments within it. On shores with unconsolidated sediments, the substrate is always moving, being eroded from one location and deposited in another. Waves usually do not reach the shore with their crests parallel to the coast but arrive at some angle less than 90°. As they break, particles of sand and water are forced up the beach at this angle in the swash but then flow under gravity back toward the sea (the backwash) perpendicular to the coastline. This constant back-and-forth motion has a net effect of moving sediments and water parallel to the coast. The movement of sand in this manner on a beach is known as beach drift; the movement of water along the coast forms longshore currents.

Photo 12.1 The two main habitats of rocky coastlines: headlands bearing the full brunt of crashing waves, and bays and beaches swept by swash and backswash. (Photo by SLW.)

The bending of the waves as they approach land concentrates their force at headlands protruding into the sea and weakens it in coves and bays (Photo 12.1). Thus, side by side, two very different habitats are created: one exposed to the full force of the crashing waves and the other sheltered and experiencing only gentle swash and backwash movement. The headlands are places of erosion often characterized by steep cliffs; the bays are places of deposition frequently characterized by sandy beaches.

Tides are, of course, the phenomena that define the limits of this biome, and their expression around the world varies considerably. The gravitational pull of the moon and the sun on Earth and its oceans causes tides. This produces a bulge in the ocean on the sides of Earth facing the moon or sun and also on the opposite sides. The moon, because it is so much closer to Earth than the sun, exerts the greatest pull; the larger, more significant bulges are associated with the position of the moon. (When sun and moon are in a straight line during new-moon and full-moon phases, the combined effect gives the greatest bulges.) The Earth rotates through the two opposite bulges once every 24 hours and 50 minutes so that, theoretically at least, every coast experiences higher water levels (high tides) twice a day and lower water levels (low tides) twice a day. In actuality, most places do have two high tides and two low tides a day; due to the shape and orientation of the coast, however, a few places, such as the shores of the Gulf of California, experience only one high tide and one low tide each day. The tidal range—that is, the difference in elevation between the high-tide mark and the low-tide mark—also varies around the world. Some places, such as the coasts of the nearly enclosed Mediterranean Sea, have a tidal range near zero. In contrast, the Bay of Fundy between New Brunswick and Nova Scotia, Canada, experiences a tidal range of 16 m (52.5 ft), the greatest on Earth. Tidal range also varies according to the phase of the moon. The highest high tides and lowest low tides occur at full moon and new moon and are called spring tides. The lowest high tides and highest low tides (or smallest tidal range)—neap tides—occur with first-quarter and third-quarter moons. Intertidal organisms must be adapted to cope with these fluctuations.

The action of waves and the flood and ebb of tides create variation in the coastal environment. The nature of the substrate—whether the materials are soft and unconsolidated as along sandy shores or hard and solid as along rocky shores—is also of great importance to the life-forms of the coastal biome. Coasts constructed of boulders too large to be moved by wave action are much like rock-bound shores in terms of the living organisms inhabiting them. Gravel and shingle beaches, on the other hand, are constantly tumbled and moved by incoming waves, and little life is possible in these conditions.

Zonation of life-forms is a characteristic feature of all parts of the coastal biome. Major life zones related to tidal influences have already been mentioned, but within each of these zones, life also tends to be arranged in vertical belts according to distance up or down the shore. Vertical zonation is especially well developed and easily recognized on rock-bound shores; in soft sediments (sandy beaches or tidal flats), differences in the physical environment are noticeable, but much of the animal life consists of small invertebrates buried in the bottom materials and requires special sampling methods and identification in the laboratory to realize zonation patterns (Raffaelli and Hawkins 1996).

Latitude is also a factor in determining the environment of living organisms in the coastal biome. On arctic and antarctic shores, the intertidal zone is scraped clean of permanent life by ice and may be inhabited only by annual algae during the few weeks of summer. In nonpolar regions, the land-derived vegetation that fringes the sea in the uppermost parts of coastal zone varies according to latitude. Salt marshes are almost exclusively restricted to temperate climates; mangroves replace them in tropical regions. Coral reefs are a feature of clear, warm waters and only barely extend out of the tropics into the subtropics along those coasts where warm ocean currents maintain water temperatures above 20° C (68° F) in winter (Nienhuis and Mathieson 1991).

Coasts are in a constant state of change, whether time is marked by hours, days, years, centuries, millennia, or eons. Some have been submerged by rising sea levels or by the weight of glacial ice. Others have emerged as the land rises relative to sea level, such as when erosion reduces the weight of landmass or when warming climates melted ice and removed the weight of ice sheets. For thousands of years, humans have used the shore to exploit its food resources, establish settlements in positions of shelter or strategic importance, and use its harbors to anchor ships of commerce. More recently, recreational use has been added to the list. All these land uses alter the coastal environment, sometimes subtly and sometimes drastically.

## MAJOR EXPRESSIONS

### Rocky coasts

Vertical zonation is evident worldwide on rocky coasts (Photo 12.2). Marine ecologists have developed several classification schemes, none of which exactly coincide or are universally accepted. One developed by Stephenson and Stephenson (1949, 1972), however, is fairly commonly used and always acknowledged in scientific literature. It identifies three zones. The supralittoral fringe, which encompasses areas the highest tides and sea spray reach, corresponds to what some call the high shore. The eulittoral zone, which lies between mean high-tide and mean low-tide marks, may also be referred to as the intertidal zone, the midshore, or the barnacle zone. The Stephensons call the third and lowest zone on the shore the sublittoral fringe. It lies more or less between the mean low tide and extreme low tide marks. Because it is submerged much of the time, others consider the sublittoral fringe the uppermost part of the subtidal zone and may call it simply the low zone of the shore (Raffaelli and Hawkins 1996).

Photo 12.2 The vertical zonation of life on rocky coasts. (Photo © Brandon D. Cole/CORBIS.)

The supralittoral fringe or sea-spray zone is itself separated into clear vertical zones. The highest parts are occupied by salt-tolerant flowering plants, **lichens**, and/or mosses of terrestrial origin (Photo 12.3). Lichens, usually *Verrucaria* spp., and cyanobacteria (*Calothris*, *Lyngbya*, etc.) occupy the lower edge of the zone. The lichens and cyanobacteria are almost universal features on rocky coasts and impart a black-colored band just above the high-tide mark. They are absent only on antarctic coasts. Annual algae are present during the winter in temperate regions and in summer at high latitudes. They provide hues of brown or green. **Perennial** seaweeds are found only where wave action is strong and the air stays humid throughout the day (Raffaelli and Hawkins 1996; Russell 1991).

The most widespread, common, and conspicuous resident animals of the supralittoral zone are periwinkles (*Littorina* spp.). Other common inhabitants of the high shore are sea slaters (amphipods of the genera *Ligia* and *Megaligia*) and, in crevices, barnacles (*Chthamalus* spp.). Quick-moving grapsid crabs such as Sally Lightfoot (*Grapsus grapsus*), hermit crabs (*Calcinus* spp.), insects, spiders, birds, and small rodents all come to the area to feed (Raffaelli and Hawkins 1996; Russell 1991).

Photo 12.3 The sea spray zone at Marwick Point, Mainland, Orkney, UK. The flowering plant is thrift (*Armeria* sp.). Fulmars nest on the rockface in the background. (Photo by SLW.)

The eulittoral zone also has distinctly different organisms that inhabit its upper and lower parts, which reflects in large part the length of time the shore is exposed to the air between high tides (or, conversely, the length of time it is submerged). An organism's ability to survive under the alternating stresses of exposure and submergence control the upper limit of its distribution. Competition from other life-forms apparently controls the lower limit. Conditions in the upper zone are most severe, and fewer life-forms live there than in middle and lower zones. Typically there are brown algae (*Pelvetia* spp., *Fucus* spp., and others) with an understory of red algae such as *Catenella* spp. or *Gelidium* spp. Crustose brown algae form a surface layer. On Atlantic coasts, lichens such as *Verrucaria mucosa* and *Lichnia pygmaea* join algae. The lower zone may have a dense canopy of fucoid brown algae (*Fucus* spp.) and below that a short turf composed of many kinds of red algae including carragheen or Irish moss (*Chondrus crispus*). These give way in the subtidal zone to kelps (*Laminaria* and other genera—see Chapter 13) (Russell 1991; Menge and Branch 2001).

The main plankton feeders in the intertidal zone are attached or sedentary filter feeders. Barnacles (*Chthamalus* spp. in the Pacific; *Balanus* spp. and *Semibalanus* spp. in the Atlantic; *Chamaesipho* spp. and *Eliminus* spp. in the Australasian region; and *Tetraclita* spp. in the tropics) occur in a zone above bivalve mollusks such as mussels (*Mytilus* spp.) in cooler waters and oysters (*Crassotrea* spp.) in warm waters. Limpets (*Patella* and other genera) are the most important grazers; whelks can be important predators (Russell 1991; Menge and Branch 2001). Sea ducks are major predators of benthic animals in temperate and subpolar regions of the Atlantic. Common Goldeneyes (*Bucephala clangula*) visit rocky shores and feed on small invertebrates living in Irish moss. Oldsquaws (*Clangula hyemalis*) consume periwinkles and other snails (*Lacuna vinita*) in rocky areas and bivalves and other invertebrates in sandy areas. Common Eider (*Somateria mollissima*) and scoters (*Melanitta* spp.) specialize on mussels (Mathieson et al. 1991). When the eulittoral zone is submerged, fishes are present and they feed a number of sea birds such as puffins (*Fratercula* spp.) and Murres (*Uvia aalge*), members of the auk family that breed in dense colonies in or above the supralittoral zone (Mathieson et al. 1991).

## Soft-Sediment Coasts

Zonation on sandy shores is less evident than on rocky shores. Particle sizes are often graded, with coarser sands high on the beach and finer particles lower down. The slope is gentle, and a sandy beach often transitions into an intertidal mudflat. Gradients also occur in the water content of the sands or muds and the length of time of exposure/submergence organisms must contend with. In soft sediments, animals have an advantage not available on hard surfaces; they can burrow into the ground to avoid unfavorable conditions. In muddy habitats in particular, oxygen can be limited at low tide, and burrows can provide access to oxygenated water as well as protection. Many organisms adopt behaviors that reflect the daily rhythms the tides impose. They may retreat to burrows, reduce activity, close shells, or vacate the area entirely when conditions are less than optimal (Raffaelli and Hawkins 1996).

The high-tide level is marked by a line of wrack or dried seaweed, shells, and driftwood (Photo 12.4). For a short distance down the beach, sands dry out completely during low tide. This uppermost part of the shore may be considered the

Photo 12.4 Debris accumulated at the high-tide mark on a soft-sediment shore. The swash and backswash of waves produces an unstable substrate. Tides alternately expose organisms to both the atmosphere and inundation by the sea. (Photo by SLW.)

supralittoral fringe. Below is the midshore region, which undergoes the greatest changes each day. The sands become saturated while submerged; although they dry out somewhat when exposed, at low tide they are still damp. The sediments of the low-shore region, on the other hand, are saturated even at low tide. This part of the beach is within the low-tide swash zone. The supralittoral fringe may have a **cover** of salt-tolerant land plants such as glassworts (*Salicornia* spp.), salt marsh grasses, and, in the tropics, mangroves (see following). Vertical zonation of these plants is sometimes apparent, with the least tolerant farthest from the shoreline and the most tolerant farther down the shore. The midshore region on sandy beaches lacks large plants, but algae and bacteria live among the grains of sand. Diatoms and dinoflagellates migrate up to the surface as the tide goes out and remain there until a few hours before tide comes in again. This vertical movement is a response, in part, to the changing intensity of light; it lets the cells photosynthesize at maximum rates yet not be swept out to sea. The low shore may have some seagrasses (seagrasses are mostly subtidal plants), and individual seaweeds attached to fragments of shells or other hard surfaces torn from the subtidal zone may also be found (Raffaelli and Hawkins 1996).

The animal life of each zone is distinct and characterized by different assemblages of small invertebrates. The actual animals present vary according to geographic location. Amphipods such as beach fleas (*Orchestia platensis*) may inhabit the supralittoral fringe on beaches lacking salt marsh or mangrove, while isopods (e.g., *Eurydice pulchra*) dominate the midshore region. Animals of the low shore are typically more diverse and may include amphipods (different than those found in the supralittoral zone), ribbon worms and bristle (polychaete) worms, crabs, and bivalve mollusks. The composition of animal life varies considerably between sandy and muddy shores. For example, ghost crabs (*Ocypode quadrata*) inhabit sandy beaches, while fiddler crabs (*Uca* spp.) are restricted to mudflats. Overall diversity is much higher on mudflats than on beaches. At low tide, numerous shorebirds—especially waders such as sandpipers and plovers—probe the sediments for food. At high tide, predators of the sea come to feed. Shrimps, prawns, and juvenile crabs and flatfish consume invertebrates. Adult flatfish, such as founders and plaice, and rays also feed on submerged sandy beaches and mudflats (Raffaelli and Hawkins 1996).

## Salt Marsh

Salt marshes develop where salt-tolerant land plants become established on mudflats sheltered enough from wave action to allow the buildup of sediments (Photo 12.5). Cordgrasses (*Spartina* spp.) are widespread and are the pioneer or foundation plants that begin the formation of the marsh habitat. Their strong network of roots binds the fine particles of the substrate together, and their stems slow water flowing off the land to increase sedimentation. The shoreline becomes stabilized, which reduces the impact of waves. The stems and roots of the cordgrass also provide sites for mussels to attach, and the mussels further stabilize and fertilize the soil to benefit plants (Pennings and Bertness 2001).

In the saturated soils of the low marsh, decomposition of dead stems and leaves depletes free oxygen in the soils. **Anaerobic** bacteria are abundant and produce toxic sulfides, which the outgoing tide flushes from the marsh. Plants of the low marsh cope with the oxygen-poor soils through morphological adaptations. Roots

Photo 12.5 Salt Marsh. (Annie Griffiths Belt/CORBIS.)

near the soil surface help oxygenate deeper roots. Special tissues (aerenchyma) that serve as air passageways run from above ground to below-ground parts to let oxygen flow to root tissues. Oxygen that leaks from the plants helps oxygenate the substrate for invertebrates. Fiddler crabs, with burrows up to 30 cm (12 in) deep, also help aerate the soil (Pennings and Bertness 2001).

Plants of the salt marsh have developed means to tolerate high salt content lethal to most other plants. They may concentrate salts in certain tissues to be able to draw freshwater up into the plant. Succulence of leaves in plants such as the sea-lavenders (*Limonium* spp.) or of stems as in the glassworts or pickleweeds (*Salicornia* spp.) is often associated with such salt accumulation. Some plants are able to exclude salts at the roots; others excrete salts through special glands in their leaves. The salt accumulated in leaves is removed when the leaves are shed. (Many of the plants of the salt marsh have close relatives, with the same adaptations, in deserts. See Chapter 4.) The most saline parts of a marsh tend to be at the midshore levels. Higher up in the marsh, rainfall and runoff dilute and wash out salts. Less evaporation occurs in the lower parts, and receding tides can carry away excess salt. Midshore is where **succulent** plants such as glassworts and sea blites (*Suaeda* spp.) are most apt to be found. Midshore salinity will be reduced and plant and animal diversity increased as plants like marsh elder (*Iva frutescens*) and rushes (*Juncus* spp.) invade and shade the surface to lower evaporation rates (Pennings and Bertness 2001).

A salt marsh has relatively few herbivores: insects, Canada geese (*Branta canadensis*), and muskrats (*Ondatra zibethica*) are among the more common. Most energy produced by marsh plants enters the detritus food chain. Fungi partially decompose dead leaves, which marsh periwinkles (*Littorina* spp.) and amphipods then shred and fragment. On the surface, bacteria decompose dead leaves and stems, and deposit feeders such as fiddler crabs consume fragments. Filter feeders such as oysters (*Crassostraea virginica*) and ribbed mussels (*Geukensia demissa*) remove particulates from the water and deposit them as feces in the sediments, thereby moving nutrients from the sea into the marsh. Predators include killifish (*Fundulus* and other genera); needlefish (*Strongylura marina*); menhaden (*Brevoortia tyrannus*) and other fishes; blue crab (*Callinectes sapidus*); wading birds such as rails, egrets, and herons; and terrestrial mammals such as raccoons (*Procyon lotor*). Blue crab predation controls the distribution of other invertebrates in the marsh. For example, ribbed mussels can thrive only in the high marsh out of the reach of the crabs, which forage when the marsh is submerged; marsh periwinkles are confined to stands of plants tall enough for them to climb out of reach of the crabs during high tide (Pennings and Bertness 2001).

Salt marshes are highly productive communities, though ones of relatively low diversity. They perform important ecological services for adjacent land and sea ecosystems, which they serve as a transition. Salt marshes protect the immediate coast from erosion. They filter nutrients and sediments, often biodegrading or accumulating toxins that may flow from land into the ocean. The marsh and the tidal streams that cross it serve as nursery areas for shellfish and finfish important to commercial fisheries. Salt marshes are more or less limited to temperate regions. The scouring of the coast by ice in polar regions prevents their establishment; mangrove forests take their place in tropical areas.

## Mangrove

Mangrove refers to any tropical, salt-tolerant woody plant that occurs only in brackish coastal waters. The term mangal is sometimes used to denote the entire ecosystem that mangroves inhabit, but that, too, is often called mangrove or mangrove forest. More than 50 different plants (in 20–27 genera and 16–19 families, depending upon the taxonomic scheme applied) worldwide are considered to be mangroves. The greatest diversity occurs in Southeast Asia; the Caribbean has the lowest diversity. Two genera are found worldwide; others are limited to either the Indo-West Pacific region or to a western region consisting of the Atlantic and Pacific coasts of the Americas and West Africa (Richards 1996; Ellison and Farnsworth 2001).

Mangroves occur in wet tropical climates and dry ones. Limited by winter water temperatures below 20° C, mangroves do occur in the subtropics where warm currents are off shore. The most poleward record is 38° 45′ S along the coast of Australia. Elsewhere they are found as far north as 32° N in Bermuda, about 31° N in southern Japan, and 29° N in the Red Sea (Richards 1996).

With the exception of a palm (*Nypa fruticans*) and a large fern (*Acrostichum aureum*) in the Indo-Malaysian region, mangroves are woody shrubs and trees. They vary in size and form according to species and habitat conditions; they range from scrubby shrubs, especially in more highly saline situations, to trees over 40 m (130 ft) tall on the Malay Peninsula. All must contend with periodic and sometimes prolonged submergence in rhythm with the tides. The substrate is usually saturated, and soils are thus depleted of oxygen. Aerial roots are a common feature of mangroves and take the form of stilt roots or flying buttresses in the red mangroves (*Rhizophora* spp.) (Photo 12.6), fluted buttress in the genus *Pellicera*, knees in

Photo 12.6 Red Mangroves (*Rhizophora mangle*) at Key West, Florida, with aerial roots exposed at low tide. (Photo © Douglas Peebles/CORBIS.)

*Bruguiera*, and rows of short vertical projections (pneumatophores) from the lateral root system in the black mangrove (*Avicennia* spp.) and the genus *Sonneratia*. Aerial roots are dotted with pores (lenticels) that lead into tiny air passages (aerenchyma) that direct oxygen down into the roots system when the structures are exposed at low tide (Richards 1996). Another peculiarity of many mangroves (reported in nine genera from six families) is vivipary or live births. Seeds germinate and develop into small plants still attached to the parent plant. The plantlets eventually drop into the water and float, often for long periods of time, until they touch a mud bottom in a new location (Ellison and Farnsworth 2001).

High salinity is handled by the same means salt marsh and desert plants use. Salts may be excluded at the roots by specialized filtration systems or may be accumulated in certain areas of the leaves. Only plants in the genus *Avicennia* have salt-excreting glands. Mangroves vary in their salinity tolerance, and thus a zonation of plants is often apparent from high shore to low shore. In Florida, for example, red mangrove (*Rhizophora mangle*) forms the seaward fringe and black mangrove (*Avicennia nitida*) grows inland from it (Richards 1996).

Though restricted to habitats with brackish water, mangroves require freshwater. Mechanisms to conserve the freshwater taken into their tissues are necessary and include succulent or leathery foliage and a thick leaf cuticle. Many mangroves have shiny, dark green foliage clustered on the ends of branches. The gloss helps reflect strong tropical light, and the open canopy created allows sunlight to dapple the surface below (Ellison and Farnsworth 2001).

Although light penetrates through the canopy, the typical mangrove forest is single storied without vines or other midsized plants. Epiphytic orchids, mistletoes, or similar plants may grow. A host of algae and cyanobacteria grow on the roots, stems, and leaves and are important food for members of the benthic and pelagic communities (Ellison and Farnsworth 2001).

Few animals other than insects consume the foliage of mangroves. A rare exception is the proboscis monkey (*Nasalis larvatus*) of Borneo. Fungi growing on the green leaves are the main food of the mangrove periwinkle (*Littorania angulifera*). The roots and trunks provide attachment sites for filter feeders. In the intertidal zone, these will be primarily barnacles and mangrove oysters; in areas near the low-tide level, where they are always submerged, sponges and mangrove tunicates (*Ascidia* spp.) dominate. Mud-dwelling crabs (*Sesmaria* spp.) consume up to 80 percent of fallen mangrove leaves. Snails are common predators (Ellison and Farnsworth 2001).

Like salt marshes, mangrove stands are important nursery areas for many marine shellfish and finfish because of the great amount of small invertebrate prey available. The tangle of roots offers protection from predators. Mangroves also provide resting and breeding sites for birds and sometimes host large rookeries of egrets, ibises, roseate spoonbills (*Ajaia ajaja*), frigatebirds (*Fregata magnificens*), and others. Also like salt marshes, mangroves protect the shore from erosion by wave action and trap sediments to build the land out into the sea.

## Coral reefs

Coral reefs are major geological features constructed by living organisms (Photo 12.7). They cover approximately 2 million km$^2$ (775,000 mi$^2$), equivalent to 0.2 percent of the ocean's surface area and 15 percent of the sea floor at depths of

Photo 12.7 Coral reef in French Polynesia with both fringing and barrier reef construction. (Photo © Wolfgang Kaehler/CORBIS.)

0–30 m (0–100 ft). They are primarily tropical in distribution and are limited to clear (that is, nutrient poor), clean, warm (minimum winter temperature above 18° C or 68° F) waters. The world's longest is the Great Barrier Reef, a string of 2,500 individual reefs that parallels the east coast of Australia for some 2,600 km (1,600 mi) between 25° and 9° S (Alonghi 1998). In the Atlantic Ocean, coral reefs are primarily a phenomenon of the Caribbean Basin. Sediments streaming out of the Orinoco and Amazon rivers inhibit reef building south of 5° N, and winter temperatures are too cold for corals to survive north of southern Florida. Their development along most tropical coasts of Africa is prevented by the cold Guinea Current, upwelling off southwest Africa, and heavy sediment loads from the Congo and other rivers. In the Indian Ocean, coral reefs are found off East Africa south of Somalia, where seasonal upwelling produces water temperatures that are too cold. Upwelling also prevents reef development off the Arabian coast. In the Persian Gulf and southern India, high salinity is a limiting factor preventing coral growth. However, reefs are common in the Red Sea and throughout the rest of Indo-Pacific region east to the Hawaiian Islands (Achitur and Dubinsky 1990; Barnes and Chalker 1990).

Coral reefs as geologic features are categorized into three types: fringing reefs, barrier reefs, and atolls. Fringing reefs form close to shore on rocky coasts. Noncoral organisms are often more important than corals in the early stages of construction. The reef builds slowly seaward; the low-tide marks limits vertical growth. A shallow reef flat develops on the dead reef between the shore and the outer, living edge of the reef. The reef flat may be exposed during low tide. Corals will begin to proliferate on the outer edge, where they are always submerged. Barrier reefs have similar origins but become separated from the shore by a lagoon as the land slowly subsides. Corals compensate for the subsidence by building the barrier reef vertically so they can stay near the surface of the sea. Atolls are formed when a volcanic island a reef was initially attached to subsides below sea level, a process first described by Charles Darwin in 1842 (Barnes and Hughes 1999).

**Disturbance** is a frequent natural occurrence on coral reefs. It can be physical destruction by storms, earthquakes, volcanic eruptions, or sudden cooling of water temperature. Diseases and predation are other natural causes of reef destruction. At any given time, a reef is a mosaic of communities in various stages of recovery (Grigg and Dollar 1990).

Reef-building organisms include certain red algae (coralline algae); formaminiferans; sponges; polycheates; mollusks; and—especially—certain cnidarians, the stony corals (*Anthozoa* and *Scleratinia*) and stony hydrocorals (most important, the *Millepora*) that excrete calcium carbonate to construct shells and exoskeletons. The corals produce intricate skeletons that assume distinctive surface patterns and overall shapes, including branches, mounds, brains, fans, and so forth (Photo 12.8). The coral animals, or polyps, use stinging cells or nematocysts to capture zooplankters but also depend on organic material that algae living within their cells produce by photosynthesis. The polyps and the algae, dinoflagellates called xooanthellae, have a symbiotic relationship much like the alga and fungus that make up a lichen. Different xooanthellae are found at different depths; each uses the photosynthetic pigments best suited to absorb the wavelengths of light penetrating to the respective part of the water column and impart their color to the host corals (Knowlton and Jackson 2001).

Though most of a given reef, in terms of its mass, is composed of dead material, coral reefs are famous for the variety of organisms that form a thin layer in and on

Photo 12.8 Corals of several growth-forms clustered near the surface on an Australian reef. (Photo © Brandon D. Cole/CORBIS.)

them. They are without doubt the richest of marine ecosystems. One scientist says reefs are

> [s]o unique in their boundless diversity, intricate relationships, and spectacular beauty, that, along with the tropical rainforest, they demonstrate the upper boundaries which be reached in the evolution of life on Earth. (Dubinsky 1990, p. v)

However, the inhabitants are still poorly known. Of an estimated million or more reef species worldwide, fewer than 100,000 have been scientifically described (Knowlton and Jackson 2001). Many are invertebrates, of course. In addition to the stony corals and hydrocorals, there are other sessile or attached cnidarians such as horny corals (Gorgonacea), soft corals (Alcyonacea), zoanthids (Zoanthidea), thorny or black corals (Antipatharia), sea anemones (Actinaria), and corallimorps (Coralimorpharia). Other attached animals of a reef may include sponges, bryozoans, and tunicates (Knowlton and Jackson 2001).

One-celled and larger leafy or turf algae also grow on the reef; without constant cropping by mobile herbivores, they would overgrow and kill the coral polyps. Ecologists divide herbivores into two groups: grazers and browsers. Grazers crop close to the surface of the reef and consume the algae and part of the reef itself. Sea urchins (*Diadema* spp., *Echinometra* spp., and others) are most significant and exhibit unexplained periodic population explosions that can kill the reef. Other invertebrates are chitons and limpets with shovellike feeding apparatuses that let them excavate the coralline algae. Parrotfish (*Scarus* spp.) also scrape off pieces of coral. Browsers, on the other hand, have teeth that allow them to cut off the algae above the hard surface of the reef. Grapsid crabs and fish such as surgeon-fish (*Prionurus laticlavius*, *Acanthurus coerulus*, etc.), blennies, and a host of others are examples of the many browsers on a reef. Many other animals feed directly on the live coral polyps. These coral eaters include polychaete worms, nudibranches, sea urchins, sea stars, hermit crabs, damsel fish (*Stegastes* spp.), and pufferfish (*Arothron meleagris*). The most notorious coral predator is the crown-of-thorn starfish (*Acanthaster planci*), outbreaks of which destroy the living coral cover of reefs in the Indo-West Pacific region (Alonghi 1998; Glynn 1990).

Fishes, too numerous to list, exhibit strong diurnal rhythms in their presence on the reef. The daytime fishes—herbivores, carnivores, and omnivores—are absent at night, and the nighttime fishes—mostly consumers of zooplankton—are absent by day. Plankton eaters undergo vertical migrations; schools of damsel-fish and others feeding above the reef during the day converge toward the reef surface shortly before sundown. By or shortly after sunset, all day fishes have left the water column and retreated to shelters in the reef. For the next 20 to 30 minutes, the reef appears deserted. During this quiet period, the fish of the night collect at the mouths of the crevices where they have passed the day. Suddenly, they rush out to forage on and around the reef through the night. Before sunrise, the nocturnal fish leave the water and another quiet period passes as diurnal fishes assemble and then emerge to take over the reef once again (Montgomery 1990). The activity of reef fishes and abundance of other organisms attracts larger marine predators such as sharks, fishes, sea turtles, sea birds, and monk seals to coral reefs.

Both horizontal and vertical zonation are characteristic of coral reefs. The top of the reef is flat and characterized by shallow water that drains off at low tide through channels cut in the reef. Sediments may accumulate on top of the dead reef material. Seagrasses grow in soft sediments; leafy algae attach themselves on hard reef substrate. The corals that grow in this part of the reef tend to be small, branched types. The outer edge of the flat is marked by a ridge that extends above the extreme low-tide mark and may be encrusted with coralline algae (*Porolithon* spp.). These algae form a honeycombed limestone that absorbs the force of waves and offers refuge to small crabs, shrimps, cowries, and other animals. Sea anemones and sea urchins may occur in dense colonies. On the outer slope of the reef, corals of different growth-forms and other animals vary according to depth. Branched corals tend to give way to massive corals that then give way to flattened, leafy forms as depth increases; sponges, gorgonians, and nonreefing building corals become more and more important with greater depth. There are marked differences in zonation patterns and species geographically and between exposed and sheltered reef structures (Barnes and Hughes 1999).

Coral reefs act as lines of defense against wave erosion of seagrass and mangrove communities. Mangroves in turn protect the reefs from land-derived sediments that would limit the ability of the zooanthellae to photosynthesize. Corals receive nutrients from the xooanthellae growing in them, and in ways not well understood, the algae also seem to promote calcification and thus the building of the reef itself. Coral reefs flourish only in low-nutrient waters because waterborne nutrients stimulate the growth of fleshy, noncalcifying algae that can cover up the corals and the growth of phytoplantkers, increased numbers of which cloud the water and diminish the light energy reaching the zooanthellae in the coral. Increased amounts of phytoplankton also support larger populations of filter-feeding animals that can overgrow the coral. Nutrient pollution is believed to be a major cause of die-off or bleaching of reef corals (Knowlton and Jackson 2001).

## HUMAN IMPACTS

The list of human impacts on coastal communities is long. They range from outright habitat destruction to the more subtle effects of global climate change (see especially Alongi 1998; Raffaelli and Hawkins 1996; Steneck and Carlton 2001; Peterson and Estes 2001). Coastal development of port facilities, industry, cities, vacation homes, and marinas has cleared salt marshes. Aquaculture and mariculture, especially for shrimp, have great significance in the removal of mangroves, although in Ecuador and parts of Southeast Asia mangroves are also cut for timber and charcoal. To protect shoreline development, seawalls, jetties, and groins have been constructed; these alter the flow of water and usually result in the erosion of intertidal habitat (Peterson and Estes 2001). Suction dredging for shellfish and trawling for demersal fish disrupt soft-sediment bottoms.

Thermal pollution from the effluent of electric power plants can quickly degrade coastal communities. The warmed water entering tropical habitats such as coral reefs can be deadly; in temperate environments, it can eliminate brown algae (fucoids). Coral reefs are also hurt by eutrophication and the increase of phytoplankton that reduce the sunlight able to reach the symbiotic algae living in the coral polyps and other animals of the reef. Other stresses include sedimentation by materials eroded from the land; oil spills and the detergents used to clean them up; and damage caused by boat anchors, recreational snorkelers, and SCUBA divers who handle corals or walk on the reefs. The increasing frequency and expanding distribution of coral bleaching (the expulsion of zooxanthellae from coral polyps' tissues) suggests an environment under stress (Grigg and Dollar 1990; Peterson and Estes 2001).

Overcollecting of coral (dead and alive); shells; and fishes for food, bait, or curios threatens reef and other coastal communities. Trampling and anchoring of fishing and tourist boats are also very damaging to coral reefs. Seaweeds are harvested for food and fertilizer without adequate regulation in many places. The near loss of large grazing animals such as green turtle, hawksbill turtle, and manatee have helped shift Caribbean coastal communities from coral dominated to algae dominated (Steneck and Carlton 2001).

**Introduced** species are another problem in many coastal communities. People transport organisms from one coast to another accidentally on the hulls or in the ballast water of ships or deliberately for aquaculture, bait, or food for other organisms. The most abundant mollusks on the rocky coasts of the Gulf of Maine are the common periwinkle (*Littorina littorea*), a native of Europe. Other conspicuous members of the community are the European green or shore crab (*Carcinus maenus*) and Asian green algae (*Codium fragile*). European and Asian sea squirts and a North Pacific kelp bryozoan are also abundant (Steneck and Carlton 2001). The green crab has also invaded the west coast of North America, South Africa, Japan, and Australia. It is often blamed for the failure of the soft-shell clam fishery in New England and Nova Scotia. The common periwinkle is implicated in the shift from fast-growing green algae to slower-growing Irish moss along New England coasts (Menge and Branch 2001).

The effects of global climate change (i.e., warming) on coastal environments are related to increased water temperatures, potential changes in oceanic circulation and upwelling zones, changes in storm paths and annual precipitation amounts, and rising sea levels. Latitudinal shifts in the distribution of fishes have already been documented off the coast of North Carolina (Peterson and Estes 2001). There is concern that mangrove and salt marsh communities may not be able to keep pace with a rapidly rising sea level and will be eliminated.

To counteract some of the degradation and destruction of coastal environments that we have caused, efforts are under way to establish marine protected areas comparable to wilderness areas, national parks, and wildlife refuges on land. Some areas are set aside as storehouses of **biodiversity** and research sites, others are protected as unique landscapes-seascapes. The sites range from those fully protected to those managed for recreation to those managed for sustainable use of natural products and services (Palumbi 2001). How successful these will be in conserving the highly productive ecosystems of the coastal biome remains to be seen.

## ORIGINS

By the late Oligocene Epoch, creatures of rocky shores had originated on the American side of the Pacific Ocean. When the Bering Strait opened some 3.5 million years ago, during the Pliocene, a **dispersal** route from the Pacific through the Arctic Ocean into the Atlantic Ocean was created. In the northwest Atlantic, some 83 percent of mollusks on rocky intertidal habitats and 50 percent of organisms on sandy and muddy shores derive from Pacific ancestors. These include many seaweeds such as kelps (Laminariales) and red algae (e.g., *Chondrus*), barnacles (e.g., *Balanus*), certain sea urchins (*Stronglyocentratotus*), and sea stars (e.g., *Asterias*) (Vermiej 2001).

Many tropical coastal organisms can trace their history back to the Eocene Epoch. Mangroves, seagrasses, and corals were spread around the shores of the Tethys Sea, the ocean that at that time more or less circled the globe in the tropical latitudes. During the Miocene Epoch, continental drift caused widespread changes in ocean currents, continental shelf areas, and the distances between continental masses that influenced the distribution of tropical coastal organisms. About 19 million to 20 million years ago, land barriers began to separate the Indian Ocean and Mediterranean Sea. The plate on which Australia and New Guinea ride also began to limit circulation between the Indian and Pacific oceans. Tropical life-forms associated today with both coasts of the Americas developed in general isolation from the Indo-Western Pacific, separated by deep oceanic waters and cold currents. A connection between the eastern Pacific and western Atlantic continued until the middle of the Pliocene Epoch, when the isthmus of Central America formed and completed the land barrier of the American continents. Extinctions in the Pliocene greatly affected tropical coastal forms on both sides of the Americas but especially in the western Atlantic, where many of the larger suspension feeders disappeared (Vermeij 2001).

In general, mangroves trace to plants that grew on the shores of the Tethys Sea at the beginning of the Tertiary Period. Continental drift, the closure of the Tethys with the suturing of Africa and Europe, and global cooling in the late Cretaceous constricted the range of these plants to the tropics. Continental shifts during the Eocene and Miocene were associated with changes in oceanic circulation and climate and caused widespread extinction in the Caribbean and eastern Pacific regions. Subsequently, the formation of the Central American land barrier about 3 million years ago isolated Atlantic populations from those of the Indo-Pacific region. Mangroves then evolved into separate species in each basin. The biogeographic history of the corals has a similar pattern. Today, the highest diversity of mangroves and corals occurs in the Indo-West Pacific Region, particularly along the coasts of Indo-Malaysia (Barnes and Hughes 1999; Ellison et al. 1999; Ellison and Farnsworth 2001).

Coastal environments around the world were affected by the changes in temperature and sea level related to the repeated glaciations of the Pleistocene Epoch. The consequences of these appear to have been north-south shifts in distribution rather than extinction (Barnes and Hughes 1999). With the warming of the Holocene epoch, some of the range shifts are continuing. Most coral reefs were drowned by rapidly rising sea level in the Holocene, so those in existence today are less than 10,000 years old (Knowlton and Jackson 2001). Coasts are in continual flux due to erosion and deposition resulting from wave action and natural population fluctuations of grazing or predatory animals. Coastal communities often assume a mosaic pattern of plant and animal assemblages reflecting such short-term disturbances. They are also disturbed, with possibly more lasting consequences, by human activities (see previous discussion).

## HISTORY OF SCIENTIFIC EXPLORATION

The study of the coastal communities can be traced in part to the fashion of seaside vacations that arose in Victorian Britain in the early nineteenth century, when people on holiday became avid collectors of shells and other coastal life-forms. Naturalists began to recognize and describe the zonation patterns so characteristic of coastal life. By the end of the nineteenth century, science was becoming professionalized, and marine biologists focused on the shores at low tide for easy access to marine communities. Marine biology stations were set up around the world, beginning in Europe in 1839. Among several established in the United States, the earliest were the Woods Hole Marine Biological Laboratory (1888) in Massachusetts, and Stanford University's Hopkins Laboratory (1892) and the Scripps Institution of Oceanography (1905) in California. The Carnegie Institution's Tortugas Marine Laboratory at Dry Tortugas, Florida, (1905–39) was one of the first with access to tropical habitats (Rafaelli and Hawkins 1996; Montgomery 1990).

The coast became a training ground for scientists and an outdoor laboratory for ecological studies. Descriptive studies of rocky shores began after World War I, culminating in Stephenson and Stephenson's 1949 classic, "The Universal Features of Zonation between Tidemarks on Rocky Coasts." After World War II, experimental and quantitative ecology became more important, and many ideas and hypotheses basic to ecology were products of the study of coastal communities (Rafaelli and Hawkins 1996). For example, after studying barnacles, Joseph H. Connell (1961a, 1961b) promoted competition and predation as having major roles in shaping local distribution patterns of organisms. Based on studies of the predatory seas star (*Pisaster* sp.), Robert T. Paine (1969) proposed the concept of a keystone species, one that has an influence on the structure of the whole community far greater than its abundance might suggest (Menge and Branch 2001). Early and influential ideas on energy and nutrient flows stemmed from work in salt marshes (e.g., Teal 1962; Odum and de la Cruz 1967) (Pennings and Bertness 2001).

Charles Darwin (1842) first described the origin and development of coral reefs and identified them as fringing, barrier, or atolls, but scientific studies of reef life really began in the twentieth century. Before the nineteenth century, there had been confusion as to whether corals were plants or animals (Falkowski et al. 1990). Observing the life and activity on coral reefs required the development of simple diving gear. In 1819 and 1837, Augustus Siebe introduced several forms of a hard-hat diving suit compressed air was pumped into. William James provided a copper or leather helmet with a window attached to a reservoir of compressed air in 1825. In 1865, greater mobility was achieved when two Frenchmen, Benoit Rouquayrol and Auguste Denayrouze, invented an air tank that could be refilled at the surface and that a diver could temporarily detach from. This foreshadowed the modern aqualung or SCUBA (Self-Contained Underwater Breathing Apparatus) developed by Jacques Cousteau and Emile Gagnan in France during World War II. SCUBA gear gave total independence to the diver. All sorts of research equipment was then adapted to underwater use (Montgomery 1990). Underwater photography became available in 1908 and was in use along with the diving helmet by William H. Longley and others by 1915 for field research in the Caribbean. Longley's studies resulted in a classic work on the comparative ecology of reef fishes (Longley and Hildebrand 1941) published after his death (Montgomery 1990). The first major research expedition to study the workings of a coral reef was the Great Barrier Reef Expedition of 1928–29. Such things were studied as the exchange of nutrients between corals and water and the role of xooanthellae, a term coined in 1883. Study of reef metabolism continued elsewhere with Howard T. Odum and Eugene P. Odum (1955) publishing the results of their work on Eniwetok Atoll in a classic study in 1955 (D'Elia and Wiebe 1990). Much work is yet to be done in understanding the complexity of reef systems, the habits of their occupants, and their vulnerability to environmental stresses—natural and manmade. Even today, much research is limited to the shallow waters conveniently reached with snorkel and SCUBA gear.

## REFERENCES

Achitur, Yair and Zvy Dubinsky. 1990. "Evolution and zoogeography of coral reefs." Pp. 1–9 in Z. Dubinsky, ed., *Coral Reefs.* Ecosystems of the World, 25. Amsterdam: Elsevier.

Alongi, Daniel M. 1998. *Coastal Ecosystem Processes.* Boca Raton: CRC Press.

Barnes, D. J. and B. E. Chalker. 1990. "Calcification and photosynthesis in reef-building corals and algae." Pp. 109–131 in Z. Dubinsky, ed., *Coral Reefs.* Ecosystems of the World, 25. Amsterdam: Elsevier.

Barnes, R. S. K. and R. N. Hughes. 1999. *An Introduction to Marine Ecology.* 3rd edition. Oxford: Blackwell Scientific.

Christopherson, Robert W. 1998. *Elemental Geosystems.* 2nd edition. Upper Saddle River, NJ: Prentice-Hall.

Connell, J. H. 1961a. "The influence of intra-specific competition and other factors on the distribution of the barnacle *Chthamalus stellatus.*" *Ecology* 42: 710–723.

Connell, J. H. 1961b. "Effects of competition, predation by *Thais lapillus,* and other factors on natural populations of the barnacle *Balanus balanoides.*" *Ecological Monographs* 31: 61–104.

Darwin, Charles R. 1842. *The Structure and Distribution of Coral Reefs.* London: Smith Elder and Co.

D'Elia, Christopher F. and William J. Wiebe 1990. "Biogeochemical nutrient cycles in coral-reef ecosystems." Pp. 49–74 in Z. Dubinsky, ed., *Coral Reefs.* Ecosystems of the World, 25. Amsterdam: Elsevier.

Dubinsky, Z. 1990. "Preface." Pp. v–vi in Z. Dubinsky, ed., *Coral Reefs*. Ecosystems of the World, 25. Amsterdam: Elsevier.

Ellison, Aaron M. and Elizabeth J. Farnsworth. 2001. "Mangrove communities." Pp. 423–442 in Mark D. Bertness. Steven D. Gaines, and Mark E. Hay, eds., *Marine Community Ecology*. Sunderland, MA: Sinauer Associates, Inc.

Ellison, A. M., E. J. Farnsworth, and R. E. Merkt. 1999. "Origins of mangrove ecosystems and mangrove biodiversity anomalies." *Global Ecology and Biogeography Letters* 8: 9–115.

Falkowski, Paul G., Paul L. Jokiel, and Robert A. Kinzie III. 1990. "Irradiance and coral." Pp. 89–107 in Z. Dubinsky, ed., *Coral Reefs*. Ecosystems of the World, 25. Amsterdam: Elsevier.

Glynn, Peter W. 1990. "Feeding ecology of selected coral-reef macroconsumers: Patterns and effects on coral community structure." Pp. 365–400 in Z. Dubinsky, ed., *Coral Reefs*. Ecosystems of the World, 25. Amsterdam: Elsevier.

Grigg, Richard W. and Steven J. Dollar. 1990. "Natural and anthropogenic disturbance on coral reefs." Pp. 439–467 in Z. Dubinsky, ed., *Coral Reefs*. Ecosystems of the World, 25. Amsterdam: Elsevier.

Knowlton, Nancy and Jeremy B. C. Jackson. 2001. "The ecology of coral reefs." Pp. 395–422 in Mark D. Bertness. Steven D. Gaines, and Mark E. Hay, eds., *Marine Community Ecology*. Sunderland, MA: Sinauer Associates, Inc.

Longley, W. H. and S. F. Hildebrand. 1941. "Systematic catalogue of the fishes of Tortugas, Florida, with observations on color, habits, and local distribution." *Papers of the Tortugas Laboratory* 3: 1–331.

Mathieson, Arthur C., Clayton A. Penniman, and Larry G. Harris. 1991. "Northwest Atlantic rocky shore ecology." Pp. 109–191 in A. C. Mathieson and P. H. Nienhuis, eds., *Intertidal and Littoral Ecosystems*. Ecosystems of the World, 24. Amsterdam: Elsevier Science Publishers.

Menge, Bruce A. and George M. Branch. 2001. "Rocky intertidal communities." Pp. 221–252 in Mark D. Bertness. Steven D. Gaines, and Mark E. Hay, eds., *Marine Community Ecology*. Sunderland, MA: Sinauer Associates, Inc.

Montgomery, W. Linn. 1990. "Zoogeography, behavior, and ecology of coral-reef fishes." Pp. 329–364 in Z. Dubinsky, ed., *Coral Reefs*. Ecosystems of the World, 25. Amsterdam: Elsevier.

Nienhuis, P. H. and A. C. Mathieson. 1991. "Introduction." Pp. 1–5 in A. C. Mathieson and P. H. Nienhuis, eds., *Intertidal and Littoral Ecosystems*. Ecosystems of the World, 24. Amsterdam: Elsevier Science Publishers.

Odum, E. P. and A. de la Cruz. 1967. "Particulate organic detritus in a Georgia salt marsh-estuarine system." Pp. 383–385 in G. H. Lauff, ed., *Estuaries*. AAAS Publication 83. Washington, DC: AAAS0.

Odum, H. T. and E. P. Odum. 1955. "Trophic structure and productivity of windward coral reef community on Eniwetok Atoll." *Ecological Monographs* 25: 292–320.

Paine, R. T. 1969. "The *Pisaster-Tegula* interaction: Prey patches, predator food preferences and intertidal community structure." *Ecology* 50: 950–961.

Palumbi, Stephen R. 2001. "The ecology of marine protected areas." Pp. 509–530 in Mark D. Bertness. Steven D. Gaines, and Mark E. Hay, eds., *Marine Community Ecology*. Sunderland, MA: Sinauer Associates, Inc.

Pennings, Steven C. and Mark D. Bertness. 2001. "Salt marsh communities." Pp. 289–316 in Mark D. Bertness. Steven D. Gaines, and Mark E. Hay, eds., *Marine Community Ecology*. Sunderland, MA: Sinauer Associates, Inc.

Peterson, Charles H. and James A. Estes. 2001. "Conservation and management of marine communities." Pp. 469–507 in Mark D. Bertness. Steven D. Gaines, and Mark E. Hay, eds., *Marine Community Ecology*. Sunderland, MA: Sinauer Associates, Inc.

Raffaelli, David and Stephen Hawkins. 1996. *Intertidal Ecology*. London: Chapman & Hall.

Richards, P. W. 1996. *The Tropical Rain Forest.* 2nd Edition. Cambridge: Cambridge University Press.

Russell, G. 1991. "Vertical distribution." Pp. 43–65 in A. C. Mathieson and P. H. Nienhuis, eds., *Intertidal and Littoral Ecosystems.* Ecosystems of the World, 24. Amsterdam: Elsevier Science Publishers.

Shramm, W. 1991. "Chemical characteristics of marine littoral ecosystems." Pp. 27–42 in A. C. Mathieson and P. H. Nienhuis, eds., *Intertidal and Littoral Ecosystems.* Ecosystems of the World, 24. Amsterdam: Elsevier Science Publishers.

Steneck, Robert S. and James T. Carlton. 2001. "Human alterations of marine communities: Students beware!" Pp. 445–468 in Mark D. Bertness. Steven D. Gaines, and Mark E. Hay, eds., *Marine Community Ecology.* Sunderland, MA: Sinauer Associates, Inc.

Stephenson, T. A. and A. Stephenson. 1949. "The universal features of zonation between tidemarks on rocky coasts." *Journal of Ecology* 38: 289–305.

Stephenson, T. A. and A. Stephenson. 1972. *Life between Tidemarks on Rocky Shores.* San Francisco: W. H. Freeman.

Teal, J. M. 1962. "Energy flow in the salt marsh ecosystem of Georgia." *Ecology* 43: 614–624.

Vermeij, Geerat J. 2001. "Community assembly in the sea. Geologic history of the living shore biota." Pp. 39–60 in Mark D. Bertness. Steven D. Gaines, and Mark E. Hay, eds., *Marine Community Ecology.* Sunderland, MA: Sinauer Associates, Inc.

# 13 CONTINENTAL SHELF BIOME

## OVERVIEW

Geologically, continental shelves are parts of continents that extend below sea level to depths of 100 to 200 m (325–650 ft) [world average is –132 m (–430 ft)]. Continental shelves slope gradually seaward toward the continental margin, where an abrupt break in slope marks the edge of a continent is and forms a steep escarpment, the continental slope. The continental slope extends to depths of 3,000 m (10,000 ft) or more, where continental rock meets the oceanic rock that composes the sea floor.

Continental shelves underlie about 7.5 percent of the ocean's surface area and hold about 0.15 percent of its total volume of seawater. The average width is 78 km (48 mi), but this number obscures the great range found worldwide. Some coastlines have no continental shelf, while in other areas the shelf extends some 1,500 km (930 mi). The width of the shelf is in part a function of the position of a coastline relative to the direction of plate movement (continental drift). If two plates are converging near the coastline, as is the case along the west coasts of the Americas, the shelf is very narrow [less than 6.4 km (4 mi)]. If, on the other hand, the coastline is on the trailing edge of a moving plate, as the east coast of North America is, shelves are typically much wider. Africa, because it was uplifted during the Tertiary Period, has a narrow shelf on all coasts (Eisma 1988; Potsma 1988).

The continental shelf biome occurs on and above the shelves from the low-tide mark (or beginning of subtidal zone) seaward to the shelf edge or continental slope. This is the neritic zone of the ocean, and it is positioned between and influenced by the coastal zone and the deep sea. Freshwater and nutrients from the land pass into the waters of the continental shelf directly from surface runoff. Many of the animals of the shelf biome spend part of their life cycle in sheltered coastal waters. The shelf also exchanges water, nutrients, and organisms with the deep sea, although the two regimes may be surprisingly cut off from each other for months at a time. Tidal action continually mixes the waters of the continental shelf, but the

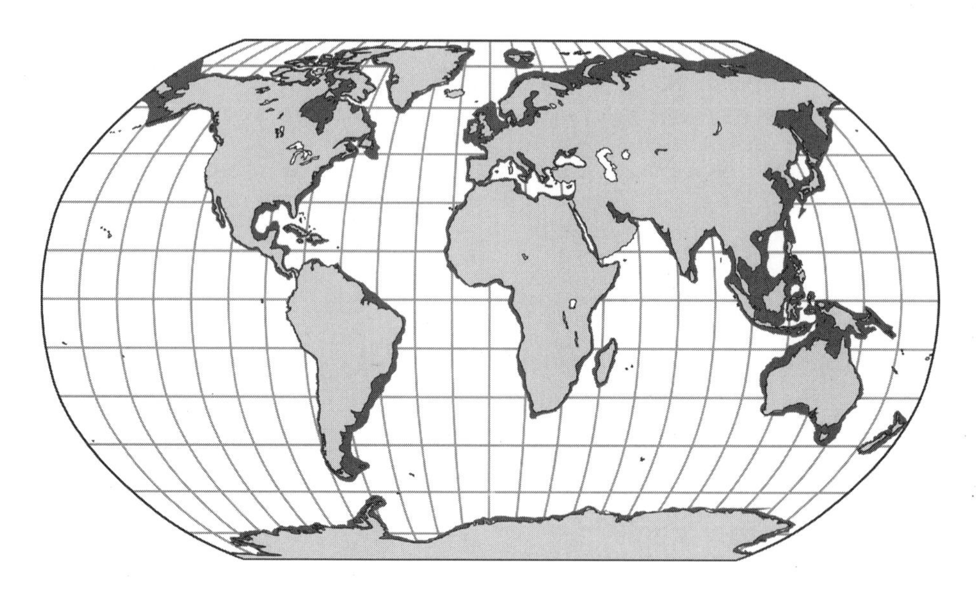

Map 13.1 World distribution of continental shelves.

waters of the deep sea become stratified in the **tropics** and during summer in the **temperate** parts of oceans. This fosters the development of distinct water masses—characterized by different phytoplankton and zooplankton—that maintain their integrity for considerable periods of time, just as in the atmosphere distinct air masses develop. The seaward edge of shelf waters is a transition zone between unstratified and stratified water masses that is called the shelf front. It is a barrier to the mixing of shelf and deep seawaters. However, the front can be dissipated or overcome when tropical storms and monsoons destroy the stratification in tropical oceans or when cooling during **midlatitude** autumn and winter destroys the stratification in temperate waters. Mixing also occurs at the edge of the shelf in regions of upwelling or downwelling. Upwelling occurs where the shelf is narrow and winds blow parallel to the coastline or offshore. Surface waters are displaced from the shelf and deeper, colder, and nutrient-rich water from beyond the shelf moves shoreward. Downwelling occurs when shelf waters are denser than those of the adjacent deep sea and water flows down off the shelf. This happens with cold water that forms in winter in places such as the Irish Sea and in warm, dry **climates** where high evaporation rates increase salinity and therefore water density, as happens in the Persian Gulf, where salinity can rise to 41 ppt (Potsma and Zijlistra 1988).

In shallower areas of the shelf, close to the coast, tides and waves mix the waters and stir up the bottom sediments. As a consequence of constant turbulence, the finest particles are held in suspension, and only sands and large particles settle out. The floor of a continental shelf is typically sand covered with scattered patches of rock surfaces exposed by currents. Tides return nutrient particles from the bottom to the water column; the result is higher productivity over continental shelves than in the deep sea. Nutrients tend to be most concentrated where currents flowing in opposite directions come into contact, such as on the Grand Banks where the Labrador Current and Gulf Stream meet. Phytoplankters, mostly diatoms and dinoflagellates, are larger in shelf waters, and food chains tend to be shorter than in the deep sea. The shorter food chains let more energy flow to predatory fishes and support larger populations than possible in the deep sea. The production of benthic organisms is probably as great or greater than that of pelagic organisms. The continental shelf biome yields 90 percent of the world's catch of shellfish and finfish (Potsma 1988; Potsma and Zijlistra 1988).

Shelf **habitats** and communities of living organisms are quite variable in terms of both location and time of year. Habitats vary according to water depth, type of substrate, and position relative to the coast or the shelf edge. For example, temperature varies seasonally and according to the amount mixing. The greatest temperature ranges are closest to shore; in waters more than 70–90 m (230–290 ft) deep, the annual range may be less than 1° C (2° F) (McLusky and McIntyre 1988). Soft sediments and rocky surfaces offer different opportunities for the benthos; in the former animals can burrow; in the latter they can attach themselves to solid foundations. Nutrients tend to be more abundant close to shore, where daily tidal mixing occurs, and at the edge of the shelf, where upwelling or a shelf front concentrates nutrients.

An especially productive habitat on continental shelves is that of an estuary. An estuary is a specific type of coastal inlet where a river enters the sea and freshwater and saltwater meet. Estuaries result from a variety of geologic processes. Many along the east coast of the United States, such as the Chesapeake Bay, are the flooded river valleys. The valleys were carved during episodes of lowered sea level in the Pleistocene Epoch and then inundated by the sea as climates warmed, ice sheets melted, and sea level rose. Where drowned valleys were enlarged by glaciers, as in British Columbia, southern Chile, Greenland, or Scandinavia, they are known as fjords. Estuaries can also be formed by the development of offshore barrier islands such as along the U.S. Gulf Coast or Albemarle and Pamlico sounds on the East Coast, or by faulting, as is the case for San Francisco Bay (Klee 1999). The estuarine environment is characterized by salinity gradients from the head of the estuary, where the river enters, down to the mouth, where it joins the ocean. Salinity gradients also occur from surface to bottom with freshwater flowing outward to the sea on top of a wedge of saltwater moving in from the ocean. These waters mix in varying circulation patterns, depending upon the relative strengths of river flow versus tidal flow. Where river flow is greater, a stratified system with freshwater on top and a wedge-shaped mass of saltwater at depth prevails. Where tidal flow is stronger than river flow, the water column becomes well mixed from turbulence related to the ebb and flow of the tide. Between these two extremes, partial mixing creates a vertical gradient from freshwater at the surface through brackish water at intermediate depths to saltier water on the bottom. A given estuary may experience different patterns according to the change of tides and to the change of seasons as evaporation rates and amounts of precipitation vary (Klee 1999; Rafaelli and Hawkins 1996).

Gradients also occur in particle size on the bottoms of estuaries. Generally, finer muds and silts are deposited at higher shore levels where slackwater at high tide lets them settle out. Coarser sands accumulate in areas of high wave energy or strong currents that keep finer particles in suspension. The nature of the substrate, as well as the degree of salinity, are important in determining the distribution patterns of estuarine life (Rafaelli and Hawkins 1996). Estuarine **ecosystems**, in particular the seagrasses meadows, are important nursery areas for coastal and oceanic shellfish and finfish.

## PLANT LIFE

Plants can live attached to the bottom in only the shallowest waters of the continental shelf, close to the coast. Seagrasses, flowering plants from six families (but none of them a true grass), grow in the subtidal zone on soft sediments and are absent only in the polar regions. They generally can occur to depths of about 15 m (50 ft) but can occur to 30 m (100 ft) in very clear water. The seagrasses host epiphytic algae. Other algae grow between the grass shoots and in the surface sediments. Large attached seaweeds, especially kelps (*Laminaria* and *Macrocystis*), thrive in cool waters on hard substrates in the **temperate zone** and sometimes form underwater forests. Seaward, in deeper waters, the euphotic zone does not extend deep enough to support attached plants, and phytoplankters, especially diatoms and dinoflagellates, are the sole producers of food. Most of the biome lies below the euphotic zone and is without chlorophyll-containing plants (Barnes and Hughes 1999; McLusky and McIntyre 1988).

## ANIMAL LIFE

Animals can be broadly categorized as benthic or pelagic. Since so much of the biome lies below the euphotic zone, with the exception of the zooplankton, animal life is dependent upon a constant rain of nutrients from above: sinking plankton, feces, or—if close to shore—materials washed from the land.

Benthic communities are composed of a wide range of organisms usually categorized as either epifauna (those animals that live on the surface of the substrate) or infauna (those animals that live in the bottom sediments). The smallest are bacteria and protozoans such as foraminiferans, and tiny ($<0.5$ mm) metazoans such as copepods, nematodes, and cnidarians. These are fed upon by juvenile shrimp, crabs, and fishes. Most inhabitants of the benthos are larger. They represent a broad array of animal groups, but most communities are dominated by echinoderms, polychaetes, crustaceans, and mollusks. Some, such as hydroids and sea anemones, are sedentary; others, such as scallops, are quite mobile. Most polychaetes, many bivalve and gastropod mollusks, and crustaceans such as lobsters are deposit feeders: they remove food particles from the bottom materials. A few, such as some of the bivalves, are filter feeders; they extract food particles from the water column. Demersal fishes such as flounders, cod, and haddock may also be considered part of the benthos (McLusky and McIntyre 1988).

Most fishes in the neritic zone are migratory and move to spawning or feeding grounds. Among them are some that reproduce and undergo their early development in freshwater, the anadramous fishes such as salmon (*Salmo* spp.) and American shad (*Alosa sapidissima*). The opposite also happens. Catadramous fishes such as eels (*Conger* spp.) reproduce at sea but spend most of their lives in freshwater. Atlantic menhaden (*Brevoortia tyrannus*) spawns at sea; larvae then move into estuaries to mature. Other fishes spawn in the seagrass communities of the coastal zone and spend their mature years at sea.

Depending upon their main food items, fish are categorized as primary, secondary, or tertiary predators. Small fish such as herring that feed on herbivorous zooplankton are primary predators. They tend to move between patches of high plankton production. The fish that feed upon herring and other primary predators are secondary predators. They include fishes such as skipjacks and mackerels that swim in schools. The tertiary predators, such as barracudas and marlin, may be solitary and more or less sedentary and let continually moving schools of prey fish come to them; or they may be very mobile—such as the tunas, travel in schools, and range widely over their habitat in search of prey (Sharp 1988). Other pelagic predators of the continental shelf include sharks, dolphins, and whales.

# MAJOR REGIONAL EXPRESSIONS

## Seagrass Meadows and Other Soft-Sediment Habitats

Seagrasses are marine flowering plants that grow in shallow, soft-bottomed bays and estuaries around most of the world. They consist of some 50–60 **species** that **reproduce asexually** by **rhizomes**. In the tropics and subtropics, several species share dominance in a seagrass meadow, but in the temperate waters of the Northern Hemisphere usually eelgrass (*Zostera* spp.) dominates. Widgeongrass (*Ruppia maritima*) may grow with the eelgrass. Many algae, both microscopic **epiphytes** and macroalgae fastened to rocks and shells on the bottom, also occur. Among the plants grasping hard surfaces are large, canopy-forming brown algae called rockweeds (*Cystoseria* spp.); kelps (*Egregia* spp. and *Laminaria* spp.); and some flowering plants, the surfgrasses (*Phyllospadix* spp. and *Amphibolis* spp.). In the tropics, large green algae occur. They, along with brown and red algae, often break off and form large drifting mats (Williams and Heck 2001).

The seagrasses and seaweeds give a complex physical structure to the seagrass meadow. Above the sea bottom is an assemblage of plant material of different heights and shapes. Seagrasses tend to grow in patches rather than as a continuous lawn, and this provides an element of horizontal structural variety to the **landscape**. In the bottom sediments, a thick mat of interwoven roots and rhizomes may extend several meters below the surface. Different seagrass species send roots to different depths, which provides structural variation within the sediments (Williams and Heck 2001).

Animals in every major group can be found below or above the sediments of the seagrass meadow. They enmesh themselves in the mat of roots and rhizomes; attach themselves or float close to the drifting mats; or, if they are mobile animals such as snails, crabs, and fish, they crawl and swim through or above the canopy. The sea grasses serve as attachment sites for invertebrates such as sea squirts or tunicates, hydroids, bryozoans, and sea anemones. Slipper shells (*Crepidula fornicata*), stacked one upon another, also cling to the seagrasses to feed upon passing plankton. At the bases of the seagrasses, where sediments are firmer, mud worms and limy tube worms can be found. These animals support a host of predators including isopods, tube-building and free-living amphipods, tiny snails, sea slugs, shrimps, crabs, and fishes. Among resident fishes in the Chesapeake Bay are sea horses, pipefish, and sticklebacks (Lippson and Lippson 1984).

Although relatively low in species diversity, seagrass meadows are known for their exceptionally high productivity and are generally considered to be very important nursery areas that provide food and protection for juvenile finfish and shellfish. The latter characteristic may be more typical of tropical and subtropical meadows than midlatitude ones. Much of the production of new biomass is unconsumed by grazers and passes to the detritus food chain and its invertebrate deposit feeders, filter feeders, and organisms of decay. Nonetheless, there are (or were) important large vertebrate grazers that depend on the seagrasses for food and habitat, including green sea turtles (*Chelonia mydas*), dugongs (*Dugong dugon*), and manatees (*Trichetus manatus*). Grazing invertebrates feed to a large degree on the epiphytic algae. Seagrass meadows support large populations of migratory waterfowl, including plant-eating ducks, geese, and swans; and invertebrate- and fish-eating diving ducks and wading birds. The seagrass meadows are vital stopover points on the major flyways (Williams and Heck 2001).

Below the depth to which light penetrates, soft-sediment communities cover 80 percent of the ocean floor. **Life-forms** inhabiting these areas range from bacteria to bottom-feeding whales. Larger invertebrates (>0.5 mm) dominate the benthic **community**, most of them members of the infauna such as polychaete worms, echinoderms, mollusks, and crustaceans. Some are completely sedentary; others are highly mobile. They all require water, so they are confined to the surface layer of the sediments or construct tubes or burrows that give them access to the top layer of sediments or to the water column itself. Filter feeders extract particles suspended in the water and deposit feces and pseudofeces that becomes food for deposit feeders (McLusky and McIntyre 1988). The most abundant animals are the polychaete worms. Some burrow into the sediments and are deposit feeders; others erect tubes that protrude from the sediment and are filter feeders. Among echinoderms are sea stars, brittle stars, sand dollars, sea urchins, and sea cucumbers. Mollusks are represented by gastropods, nudibranches, and bivalves such as clams and mussels. Crustaceans include ostracods; amphipods; isopods; and decapods such as true lobsters, true crabs, and spiny lobsters. Even cnidarians such as anemones can be attached to buried worm tubes. The benthic community provides important ecological services to the oceans and to humans. As they filter water and sediments, the animals recycle nutrients and detoxify pollutants. Burrowing, digging, and feeding activities stir up the sediments to resuspend sediments and oxygenate the surface of the bottom. Benthic organisms are also essential food for larger invertebrates such as rays, horseshoe crabs, octopi, and shrimp; demersal and pelagic fishes; and marine mammals such as sea otter (*Enhydra lutris*), walrus (*Odobenus rosmarus*), and gray whale (*Eschrichtius robustus*). Crabs, shrimps, and bivalves constitute major commercial fisheries, as do some of the predatory fishes. Soft sediments have an infauna that is necessarily absent on hard surfaces. The richest benthic communities are found in shallower waters less than 55–60 m deep (180–195 ft) (Lenihan and Micheli 2001).

Twenty-five percent of the world's continental shelf area lies in polar latitudes. These regions support fewer species than other areas of continental shelf and have shorter food chains (Photo 13.1). The greatest abundance and biomass of invertebrates are found in soft sediments off Antarctica due to the high production of phytoplankton (primarily diatoms) and the high rate of sinking of organic material in the downwelling currents. Little disturbs the bottom sediments or returns nutrients back to the surface. No crabs, skates, rays, sharks, large fish, or bottom-feeding mammals live there. Diatoms grow on the underside of the ice, through which light penetrates, and on the sea bottom in shallow waters. Production is so great that sponges with silica-based skeletons growing in complete darkness can reach diameters of 3 m (10 ft). The same kinds of invertebrates can be found throughout the waters surrounding the Antarctic continent; many are found nowhere else. Especially characteristic of the benthic **fauna** are sponges (approximately 400 different

Photo 13.1 Seal hauled out on sea ice. The animal represents the top of the short food chain found in the rich waters of Antarctica's continental shelf. (Photo by Richmond Woodward, used with permission.)

kinds) and cnidarians. Crustaceans in the form of krill, the food of penguins and whales, are extremely abundant in the pelagic zone (Lenihan and Micheli 2001).

The fishes of Antarctic shelf waters are mostly **endemic** to the region, and more than half belong to the family Notothenioidae. Notothenioids fill many of the niches occupied by fishes from numerous other families in temperate regions. Antarctic fishes possess some unique adaptations to their cold-water environment and cannot tolerate warm waters. Members of the family Channichthyidae, for instance, have pale whitish blood without hemoglobin and relatively few red blood cells. These two factors keep the blood thin at low temperatures so less energy is required for circulation. The water temperature in coastal areas is below the normal freezing point of marine bony fishes (–0.8° C or 29° F), therefore the freezing point of blood and other fluids is reduced by the liver's year-round production of glycopeptides, or natural antifreezes. Antarctic fishes lack swim bladders and maintain neutral buoyancy by minimizing body weight through a lack of scales, the substitution of cartilage for bone, and the storage of fat. This weight reduction also allows them to move rapidly up and down in the water column with minimum energy expenditure (di Prisco 1997).

Arctic marine communities are unlike those of the Antarctic. Rather than being more or less uniform throughout the polar waters, four different environments and assemblages of organisms are recognized in the Beaufort Sea, the Bering/Chukchi seas, the Russian Arctic, and the eastern Canadian Arctic. Prime examples of short food chains are found in the Bering/Chucki region, where high phytoplankton production supports the clams that are fed upon by walruses; the amphipods consumed

Photo 13.2 Scuba diver in kelp forest off southern California. (Photo © Jeffrey L. Rotman/CORBIS.)

by gray whales; and a great abundance of invertebrates fed upon by Alaskan king crabs (*Paralithododes camtschaticus*) and snow crabs (*Chionectes opilio*), which constitute major commercial fisheries (Lenihan and Micheli 2001).

## Kelp Forests and Other Hard Surface Habitats

The hard surface substrates on the continental shelf host those benthic organisms that can attach to rock. Two worldwide subdivisions of the subtidal or shelf region have been made. The shallower, more horizontal places in the infralittoral zone are dominated by large erect or turf-building algae attached to the bottom. Steep slopes are dominated by invertebrates, which cannot tolerate the sediments that can accumulate on the flatter surfaces. The most notable of these communities are the kelp forests of cool waters in the temperate zone and cold upwelling zones in the subtropics (Photo 13.2). Kelps are large brown algae whose life cycle involves alternating between a tiny **sexually reproducing** gametophyte and a large asexual sporophyte. The sporophyte consists of a holdfast, stipe, and blade or lamina. The stipes of some kelps are rigid, while in others they are quite flexible. Some have very long blades supported by gas-filled bladders and reach the surface of the water. They are quite visible as they slosh back and forth with the waves. The giant kelp (*Macrocystis pyrifera*) off the coast of California attains lengths of more than 50 m (160 ft). *Ecklonia maxima*, a kelp that grows along the coast of southwest Africa, reaches lengths of 20 m (65 ft). Many kelp are **annuals**; in others the holdfast is **perennial**, but the stipe and blade are shed (Foster 1991; Barnes and Hughes 1999; Witman and Dayton 2001).

Kelps grow to depths of 20–40 m (65–130 ft) and on shelves with gentle slopes may be found 5–10 km (3–6 mi) offshore. Buoyant kelp blades form an upper surface canopy and give a three-dimensional character to these communities that makes them look like underwater forests. Off central California, the surface canopy consists of the giant kelp (*Macrocystis pyrifera*). Beneath the surface is an understory of *Pterygophora californica* and *Laminaria* spp., kelps with stiff stipas and blades that form a second canopy about 2 m (6 ft) above the bottom. A third layer of leafy red and brown algae and coralline algae is common beneath the *Laminaria*. On the bottom, a turf of filamentous and encrusting algae is common. A crust of corralline algae covers spaces on the rocks between kelp holdfasts (Foster 1991).

The vertical **zonation** previously described is mirrored in a seaward, horizontal zonation of algae reflecting increasing water depth. In shallow waters, the large *Laminaria* and articulated coralline algae are characteristic. At intermediate depths, the surface-canopy kelps dominate. And in the deeper waters of the shelf, red algae and sedentary invertebrates are the most common life-forms (Foster 1991).

Kelp forests are extremely productive and home to a large number of invertebrates and fishes. Numerous mobile invertebrates, such as polychaete worms, brittle stars, sea urchins, abalone, spider crabs, and snails, live among the holdfasts. In one site off California, over 300 kinds of invertebrates were reported and 125 fishes. Among the fishes, temperate families such as rockfish (Scorpaenidae) and sculpins (Cottidae) are important. Marine birds such as cormorant (*Phalacrocorax* spp.), brown pelican (*Pelecanus occidentalis*), and gulls are typical top carnivores, as are sea otter (*Enhydra lutris*) and harbor seal (*Phoca vitulina*) (Foster 1991; Barnes and Hughes 1999).

Sea urchins are the major consumers of kelp. These echinoderms undergo population explosions from time to time and can denude the forest. Some of these outbreaks appear to be natural; others are undoubtedly the consequence of human **disturbance** of the kelp ecosystem, particularly the overharvesting of the predators of sea urchins such as cod and sea otter (Barnes and Hughes 1999).

The kelp forests of the temperate zone are in a sense counterparts of the coral **reefs** of shallow tropical waters (see Chapter 12 for a description of coral reefs). Both are highly productive communities dominated by sedentary organisms (Barnes and Hughes 1999).

A second benthic habitat on rocky substrate occurs at depths below the infralittoral zone. This part of the continental shelf is dominated by a suspension-feeding epifauna of invertebrates, both sedentary and mobile. Particularly conspicuous are sponges, anthozoans, octocorals, gorgonians, bryozoans, ascidians, brachiopods, sea cucumbers, and nonreef-forming corals. Although not as tall as kelp forests, this community does possess a vertical structure due to the presence of so many erect animals. There are trees and fans of cnidarians such as the gorgonians and hydrocorals, stalked ascidians, and mounding and finger-shaped sponges. The vertical relief is sufficient to alter water movement and provide refuge from predators (Witman and Dayton 2001).

## Estuaries

No two estuaries are alike, but the Chesapeake Bay can be used as a model for the life-forms and their distribution patterns in temperate estuaries. Freshwater habitats are found at the heads of the bay and the estuaries of its major tributaries. Downstream, the water becomes brackish to moderately salty with a salt content up to 18 ppt. In the lower parts of the bay, salinity ranges from 18 ppt to 30–35 ppt, the salinity of the ocean. The distribution of communities reflects this gradient, the nature of the substrate, and increasing depth. Freshwater marshes may occur at the head of the estuaries. Salt marsh replaces them in the upper intertidal zone. In shallow subtidal and slightly brackish water (10 ppt or less), weed beds with plants such as redhead grass (*Potamogeton perfoliatus*) and sago pondweed (*P. pectinatus*) dominate in association with widgeongrass (*Ruppia maritima*), a seagrass. In more brackish water, seagrass meadows with widgeongrass and eelgrass (*Zostera maritima*) occur (Lippson and Lippson 1984).

Oyster reefs are another habitat type in estuaries. In areas that are warm year round, oysters (*Crassostrea virginica*) can inhabit the intertidal zone, but where winter temperatures can be low, oysters are restricted to the subtidal zone of estuaries. They are concentrated in brackish waters by an intolerance of freshwater and by the diseases and predators that are more common in saltwater. Oysters first establish themselves on hard mud and firm, sandy bottoms and then become a hard, immobile platform that oyster spat and other sessile life-forms attach to. The oyster reef community can be very diverse and include sponges, anemones, comb jellies, bryozoans, various worms, mollusks, crustaceans, and fishes. The Chesapeake Bay oyster beds are home to mollusks such as oyster drills (*Urosalpinx cinerea*), slipper shells, mussels, soft-shelled clam (*Mya arenaria*), giant whelk (*Buscyon cavica*), and oyster snail (*Odostomia impressa*). Among crustaceans are barnacles, amphipods, and crabs—including the commercially important blue crab (*Callinectes sapidus*). Sea stars (*Asteria* sp.) represent the echinoderms. The somewhat bizarre-looking oyster toadfish (*Opsanus tau*) is a resident fish (Lippson and Lippson 1984; Ray et al. 1997).

At depths below where plants and algae can grow rooted in or attached to the bottom, the benthic animals become fewer and the pelagic community more approximates that of the ocean. Many kinds of fish depend upon estuaries at some point in their life history for feeding or spawning, or find nursery habitat in salt marsh and seagrass meadows. Ray et al. (1997) estimated over 150 Atlantic fish rely upon the Chesapeake Bay and other east coast estuaries for their existence. These include certain sharks, rays, eels, sturgeon, herring, alewife, killifish, blennies, gobies, menhaden, striped bass, flounders, and many more. Anadramous species such as Atlantic herring (*Clupea harengus*) and shad (*Alosa* spp.) move up the estuary from the ocean in spring to spawn in freshwater. The juveniles move back into brackish water to grow and then spend most of their adult life in the ocean. Striped bass or rockfish (*Morone saxitilis*) is semianadromous in that adults live in the ocean or lower bay but spawn in brackish water. Its close relative, the white perch (*M. americana*), spends its entire life in brackish waters of the estuary. Menhaden (*Brecoortia* spp.) spawn in the open sea, but juveniles and adults enter the estuary and can even be found in tidal freshwater. The American eel (*Anguilla rostrata*) is truly catadraomous; it spawns in the Sargassum Sea, but its young swim to freshwater, where it spends most of its adult life. Freshwater fishes also can be found in an estuary near its head. In the Chesapeake Bay drainage system, these include sunfishes, catfishes, and minnows (Lippson and Lippson 1984).

Oyster reefs provide many keystone services within the estuary and provide habitat for coastal animals. The oysters filter the waters to maintain water quality and act as nutrient traps by depositing organic matter that becomes food for the benthic organisms. Estuaries as a whole are highly productive systems that are important sources of food for ocean life and humans as well. They are easily overwhelmed by nutrient pollution; sedimentation; and accumulation of toxic materials derived from agriculture, urban and industrial development, clearing of salt marsh, and ship traffic.

## ORIGINS

Although different phyla dominate at different latitudes, the ancestors of all benthic organisms can be traced to the tropical Tethys Sea, which circled the globe until the Miocene Epoch. During the Miocene, life-forms diversified poleward (Lenihan and Micheli 2001). For the most part, the contemporary habitat, the world's present continental shelves, came into being only with the breakup of Pangaea; those of the Atlantic Ocean, Antarctica, Australia, India, and eastern Africa are less than 200 million years old. Since then, they have been exposed to many fluctuations in sea level related to eustatic forces causing land masses to rise and fall and, even more importantly, to alternating periods of cooler (glacial) and warmer (interglacial) climate on Earth that began in the late Pliocene Epoch and continued through the Quaternary Period. At present, about 16 percent of continental rock is submerged beneath the sea and thus designated continental shelf. As recently as the last glacial period of the Pleistocene, when sea level was lowered approximately 135 m (400 ft), only 5 percent was submerged. Although life probably first evolved on continental shelves more than 3.5 billion years ago and many shelves, geologically, are nearly 200 million years old, contemporary communities of plants and animals may be only 10,000 years old or even younger (Eisma 1988).

Shallow-water tropical life-forms—including plant-eating sirenians, most major groups of bony fishes, chelonian sea turtles, sea urchins, and seagrasses, are known from the Eocene (49 million–41 million years ago). Many seem to have originated in the seas around tropical America and then spread eastward across the Tethys Sea to the Indo-West Pacific region in the early Oligocene. For the last 25 million years, the tropical regions of the ocean have been more or less isolated from one another and the evolution of geographically distinct forms has been a major trend. The movement of tectonic plates created a barrier between the Indian Ocean and the Mediterranean Ocean approximately 19 million–20 million years ago and constricted the connection between the Indian and Pacific oceans by the middle of the Miocene. The tropical Atlantic was isolated from the Indo-West Pacific region by the Miocene and from the Pacific during the Pliocene, when the Central American isthmus formed. The Indo-West Pacific region was to accumulate the greatest number of species, while tropical American seas suffered greatly from extinctions during the Pliocene—probably due to changes in oceanic circulation with the completion of the Central American isthmus (Vermeij 2001).

Cold-water life-forms derived from tropical and warm-temperate ancestors. Cold-water species in the waters off Antarctica are known since at least the late Cretaceous. Antarctica had reached its present position about 65 million years ago and separated from Australia 38 million years ago (at the transition between the Eocene and Oligocene epochs) and from South America 22 million–25 million years ago (at the transition between the Oligocene and Miocene epochs). After the middle of the Miocene Epoch, continental ice sheets formed every 1 million to 3 million years. The last episode of cooling began 2.5 million years ago, during the Pliocene. As a consequence of long isolation or near isolation from other cold-water areas by virtue of intervening tropical waters, the fishes of Antarctic waters are largely endemic and possess some unique adaptations to cold (di Prisco 1997).

Some cold-water life-forms of the Pacific, especially gastropods such as cockles and razor clams, originated in the temperate waters of the Northwest Pacific in the late Eocene or early Oligocene—the first cool interval of the Cenozoic in the Northern Hemisphere—and then spread eastward. Others evolved on the American side of the Pacific after the appearance of kelps in the middle to late Miocene and dispersed westward. Sea cows are among those forms originating in the Northeast Pacific. Carnivorous seals apparently evolved in warm-temperate waters off California in the late Oligocene or early Miocene. Otters are known from the late Miocene and sea otters from the late Pliocene or early Pleistocene. Marine gulls arose during the Pliocene, and the Pleistocene saw the appearance of gray whales and walrus in the northeastern Pacific. Adaptation to cold took place on the American side of the Pacific regardless of whether origins were in the Northwest or Northeast, and cold-adapted species then spread westward across the Pacific (Vermeij 2001).

The cold-water life of the Atlantic arose earlier than that of the Pacific, and ancient forms are found on sandy and muddy bottoms on both sides. The Atlantic remained isolated from the Pacific until the late Miocene or early Pliocene, at which time the Bering Sea opened to allow Pacific forms to pass into the Arctic Ocean and then the Atlantic. Pacific forms invading the North Atlantic at this time included many seaweeds such as kelps (Laminariales) and red algae (*Chondrius* spp.), eelgrass (*Zosteria marina*), soft-shelled clams (*Mya* spp.), sea urchins (*Strongylocentrotus* spp.), and auks (Alcidae). A wave of extinctions followed; the American side of the Atlantic suffered the most (Vermeij 2001).

## HUMAN IMPACTS

Major human impacts on soft-sediment communities are related to modern fishing practices and nutrient pollution from the land. The otter trawls used for collecting demersal fish, crabs, and shrimps and the dredges used to harvest clams (and also gravel) are very damaging to the bottom environment. Sediments are resuspended, tubes and other structures made by benthic organisms are destroyed, animals are crushed or mutilated. Otter trawls penetrate 1–16 cm (0.5–6 in) into the sediments; dredges penetrate up to 30 cm (12 in) (Lenihan and Micheli 2001).

As in the intertidal zone, excess nitrogen and phosphorus derived from agricultural and urban land uses can cause algal blooms far in excess of what zooplankton can consume. Phyoplankter cells settle to the bottom, where they are decomposed by oxygen-consuming bacteria. The processes of decomposition can severely reduce oxygen in the sediments and bottom of the water column with negative effects on the benthic animals.

The kelp forests of the Northern Hemisphere have been affected by overfishing and overhunting of species such as Stellar's sea cow (*Hydrodamalis gigas*) and sea otter in the North Pacific and lobsters in the North Atlantic. The sea otter, by preying heavily upon sea urchins, prevents the latter from overgrazing the kelp beds. In the absence of otters, Alaskan subtidal areas suffered from dense populations of sea urchins (*Strongulocentrus* sp.), limpets, and chitons. The large, fleshy kelps and other algae disappeared, and fish became uncommon (Witman and Dayton 2001). The sea otter, once near extinction, has been an internationally protected species since 1911. Severely depleted kelp beds recovered as otter populations increased and their distribution area expanded. Stellar's sea cow was hunted to extinction by 1768. Largest of the sirenians, it was a grazer of kelp; the sea otter's control of urchin populations may have helped maintain its pastures (Barnes and Hughes 1999).

Fishing has especially targeted large top predators, and the reduction of their numbers can have effects throughout the ecosystem. An example is the Atlantic cod (*Gadus morhua*), a large, bottom-feeding fish that was the mainstay of aboriginal, colonial, and modern commercial fisheries. Cod have declined in numbers and size and are now essentially gone from the great fishing banks of the northwest Atlantic. In the early 1800s, the average size of fish caught was 45–73 kg (100–160 lb). A century later, fish weighing more than 45 kg were exceptionally rare. Numbers nosedived between 1989 and 1992, when U.S. fisherman gained exclusive use of the banks and fished them intensely. Today, the dominant predatory fish are small fishes such as sculpin (*Myoxocephalus octodecemspinosus*) or rock gunnel (*Pholis gunnelis*). The loss of the cod led to an increase in numbers and expansion in distribution of some of its prey species such as lobster and sea urchin. The increase in sea urchins led to a decline in kelp. However, Japanese fishermen began to exploit the urchins for their roe, and as a result, kelp and other macroalgae have again increased. The kelp beds are important nursery areas for some of the larger finfish, so this may be a positive change (Steneck and Carlton 2001).

Estuaries are severely affected by nutrient pollution, the addition of excess nitrogen and phosphorus. Much of this comes in runoff from the land as rivers collect fertilizers lost from cropland, improved pastures, suburban lawns, and wastewater inadequately treated or lacking any treatment at all. A large proportion (up to one-half), however, comes as fixed nitrogen precipitated from the atmosphere. The original sources of the atmospheric pollutants are primarily industrial and vehicular exhausts. Seagrasses in particular are hurt. Increased nutrients (eutrophication) stimulate the growth of one-celled algae that cloud the water or cover the seagrass blades, either way reducing the light that can reach the plants. The rate of nutrient accumulation is increased when oyster beds are destroyed. The huge oyster reefs of the Chesapeake Bay may have been able to filter plankton and other particulates from the entire Bay every three days under the more natural conditions that existed at the beginning of the twentieth century (Peterson and Estes 2001). Loss of seagrasses is a critical worldwide phenomenon. In the most important estuaries of the United States, up to 80 percent of salt marsh and seagrass meadow has been lost. This means the loss of important nursery areas for shellfish and finfish, as well as feeding grounds for marine fishes, sea turtles, and waterfowl.

The number of confirmed extinctions of marine life from continental shelf areas in historical times is relatively low, but most were top predators or grazers so the impacts of their losses probably were felt throughout the ecosystems involved. Three mammals, five birds, and four snails have been lost forever. The mammals are the Stellar's sea cow (*Hydramalis gigas*), exterminated from the north Pacific in 1768; the sea mink (*Mustela macrodon*), last reported from the northwest Atlantic in 1880; and the West Indian monk seal (*Monachus tropicalis*), which had disappeared from the Caribbean Sea and Gulf of Mexico by 1952. The birds now extinct are

Palla's Cormorant (*Phalacrocorax perspicillatus*), gone from the Northwest Pacific by 1850; the Great Auk (*Alca imperius*), lost throughout its range on both sides of the Atlantic by 1844; the Labrador Duck (*Camptorhyncus labradorius*), exterminated from the northwest Atlantic by 1875; Auckland's Merganser (*Mergas australis*), a native of the southwest Pacific not sighted after 1902; and the Canary Island Oystercatcher (*Haematopus meadewaldoi*), extinct in the northeast Atlantic by 1913 (Carlton et al. 1999). Among species on the verge of extinction are the white abalone (*Haliostes sorensi*) off California and the barndoor skate (*Raja laevis*), native to the western parts of the North Atlantic. Numbers of green sea turtle (*Chelonia mydas*), hawksbill turtle (*Eretmochelys imbricata*), and manatee (*Trichecus manatus*) are so reduced in the Caribbean that they have lost their dominant roles in the seagrass ecosystems (Steneck and Carlton 2001).

**Introduced** species are also part of the human impact on ecosystems of the continental shelf. The Gulf of Maine has more than 50 **exotic** species, and the San Francisco Bay has more than 164. Algae, cnidarians, mollusks, and crustaceans compose the list. The green alga *Codium fragile* is now a dominant seaweed in many places along the Atlantic coast of the United States, where exotic invertebrates found on hard-surface subtidal habitats include the European shorecrab *Carcinus maenus*, introduced in the 1820s; the Asian sea anemone *Diadumene lineata*, introduced in the 1890s; the European flat oyster *Ostrea edulis*, introduced in the 1940s; and the Asian shorecrab *Hemigraspus sanguineus*, introduced in 1988. The European shorecrab also now inhabits soft-sediment habitats, as does the European opossum shrimp (*Praunus flexuosus*), first reported in the 1960s. Most exotics were accidentally introduced in or on the hulls of ships. Elsewhere in the world, the American combjelly (*Mnemiopsis leidyi*) is now found in the Azov, Black, and Mediterranean seas; the Japanese seastar (*Asterias amuriensis*) occurs in Australian waters; and the Caribbean algae *Hypnea musiformis* and *Caulerpa taxifolia* are problems off Hawaii and in the Mediterranean (Steneck and Carlton 2001).

## HISTORY OF SCIENTIFIC EXPLORATION

Some of the earliest studies in quantitative **ecology** (Petersen 1918) were conducted in soft-sediment communities on the continental shelf. The concept of describing species composition in terms of relative abundance rather than simply presence or absence was a result of this work (Lenihan and Micheli 2001). The scientific study of seagrass communities and kelp beds had to await the invention of SCUBA during World War II. Research became especially prolific in the 1970s and made major contributions to ecological thought. Among important outcomes were comprehensive analyses of detritus food chains, better understandings of relationships between the complexity of habitat structure and the abundance and diversity of species, and the intricate nature of interactions among native and nonnative species. Contemporary research focuses on the needs of marine habitat restoration, especially seagrass beds (Williams and Heck 2001). Research on rocky subtidal substrates was long neglected in favor of work on rocky intertidal habitats, but some work has been done. In the 1960s, the focus of attention was algae and invertebrate zonation. In the 1970s and 1980s, the pressures of grazing and predation received most of the scientific scrutiny, and in the 1990s phenomena associated with recruitment dominated research (Witman and Dayton 2001).

## REFERENCES

Barnes, R. S. K. and R. N. Hughes. 1999. *An Introduction to Marine Ecology.* 3rd edition. London: Blackwell Scientific.

Carlton, J. T., J. B. Geller, M. L. Reaka-Kudla, and E. A. Norse. 1999. "Historical extinctions in the sea." *Annual Review of Ecology and Systematics* 30: 515–538.

Di Prisco, Guido. 1997. "Physiological and biochemical adaptations in fish to a cold marine environment." Pp. 251–260 in B. Battaglia, J. Valencia, and D. W. Walton, eds., *Antarctic Communities: Species, Structures and Survival.* Cambridge: Cambridge University Press.

Eisma, D. 1988. "An introduction to the geology of continental shelves." Pp. 39–91 in H. Postma and J. J. Zijlstra, eds., *Continental Shelves.* Ecosystems of the World, 27. Amsterdam: Elsevier.

Foster, M. S. 1991. "Open coast intertidal and shallow subtidal ecosystems of the Northeast Pacific." Pp. 235–272 in A. C. Mathieson and P. H. Nienhuis, eds., *Intertidal and Littoral Ecosystems.* Ecosystems of the World, 24. Amsterdam: Elsevier.

Klee, Gary A. 1999. *The Coastal Environment. Toward Integrated Coastal and Marine Sanctuary Management.* Upper Saddle River, NJ: Prentice Hall.

Lenihan, Hunter S. and Fiorenza Micheli. 2001. "Soft-sediment communities." Pp. 253–288 in Mark D. Bertness, Steven D. Gaines, and Mark E. Hay, eds., *Marine Community Ecology.* Sunderland, MA: Sinauer Associates, Inc.

Lippson, Alice Jane and Robert L. Lippson. 1984. *Life in the Chesapeake Bay.* Baltimore: The Johns Hopkins University Press.

McLusky, D. S. and A. D. McIntyre. 1988. "Characteristics of the benthic fauna." Pp. 131–154 in H. Postma and J. J. Zijlstra, eds., *Continental Shelves.* Ecosystems of the World, 27. Amsterdam: Elsevier.

Petersen, C. G. J. 1918. "The sea bottom and its production of fish food. A survey of the work done in connection with valuation of the Danish waters from 1883–1917." *Rap. Danish Biol. Stat.* 25: 1–62.

Peterson, Charles H. and James A. Estes. 2001. "Conservation and management of marine communities." Pp. 469–507 in Mark D. Bertness. Steven D. Gaines, and Mark E. Hay, eds., *Marine Community Ecology.* Sunderland, MA: Sinauer Associates, Inc.

Potsma, H. 1988. "Physical and chemical oceanographic aspects of continental shelves." Pp. 5–37 in H. Postma and J. J. Zijlstra, eds., *Continental Shelves.* Ecosystems of the World, 27. Amsterdam: Elsevier.

Potsma, H. and J. J. Zijlstra. 1988. "Introduction." Pp. 1–4 in H. Postma and J. J. Zijlstra, eds., *Continental Shelves.* Ecosystems of the World, 27. Amsterdam: Elsevier.

Raffaelli, David and Stephen Hawkins. 1996. *Intertidal Ecol*ogy. London: Chapman & Hall.

Ray, G. C., B. P. Hayden, M. G. McCormick-Ray, and T. M. Smith. 1997. "Land-seascape diversity of USA East Coast coastal zone with particular reference to estuaries." Pp. 337–371 in Rupert F. G. Ormond, John D. Gage, and Martin V. Angel, eds., *Marine Biodiversity, Patterns and Processes.* Cambridge: Cambridge University Press.

Sharp, G. D. 1988. "Fish populations and fisheries. Their perturbations, natural and man-induced." Pp. 155–202 in H. Postma and J. J. Zijlstra, eds., *Continental Shelves.* Ecosystems of the World, 27. Amsterdam: Elsevier.

Steneck, Robert S. and James T. Carlton. 2001. "Human alterations of marine communities: Students beware!" Pp. 445–468 in Mark D. Bertness, Steven D. Gaines, and Mark E. Hay, eds., *Marine Community Ecology.* Sunderland, MA: Sinauer Associates, Inc.

Vermeij, Geerat J. 2001. "Community assembly in the sea: geologic history of the living shore biota." Pp. 39–60 in Mark D.Bertness, Steven D. Gaines, and Mark E. Hay, eds., *Marine Community Ecology.* Sunderland, MA: Sinauer Associates, Inc.

Williams, Susan L. and Kenneth L. Heck, Jr. 2001. "Sea grass community ecology." Pp. 317–337 in Mark D. Bertness, Steven D. Gaines, and Mark E. Hay, eds., *Marine Community Ecology*. Sunderland, MA: Sinauer Associates, Inc.

Witman, Jon D. and Paul K. Dayton. 2001. "Rocky subtidal communities." Pp. 339–366 in Mark D. Bertness, Steven D. Gaines, and Mark E. Hay, eds., *Marine Community Ecology*. Sunderland, MA: Sinauer Associates, Inc.

# 14 THE DEEP SEA

## OVERVIEW

The deep sea biome begins at the edge of the continental shelf and involves all parts of the water column and sea floor at depths greater than 200 m (650 ft). Some 80 percent of the biome occurs at depths greater than 2,000 m (6,500 ft) and 60 percent at depths below 4,000 m (13,000 ft) (Barnes and Hughes 1999). Although it covers nearly two-thirds of Earth's surface, the deep sea remains largely a great unknown. In the past two decades or so, rapid advances in our understanding of the deep-sea environment and deep-sea communities have been made thanks to technological advances such as submersibles and remote-sensing methods. Long-held ideas are being scrapped as new information is acquired. The student is cautioned that any book is probably out of date; the most accurate information is best found in science journals and the Web sites of professionals in the field.

It is known now that benthic and pelagic communities of the deep sea are basically the same as those of the continental shelf and do not contain only a few unique or monstrous forms, as once believed. Many, many different forms of life occur here, though none is especially abundant. One estimate puts the total number of **species** at over 10 million (Grassle and Maciolek 1992), but no one really knows.

## THE DEEP-SEA ENVIRONMENT

The average depth of the ocean is 3,800 m (12,470 ft) below sea level. The greatest depth, –11,033 m (–36,198 ft), occurs in the Mariana Trench in the western Pacific Ocean just east of the Mariana Islands. Most of the bottom of the deep sea consists of a flat abyssal plain covered by soft sediments. Close to the continents, these sediments may be coarse sands derived from land, but elsewhere fine muds or oozes are the norm. These finer materials are biological in origin and form as dead formaniferans, diatoms, and radiolarians settle down from the surface zone. The

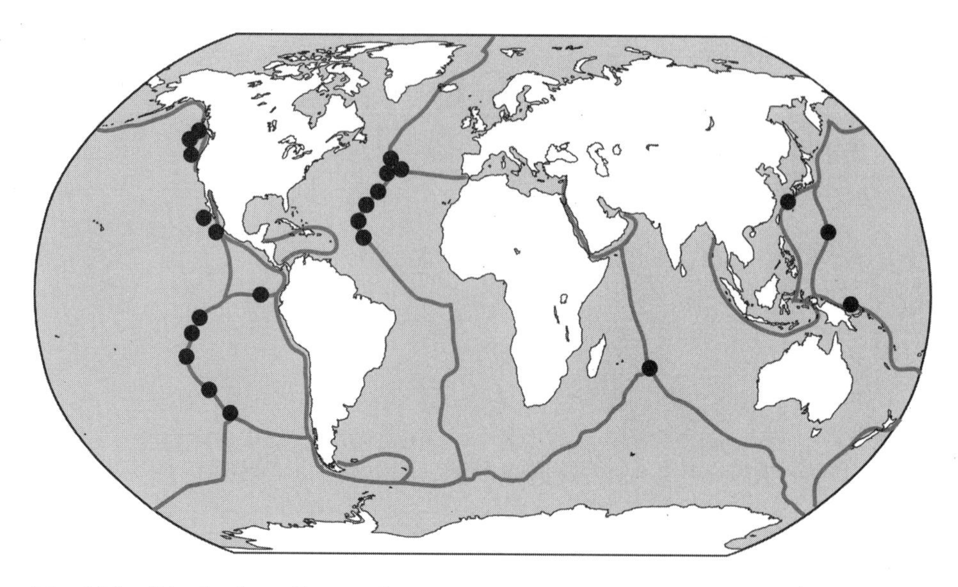

Map 14.1 Distribution of known deep-sea vents.

rate of accumulation varies from a few millimeters to a few centimeters every 1,000 years. On older parts of the seafloor, these oozes can be kilometers thick. The abyssal plain is interrupted by the world's longest mountain range, the 40,000-km (25,000-mi) chain of midoceanic ridges that wraps around the Earth. Midoceanic ridges are places where the oceanic plates are separating and volcanic activity is producing new seafloor. Other volcanically active parts of the sea bottom occur or occurred at so-called hot spots, where plumes of magma ascend from the earth's mantle far from plate boundaries to form seamounts and oceanic islands. Seamounts may occur in association with midoceanic ridges or in chains formed by ocean plates moving over hot spots. The Pacific alone has an estimated 30,000. Midoceanic ridges and seamounts provide unique **habitats** for deep-sea marine life because they offer hard surfaces for attachment and, in places, hydrothermal vents (see following). They also alter the flow of bottom currents and thus can affect nutrient distribution and the **dispersal** of larval forms. Even small features affect animal distribution because they can lift filter feeders above soft sediments that would clog their filter apparatuses. Other hard surfaces on the sea bottom are provided by manganese nodules, shipwrecks, and whale carcasses (Barnes and Hughes 1999; Etter and Mullineaux 2001; Van Dover 2000).

Below 200 m (650 ft), the ocean waters are cold and dark. Salinity is constant, but pressure increases with depth. Bottom currents usually move very slowly; however, benthic storms do occur. At –3,000 m (–10,000 ft) the temperature is 1–2° C (34–35° F), salinity is 34.8 ppt, and pressure is equivalent to 300 times atmospheric pressure at the sea surface. The only light comes from bioluminesence (light produced by living organisms) and the radioactive decay of potassium-40 (Van Dover 2000).

Deep-sea water is well oxygenated since it originates from cold surface waters that sink in the North Atlantic and Antarctic. Because of the sparseness of living organisms on the sea bottom, these waters lose little of the oxygen acquired in contact with the atmosphere, even though they may remain on the bottom for hundreds of years before rising again in the Pacific and Indian oceans (Van Dover 2000).

## PLANKTON

The deep sea biome lies below the level at which phytoplankters can actively produce food. However, abyssal olive-green algae have been reported to depths of –5,000 m (–16,000 ft). These apparently are the resting stages of diatoms—once diatoms sink below the euphotic zone they change into an inactive state. Other phytoplankton known from depths of –1,000–3,000 m (–3000 to –10,000 ft) are coccolithophores and **cyanobacteria**, but these also are probably not photosynthesizing (Barnes and Hughes 1999). Without producers, few living zooplankters are found. Fecal pellets and dead zooplankters, on the other hand, are constantly sinking through the water column and provide food for filter feeders. This detritus is not uniformly distributed but reflects the patchy distribution of plankton in surface waters.

## BENTHIC ANIMALS

The benthos depends on food particles settling down from higher parts of the water column or on the large numbers of bacteria in the sediments that decompose the detritus. (Chemosynthetic bacteria in symbiotic relationships with animals occur at hydrothermal vents. See following.) Other than fecal pellets and dead, unconsumed zooplankters, the detritus may consist of dislodged and sunken seaweeds such as kelps and sargassum, the carcasses of fish and marine mammals, and even sunken logs and other debris from the land (Van Dover 2000).

Echinoderms such as sea stars, brittle stars, sea urchins, and primitive sea lilies or crinoids are the most conspicuous inhabitants of the deep sea, but many other groups are well represented. Typically, deep-sea communities on soft bottoms contain a few individuals of a diverse group of small (<1 mm) organisms that are deposit feeders and consume bacteria, organic particles, and the even smaller organisms (42–300 μm) such as nematodes and foraminiferans. They include many kinds of polychaete worms, crustaceans, and mollusks. Some 250 different organisms per square meter have been recorded. The megafauna consists of highly mobile forms such as demersal fishes, cephalopods, lobsters, and large amphipods; and more sedentary animals such as glass sponges, soft corals, gorgonians, sea pens, sea anemones, and huge protozoans (foraminiferans and xenophyophores). Many benthic organisms, such as glass sponges and tube-building foraminferans, become habitat themselves since they provide attachment sites for other organisms (Etter and Mullineaux 2001). At least some benthic animals seem to have depth limits. Lobsters and their relatives are found only above 5,000 m (16,000 ft). Below 8,300 m (27,000 ft) animal life is essentially limited to deposit (ooze) feeders, 90 percent of which are holothurians (Echinodermata) (Barnes and Hughes 1999). Interesting trends among the deep-sea benthos are gigantism among predators and miniaturization among deposit feeders. Examples are giant scavenging amphipods 0.33 m (1 ft) in length and baseball-sized protozoans (Barnes and Hughes 1999; Van Dover 2000).

On the tops of seamounts and at other locations with hard bottoms and relatively fast currents, large suspension feeders are prevalent. These include animals such as sponges, hydroids, anemones, crinoids (or sea lilies), tunicates (or sea squirts), corals, and sea pens. Smaller suspension feeders such as tube-building polychaetes, solitary corals, and crustaceans are also present. These communities may be quite dense on the tops of seamounts since nutrients are concentrated in these locations by ocean currents (Etter and Mullineaux 2001; Van Dover 2000).

## NEKTON

The free-swimming nekton of the deep sea includes gelatinous filter feeders such as medusas and ctenophores, prawns, squids, fishes, and whales. The upper waters [at depths of 250–700 m (800–2,000 ft)] of the deep sea biome are inhabited by fishes typically countershaded with black backs, silvery flanks, and a line of light-producing organs called photophores along their bellies. All these features camouflage the fish against predators lurking below. The silvered sides make them invisible in the diffuse light a predator would see when looking toward the surface. A string of lights along the belly helps breaks up the outline of the body. Squids (*Histroteuthis* spp.) in this zone also have photophores. Organisms in these upper waters undergo daily vertical migrations; they move upward during the night to feed in the food-rich surface zone and descend during the day to avoid surface-water predators that could easily see them (Robinson 2001; Angel and Martin 1977).

At depths greater than –700 m (–2,000 ft), vertical migrations are no longer practicable, and organisms must rely on what sinks from above or swims by them. Fishes in this zone tend to be uniformly colored a dark bronze or black. Ninety percent of life-forms have some type of bioluminescence, which serves the functions that color does higher in the water column. It is used to distract predators (for example, squids produce bioluminescent ink), to lure prey [e.g., the dragon fish (*Idiacanthus* sp.)], or for recognition of individuals of the same kind so that they can find mates. Deep-sea fish have eyes adapted to the extremely low light levels still available at depths down to –1,000 m (–3,000 ft), and prey have evolved camouflage and transparency to trick their predators. Below 1,000 m, hunting by sight becomes impossible, and taste and smell become more important. Fishes living below 1,000 m are blind or have primitive eyes that may detect light but cannot focus (Robinson 2001). The greatest depth at which fish have been observed is –8,730 m (–27,460 ft) (Gradwohl 1996).

Swimming animals have greater energy needs than attached or floating organisms. To stay deep within the water column, fishes inhabiting the deep sea lack swim bladders or other buoyancy devices but must keep swimming to prevent sinking. With increasing depth, less food is available to the nekton, and energy conservation mechanisms show up in morphological and behavioral adaptations. Fishes tend not to pursue prey but stalk it or wait in ambush. Feathery bristles and exceptionally long antennae help keep them afloat. Body tissue is reduced and often takes on a gelatinous consistency. The only muscles preserved are those that operate the jaws, so bizarre body shapes develop that can make a fish look like a large mouth with fins and tail (Barnes and Hughes 1999; Robinson 2001).

## HYDROTHERMAL VENTS

Unique habitats occur at widely separated locations in the deep sea where superheated water rises from the sea floor. These are called hydrothermal vents, and most are associated with midoceanic ridges. Midoceanic ridges form where tec-

Photo 14.1 A black smoker on the East Pacific Rise. (Photo © Ralph White/CORBIS.)

tonic plates are separating and new oceanic crust is being formed as magma wells up from Earth's mantle. Heat also escapes from the interior of the planet. Young oceanic crust tends to be porous, and cold seawater percolates into it. As it approaches the rising magma, the water is heated and its density lessens. The heated water then rises back to the seafloor and issues forth as hot springs or geysers. Older crust becomes sealed by mineralization and, to some extent, by a coating of sediments so hydrothermal activity is limited to areas of recent volcanism (Van Dover et al. 2001).

When water temperatures are around 400° C (750° F), metal sulfides from the oceanic crust dissolve in it. The hot water that spews forth is acidic and rich in iron, copper, zinc, and sulfide ions. When it mixes with the very cold and alkaline seawater, the metal sulfides quickly precipitate out of the water to form a plume of black particulates. Chimneylike structures known as black smokers build up from the deposits (Photo 14.1). They may grow to heights of 10–20 m (30–65 ft) and have several vents through which the particulates escape. These structures, typical of the East Pacific Rise, are rather fragile and easily toppled in earthquakes or in collisions with submersibles (Van Dover 2000).

White smokers also occur; they form at water temperatures of 100–300° C (212–570° F) and precipitate silica, anhydrite (calcium sulfate), and barite (barium sulfate) as white particles (Van Dover 2000).

Structures built at hydrothermal vents take on various shapes in addition to chimneys. Beehives develop when a set of horizontal passages vent a mound developed on the side or top of a chimney. Rows of vertical vents can develop into buttresslike features like "Godzilla," which rises to a height of 45 m (150 ft) on the Juan de la Fuca Ridge. Sulfide mounds up to 200 m (650 ft) in diameter are found on the Mid-Atlantic Ridge. Hydrothermal structures often occur in vent fields that range in area from several hundred to several million square miles (Van Dover 2000).

Hydrothermal vents last only for a limited time, although how long one is active varies considerably. On fast-spreading [more than 90 mm (3.5 in) per year] ridges such as the East Pacific Rise, a vent's life span may be less than 20 years. The TAG mound on the slow-spreading [10–50 mm (0.2–2 in) per year] Mid-Atlantic Oceanic Ridge has been active on and off for 40,000–50,000 years (Van Dover 2000). The frequency and spacing of vents also is variable. On the East Pacific Rise, they are abundant and closely spaced (tens of meters to tens of kilometers apart). On the Mid-Atlantic Ridge, they are uncommon and widely spaced (hundreds of kilometers apart) (Etter and Mullineaux 2001).

Life at hydrothermal vents has astounded and fascinated scientists since they first discovered them in the late 1970s. Microbes are able to survive at temperatures up to 115° C (240° F), and invertebrates can survive at temperatures up to 40° C (104° F). **Photosynthesis** is impossible at these depths, but some bacteria can produce food (organic carbon) for zooplankters by chemosynthesis. They use chemical energy from sulfides to combine carbon dioxide, oxygen, and water and form carbohydrates (and, to complete the equation, sulfates) (Van Dover 2000). These bacteria occur in several habitats associated with hydrothermal vents. They may live free in the water column held in suspension hundreds of meters above the vent by the outflow of warm water or in the plumes of black smokers. They may occur as mats or colonies on the surfaces of the seafloor and chimneys. Chemosynthetic microorganisms also may live within the pores of sulfide structures and below the surface of the seafloor. Most importantly, however, they live in symbiotic relationships with invertebrates. Host organisms have a variety of biochemical, physiological, and morphological adaptations to accommodate the bacteria they receive carbohydrates from. Most bacteria live within the tissues of their hosts. They live in the gills of clams, mussels, and scallops. The tubeworm (*Riftia pachyptila*) has neither mouth nor gut but does have special structures to house bacteria and depends exclusively on them for food. The tubeworm in turn ensures the bacteria a steady supply of the inorganic compounds they need to produce the food. Other chemosynthetic bacteria live on the outside of organisms. Examples of hosts for these bacteria are the vent shrimps (*Rimicavis exoculata*) that swarm around Mid-Atlantic Oceanic Ridge vents, Pompeii worms (*Alvinella pompejana*), and vent limpets (*Lepetodrilis fucensis*) (Van Dover 2000; Etter and Mullineaux 2001).

Most animals at hydrothermal vents are not found in other marine environments, although predators and scavengers from other parts of the abyssal zone may visit. Different assemblages occur at different types of vents, but in all, the bulk of living matter is made up of invertebrates with symbiotic bacteria. Three groups are dominant: annelids (segmented worms), mollusks, and arthropods. Free-living bacteria provide food for certain limpets, polychaete worms such as spaghetti worms and dandelions, and small crustaceans. These are eaten by shrimps, crabs, and fish, as are the tubeworms and bivalves with symbiotic bacteria (Van Dover 2000; Etter and Mullineaux 2001).

## HUMAN IMPACTS

Despite their distance from most human activities, and perhaps because of our ignorance of the marine environment, deep-sea habitats and organisms are not free of human impact. The deep sea is used as a disposal site for dredge sludge; sewage; and hazardous wastes, including radioactive materials. Mining of manganese nodules and hydrothermal deposits is ongoing; these activities tend to stir up sediments that resettle elsewhere at rates greater than the slow natural accumulation rates of sediments. Overfishing has occurred: orange roughy (*Hoplostethus atlanticus*), a fish that gained popularity in America and Europe when haddock, sole, and flounder fisheries disappeared in the 1980s, was soon depleted from seamounts around Australia and New Zealand. Collecting of the more intriguing life-forms can affect some deep-sea and hydrothermal communities at least short term (Etter and Mullineaux 2001). **Climate** change significant enough to stop the great conveyor belt of cold, oxygenated waters from polar regions to the sea bottom would have major effects on life in the deep sea. As with so much about the deep sea biome, we really don't know yet a lot about the long-term consequences of human activities.

## ORIGINS

Deep-sea animals seem to have evolved in the deep sea or migrated to it from cold, shallow waters such as those surrounding Antarctica. Organisms on deep seamounts are members of wide-ranging species. Those on shallow seamounts are similar to animals inhabiting nearby continental shelves (Etter and Mullineaux 2001; Van Dover 2000).

Since the discovery of hydrothermal vents in the deep sea, more than 500 new species have been described. Most invertebrates seem to be recent (mid- to late Cenozoic) immigrants from the surrounding deep sea. For example, the squat lobster (*Munidopsis lentigo*) is a black-smoker specialist but closely related to another squat lobster (*M. subsquamosa*) that thrives in high densities at the edge of vent fields. A few animals are relicts from the Mesozoic or possibly even Paleozoic eras. The vent barnacle (*Neolepas zevinae*), the most primitive of living stalked barnacles, is a so-called living fossil from the Mesozoic. Another example, the vent limpet (*Neophalus fretterae*), may have first lived in shallow seas and migrated to the deep seas, where it was able to survive the extinctions that eliminated its relatives. Overall, diversity at vents is relatively low, as is often true of high-nutrient ecosystems (e.g., estuaries and salt marshes). Animals at Atlantic vents are a subset of those found at vents at the Pacific, as one might expect since the Atlantic is considerably younger than the Pacific. Dispersal into the Atlantic may have come along the South Atlantic ridge system that is a continuation of that of the Indian Ocean. A large number of life-forms are **endemic** to the Atlantic (Van Dover 2000). The first report from a deep-sea hydrothermal vent 1,600 km (1,000 mi) east of Madagascar revealed that the most common organism was a shrimp like that of North Atlantic vents. The other animals—crustaceans, mussels, anemones, and large snails—appear to be similar to those of the West Pacific (Van Dover et al. 2001).

Part of the fascination science has with hydrothermal vents is the presence of chemosynthetic microorganisms. These bacteria, which do not depend on sunlight, may be models for the earliest forms of life on Earth. Their survival in harsh, dark environments has also made the discovery of life on other planets more likely—at the moment, Europa, a moon of Jupiter, is a promising candidate (Delaney 2001).

## HISTORY OF SCIENTIFIC EXPLORATION

The science of oceanography and exploration of the deep sea began with the voyage of the British research vessel HMS *Challenger*, 1872–76. The naturalists on board collected specimens of thousands of animals previously unknown to science and amassed data on currents, tides, water chemistry, and water temperatures in all oceans except the Arctic. They invented new instruments and devices to sample the sea and its inhabitants. They also attempted to map the seafloor using the technology of the day, a heavy weight fastened to a hemp line (Barnes and Hughes 1999; Tharp 2001).

The history of deep-sea exploration is in many ways a history of technology, and military needs have driven much of it. Otis Barton designed the bathysphere, a steel sphere attached by cable to a mother ship, in the early 1930s. In it, he and William Beebe descended a record 914 m (3,000 ft) off the coast of Bermuda on August 15, 1934, far surpassing the deepest helmet dive of 18 m (60 ft). In the 1940s, Auguste Piccard developed another submersible that he called the bathyscape. After refinements, his son, Jacques Piccard, and U.S. Navy Lt. Donald Walsh descended 10,915 m (35,810 ft) into the Mariana Trench in the bathyscape *Trieste;* they are still the only humans to have seen the deepest part of the ocean. The use of deep-diving, nuclear-powered submarines encouraged the U.S. Navy to support research and technological development to map the deep-sea world. The urgency to explore this largely unknown realm became clear with the sinking of the nuclear submarine *Thresher* in 2,500 m (8,400 ft) of water off the coast of New England in 1963. The following year, the three-person submersible *Alvin* was constructed for the Navy to be operated by the Woods Hole Oceanographic Institution. *Alvin* carried researchers to depths down to –4,500 m (–14,763 ft) and played a major role in the exploration of the deep sea. Discoveries led to new understandings of the configuration of the seafloor, **plate tectonics**, and deep-sea life. In 1975, Robert Ballard and other members of project FAMOUS (the joint French American Mid-Ocean Undersea Study) witnessed new oceanic crust being produced and confirmed the notion of seafloor spreading. In 1977, John Corliss and Robert Ballard, both members of FAMOUS, while in *Alvin* discovered hydrothermal vents 3 km (2 mi) beneath the sea surface on the Galapagos Rift and saw the first giant tube worms and evidence of chemosynthetic microbes. As recently as December 2000, Deborah Keeley saw through the windows of *Alvin* the "Lost City," the first hydrothermal structure known to be constructed on ancient oceanic crust and formed of carbonates and silica. It had formed on the side of a 3,650-m (12,000-ft) mountain about 15 km (9 mi) from the North Atlantic mid-oceanic ridge (Tharp 2001).

A number of other submersibles have been constructed. One-person vehicles such as *DeepWorker* operate to depths of 2,000 ft. Robotic vehicles were developed in the 1980s, and remotely operated vehicles (ROVs), both tethered and untethered, are now in wide use. Japan's *Kaiko*, an untethered ROV, can descend to depths of –11,000 m (–36,000 ft). The current record for a manned vehicle, 6,527 m (21,414 ft), is held by the three-person submersible *Shinkai.* (Tharp 2001; Gradwohl 1996).

World War II saw the development of accurate echo sounders that were later used to map the ocean floor. Scientists at the Woods Hole Oceanographic Institution collected data for the U.S. Navy beginning in 1952. The crest of the midoceanic ridge was located through all oceans but the Arctic. Since the exact depth values were classified information until 1961, the first world map of the seafloor was drafted with a pictorial representation of the vast undersea mountain chain that became the standard way of showing the ridge for years to come (Tharp 2001). In the early 1980s, a new technology was used to map the seafloor. The U.S. Navy launched the GEOSTAT satellite to a position 800 km (500 miles) above Earth. It used a multibeam acoustical system for mapping the seafloor. When that information was declassified in 1995, workers at Scripps Institute for Oceanography and National Oceanographic and Atmospheric Agency (NOAA) produced a new (though still low resolution) map of the world. The Global Ocean Observation System (GOOS) is a newer satellite system monitoring the sea surface, including the distribution and abundance of phytoplankton (Tharp 2001).

Mapping the seafloor was important to the military and also to oil and gas producers and other mining operations. The geologists employed by these industries used drilling methods to core the seafloor and gain knowledge of the resources that may lie within the sediments or on the seafloor surface. Their information was useful to scientists studying the history of the Earth and past climate changes. The Ocean Drilling Program is an ongoing international project. The famous ship *Glomar Challenger* operated from 1968–83 and was involved in an unsuccessful attempt to drill through the oceanic crust and sample the mantle below. The *Resolution*, built for oil exploration in 1978, was refitted in 1984 and has continued the work of the *Glomar Challenger* (Tharp 2001).

Research on life in the deep sea remains a high priority, and new techniques as well as discoveries arise every year. The United States, Russia, and Japan sponsor major research programs. At the same time, there are efforts to conserve the known and unknown in marine reserves. The Law of the Sea Convention sponsored by the United Nations provides an international legal framework governing the seas and includes provisions regarding environmental protection and research. It gained enough signatories to go into effect in 1993 (Cousteau 2001).

## REFERENCES

Angel, Martin and Tegwyn Harris. 1977. *Animals of the Oceans, the Ecology of Marine Life.* New York: Two Continents Publishing Group.

Barnes, R. S. K. and R. N. Hughes. 1999. *An Introduction to Marine Ecology.* 3rd edition Oxford: Blackwell Scientific.

Cousteau, Jean-Michel. 2001. "The future of the ocean." Pp. 170–180 in Sylvia A. Earle, ed., *National Geographic Atlas of the Oceans.* Washington, DC: National Geographic Society.

Delaney, John. 2001. "Antarctica and the Europa connection." Pp. 150–151 in Sylvia A. Earle, ed., *National Geographic Atlas of the Oceans.* Washington, DC: National Geographic Society.

Etter, Ron J. and Lauren S. Mullineaux 2001. "Deep-sea communities." Pp. 367–394 in Mark D. Bertness, Steven D. Gaines, and Mark E. Hay, eds., *Marine Community Ecology.* Sunderland, MA: Sinauer Associates, Inc.

Gradwohl, Judith (curator). 1996. "Ocean Planet, A Smithsonian Traveling Exhibition." National Museum of Natural History. http://seawifs.gsfc.nasa.gov/ocean_planet.html.

Grassle, J. F. and N. J. Maciolek. 1992. "Deep-sea species richness: Regional and local diversity estimates from quantitative bottom samples." *American Naturalist* 139: 313–341.

Robinson, Bruce. 2001. "Water column life." Pp. 98–103 in Sylvia A. Earle, ed., *National Geographic Atlas of the Oceans.* Washington, DC: National Geographic Society.

Tharp, Marie. 2001. "Ocean cartography." Pp. 32–44 in Sylvia A. Earle, ed., *National Geographic Atlas of the Oceans.* Washington, DC: National Geographic Society.

Van Dover, Cindy Lee. 2000. *The Ecology of Deep-Sea Hydrothermal Vents.* Princeton: Princeton University Press.

Van Dover, C. L., S. E. Humphris, D. Fornari, C. M. Cavanaugh, et al. 2001. "Biogeography and ecological setting of Indian Ocean hydrothermal vents." *Science* 294 (5543): 818–823.

# IV HUMAN-DOMINATED BIOMES

## INTRODUCTION

Earth is the home of mankind. We humans have been modifying our **habitat** and interacting with our fellow inhabitants of the planet for at least 2 million years. No **ecosystem** has escaped our direct or indirect effect. The preceding chapters on terrestrial and aquatic **biome**s give many instances of this. Some alterations of nature are very subtle—a change in the **species** composition of stands in the tropical rainforests, for example. Others are quite obvious, such as the extension of prairies into humid **climate** regions in eastern North America or the heathlands covering once-forested Britain. The most complete transformations are usually credited to agriculturalists and city dwellers, who have altered natural systems and created new ones that they themselves manage and control. The resulting **agroecosystems** and urban ecosystems are the subjects of this section of the book.

The case can be made, however, that ecologically, humans dominate the entire biosphere. Although much of our current attention focuses on the last 500 years, this dominion is not recent; it goes back in time to whenever humans first arrived in a previously uninhabited area and became new members of the local ecosystems. Fire was the first environmental management tool people used to make their surroundings more habitable and more productive. Evidence from dated layers of charcoal shows that fire has been so used for at least 800,000 years perhaps even for 1.5 million years. Indeed, some have stated that the original home of our **genus**, *Homo*, in East Africa was a human-created tropical **savanna**. This may be too ambitious a claim, but it is widely accepted that the control of fire enabled *Homo erectus* to spread out of Africa north and west into Europe and east across Asia, where charcoal and other evidence at Zhoukoudian, China, dates to 400,000 years ago (Simmons 1996).

**Vegetation** was burned for a variety of reasons. Fire selects for those plants that, among other traits, can resprout from underground structures. These include shrubs and **perennial forbs**. In the destruction of the preexisting above-ground parts of plants, fire is instrumental in mineralizing organic matter—that is converting it to the inorganic chemicals that constitute plant nutrients. The flush of new growth that emerges from **rhizomes** or bulbs or root crowns quickly recycles these nutrients and produces tender, nutritious forage for grazing and browsing animals. Post-burn vegetation is often better quality feed than older plants, and growth rates are faster so the quantity of accessible and palatable forage increases. Undoubtedly, such pasture improvement and the consequent attracting of game animals was a major reason that early peoples burned their habitats frequently. However, fire was also used as a hunting technique to drive game and to clear undergrowth to increase visibility and defend against predators, including other human beings.

Repeated burning by hunter-gatherers for hundreds and even thousands of years changed the structure and composition of vegetation. Fire-adapted plant species were selected for. These include plants that can resprout if their above-ground foliage and buds are destroyed, as previously mentioned; those with thick bark such as cork oak (*Quercus suber*) or sequoia (*Sequoia gigantea*) or protective outer layers of dead or expendable plant tissue such as many palms or **rosette** plants such as yuccas (*Yucca* spp.); and plants that require the high temperatures generated by a hot fire to open their seed capsules and disperse their seeds. An example of the latter from the Boreal Forest of North America is the jack pine (*Pinus bansksiana*); it requires a minimum temperature of 120° C (250° F) to unseal its cones and release seeds onto the fresh deposit of ash that lies open to the sun. Burning selects against those plants that cannot tolerate fire. In forested areas, woody species that have thin bark or that do not sprout will be eliminated. In temperate grasslands, perennial forbs tend to be selected over grasses, and in drier grasslands, **annuals** increase at the expense of perennial grasses (Simmons 1996). Hunting and gathering peoples from the Boreal Forest to the dry forests of Australia still use fire to modify their environments. Large parts of several terrestrial biomes may be fire communities induced or expanded by the activities of past hunter-gatherers. These include Tropical Savannas, Temperate Grasslands, and Mediterranean Woodland and Scrub. (Today, fire suppression rather than initiation has affected chaparral, grassland, and forest in the United States. Prevention and the rapid extinguishing of natural fires have allowed for the accumulation of litter and the development of dense undergrowth, both of which represent fuel. Under the natural fire regime, both litter and undergrowth were regularly eliminated and thereby kept to a minimum so fires were not hot enough to destroy the **soil** or the seed bank held within. Tree seedlings were killed on the prairies, which maintained grasslands even in humid climate regions. Fires died out when they encountered an area that had burned in recent years and didn't have enough fuel to sustain the flames. Now we are in a period of huge, excessively hot fires that rage through vast, densely vegetated areas of chaparral and **needle-leaved** forest. Aspen and other trees dominate what was once tall grass prairie.)

Hunting and gathering peoples also must have affected animal life. Selective killing may have altered the gene pools of game animals, even though their reproduction was not controlled, as was the case in agricultural societies. Range contractions and outright extinctions might also have been a result of human hunting in the past. Most intriguing and controversial is the loss of many of Earth's largest mammals and flightless birds about 11,000 years ago, the so-called Pleistocene extinctions (Martin and Klein 1984). The wave of extinction at the end of the Pleistocene appears to coincide with the spread of *Homo sapiens* out of the Old World to the Americas, Australia, and Oceania. In the United States, it has been associated with the rise of Clovis culture and a hunting economy whose mark in the archeological record is a distinctive bifacial stone point. However, the end of the Pleistocene was also a time of global climate change, and some scientists argue that warming climate and resulting changes in habitat, particularly vegetation, triggered the demise of large grazing and browsing animals. Yet a third hypothesis receiving attention recently is that lethal diseases spread from species to species and from continent to continent, perhaps with humans and their domesticated dogs, near the end of the Pleistocene. The populations of large animals with slow reproduction rates were unable to recover from the high death rates that ensued. Since animals such as mammoths (*Mammuthus* spp.), mastodons (*Mammut* spp.), woolly rhinos (*Coelodonta* spp.), and a host of others survived earlier warming periods at the onset of previous interglacials, it seems likely that the presence of skilled hunters played at least a contributing if not primary role. *If* early peoples did play a significant role in the loss of the megafauna, this would have been a major human impact on Earth's terrestrial biomes. In North America, two-thirds of mammals over 50 kg (100 lb) were lost. Among them were dominant herbivores, including 3 genera of elephants, 6 genera of edentates (Order Xenartha, mainly giant ground sloths), and 15 genera of **ungulates** (Simmons 1996). Grazing, browsing, and rooting by these large animals would have influenced the vegetation structure, and the release of these pressures when the animals disappeared would forever alter the structure and composition of the vegetation.

It has been customary to consider the vegetation of North America before 1492 as pristine—in other words, unaffected by human activity. Land and wildlife management policies have been developed with the supposed pre-Columbian vegetation as the ideal to be maintained or restored. However, here, as throughout Earth's habitable lands, there is no such thing as pristine. Humans have been a key part of Earth's ecosystems and influenced their composition, structure, and functioning for as long as they have occupied given biomes. Indeed, most of the world's vegetation does not reflect present climate but is the product of past and present **disturbance** (Head 2000).

Humans have affected life on Earth in a number of other important ways that will receive attention in the following chapters. People are agents of evolution and select for those traits in plants and animals that are particularly useful or attractive to humans. Domestication—the process by which wild plant and animal species are genetically transformed into crops and ornamental plants, livestock, and pets—has modified and even created species that became integral parts of human-dominated ecosystems. Selective breeding has enhanced the traits desired by people and changed the species into organisms tolerant of or entirely dependent on human activities. Other species have been able to evolve to take advantage of habitats regularly disturbed by humans, the common weeds of agricultural fields and city streets being prime examples.

Human selection may also alter wild species. Forests are selectively logged, which removes mature plants from the breeding population. Some forests and woodlands are managed to the degree that they are essentially plantations. While such agroforestry may be commendable from the viewpoint of economic and even ecological sustainability relative to other types of land use, they are nonetheless much reduced in species richness when compared to more natural systems. Male game animals bearing trophy-worthy antlers and horns are selectively hunted and removed from the breeding population. Behavioral, demographic, and physiological changes occur among hunted populations. Animals that are normally diurnal may, for example, become nocturnal in their habits. The sex and age structure of populations may change as few adults reach old age and, if a trophy animal, more males are shot than females. Sexual maturity may occur at an earlier age, and birth rates may increase.

People have also been agents of **dispersal** by carrying plants, animals, and microorganisms (most importantly, pathogens) with them across natural biogeographic barriers to new regions where many established viable populations. Such human-enabled dispersal has been both deliberate and accidental. This transport of species to great distances from their native ranges is causing a homogenization of Earth's biota, so that the same species may be found everywhere. **Introduced** (or **alien** or exotic) species are a major problem today, particularly on oceanic islands. On the continents, aliens are generally concentrated in the human-dominated agroecosystems and urban ecosystems and rarely invade the interiors of natural areas. However, these natural areas themselves are being greatly reduced, so introduced species may continue to expand their distribution.

With the near-global adoption of agriculture beginning about 9,000 years ago, transformation of woodland and forest to field and pasture began. Villages, towns, and then cities arose. Wildness became viewed as uncivilized. "Around the globe, the physical environment remodeled by human action has long and overwhelmingly been preferred to nature raw" (Meyer 1996, 15). Natural habitats were modified and reduced and also became fragmented. Fragmentation creates habitat islands. Small areas support fewer species and often fewer individuals of a given species, so native species become more vulnerable to extinction. Accessibility between similar patches becomes increasingly limited as distance between patches expands, and the arrival of dispersing organisms or seeds or spores that might replenish declining populations becomes less likely.

It has been estimated that human activities have transformed one-third to one-half of the land surface of Earth. Between 10 and 15 percent is now in row-crop agriculture or urban-industrial land uses. Six to 8 percent is used as pasture. Domestic animals graze grasslands and deserts; forests and woodlands are harvested for a variety of products including fuel wood, timber, and fruits. Add to these obvious changes the subtle and largely unknown influences of global climate change, introduced species, and accelerated rates of extinction in the past 400 years, and one can see that today the human impact is all pervasive on the vegetated surfaces of Earth (Vitousek et al. 1997). In fact, as the human population continues to grow and urbanization increases, it has been predicted that the influence of people on terrestrial ecosystems will be greater than that of climate change in all biomes except the Boreal Forest and the Tundra (Walker et al. 1999).

Impacts on aquatic biomes are more difficult to ascertain, but approximately 60 percent of the human population lives within 100 km (65 mi) of a coastline. Coastal wetlands have been altered or destroyed. Humans use 8 percent of the energy produced in the ocean, but a disproportionate amount comes from the most productive areas: 25 percent of the energy produced in upwelling areas and 35 percent of that of ecosystems on temperate continental shelves is captured by humans. In 1995, 22 percent of marine fisheries were considered overexploited or already depleted. Where freshwater systems are involved, about one-half of accessible water is directed to human use (Vitousek et al. 1997).

There is a longstanding philosophical debate as to the relationship between humans and nature. Is mankind part of or outside of nature? For purposes of describing biomes and identifying the impacts of the dominant humans, it is a convenient device to consider modern humans as separate from nature. Thus hunter-gatherers are generally viewed as living in nature and totally dependent upon natural ecosystems; agricultural peoples live in a modified natural environment, in cultural **landscapes** that exhibit the interplay between natural and cultural processes; and urban peoples live in an environment constructed separate from nature. Of course, urban-industrial societies are not separated from nature, but like hunter-gatherers, they are totally dependent on the resources of the entire biosphere—land, air, and water. They put themselves in danger when they forget that. Also true is the fact that their effects extend far beyond the urban fringe and affect life throughout the world. They are the dominants among the single most dominant species on the planet and determine what nature is and will be.

> There is no clearer illustration of the extent of human dominance of Earth than the fact that maintaining the diversity of "wild" species and the functioning of "wild" ecosystems will require increasing human involvement. (Vitousek et al. 1997, 499)

The two biomes described in this section are valid as separate entities in terms of the degree of human transformations each encompasses. These transformations have enabled the human population to increase enormously and become the global dominant. Agroecosystems and urban ecosystems exhibit human qualities of creativity, societal value, and often great beauty. But they do not remove humanity from the rest of life contained in the other biomes of land or water.

## REFERENCES

Head, Lesley. 2000. *Cultural Landscapes and Environmental Change.* New York: Oxford University Press.

Martin, Paul S. and Richard G. Klein, eds., 1984. *Quaternary Extinctions: A Prehistoric Revolution.* Tucson: The University of Arizona Press.

Meyer, William B. 1996. *Human Impact on the Earth.* Cambridge: Cambridge University Press.

Simmons, I. G. 1996. *Changing the Face of the Earth.* 2nd edition. Oxford: Blackwell Publishers.

Vitousek, Peter M., Harold A. Mooney, Jane Lubchenko, and Jerry M. Mellilo. 1997. "Human domination of Earth's ecosystems." *Science* 277(5325): 494–499.

Walker, B. H., W. L. Steffen, and J. Landridge. 1999. "Interactive and integrated effects of global change on terrestrial ecosystems." Pp. 329–375 in B. Walker, B. Steffen, J. Canadell, and J. Ingram, eds., *The Terrestrial Biosphere and Global Change: Implications for Natural and Managed Ecosystems.* Cambridge: Cambridge University Press.

# 15 AGROECOSYSTEMS

## OVERVIEW

Agriculture involves the human manipulation of plants, animals, soils, and microclimates for the purpose of directing as much of their production as possible to human beings. The artificial ecosystem created, the agroecosystem, is perpetuated only with continual human management. The dominant life-forms in an agroecosystem are a small group of domesticated plants (crops) and/or animals (livestock). These species have been genetically altered from wild progenitors by selection for traits that increase their usefulness to people and adapt them to the agricultural environment where they are produced. Out of an estimated 10,000–50,000 edible plants on Earth, just 30 provide 90 percent of the world's food (measured in terms of caloric intake). Sixty percent of the world's cropland is devoted to just three: wheat (*Triticum* spp.), maize or corn (*Zea mays*), and rice (*Oryza sativa*). Of 15,000 species of mammals and birds, 14 account for 90 percent of the global livestock production. Four species dominate: cattle (*Bos taurus, B. indicus*), water buffalo (*Bubalus bubalis*), sheep (*Ovis aries*), and goat (*Capra hircus*) (Wood et al. 2000).

Agroecosystems differ from natural ecosystems in important ways. As the previous figures suggest, they are much simpler than most terrestrial biomes. Fewer species are involved; while there are exceptions such as traditional agricultural systems in the humid **tropics** or the fish pond-dyke system of southeastern China, this usually results in a less complex structure both in vegetation stratification and food chains. In **monocultures** (one crop species or **variety** per field), only one layer of plant growth occurs; in many **intercropping** schemes, where two species are grown in the same field, two nonoverlapping layers may occur. Only in the **multicropping** schemes found in **kitchen gardens** and with **vegeculture** practices in the humid tropics does a multilayered **canopy** mimic natural ecosystems. Since all plant production is purposefully directed toward human consumption, there are only two tropic levels (producers = crop plants, and herbivores = humans), or three

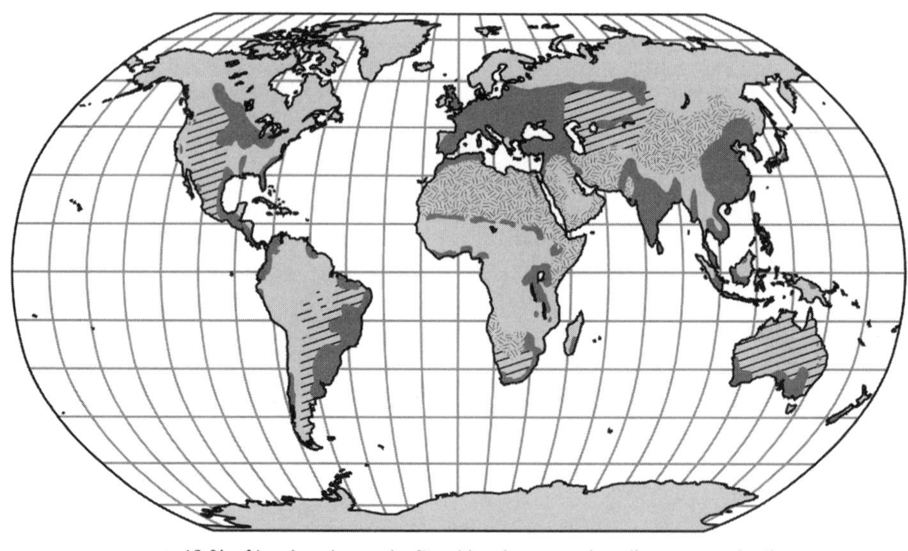


Map 15.1 Dominance of crop and livestock production as land use practices worldwide.

(producers = grasses and other forage plants, herbivores = livestock, and carnivores = humans).

Food or other products are removed from the agroecosystem once they have been produced. The nutrient cycles that would occur in a natural ecosystem are interrupted, and nutrients must be added frequently to maintain the system. Such systems are called open systems: inputs are continually needed to replace the flow of outputs. (Natural ecosystems are considered more or less closed systems, with nutrients being recycled to the plants by fungi and bacteria that break down accumulated organic detritus.) The farmer then must manage the agroecosystem by supplying some form of fertilizers. To increase production, the farmer may also manage the system by removing competitors, (weeds, in the case of crops; wild grazing animals in the case of livestock), predators, and pests such as insects and disease-causing organisms that could reduce yields—or, in commercial agriculture, market value. Such protection demands additional inputs of labor and pesticides. In crop systems, the labor input—either from draft animals or people—includes preparing the soil for planting; weeding; eliminating as much as possible insects, birds, rodents and other crop robbers; and harvesting the crop. This labor is often referred to as an energy subsidy, a necessary support to the enterprise of farming. In mechanized agriculture, the labor subsidy is part of a broader energy subsidy that may include fossil fuels for running tractors and other equipment, chemicals for nutrients and pesticides, and electricity to run irrigation pumps (Tivy 1990). Often much more energy is applied to the cropland (input) than is produced by the crop plants (output)—which makes the system inefficient, especially in comparison to solar-powered natural ecosystems. The energy inefficiency of agriculture is rewarded, however, by a greater production of food, fibers, medicinal herbs, and industrial raw materials than wild ecosystems could provide.

While climate and soils are major influences on the types of crops and livestock produced, often of equal or even greater importance are a complex of cultural, technological, and economic factors. Farmers work within the context of traditions that help dictate such things as food preferences, agricultural methods, skills, and scheduling: in other words, culture provides a sense of what to do and how to do it. Available technology also influences agricultural practices and, to some extent, the types of plants and animals that are raised. Local economic conditions can determine what technologies are used and the types and amounts of energy subsidies possible, while regional and global economic patterns influence the profitability of agriculture in a particular location and create regional specializations on particular commercial products.

The influences of tradition, economy, and nature result in a range of agricultural systems from those with low labor/energy subsidies to those with extremely high inputs. Generally speaking, at the lower end are the more complex systems involving an array of crops and livestock that tend to mimic nature and are still found among traditional farmers in the humid tropics. At the high end are industrialized monocultures and factory-style livestock production such as found in the United States.

Low-input cropping systems often practice a polyculture in which several different crops are grown. The crop varieties either are so-called **landraces**—those crops that evolved over a long period of time in response to the local climatic and soil conditions and folk selection (Sauer 1993)—or local **cultivars**—varieties deliberately selected and modified by the indigenous peoples. The result is a relatively high genetic diversity among the crop plants. Different crops may be grown together in the same field, a method known as mixed cropping, and develop a multilayered canopy. This happens when tree crops, shrubs, and herbs are produced in the same plot. Different crops are planted and harvested at different times in the agricultural year. The inputs are primarily labor, either human or draft animal, or both. Fertilizers are in the form of crop residues and manures. The main purpose of such agriculture is **subsistence**, although some cash crops or livestock products may also be produced (Wood et al. 2000).

In contrast, high-input, industrial agriculture is characterized by monocultures of high-yielding varieties produced in scientific laboratories and field research stations. These varieties often require high amounts of chemical fertilizer and sometimes irrigation to produce high yields. They tend to be genetically uniform so that all individuals can be planted and harvested simultaneously by machines. The lack of genetic diversity imposes a vulnerability to disease, so these crops require high inputs of pesticides as well. (The infamous Irish potato famine of the 1840s is a prime example of what can happen when disease strikes a crop with little or no genetic diversity, even under subsistence conditions.) The counterpart in livestock production is feedlot beef cattle production, production of hogs or poultry confined in climate-controlled structures, or intensive dry-lot dairying. In all, large numbers of animals are housed in corrals or buildings, and high-quality feed grains and high-protein meal are carried to them (Wood et al. 2000). This form of agriculture is strictly commercial and occurs where uniformity of an abundant and relatively low-cost product is required to satisfy the demands of consumers in developed regions of the world.

Agroecosystems can be categorized in several ways. One basic way is to distinguish among crop cultivation; pastoralism or livestock production; and mixed farming, wherein crops and livestock are equally important and are produced in an integrated system. Within each, there are differing degrees of intensification—the increased used of external inputs such as labor, fertilizers, pest-control measures, and/or water. Some of the more widespread agroecosystems are described later within a regional organization that more or less reflects the terrestrial biomes.

## PLANTS

In comparison to their wild ancestors, the world's major crop species are characterized by enlargement of those parts people harvest. For example, the common field bean (*Phaseolus vulgaris*) has a seed five to eight times larger than wild species of bean (Tivy 1990). Other attributes people like or need also increase; examples include protein content, sugar content, oils, aromatics, or drugs. Those features that served as defenses in wild species are often reduced—traits such as toxic and bitter compounds or thorns. The dispersal capability of plants is also reduced. This is most noticeable (and extreme) in the case of domesticated grasses such as wheat, barley, and maize, where a nonshattering flower stem (rachis) has developed that prevents the ripe seed from falling to the ground. These plants are totally dependent upon humans' gathering and planting their seeds.

Crops are classified according to the degree of human intervention in their evolution. Within each species (for example, maize, peppers, or brussels sprouts), a large number of different types have arisen. Those types that evolved primarily in response to natural selection from local climatic conditions and soils, as well as the disturbance and cultivation regimes of traditional agricultural systems, are called landraces. These developed slowly over time as the crops were dispersed to new regions or new elevations. Many are highly localized but finely tuned to specific soil types and have developed resistance to drought, low nutrients, or to the specific pests or diseases they coevolved with. Populations maintain high genetic diversity that lets them continue to evolve to meet changing circumstances (Fowler and Mooney 1990). Cultivars or varieties are more directly the product of deliberate selection for characteristics people want to propagate in future generations. Traditional agriculturalists saved the seeds from good individuals and planted only those seeds the next season. Over generations, new properties such as increased sweetness or bigger seed heads evolved. Considerable genetic variation, and therefore variation among individual plants in the field, was nonetheless maintained. Modern crop varieties are produced by scientific crossbreeding and have only a narrow range of variability and hence breed true. A clone—a group of genetically identical individuals **reproduced vegetatively** or **asexually** through grafting, as in apples, or planting pieces of the plant stem, as in manioc or bananas or sugar cane—is a type of variety. In clones, genetic variation happens only by mutation.

Crop plants can be divided into categories in other ways. First are the cereals or grains, grasses grown for their seeds. The small grains include wheat (*Triticum aestivum*) and rice (*Oryza sativa*). The large or coarse grains include maize (*Zea mays*), sorghum (*Sorghum bicolor*), and millet (*Pennisetum* and other genera). Cereals are easily stored, an important consideration in the provision of a year-round food supply, and are staple foods for much of humanity. A second important and widespread group are the seed **legumes** or pulses, such as peas (*Pisum sativa*), beans (*Phaseolus* spp.), and peanuts (*Arachis hypogea*) whose protein-rich seeds are encased in a pod. Vegetables are the plants of which leaves and young stems are consumed. These include spinach (*Spinacia oleracea*) and lettuce (*Latuca sativa*), but also plants that supply nicotine and caffeine (i.e., tobacco and tea). Fruits are those plants where the fleshy material containing or encasing the seeds is eaten or used. The fifth group is the roots and tubers whose swollen underground roots and stems are the focus of human attention. White potato (*Solanum tuberosum*) and beet (*Beta vulgaris*) are the chief tubers of **midlatitude** diets, but a number of others are important in the high Andes and in the humid tropics. The people of the Andes cultivate oca (*Oxalis tuberosa*), ullucu (*Ullucus tuberosum*), and other tubers along with an enormous variety of potatoes. In the humid tropics, manioc or cassava (*Manihot esculenta*) is the staple crop in many places, and yam (*Dioscorea* spp.) is also very important. Sweet potato (*Ipomoea batatas*) and taro (*Colocasia esculenta*) are among other tropical root crops significant in local diets (Tivy 1990; Gade 1999).

Industrial crops are a group of domesticated plants that provide raw materials for processing into usable products. These include oil plants such as rapeseed or canola (*Brassica napus*), soybean (*Glycine max*), oil palm (*Elaeis guineensis*), and cotton (*Gossypium* spp.). Cotton, of course, is also an industrial fiber plant, as are flax (*Linum usitatissimum*), jute (*Corchorus* spp.), sisal (*Agave sisalana*), and hemp (*Cannabis sativa*). Sugar cane (*Saccharum officinarum*) and sugar beet (*Beta vulgaris*) provide most of the world's sweeteners; coffee (*Coffea arabica*), tea (*Camellia sinesis*), and cocoa (*Theobroma cacao*) are important beverage crops.

Most crop plants are annuals or perennials, such as cotton, that are grown as annuals. The main exceptions are the tree crops—such as oil palm, cocoa, citrus (*Citrus* spp.), coconut (*Cocos nucifera*), and rubber (*Hevea brasiliensis*)—and **evergreen** shrubs such as coffee and tea.

Weeds are also a regular component of agroecosystems, and many evolved in the open, frequently disturbed habitat of the agricultural plot (Harlan 1995). Typically weeds are sun-loving plants with excellent dispersal capabilities and high reproduction rates. Worldwide, the most common weeds in agroecosystems are nut grass (*Cyperus rotundus*), Bermuda grass (*Cynodon dactylon*), barnyard grass (*Echinochloa crus-galli*), goosegrass (*Eleusine indica*), Johnson grass (*Sorghum halepense*), lalang (*Imperata cylindrica*), water hyacinth (*Eichornia crassipes*), and purslane (*Portulaca oleracea*) (Tivy 1990).

## ANIMALS

Domesticated mammals are mostly ruminants; they tend to be smaller than their wild relatives and have higher reproductive rates. Many display the characteristics of the juveniles of wild ancestors, features such as baby faces with short muzzles, curly or fuzzy hair rather than straight, and multicolored or patterned coats rather than solid colors (Simmons 1989). Most prominent in agroecosystems are social or herd animals, such as cattle, sheep, goats, water buffalo, yaks (*Bos grunniens*), horses (*Equus caballus*), reindeer (*Rangifer tarandus*), llamas (*Lama glama*), alpacas (*L. pacos*), and camels (*Camelus bactrianus* and *C. dromedarius*). These animals are grazers or

Photo 15.1 The long-haired Highland breed of cattle. It was developed to flourish in the harsh environment of the Scottish highlands, seen in the background. (Photo by SLW.)

browsers that consume plant material people cannot digest. Among birds, ducks and geese are primarily consumers of plants. However, the agricultural **community** may also contain carnivores and scavengers such as chickens that, when free roaming, feed on insects and grubs; and pigs (*Sus scrofa*) that search out roots, nuts, feces, and carrion.

Landraces developed among domestic animal species, just as they did among crop plants. They adapted to local climates, wild forage, and living pretty much on their own. Landraces typically show the ability to convert low-quality grass and browse to milk, meat, and fiber; they also demonstrate good mothering skills often lost in modern industrial **breeds**. However, instead of talking about cultivars and varieties, the breed is the basic unit for livestock deliberately bred by humans. A breed is a group of animals carefully selected and bred to reduce genetic variation to ensure the production of specific desirable traits. The outcome is a "group readily distinguished from other members of the species and . . . consistently within a relatively narrow range of parameters: when bred to one another, members of the breed reproduce this distinguishing type" (Bixby et al. 1994, p. 8). The most recognizable traits are external features such as color, pattern, or size, but behavior (e.g., docility, care of offspring), tolerances to particular environmental stresses, growth, and production rates are also selected and may be distinguishing characteristics (Photo 15.1). Because variation within a breed is minimized, maximum preservation of genetic diversity in any given livestock species requires conservation of as many breeds as possible.

Other animals of significance in agroecosystems are pests such as leaf-chewing insects, leaf miners, and plant-sucking insects; grain- and fruit-eating birds; voles, mice and other rodents; and predator animals, usually avian or mammalian, that prey on crops or livestock or eat carrion. Hawks and eagles are common birds of prey in agroecosystems, while vultures perform the service of removing dead animals. Raccoons and white-tailed deer are common crop predators in the much of the United States, while the coyote is a widespread livestock predator—probably more maligned than facts warrant because they eat more fruits and small rodents than large animals.

## MAJOR TYPES OF AGROECOSYSTEMS

### Pastoral Systems

Pastoralism is an extensive form of agriculture that requires few inputs. These agricultural ecosystems use natural or seminatural vegetation, which is called rangeland when so used, to produce livestock. Domesticated animals provide people with meat, milk, hides, fat or tallow, wool or other fiber, and, in many places, dung for fuel. In most of the subsistence pastoral economies that remain in the developing countries of the world, milk is the main product, although the animals are multipurpose. In the commercial ranching operations associated with industrialized countries, meat or wool is the main product (Vasey 1992; Tivy 1990).

Almost 50 percent of the total land area of Earth is used as rangeland. A large proportion of pastoral agriculture takes place in deserts and grasslands, especially in the Old World, on land that is unsuited (ecologically or economically) to the cultivation of crops. About half occurs in tropical and subtropical climate regions and half in temperate and cold climate regions. Most livestock are raised under free-range conditions and consume primarily grasses and leguminous shrubs and trees. Cattle are the preferred animal throughout the world. Sheep dominate in more arid regions. Goats are found everywhere but are most prominent in the agricultural economies of semiarid and tropical Africa and Asia (Tivy 1990). Camels gain importance in the driest areas, where they are herded in mixed flocks with sheep and goats. Other specialized animal use occurs in Arctic Eurasia, where reindeer are herded; in high Himalayan countries where yaks are grazed along with horses, cattle and sheep; and in the Andes, where llamas and alpacas were the traditional livestock and are now raised together with cattle, sheep, and goats.

**Nomadic Herding**. True nomadism is rare, but when it occurs people continually move their flocks in search of forage and water. The movements have no established pattern; instead they have a built-in flexibility to exploit the unpredictable flushes of grasses and foliage in a desert environment. Movements are confined to recognized tribal territories, however. The overall economy usually involves an exchange of goods (animal products for crop products) and services (e.g., transport of goods or manuring of fields in exchange for food, forage, and water) with permanently settled farming people at desert oases or along **exotic** streams. The animal complex may contain camels, cattle, donkeys, goats, and sheep. The Bedouin tribes of Southwest Asia, for whom camels are the dominant livestock species, exemplify this system (Vasey 1992; López-Bellido 1992; Tivy 1990).

Forms of seminomadism are more prevalent today than true nomadism is. Pastoralists move their flocks seasonally between summer and winter pastures along well-established migration routes, and they have permanent settlements where some members of the family or community may raise crops. This type of nomadism is also referred to as transhumance. Two general patterns occur: one horizontal, the other vertical. A horizontal pattern is developed in desert areas where movement is to the desert rangeland during the wettest time of year and to the home base at an oasis or some other more humid habitat during the driest part of the year. A variation of this occurs in the Sahel, where movement is scheduled according to the activity period of the tsetse fly (*Glossina* spp.), which transmits trypanosomias, a disease that affects livestock and people. Another variation can be found among reindeer herders in northern Eurasia. In the summer, the herds are taken to the Arctic tundra and in winter to settlements at the edge of the Boreal Forest (Vasey 1992).

In mountainous areas outside the tropics, people commonly practice a vertical movement of herds and flocks from a winter range in valleys to mountain meadows in summer. Sheep tend to be the dominant animal involved in this type of transhumance; although where forage quality and water availability permit, such as in Scandinavia or Switzerland, cattle are involved. In Central Asia, Kazaks and others congregate in winter settlements in the valleys where they have their permanent houses and then disperse to mountain pastures in summer, where they live in yurts (Vasey 1992). Transhumance is still practiced in the United States, especially in the Rocky Mountains and the Basin and Range Physiographic Province, although sheep may be trucked between their seasonal rangelands instead of herded.

**Commercial Ranching**. Ranching is a sedentary form of pastoralism primarily associated with European cultures in the Americas and Australia (Photo 15.2). In the United States, ranchers graze cattle and sheep on vast tracts of land, part of which they own and much of which is public lands they lease from the federal government. The rancher has a permanent base at his homestead, but along with ranch

Photo 15.2 Cattle ranching in the western United States. Commercial ranching depends on wild plant production, although wells artificially increase the availability of water. (Photo by SLW.)

hands may spend days at a time at line camps during calving and roundup times. As in nomadic pastoralism, the plants in the system are wild grasses and shrubs. The livestock is produced for sale as live animals either to finishing establishments, where they are fattened for slaughter, or to breeding stations, where they may be crossbred with other breeds to produce large, fast-growing hybrid calves or lambs (Vasey 1992).

## Traditional Rain-Fed Cropping Systems

**Tropical Agroecosystems.** Shifting agriculture or the use of temporary agricultural plots is still widely practiced in the forest and savanna regions of tropical South America, Africa, and Asia. This is a very extensive form of agriculture that minimizes the labor subsidies of weeding and fertilizing by abandoning plots once crop yields decline and/or weeds and secondary forest encroach. The abandoned plot is left in bush or tree fallow (i.e., left unplanted but overgrown with shrubs or trees) for a number of years before being reused. The typical method of preparation for planting involves clearing a patch of forest or woodland and burning the debris during the dry season. Hence the agricultural method is often called slash and burn. The resulting ash is high in many plant nutrients; any unburned material contributes nitrogen to the soil as it rapidly decays. A specialized type of slash and burn agriculture, chitemene, is found in East Africa. In chitemene, branches are cut from trees over a wide area surrounding the plot and then carried to the open area to be burned.

The planting of a mixture of crops begins at the start of the rainy season (Photo 15.3). A digging stick or hoe is used, which reduces disturbance of the soil. The crops may include several **growth-forms**—annual herbs, shrubs, and trees—and therefore produce a complex, multilayered, **closed-canopy** vegetation structure. The different crops typically have different growth periods, different nutrient requirements, and different food value, so they are complementary in the agroecosystem and the human diet and yield a year-round supply of food. Vegeculture, the cultivation of root crops propagated from stem or rhizome cuttings instead of seed, is characteristic in rainforest regions. Common crops of vegeculture are manioc, yams, maize or rice, bananas, beans, and various vegetables. In Africa, up to 50 species have been reported in the same plot (Vasey 1992; Tivy 1990). Where this agroecosystem occurs in tropical rainforests, animal protein usually derives from fish and wild game.

The typical time a particular plot of ground is cropped ranges from one to three years, although longer periods do occur, and tree crops may be harvested long after the planting of annual crops has ceased. The fallow period is also quite variable, but 10–15 years is common. By then, enough secondary woody growth has invaded that the plot may be cleared and burned again.

In drier parts of humid Africa, at the savanna forest edge, fewer crop species are able to tolerate the long dry season, and cereals become more important. The fallow is also necessarily longer, and grasses rather than shrubs and trees often invade the resting plot. To reuse the land after the fallow period, plows are needed, and these are usually pulled by oxen or water buffalo. Cereals and manioc characterize fields distant from the settlement. Diversity increases in fields closer to the farm compound and is maximized in kitchen gardens next to homes. Kitchen gardens have multistoried plantings of bananas, fruit trees, oil palm, maize, and a variety of

Photo 15.3 An Indonesian farmer planting cassava (manioc) in a recently cleared and burned plot. Rice, bananas, and eggplants will also be grown during the few years this temporary plot will be in use. (Photo © Michael S. Yamashita/CORBIS.)

vegetables. They may also house fish ponds, poultry, and small livestock such as sheep and goats (Juo and Ezumah 1992).

Mounds and ridges are common features of the agricultural landscape in the savanna region of Africa. At the end of a rainy season, sod is cut with a special hoe and stacked to create a raised platform where planting will commence once the rains begin again. The mound creates a well-drained surface above waterlogged soils created either by a raising water table or by floods resulting from rising streams in the wet season. White yam (*Dioscorea rotundata*) is planted at the top of the mound and a pole inserted for this plant to grow up. A number of other crops are planted together on the mound around the yam. Major components of the mound community are maize or upland rice and cassava. They are joined by minor root crops such as cocoyam (*Xanthosoma sagittifolium*) and sweet potato, vegetables such as African spinach (*Amaranthus* spp.), and beans, okra, and melons. Wetland rice may be planted in the saturated soils between mounds (Juo and Ezumah 1992).

Permanent field cultivation systems also exist in the humid tropics and subtropics. People have long taken advantage of nutrient-rich soils such as those produced by annual deposits of alluvium on floodplains or those derived from basic volcanic rock materials. Kitchen gardens fertilized by plant detritus and household wastes are also used on a continual basis. In prehistory, raised fields in Amazonia and other parts of tropical America and the *chinampas* or floating gardens near present day Mexico City supported dense human populations on maize-based diets. Today, however, the major forms of continuous cropping in the humid tropics are based on wetland rice (*Oryza sativa*). About half the world's production takes place in South and Southeast Asia. Wetland or paddy rice can grow in saturated soils because its

Photo 15.4 Paddy rice production. This is an ancient and intensive form of permanent agriculture. The crop with the white flowers growing here in Sichuan Province, China, is lotus. (Photo by SLW.)

roots have pore spaces that hold air. Hundreds, if not thousands, of varieties of rice were developed within three traditional groups, *japonica*, *javanensis*, and *indicus*. Some were tall, others short; some were early maturing, some late, others intermediate. They were adapted to various local environmental factors as well as taste and texture preferences. *Japonica* is a midlatitude form than does not tolerate the short days of the tropics. It is common in China, Korea, and Japan. The other two groups are tropical in distribution. Today, modern high-yielding strains are increasingly replacing the traditional varieties.

The wetland rice agroecosystem is typically a high-input, labor-intensive form of agriculture (Photo 15.4). Water control is paramount. In Asia, 80 percent of rice production depends upon precipitation and runoff from higher slopes (Tivy 1990). To maintain proper water levels, terraces and shallow ponds or paddies have been constructed and linked by a series of canals and drainage ditches, all of which require constant maintenance. Nitrogen-fixing organisms such as **cyanobacteria** and water fern (*Azolla pinnata*), which has algae living symbiotically with it, enrich the paddy. Rice is commonly grown in seed beds and then transplanted by hand in the prepared paddy. The water level is raised to keep pace with the growing rice.

Throughout most of tropical and temperate Asia, one crop of rice per year is the norm. However, it is often part of a crop rotation system. In summer, paddy rice is grown, and in the dry winter months, wheat and barley, legumes, vegetables, or even cotton and tobacco are produced in the fields.

Dryland farming in the tropics is found in the drier parts of savanna regions. Dry farming refers to techniques that allow rain-fed crop production in areas of low precipitation. The techniques involve conserving moisture by reducing evaporation or transpiration and, sometimes, improving the infiltration of rainwater by slowing the movement of runoff. Crop plants are widely spaced to reduce the number of plants, and weeds are removed through tilling. Both practices reduce the loss of soil moisture via the plants themselves. Sometimes a mulch of plant residue is used to lower evaporation from the soil surface. Ridge tying and furrow dams slow the movement of surface water flowing across the field and direct it into the soil. The typical practice in dry farming is to leave fields fallow for a year or more to allow soil moisture to build up while there are no plants at all on the plot. This usually results in a grass or bush fallow, which is then burned before the plot is reused (Tivy 1990; Vasey 1992).

Cereals are the main crops of dry farming. In the Sahel of Africa, where summer is the rainy season, the main grains are millets and sorghums. Farther north where a mediterranean pattern of winter rains occurs, wheat and barley are produced. Dry farming in both instances is often integrated with animal husbandry, either with animals owned by the farmers or with the animals of seminomadic pastoralists that arrive to graze the stubble during the nongrowing and fallow periods. As they graze the harvested fields, both resident and migratory livestock fertilize the cropland with their manure. Resident animals may also provide draft power for the agroecosystem (Tivy 1990; Vasey 1992; Fussell 1992).

In West Africa, the main crops are pearl millet and sorghum. They are grown mixed together or intercropped with cowpeas (*Vigna unguiculata*) and maize. Crop rotation may also be practiced. Peanuts, a cash crop, may be planted after cereals to help enrich the soil, or cereals may be planted after a harvest of cotton, another cash crop. In some instances, organic matter derived from weeding is collected between the rows of cereals to act as a mulch and to decay. The next season, a cereal will be planted in these now nutrient-enriched ridges or mounds of compost. A variation on this practice is to burn the weed and crop residue in the fields (Fussell 1992).

Décrue farming is a specialized drylands agricultural method that takes advantage of the flood regime of exotic rivers. Planting occurs on the floodplain as waters recede at the end of the rainy season. The floodwaters are the only source of soil moisture and early annuals that can complete their growth cycles before the root zone dries out are the preferred plants. Décrue has long been practiced on the inland delta of the Niger River, where the dominant crops are drought-resistant millets and sorghums. In complementary fashion, African rice (*Oryza glaberrima*), a floating rice, is planted as waters rise during the rainy season—a practice known as crue (Harlan 1995; Vasey 1992; Fussell 1992; Harris 1976).

**Midlatitude Agroecosystems.** Mediterranean agriculture around the world incorporates several strategies to deal with the fact that rain falls during the cool season and summers are dry. One system grows temperate crops, especially wheat and barley, during the winter growing season and lets sheep graze the stubble during the summer fallow. The wheat tends to be the hard durum types (*Triticum durum*) used for pasta. Another system depends upon subtropical trees that are resistant to drought but sensitive to frost. Typical mediterranean crops in this agroecosystem are olive (*Olea europaea*), grape (*Vitis vinifera*), and almond (*Prunus dulcis*). Most of the olives are processed into oil, and most of the grapes are processed into wine or raisins, although table grapes can also be important. With

Photo 15.5 European mixed farming, best exemplified in the United States by dairy farming. The farmer produces and stores hay and silage to feed his livestock through the winter months. (Photo by SLW.)

irrigation, citrus, melons, bananas, pineapples, avocados, and temperate fruits such as apples and peaches may also be produced. Today, intensive irrigated agriculture—complete with heaters and fans to ward off frosts—is the rule. Early vegetables and fruits such as strawberries, lettuce, and potatoes are grown along with tree crops (Tivy 1990).

Rain-fed mixed farming is the traditional mode of European agriculture in humid temperate regions, and it was transferred to North America during the colonial period (Photo 15.5). Mixed farming in this sense refers to the complete integration of crops and livestock into the agroecosystem and the human diet. Crops today are a mix of European and American natives on both continents. Prehistoric Europeans developed small grains such as wheat and rye (*Secale cereale*) and most of our livestock: cattle, sheep, goat, pig, and poultry (with the exception of the turkey, which is North American in origin). Native American agriculturalists contributed maize and the white potato as primary crops in the system. Crops are grown in monocultures, one crop to a field. The plow and, originally, draft animals such as oxen and horses are integral to the system. Livestock was stabled at night, a practice that concentrated the manure and made it easier to collect and transfer to the fields. The hay meadow and the production of silage crops is as important as the production of cereals and other foodstuffs (Vasey 1992). A long-period rotation among field, pasture, and woodlot may occur in southeastern North America, discernible at the scale of human generations (Hart 1998).

In the humid temperate regions of eastern Asia, paddy rice is the main crop, often grown in rotation with a winter crop of wheat, soybeans, or some other crop). While the paddy is flooded, carp may be introduced and fed plant debris or manufactured

Photo 15.6 Monoculture of wheat. This is an extensive form of agriculture that originated in the Middle East and Europe and allows a few people with access to plows and land to produce surplus grains. (Photo by SLW.)

feed pellets. Ducks also may be raised on the paddy. On the dikes separating ponds, farmers grow a variety of vegetables for human consumption and fodder plants for livestock. Pigs kept in pens near the farmer's house provide manure; traditionally, human night soil was collected, fermented in vats, and applied to fields as well. A water buffalo may be kept as draft power to plow the paddy between plantings, and it also contributes manure to the system. The nutrient-rich sediments that accumulate on the bottom of the paddy are applied to the dikes as fertilizer in what is one of the most intensive forms of midlatitude agriculture.

Dryland farming techniques are prevalent in the semiarid regions of the midlatitudes. As in the tropics, they include leaving fields fallow for one or two years to recover soil moisture. At any given time, only one-third to one-half of the land is cropped. The fallows used to be plowed to kill weeds and prevent water loss by transpiration, but excessive soil erosion from wind and water has led to the practice of leaving the stubble and crop residues in the field during fallow years to better bind the soil and act as a mulch. The stubble also traps snow in winter and in spring to direct meltwater downward along the root system into the soil, thereby increasing infiltration. The dominant crop in North America and Eurasia is winter wheat (*Triticum aestivum*) or bread wheat (Photo 15.6), but industrial crops such as sugar beet, rapeseed, and sunflower are also extremely important. White potato and buckwheat (*Fagopyrum esculeritum*) have regional importance, especially in central Europe.

## Irrigation Agriculture

Irrigation refers to the transport of water to cropland through man-made features and its subsequent delivery to the root zone of crops. This is more a technique of farming than a type of agroecosystem because just about any crop can be grown under well-managed conditions of irrigation. It is a practice formerly limited to arid and semiarid areas, but more and more frequently commercial agriculture is using it in humid regions as a safeguard against unpredictable droughts during the growing season. Water for irrigation in arid lands comes from two major sources: surface waters and groundwater (Vasey 1992; Tivy 1990). Techniques for distributing water from each of these sources are briefly described.

In most deserts, the primary surface water supply is exotic rivers. To convey this water to cropland, a network of canals and/or aqueducts is constructed. Since river levels usually fluctuate seasonally and may be at their lowest when water is most needed in the fields, rivers are frequently dammed to create reservoirs and allow for the regulation of water flow.

Groundwater is the product of precipitation, an accumulation of water that has percolated through the ground and formed a saturated layer of soil, regolith, or bedrock below the surface. When the groundwater readily moves through the substrate and is retrievable, the water-bearing material is known as an aquifer. While it may seem that the supply should be limited in dry areas, groundwater derives not only from local precipitation but slowly flows through aquifers originating in well-watered uplands. Indeed, not all groundwater represents contemporary accumulation, but vast deposits were formed in many places during the pluvial periods of the Pleistocene epoch, when many of today's drylands were considerably more humid than at present. This is now considered fossil groundwater, and it is mined as a nonrenewable resource in the Sahara, on the Arabian Peninsula, and in the central and western United States.

Groundwater and often river water must be lifted to the surface. This can be accomplished quite simply by lowering buckets into a well, but a variety of other means have been devised. In the Old World deserts, a *shaduf* or reverse waterwheel has been in use for millennia. Another ingenious method for reaching groundwater developed first in Iran in ancient times. The goal was to extract the groundwater in alluvial fans at the base of desert mountains and transport it to cultivated fields in the valleys. A narrow, essentially horizontal tunnel was built between the valley floor and the fan to tap the resource and let water flow underground, by gravity, to the farms. Vertical shafts were dug down to the tunnel at intervals of 30–45 m (100–150 ft) to permit removal of excavated material and to let air and light into the tunnel for maintenance workers. Visibly raised piles of debris encircle the shaft holes and stretch in straight lines along the desert surface for kilometers until water gushes forth beside a bright green field. This technology spread from Iran, where structures are known as qanats, across North Africa, where the Arabic name foggarra is used, and into western China, where they are known as karezes. Many are still in use today (Vasey 1992).

More modern systems use power-driven pumps to draw water out of rivers or up from the ground and empty it into canals for distribution to fields. Application onto the field traditionally took one of two forms; both required the preparation of the field, particularly leveling it to allow gravity flow. In one method, basin irrigation, the entire field is flooded with a sheet of water that is then allowed to sink into the ground. In the other, furrow irrigation, the water is directed into deep furrows or trenches from where it percolates into adjacent ridges planted with crops.

Much water can be lost by evaporation from irrigation canals and aqueducts, as well as from basin and furrow applications. Many commercial growers have switched to sprinkler irrigation systems and, increasingly, to drip or trickle irrigation methods. Sprinklers spray water under pressure and eliminate the need to level the field. Rotating sprinkler heads or fixed or movable perforated pipes are used. Center pivot sprinklers leave their indelible imprint on the land as they create large circles of green crops on the desert floor. Drip irrigation involves the use of perforated hoses laid on the ground to deliver water drop by drop in a very controlled fashion to the root zone of individual plants. It results in less water loss from evaporation or excessive application (Vasey 1992; Tivy 1990).

Watering crops with irrigation systems is a major way humans modify local climates to create desired agroecosystems. Water is used to control climate in other ways, too. In greenhouses, water is applied for evaporative cooling. In hot climates, sprinklers can apply water to the canopy of citrus groves to cool the area. Water can also be used to prevent killing frosts in late spring or early fall in temperate climates, either by flooding crops such as cranberries or spraying fruit orchards when freeze warnings are issued (Tivy 1990). The water that encases the plants will not go below freezing, and often—because of water's heat-storage capacity—temperatures at the plant surface do not get as low as 0° C (32° F), even though air temperature might be lower than that.

Irrigation greatly increases the production of cultivated crops in arid and semiarid regions. However, it is an expensive endeavor. In simpler forms, the expense is in human labor; in mechanized forms, the expense is monetary, which often prevents irrigation from being an option for the small landholder.

## Modern Industrial Agriculture

Modern industrial agriculture differs from traditional forms by the use of crops and livestock developed in scientific breeding programs or through biotechnology. Crops, herds, or flocks consist of genetically identical or nearly identical individuals. Machinery—including tractors, automatic feeders, and electric or gas-driven irrigation pumps—replaces human and animal labor. Massive inputs of fertilizers produced by the chemical industry and pesticides produced by the petrochemical industry are the rule. This translates to enormous energy subsidies that do not end at the farm. More energy is used in processing, distributing, and preparing food and other agricultural products. Since considerable capital investment is needed, operations are strictly commercial and most often large scale. Industrial agriculture has increased production, whether measured per man-hour of labor or per unit of land.

**Industrial Livestock Production.** Instead of depending upon wild vegetation for the bulk of livestock feed, as in extensive pastoral systems, modern livestock raising is an intensive form of agriculture. Several breeds of beef cattle are used, among them Hereford, Black Angus, Simmentel, and Charolais. Animals graze in improved, cultivated pastures, where they feed upon improved, high-yielding varieties of grasses and forbs cultivated by the stockman. The most common improved forage crops are perennial ryegrass (*Lolium perenne*), timothy (*Phleum pratense*), meadow fescue (*Festuca pratense*), tall fescue (*F. arundinaceae*), and the legume clover (*Trifolium* spp.). Alfalfa (*Medicago* spp.) is grown for a high-quality hay and maize for silage, the enriched product of fermentation processes. The most intensive grazing systems involve either continuous stocking of pastures where the number

Photo 15.7 Steers being fattened in a feedlot prior to slaughter. This finishing of livestock is a form of industrialized agriculture. (Photo by SLW.)

of animals are kept in balance with the carrying capacity of the land or rotational grazing in which pasture is heavily grazed for a period of time and then the animals are removed to permit recovery (Tivy 1990).

The extreme form of industrial livestock farming is factory-style production that confines animals to dry lots (outside areas without grass or other vegetation) or to specially constructed structures. Feed is brought to the animals. Beef cattle bred on the open range are finished in dry lots, where feed and water are provided in mechanically filled troughs to tens of thousands of animals per operation (Photo 15.7). They are fattened for six months before being shipped to slaughterhouses. Maize, grain sorghum, barley, and alfalfa hay are dominant feed crops purchased by feedlot owners. Cattle can also be fed a wide range of waste products from other forms of industrial agriculture, including cottonseed cakes, begasse from sugar processing, spent mash from distilleries, sugar beet tops, citrus pulp, and even the remains of other cattle—the way mad cow disease was spread in Great Britain (Hart 1998).

Dry lot dairying operations in the western United States handle hundreds of milk cows fed on irrigated alfalfa hay produced elsewhere. Most dairy cattle around the world are the same breed, the black and white Holstein. This is a highly specialized animal that depends upon high-quality feed and intensive management to yield prodigious amounts of milk. The cows are brought into milking barns and attached to milking machines twice a day. At those times, they are fed measured quantities of concentrated feed. Cows are often worn out after two or three lactations and are then sold to become hamburger (Hart 1998; Bixby et al. 1994). Dry lot dairies are located near large cities so they can supply the inhabitants with large quantities of fresh milk.

Industrial breeds of poultry and hogs are sheltered in climate-controlled structures. Egg farmers may have as many as 4 million laying hens confined in small cages. The White Leghorn breed is used almost exclusively. Each hen produces three or four eggs every four days. Immediately after they are laid, the eggs roll onto conveyor belts for sorting and packing. Broiler chickens are produced factory style by farmers caring for 10,000 or more birds at a time. The farmer purchases day-old chicks, mostly Cornish × White Plymouth Rock hybrids, from a hatchery and grows them to a finished weight of 2 kg (4.5 lb) in seven weeks. At the farm, each bird is allocated about 0.3 $m^2$ (1 $ft^2$) of space on the floor of a well-ventilated shed. Feed and water troughs suspend from the ceiling so they can be raised out of the way when the finished birds are caught for transport to a processing plant. The poultry farmer is a link in an integrated agribusiness. He buys chicks and feed from a corporation that also dictates management of the birds, owns the processing plant, and is responsible for marketing and distributing the products to retail outlets (Hart 1998).

Industrial pork production operates in a fashion similar to the poultry industry. In North America, it relies on three swine breeds: Duroc, Hampshire, and Yorkshire. The hogs are confined in long, closed sheds ventilated with a series of large fans to keep temperatures close to 18° C (65° F), the temperature at which hogs are most productive. The improved breeds maintained in dense populations are highly susceptible to diseases, so extreme care is taken to prevent the introduction of pathogens. Hogs reach market weights of 100–120 kg (225–260 lb) in six months. Hogs produce huge amounts of urine and fecal matter, and waste management is an integral part of the operation. Manure lagoons in which solids are allowed to settle out are key parts of the industrial hog landscape. Due to strong odors, operators try to locate in sparsely populated areas. The hog sheds and lagoons are increasingly common additions in regions where center-pivot irrigation leaves space available between adjacent circular fields of forage crops (Hart 1998).

While sheep production has not been industrialized, modern meat and wool producers depend upon just a few breeds. In the United States, sheep have been developed for meat production, and the black-faced Suffolks are favored because they produce large lambs. Most other breeds are maintained in small numbers for specialty wools (Bixby et al. 1994).

Most goats raised commercially in the United States are Angoras and produce mohair (cashmere, the other fiber from goats, is the winter undercoat of goats and is not restricted to a particular breed) (Bixby et al. 1994).

**Industrial Crop Production.** The industrial production of crops involves huge monocultures of hybrid or high-yielding varieties of crop plants developed since the 1960s. These crops respond well to irrigation and to high amounts of chemical fertilizers; they yield more product per unit of land and per worker than traditional crops. Seeds are produced at agricultural research stations or by major corporations, and farmers must buy new seeds each year. (In traditional farming, seeds are saved from the harvest to be planted the following year.) Plant varieties are selected for higher yields, uniformity of size, simultaneous maturity, appeal to the consumer, and ease of mechanical harvest. Food crops, fodder crops, and industrial crops may all be grown in this manner. Two of the most widespread crops worldwide are soybean and fodder maize.

Vast stands of genetically identical plants are highly vulnerable to disease and other pests, so large amounts of chemical pesticides are applied annually. As the pests evolve immunity to the chemicals, newer, more toxic herbicides and insecticides need to be developed. Today it is possible to extract genes that introduce immunity to disease or tolerance to salts or drought from one variety to another of the same crop and from one species to another. Such genetic engineering holds the possibility that crops may be developed in the future that do not require the input of pesticides frequently added to industrials agroecosystem today.

Large-scale crop production by industrial means is characteristic in the United States, Canada, Australia, Argentina, and Brazil. In the developing world, the high-yielding crop varieties of the **Green Revolution** (see p. 366)—wheat, rice, and potatoes—are produced by growers who can afford irrigation systems, chemical fertilizers, and industrial pesticides—usually larger landholders or foreign corporations. Around the world, the new varieties are replacing the lower-yielding but genetically diverse landraces and cultivars of traditional agriculture.

**Alternative Agriculture.** In response to some of the negative impacts of large-scale industrial agriculture (see following), some farmers, especially in developed countries, are experimenting with modern forms of low-input, small-scale farming or organic farming. Integrated pest management, the planting of diverse crops and varieties (and sometimes using so-called heirloom varieties), intercropping, and crop rotation are common practices to encourage the biological control of pest species. Manures and composts are preferred over chemical fertilizers and mulching and weeding over the application of herbicides (Vasey 1992). Free-ranging, grass-fed traditional livestock breeds may also part of these agroecosystems (Bixby et al. 1994). Due to the higher cost of farming in this manner, growers must often target specialty niches in the market to sell their products. Farm-fresh, organically grown vegetables and eggs, specialty cheeses, and colored wool for handspinning are among economically viable products. Consumers are frequently people in the middle and upper socioeconomic classes who are vegetarians and/or espouse environmentally sound land-use practices.

## IMPACTS OF AGRICULTURE

Early agriculturalists were agents of evolution who created new habitats to which wild species (especially those we categorize as weeds) adapted and nurtured the development of landraces of crops and livestock numbering in the thousands. An estimated 5,000 potato landraces alone have evolved, 3,000 of them cultivated in the Andes. Tens of thousands of types of wheat, maize, and rice also developed during the millennia since they were first domesticated (Fowler and Mooney 1990). At the end of the Pleistocene, possibly 10 million people lived on Earth supported by hunting and gathering economies. Today, the human population exceeds 6 billion, and almost everyone is supported by agriculture in one form or another. Agricultural systems are one of the great achievements of mankind and also one of the major ways humans have modified the planet.

Agriculture has created the rural landscapes that many people prize as open space (Photo 15.8). The mosaic of field and wildland and the farmsteads and other structures associated with farming—often with distinct cultural or regional styles—

Photo 15.8 The treeless landscapes of Scotland and England, largely the product of sheep husbandry practices. (Photo by SLW.)

have continually changed to reflect new technology, changing crop and livestock preferences, the migrations of ethnic groups, and the economic conditions of the time. With the decline of small-scale farming today, many view rural landscapes as a disappearing part of a culture's or region's history that should be preserved for future generations.

The development of European mixed farming—where livestock were kept near humans, often sheltered in the same building—has been implicated in the rise of many infectious diseases, such as smallpox, measles, and influenza, **endemic** to the Old World until A.D. 1500. The disease-causing microbes originally had livestock hosts but were able to evolve and infect humans. When Europeans entered and colonized the Americas, they carried these diseases with them and spread them to indigenous peoples who had no prior exposure and hence no immunity to them. Native human populations were decimated, and European diseases became both an unintentional and sometimes a deliberate tool in vanquishing the original inhabitants of the land. It has been argued that such diseases are a major reason Europeans so successfully colonized temperate regions around the world beginning in 1492 (Crosby 1972, 1986; Diamond 1997).

The negative environmental impacts of traditional agriculture, both crop cultivation and pastoralism, have been both extensive and significant. Foremost among them are the reduction, fragmentation, modification, and destruction of wild vegetation with consequences to the wild animal life and other elements of the region's biomes. Deforestation, drainage of wetlands, overgrazing, and trampling of grasslands are all factors. Clearing of the natural vegetation provided open, frequently disturbed patches of land in which the soil was exposed to erosion by water and wind. Indeed, topsoil erosion is considered by many to be the most deleterious impact of agriculture throughout the world (Tivy 1990). Soil from the Sahel blows on the tradewinds to the Caribbean Sea, and soil microorganisms carried aloft are implicating in the bleaching of coral reefs there. The windblown soil from interior China can interfere with astronomical observations in Hawaii and sometimes reaches all the way across the Pacific Ocean to the United States. It was the famous Dust Bowl of the 1930s in the United States that first made people aware of the great losses of topsoil by wind erosion (Tivy 1990).

Modern agricultural practices expanded and intensified the impacts of traditional farming and pastoralism and added significant new problems. The waste products of industrial farming are a major source of air, water, and soil pollution. These waste products include massive accumulations of the excreta of livestock and excess inorganic fertilizers and organic pesticides. Water-soluble nitrates that are not taken up by crop plants are leached into the groundwater and become a major source of water pollution in lakes, rivers, and bays. Less soluble nutrients such as potassium and phosphorous get washed into waterways attached to soil particles eroded from agricultural fields. Many of the synthetic organic herbicides, fungicides, and insecticides applied to cropland are nonbiodegradable and pollute both soil and water. Attached to dust and other particulate matter, these toxic compounds have been distributed around the world in the global wind system and can be found in the soils of remote oceanic islands and even the Antarctic ice sheet (Tivy 1990). As these chemicals pass up the food chains of land and sea, they become biologically magnified, or more highly concentrated, in the tissues of predatory species and affect nervous and reproductive systems. The negative impacts of the chlorinated hydrocarbon DDT, documented by Rachel Carson in her 1962 book *Silent Spring*, became widely appreciated after use of the pesticide was banned in the United States and populations of top-of-the food-chain birds such as brown pelican (*Pelecanus occidentalis*) and bald eagle (*Haliaeetus leucocephalus*), once on the verge of extinction, began to recover.

Industrial agriculture has created what is seen by some as a crisis in agriculture itself, namely a decline in genetic diversity. The widespread adoption of a few high-yielding varieties of crop plants and industrial livestock breeds are pushing the thousands of landraces, old crop varieties, and traditional breeds of livestock toward extinction. With them go a storehouse of valuable genes that could be used in future breeding programs (Harlan 1995; Bixby et al. 1994; Fowler and Mooney 1990). Biotechnology is able only to transfer genes, not create them. The sources of new genetic information lie in the multitude of old crop varieties, minor livestock breeds, and their closest wild relatives. Seed banks, sperm banks, and *in situ* preservation programs on farms throughout the world are necessary adjuncts to modern agriculture. The IBPGR (International Board of Genetic Resources, an agency of the United Nations) places highest priority globally on collecting and preserving rapidly disappearing varieties of wheat, *Phaseolus* beans, manioc, sweet potato, coffee, and tomato (Fowler and Mooney 1990).

The environmental impacts of prolonged or mismanaged irrigation can be severe. Without proper drainage, the local water table (the top of the saturated or groundwater zone) may be raised into the root zone of plants to waterlog the soil. This alone is detrimental to the production of most crops, but in arid and semiarid climates, evaporation of excess soil moisture causes dissolved salts to precipitate out at or near the surface. In basin irrigation, when fields are flooded and the water is contained so that it can infiltrate into the ground, much is also lost to evaporation with the same effect. Without adequate flushing, an accumulation of toxic salts develops in the soil, and the land is no longer usable for agriculture. This process is called salinization, and it is both an ancient and a contemporary problem. Evidence shows salinization had become a problem in Mesopotamia by 2400 B.C. After this time, the area planted to salt-intolerant wheat began to decline and by 1700 B.C. wheat was no longer planted in the southern part of the valley. By the late twentieth century, it was estimated that 50 percent of the irrigated land in Iraq was affected by increased salinity and/or waterlogged soils (Goudie 1981). Salinization is also a major problem in the Indus Valley, western China, Central Asia, northern Mexico, and the southwestern United States (for example, around the Salton Sea in California's Imperial Valley) (Meyer 1996).

Removal of water from groundwater supplies and from rivers also has deleterious effects. Land subsidence is associated with withdrawals of groundwater at rates exceeding recharge of the aquifer and is a common consequence of the use of fossil groundwater, which also depletes a nonrenewable resource. Diversion of water from exotic rivers to adjacent irrigated lands diminishes the supply of water available to downstream users. Drain water may be returned to rivers, but it will contain high concentrations of dissolved salts that then increase the salinity of formerly freshwater streams. The volume of the Colorado River has been so reduced that the Colorado no longer reaches the Gulf of California, and the salinity of that body of water has increased to the detriment of its marine life. The river itself had become so saline in the lower reaches that the United States had to construct a large desalination plant to provide Mexico with the freshwater guaranteed by treaty.

Perhaps the greatest modern catastrophe associated with irrigation, however, has happened beside the Aral Sea. Between 1960 and 1989, the sea lost 60 percent of its surface area and two-thirds of its volume as river water flowing into the sea was diverted to irrigate cotton. Salinity of the sea increased and an important **fishery** was destroyed; ports and settlements were left far from the water's edge; but even more significantly, the dry seabed was left exposed to the wind. Blowing dust and salts, laced with pesticides and sewage, have damaged human health over a wide area. High rates of cancer, infant mortality, and birth defects are a fact of life in the region today (Meyer 1996).

## ORIGINS

The origin of agriculture is a much debated topic. Numerous theories have been put forth to explain the adoption of agriculture by early peoples on all the inhabited continents but Australia. Among the general themes, not necessarily mutually exclusive, are that agriculture developed as: (1) an invention that occurred in one or more localized centers and then dispersed to other parts of the world; (2) a necessity forced upon people by population growth beyond the capacity of natural environments to support them; (3) a result of environmental changes near and after the end of the Pleistocene Epoch that brought about new spatial and ecological relationships among humans and wild species of plants and animals; and (4) an outgrowth of the coevolution among certain wild species, human land use practices, and human societal changes (Rindos 1984; Vasey 1992; Reed 1977).

The general consensus today is that domestication had to precede agriculture itself. In other words, at least some species already adapted to disturbed habitats near human settlements and bearing traits desired by humans had to be in existence before they were deliberately propagated. Independent invention and subsequent diffusion has lost favor. Coevolution, especially between plants and people, has been effectively argued by Rindos (1984) as being no different that the coevolution of certain plants and the wild animals that are their predators and dispersers. One suggestion, known as the "dump-heap hypothesis," (see later discussion) proposes that the precursors of crops were adapted to the middens and latrine sites built up near human habitations. In these open, nutrient-rich places, they evolved larger seeds that, along with proximity, attracted human collectors. People provided some protection to the seedlings of useful plants they could recognize. As humans came to be reliable collectors, protectors, and dispersers—carrying seeds (as food) with them when they changed habitation sites, the plants lost any natural defenses such as thorns and reduced natural dispersal mechanisms (i.e, mature seeds and fruits did not fall readily from the stalk) and became even more attractive to people.

Evolution could be rapid among self-fertile plants, but among self-sterile plants and **sexually reproducing** animals, the domestication process was probably slow and probably required the isolation of evolving populations from their wild relatives. Migration out of the range of wild progenitors would have been one way to prevent crossbreeding. A group of domesticates, including hemp and tomato, are identified as camp followers or hitchhikers that as weeds accidentally moved with people only to evolve the characteristics of crops after they had become spatially and genetically isolated from their wild relatives. Although it is harder to extend these scenarios to animals, the pig and the dog, camp scavengers, are apt to have domesticated themselves this way.

One way or another, population growth is involved in the adoption of agriculture. It is well known that hunting and gathering peoples know how plants grow; but at least in historic times, most groups have resisted becoming farmers until forcibly settled and presented no alternative. The debate centers around whether the population growth and sedentism that began in the late Pleistocene/early Holocene epochs occurred before agriculture and required a change in life ways to maintain high population densities (e.g., Cohen 1977) or whether they are a consequence of early agriculture.

Changes in subsistence patterns at the end of the Pleistocene apparently were related to environmental changes, such as the extinction of the largest animals and vegetation changes due to climate change and also to the widespread use of fire by hunting and gathering peoples. Archeological evidence shows that human diets became less specialized (the broad spectrum revolution), and sedentary village life increased (Harris 1977).

Dependence upon domesticated plants and animals began everywhere but in Australia between 8000 B.C. and 4000 B.C. Some of the oldest evidence for the cultivation of seed crops (grains) and animal husbandry comes from Southwest Asia, where domestic wheat and barley and domestic sheep and goats were first raised. Millet was first cultivated in northern China sometime between 6000 B.C. and 5000 B.C., and pigs were also being raised (Boyden 1992). Rice was being grown in China by about 4800 B.C. and in India by 2300 B.C. (Hanfei 1992). Dry farming of wheat and barley developed in the eastern Mediterranean around 3000 B.C., coinciding with the appearance of plows (López-Bellido 1992), and from there spread into western and northern Europe during the next 1,000 years. In the New World, maize was being cultivated on the Mexican Plateau by 6000 B.C. to 5000 B.C. Squash, beans, avocado, and chilies were being grown by 3000 B.C. In South America, manioc was being cultivated by 2500 B.C. and potatoes and other root crops by 1000 B.C. (Boyden 1992).

Agricultural practices, crops, and livestock spread from one place to another through human migrations, conquest, and trade. For example, the expansion of Islam after A.D. 700 brought irrigation to the Mediterranean world and the intensive cropping of fruits and vegetables. New crops—including sugar cane, alfalfa, cotton, and rice—were introduced to Europe then. The greatest exchange of domesticated species, however, probably occurred during Europe's so-called Age of Discovery. Maize was introduced to Europe in A.D. 1500 and the white potato in 1570 (Boyden 1992). Other New World crops, tomatoes for example, also became part of national dishes throughout Europe. Maize and other American crops were also introduced to Africa. It was a two-way exchange across the Atlantic, however; wheat, barley, rye, peas, lettuce, cabbages, eggplant, and other crops, as well as all our common barnyard animals, were brought to the Americas by European colonists. African foods such as yams, okra, and watermelon also found their way across the Atlantic as adjuncts of the Atlantic slave trade.

Landraces of both crops and livestock dominated agroecosystems until the eighteenth century, when selective breeding of livestock began in Europe. Even so, most breeds remained multipurpose by providing milk, meat, hides, and, at least for the larger species, draft power until the twentieth century. Improved or higher yielding varieties and breeds are mostly products of the post–World War II world, although Japanese scientists were selecting for specific traits in rice and wheat by 1910 (Vasey 1992). In the 1940s, Norman Borlaug produced a high-yielding strain of wheat at a research station in Mexico funded by the Rockefeller Foundation. Borlaug received the Nobel Peace Prize for his development of a miracle crop and ushered in a new era in crop breeding that would become known as the Green Revolution. It was the beginning of a concerted international effort to create high-yielding strains of other staple crops that would feed the burgeoning populations of the developing world (Brookfield 2001; Fowler and Mooney 1990). The hybrid rice IR-8, produced by the International Rice Research Station in Baños, Philippines, was introduced in 1966.

At about the same time, livestock selection turned toward specialty breeds that were good at producing just one product and responded well to industrial agroecosystems. The 1970s saw the beginning of factory-style poultry production, initially for broilers and eggs. Selection for uniformity and high rates of production in an industrial setting took over livestock breeding programs. As a consequence, one or just a very few of the many traditional breeds of each livestock species is now dominant globally. Agricultural is currently undergoing what has been termed a livestock revolution as demand for meat has grown in response to rising incomes in the world's developing countries. Related to this is an increase in demand for, and therefore production of, feed grains and high-protein meal (Wood et al. 2000).

The machinery of agriculture—plows, cultivators, harvesters—was originally wooden and horse, ox or mule drawn. All-iron plows were in use in England by the 1770s, and John Deere introduced the steel plow in 1839. McCormick's reaper appeared in 1831 and replaced the hand cutting of cereals or use of earlier stationary reapers. The reaper was soon followed by hay balers, combines, and hay mowers. Early steam-driven tractors date to the 1800s, and the first gasoline tractor was introduced in 1893. However, widespread adoption of tractors in the United States did not take place until after the 1940s. Mechanical harvesters for fruits and soft vegetables have been available since the 1960s (Vasey 1992).

## HISTORY OF SCIENTIFIC RESEARCH IN AGRICULTURE

Up until the nineteenth century, crop and livestock breeding and the development of agricultural methods were in the hands of the farmers themselves. The early interests of scientists lay in trying to explain the successes that took place in field and pasture. A prime example is Charles Darwin's 1883 publication of *The Variation of Animals and Plants under Domestication.* Before the twentieth century, breeding programs involved field-growing plants and selecting seed from individuals with desirable traits to be the seed stock for the next generation. The process continued until a fairly consistent new variety was produced. In the twentieth century, the genetic theories of Gregor Mendel were rediscoverd and put to use in the crossbreeding of plants to obtain desired properties. This involved manipulation and control over pollination and lent itself to more rigorous scientific methods than the selection process had. T. A. Knight is often considered the prototype of the scientific breeder. He published on heredity in peas and wheat, developed useful field varieties, and in the 1820s developed the first hybrid wheat in Europe (Vasey 1992). In the United States, Luther Burbank (1849–1926) was an early scientific breeder of crops.

Scientific interest in crops, in particular, gave rise to explorers who went around the world collecting and cataloging crop varieties and their traits. One of the most important of the pioneers in this endeavor was Nikolai Ivanovich Vavilov, whose collections began in 1916. His collections are housed today at the Vavilov Institute of Plant Industry in St. Petersburg, Russia, and represent one of the world's major stores of crop genetic information (Harlan 1995; Vasey 1992).

Livestock, too, had initially been developed by selecting the breeding stock. Back-crossing, which usually involves mating a male to his daughters and granddaughters, became common practice in eighteenth century in western Europe and led to the development of many of the standardized breeds of cattle and sheep still important today (Vasey 1992).

A major influence of early science in agriculture came from chemistry. Chemists evaluated fertilizers and suggested new compounds. An important early work was Justus von Liebig's 1840 book *Organic Chemistry in Its Applications to Agriculture and Physiology,* published in both German and English (Vasey 1992). In 1903, scientists developed the electric arc process for fixing nitrogen, and during World War I the Haber ammonia synthesis process was developed. These heralded the era of synthetic fertilizers (Vasey 1992).

Soil science and soil classification systems are direct outgrowths of scientists' interest in understanding and improving agricultural productivity. Russian scientists are credited with being first in these fields (Vasey 1992).

Until the 1970s, the breeding and distribution of seeds was guided by national governments, government-funded universities and institutes, and nonprofit international research organizations. In 1862, the U.S. federal government established the land grant system to help states fund agricultural colleges. (Segregated black agricultural colleges were established after the Civil War in 1890.) In 1887, the U.S. Congress passed the Hatch Act, which created agricultural experimental stations at the land grant colleges (Vasey 1992). Since the 1970s, the role of government has faded, and multinational corporations now control the entire spectrum of breeding, producing, processing, and marketing. Many of these companies have their roots in the chemical and pharmaceutical industry. By 1990, Royal Dutch Shell was the world's largest manufacturer of pesticides and one of the largest seed companies. Corporations such as Monsanto and Ciba Geigy sell farmers a complete package of seed, fungicides, and herbicides and breed resistance to herbicides into the crop seed. Among other major global seed houses were Occidental Petroleum, Upjohn, Unilever, and Volvo (Fowler and Mooney 1990).

Biotechnology now circumvents the need to crossbreed plants in elaborate pollination schemes. Recombinent DNA procedures remove genes from the nucleus of one cell and implant them in the nucleus of another cell, thus allowing the transfer of genetic material between species that cannot interbreed. Tissue culture allows cloning millions of plants from a single one exhibiting desirable traits. A few cells, usually from the growing tip, are grown in the laboratory on a sterile medium enhanced with growth promoters and nutrients. From these undifferentiated cells, millions of plant embryos develop. These can now be encased in herbicides, fungicides, and fertilizers and sold as artificial seeds (Fowler and Mooney 1990; Vasey 1992). The new varieties are often patented as inventions and are the property of the multinational corporations. The source of new genes to impart resistance to any variety of environmental hazards remains, however, in the old varieties still propagated by small farmers primarily in the underdeveloped world or in the wild relatives living in the world's remaining natural areas.

While much research in agricultural science continues to be directed at improving industrial crops, there is also a major focus on developing sustainable practices for agriculture and land management. Intercropping, integrated pest control, no-tillage cultivation, and grazing by geese or sheep as an alternative to herbicides are all part of mainstream agriculture today (Brookfield 2001).

Studies of the origins of crops began with the Swiss botanist and geographer Alphonse DeCandolle, who in 1882 published *L'Origine des Plantes Cultivées.* Vavilov proposed eight primary centers of origin in this 1926 paper "Studies on the origin of cultivated plants." Anthropologists, botanists, and geographers tackled uncovering the origins of agriculture as well as domestication. The anthropologist V. Gordon Childe (1951) introduced the concept of a "Neolithic Revolution" that marked the transition from food collecting in the Old Stone Age to food producing in the New Stone Age. Among other early and influential studies was Carl O. Sauer's 1952 *Agricultural Origins and Dispersals* (New York: American Geographical Society). Sauer, a geographer, reasoned that agriculture must have begun among sedentary people with an ample and abundant food supply that would give them the leisure to domesticate and experiment with crops. The botanist Edgar Anderson, in his 1954 book *Plants, Man and Life* (London: A. Melrose), proposed the "dump-heap" hypothesis. More recently, anthropologist Mark Cohen (1977) championed the idea that agriculture developed under stress of population pressure and David Rindos (1984) emphasized biological factors and proposed agriculture as a product of coevolution between plants and human societies. Jonathan D. Sauer (1993) compiled the histories of crop species—the work of many, many scientists. Among those concentrating on animal domestication have been Charles A. Reed (1969) and Juliet Clutton-Brock (1981). Theories continue to be revised as new evidence becomes available. Much interesting work still needs to be done and is being done to unravel agriculture's past. Current research is concentrated within the interdisciplinary field of cultural ecology, which blends anthropology and geography (Brookfield 2001).

## REFERENCES

Bixby, Donald E., Carolyn J. Christman, Cynthia J. Ehrman, and D. Phillip Sponenberg. 1994. *Taking Stock: The North American Livestock Census*. Blacksburg, VA: The McDonald & Woodward Publishing Company.

Boyden, Stephen. 1992. *Biohistory: The Interplay between Human Society and the Biosphere*. Park Ridge, NJ: Parthenon Publishing Group.

Brookfield, Harold. 2001. *Exploring Agrodiversity*. New York: Columbia University Press.

Childe, V. Gordon. 1951. *Man Makes Himself*. New York: The New American Library.

Clutton-Brock, Juliet. 1981. *Domesticated Animals from Early Times*. Austin, TX: University of Texas Press.

Cohen, Mark Nathan. 1977. *The Food Crisis in Prehistory; Overpopulation and the Origins of Agriculture*. New Haven: Yale University Press.

Crosby, Alfred W. 1972. *The Columbian Exchange: Biological Consequences of 1492*. Westport, CT: Greenwood Press.

Crosby, Alfred W. 1986. *Ecological Imperialism: The Biological Expansion of Europe, 900–1900*. Cambridge: Cambridge University Press.

Diamond, Jared. 1997. *Guns, Germs, and Steel: The Fates of Human Societies*. New York: W.W. Norton & Company.

Fowler, Cary and Pat Mooney. 1990. *Shattering; Food, Politics, and the Loss of Genetic Diversity*. Tucson: The University of Arizona Press.

Fussell, L. K. 1992. "Semi-arid cereal and grazing systems of West Africa." Pp. 485–518 in C. J. Pearson, ed., *Field Crop Ecosystems*. Ecosystems of the World, 18. Amsterdam: Elsevier.

Gade, Daniel W. 1999. *Nature and Culture in the Andes*. Madison: The University of Wisconsin Press.

Goudie, Andrew. 1981. *The Human Impact: Man's Role in Environmental Change*. Cambridge, MA: The MIT Press.

Hanfei, Dong. 1992. "Upland Rice systems." Pp. 183–203 in C. J. Pearson, ed., *Field Crop Ecosystems*. Ecosystems of the World, 18. Amsterdam: Elsevier.

Harlan, Jack R. 1995. *The Living Fields: Our Agricultural Heritage*. Cambridge: Cambridge University Press.

Harris, David R. 1976. "Traditional systems of plant food production and origins of agriculture in West Africa." Pp. 311–356 in Jack R. Harlan, Jan M. J. DeWet, and Ann B. Stemler, eds., *Origins of African Plant Domestication*. The Hague: Mouton Publishers.

Harris, David R. 1977. "Alternative pathways toward agriculture." Pp. 179–243 in Charles A. Reed, ed., *Origins of Agriculture*. The Hague: Mouton Publishers.

Hart, John Fraser. 1998. *The Rural Landscape*. Baltimore: The Johns Hopkins University Press.

Juo, A. S. R. and H. C. Ezumah. 1992. "Mixed root-crop systems in wet sub-Saharan Africa." Pp. 243–258 in C. J. Pearson, ed., *Field Crop Ecosystems*. Ecosystems of the World, 18. Amsterdam: Elsevier.

López-Bellido, L. 1992. "Mediterranean Cropping Systems." Pp. 311–356 in C. J. Pearson, ed., *Field Crop Ecosystems.* Ecosystems of the World, 18. Amsterdam: Elsevier.
Meyer, William B. 1996. *Human Impact on the Earth.* Cambridge: Cambridge University Press.
Reed, Charles A. 1969. "The pattern of animal domestication in the prehistoric Near East." Pp. 361–380 in Peter J. Ucko and G. W. Dimbleby, eds., *The Domestication and Exploitation of Plants and Animals.* Chicago: Aldine Publishing Company.
Reed, Charles A., ed., 1977. *Origins of Agriculture.* The Hague: Mouton Publishers.
Rindos, David. 1984. *The Origins of Agriculture, An Evolutionary Perspective.* New York: Academic Press.
Sauer, Jonathan D. 1993. *Historical Geography of Crop Plants, A Select Roster.* Boca Raton, FL: CRC Press.
Simmons, I. G. 1989. *Changing the Face of the Earth: Culture, Environment, History.* Oxford: Basil Blackwell.
Tivy, Joy. 1990. *Agricultural Ecology.* Essex, England: Longman Scientific & Technical.
Vasey, Daniel E. 1992. *An Ecological History of Agriculture, 10,000 B.C.–A.D. 10,000.* Ames, IA: Iowa State University Press.
Wood, Stanley, Kate Sebastian, and Sara J. Scherr. 2000. *Pilot Analysis of Global Ecosystems: Agroecosystems.* Washington DC: International Food Policy Research Institute and World Resources Institute. Also available at http://www.wri.org/wr2000.

# 16 URBAN ECOSYSTEMS

## OVERVIEW

The urban environment is the **biome** humans and their activities most dominate. Under this designation are towns and cities, including their suburbs, where people are densely settled and have significantly and obviously altered the natural patterns of plant and animal life characteristic of the region. An urbanized area is defined as the built-up area where urban land uses (residences, industries, stores, administrative offices, transportation, and utilities networks, and the like) prevail regardless of political boundaries. A city or town, strictly speaking, is a political unit with its own government responsible for such things as schools, public water and sewerage systems, and local roads. In popular usage, "city" implies a larger settlement than "town," and "village" refers to a rural cluster of houses whose inhabitants engage largely in agricultural occupations. Other terms used in urban **ecology** and geography to refer to urban settlement types are metropolitan area and agglomeration, both of which refer to one or more central cities and adjacent areas that either constitute a continuous built-up (i.e., urbanized) area or are functionally connected to one of the central cities. The U.S. Bureau of the Census uses its own set of terms defined according to the size of the human population; these may be considered statistical units. For example, an urban place is defined as any settlement with more than 2,500 people, and a Metropolitan Statistical Area (MSA) is defined as one or more central cities with a combined population of at least 50,000 plus those contiguous counties in which more than 75 percent of the inhabitants are engaged in nonagricultural employment. Other countries use different numbers for their official statistical units.

Urbanization is the process by which an area changes from a region dominated by small, agriculturally based settlements to one dominated by large, dense settlements based upon an urban economy of manufacturing or services. In this process, people also change their values, attitudes, and behavior as they adjust to the new urban setting. It may be obvious that the density of housing and other elements of the built environment will influence the kinds of nonhuman **species** that share an urban area with people, but urban taste also influences the species composition and distribution patterns of plants and animals in the urban environment. The latter can be conspicuous in terms of what is socially acceptable in the landscaping of front yards in neighborhoods of different socioeconomic classes.

Urban geographers have modeled the physical structure and development of cities, and the distribution patterns of plants and animals generally reflect this layout. The urban core is a densely built-up area referred to as the downtown or central business district in the United States and the city center in Europe. Typically, the center is a totally built environment dominated by hardscape—building surfaces (skyscrapers in the modern city), sidewalks, and streets—and plant life is represented primarily by ornamental trees lining streets or container gardens on sidewalks and buildings. Outward from the center, population density and the compactness of the built-up area decreases. Residential areas with tree-lined streets and gardens come to predominate. Closer in, these may be apartment buildings and townhouses; farther out, single-family homes become more and more prevalent, and lot sizes become larger as distance from the downtown area increases. Beyond the urbanized area, the effects of city life can still be felt in what is termed the urban fringe. In the United States, this zone consists of subdivisions with low-density housing on large lots (0.2–4 ha or 0.5–10 ac). The people who live in the urban fringe live in a rural setting but maintain urban values and behaviors and may commute 80 km (50 mi) or more to urban jobs. They are attracted to the open space and aesthetic and recreational aspects of the countryside but not to an agricultural lifestyle. Their influence may extend even farther from the urban center into national forests, nature preserves, and national parks through the effects of their pets, garden escapes, and vehicles (Theobold 2001).

Not all parts of the world share the patterns typical in the United States. In Russia, for example, rural land uses and values dominate the urban fringe of large cities. Instead of urban people moving out to the edge of the city, rural folk have moved toward the city because farming is more profitable there than in more remote areas that lack reliable transportation (Ioffe and Nefedova 2000). In the Soviet period, and still today, city dwellers also use the urban fringe to grow food. Weekend trains leaving Moscow carry many people taking their pitchforks and shovels out to their small plots of lands.

The urban environment (Photo 16.1) differs from natural and agricultural ones in several key ways, in addition to the prevalence of the artificial hard surfaces previously mentioned. Urban temperatures are higher and relative humidity is lower than in the surrounding countryside. Precipitation is higher, but because concrete and asphalt cover much of the ground, evaporation and runoff are also higher than in rural or natural areas. Pollution of air and water affects urban **life-forms** and prevents some native species of the region from surviving but selects for those that can tolerate or adapt to it. **Soils** in urbanized areas are altered by compaction, removal of topsoil, or the addition of nutrients, and it may even be man-made. Each type eliminates some plants and offers unique opportunities for others. Repeated **disturbance** by humans is a final major factor in the urban environment that determines which life-forms can succeed. The physical structures of the city are built, demolished, and rebuilt; lawns are mowed, shrubbery is

Photo 16.1 Varied urban environments from roadways to gardens, parks, and high-rise buildings, can be seen in this view of São Paulo, Brazil, one of the world's largest cities. (Photo © Jonathan Blair/CORBIS.)

pruned, gardens are tilled; pesticides are applied; and foot traffic, vehicular traffic, and noise disrupt movements and other activities of nonhuman organisms, especially animals.

Plant and animal life of urbanized areas consist of a group of organisms able to survive the urban environment and the fashions, tastes, and land-use practices of the humans who are the dominant species of the biome. The majority of nonhuman organisms are native to the local natural biome and vary accordingly from region to region and continent to continent. However, a somewhat consistent group of weedy plants and animals are found in cities. Indeed, one of the characteristic elements of urban life is the high proportion of nonnative (**alien**, adventive, or **exotic**) species. For plants, these most often are garden escapes or weeds accidentally **introduced** in the past in agricultural commodities or ship and railroad ballast. The counterparts among animals are **feral** (escaped from domestication) pets or livestock (pigeons, for example) and accidental world travelers such as house mice (*Mus musculus*), common or Norway rat (*Rattus norvegicus*), cockroaches (Family Blattidae), and silverfish (*Lepisma saccharina*).

Urban ecology and **biogeography** are in their infancy. Much of the information available occurs in case studies and most deals with cities in Europe and North America. For the most part, the work has been descriptive—a matter of inventorying what plants and animals are found in which cities and in which **habitats** within a city. Ecological interactions among species in terms of energy flow, nutrient cycles, or **successional** communities have not been extensively examined; what work has been done has largely addressed managing pests or creating and conserving wildlife habitats to bring nature into the city. Due to the imbalance in geographic treatment, no attempt will be made here to distinguish among regional expressions of the urban biome. The general descriptions of urban habitats, plants, and animals that follow are strongly biased toward North American and European cities. It should be remembered that most people live in very different kinds of city environments at different levels of economic development, and they possess cultures with different values and histories with different connections to the rest of the world. All these human factors influence the composition and patterns of nonhuman life in ways yet to be fully explored.

The human population is becoming increasingly urbanized. It is estimated that by 2020, more than half of the people on Earth will live in cities, and urbanized areas will continue to expand to accommodate them. The urban biome and the lifeforms in it may well represent the future of life on the planet (Sukopp et al. 1979).

## THE URBAN ENVIRONMENT

The urban environment differs from natural and agricultural ones in several key ways. Built-up areas alter the local **climate**, especially temperature; city soils are usually strongly compacted or are obliterated altogether—covered by pavement, buildings, rubble, or fill. Air and water pollution can be severe. Frequent disturbance is another factor; urban plants and wildlife must be able to tolerate such conditions or be able to take advantage of the niches urban types of disturbance create.

Local climate is greatly influenced by the preponderance of man-made structures in an urban area and the great use of fossil fuels and electricity by the city's human inhabitants. The buildings, concrete, and pavement of city centers absorb more solar radiation than vegetated land does. Especially at night, this energy is reradiated as heat to the atmosphere and—together with the heat generated by electricity, air-conditioning or heating, and vehicles—is responsible for a warming known as an urban heat island effect. Mean annual temperatures over large towns are 0.5–1.5° C (0.9–2.7° F) higher than air temperatures above the surrounding countryside. The daily temperature differences are slight during the day but rapidly increase after sunset and peak some two or three hours later. Temperatures in London may be 5° C (9° F) higher than in nearby rural areas, and New York City averages 2–3° C (4–5° F) warmer than surrounding wooded areas. The greatest difference (up to 10° C or 18° F) is found in narrow streets, courtyards, and other similarly confined areas in the urban core. The difference lessens across the gradient from city center to outlying residential areas [This pattern does not always hold true. Phoenix, Arizona, with its irrigated lawns and other plantings, is actually cooler in summer than the surrounding desert] (Pickett et al. 2001). The effect on plant life can be significant. The active growing season for plants in central London, for example, is 3 weeks longer than in nearby rural areas, and the frost-free period is 10 weeks longer because the first killing frost of autumn is delayed. As a result of urban heat islands, some plants have been able to extend their northern limits into large cities such as London (Gilbert 1989).

Other urban effects on local climate include a 20–30 percent reduction in wind speed (caused by friction of air moving against building surfaces), an eddying effect of winds at the base of tall buildings, lower relative humidity (on average about 6 percent), 5–10 percent greater cloudiness, and 5–15 percent less sunshine. Due to the high amount of particulates that serve as condensation nuclei in the air above cities, some 5–15 percent more precipitation falls in cities than in surrounding rural areas, but much of it is immediately lost through surface runoff and evaporation (Gilbert 1989; Adams 1994). Since air pollution is likely to increase during the work week, the likelihood of rain is greatest at the end of the week and on weekends (Collins et al. 2000).

Increased runoff, much of which passes through storm sewers, leads to a lowering of the water table and can cause an increased rate of downcutting in urban streams. As the streams deepen their channels in the floodplain, riparian vegetation suffers because its root zone dries out (Pickett et al. 2001).

The main ingredients of urban air pollution are sulfur dioxide ($SO_2$) and smoke from factories, and nitrogen oxides ($NO_x$) and carbon monoxide (CO) from vehicular exhaust. $SO_2$ is toxic to plant cells and acidifies the soil. When mixed with water vapor in the atmosphere, $SO_2$ forms sulfuric acid and becomes a major ingredient of acid precipitation. **Conifers** are especially sensitive, and most species had disappeared from English manufacturing towns by 1950 (Gilbert 1989).

Smoke causes severe damage to **evergreen** plants, but **deciduous** trees and shrubs can also be hurt. In 1930s and 1940s in Western Europe, it was noticed that evergreen privet (*Ligustrum lucidum*) had become deciduous in habit, and normally deciduous plants shed their leaves earlier in the season. Due to the haze and reduction in light levels, it was impossible to grow some sun-loving species, while others could be used as ornamentals only if raised outside the city and then transplanted to urban gardens. Bulbs such as crocuses had to be replanted each year because there was not enough sunlight for them to replenish their nutrient stores after blooming. While smoke pollution has been greatly reduced in Western Europe and North America, it is still a major problem in cities in Eastern Europe, Asia, Africa, and Latin America (Gilbert 1989). The nineteenth- and early twentieth-century effects of this pollution foreshadowed what is occurring in many cities of the developing world today.

Automobile exhaust contains unburned hydrocarbons, nitrogen oxides, and carbon monoxide. These react with ultraviolet radiation to produce photochemical smog, a mixture of ozone and peroxyacetyl nitrates (PANs). The reddish-brown gas nitrogen dioxide ($NO_2$), ozone, and PANS all damage plants. Concentrations of $NO_2$ in North American urban areas may be 10–100 times greater than in rural areas. $NO_2$ is another component of acid deposition and contributes significantly to the eutrophication of surface waters.

Urban soils show great variability both vertically and horizontally. Since soil science developed as an outgrowth of agricultural science, no standard classification system has yet been devised for them. On soil maps, they may be labeled simply as urban soils or as man-made soils, disturbed soils, or urban fill. Some urban soils are actually enriched with nutrients by human use and have a high **humus** content. These so-called hortisols (Sukopp et al. 1979) generally occur in areas gardened or used for dumping sewage, kitchen scraps, and other household trash for years and years. Most disturbed soils and urban fill, however, are primarily mineral in content. They tend to be alkaline due to the **weathering** of calcium from bricks, cement, plaster, and mortar; they harbor calcium-loving plants such as Jacob's ladder (*Polemonium caerulea*) and centaury (*Centaurium erythraea*). Urban soils are often polluted with lead from paints and vehicular exhaust. Boron, copper, and zinc are other common contaminants. These appear to have a greater impact on human health than on plant life, however (Gilbert 1989).

Disturbance is a constant in the urban environment, much of which is carefully manicured in accordance with the aesthetic values of the human inhabitants. Traffic, construction, demolition, digging for gardening and landscaping, and noise are all significant agents of disturbance that shorten periods of habitat stability and deter many species in the natural biome of the region from becoming established in urbanized areas.

## URBAN HABITATS

Plants and animals have a multitude of habitats within a city. Gilbert (1989) has grouped them into three categories: encapsulated countryside, created habitat, and unofficial wild places. Stearns (1971) correlated these with three vegetation types: residual, managed, and ruderal, respectively. Encapsulated countryside is a remnant of the preurban rural **landscape**—farmland, woodlot, meadow, or wetland—that has been surrounded by urban development but by chance has survived more or less intact. Created habitat includes most of the urban area that is carefully tended by humans: lawns, backyards, private gardens—be they on ground, stoop, balcony, or rooftop, or public parks, arboretums, botanic gardens, and cemeteries. (To be really complete, one should also include the insides of buildings with their gardens of houseplants and menageries of pets.) Common limiting factors to life in created habitats include fashion, taste, lawnmowers, herbicides and insecticides, foot traffic, and dogs. Finally, the wild places, or what Hough (1994) has called the "fortuitous urban landscapes," are largely unnoticed and ignored; they are where the plants and animals truly adapted to independent living in an urban environment can be found. These habitats and **microhabitats** are found throughout the built-up area and range in size from cracks in pavement or brickwork to railroad rights-of-way; they include the berms and median strips of divided highways, vacant lots, underground sewerage systems, bomb craters, and industrial **brownfields**. Ecological succession occurs in some of these places, and plant communities rich in alien (nonnative) species often develop (Gilbert 1989; Schmid 1975). Created habitats tend to reduce diversity, but the wild places encourage diversity.

> It is a paradox of our cities and urban regions that the landscapes we ignore are often more interesting and complex, and have a greater sense of place, than the ones we admire as the expressions of civic pride and good urban design. (Hough 1994, 42)

Indeed, cities overall may have a higher number of species in comparison to surrounding rural areas because of their diversity of habitats, particularly small-scale ones. Also, they have served as accumulation areas for immigrant species deliberately or accidentally introduced by people—particularly those species that can adapt to living in frequently disturbed, man-made landscapes (Sukopp and Werner 1983).

## PLANTS

Nonnative species and horticultural varieties—plants selected and bred for their value as garden crops or ornamentals—are important components of the plant life of urban areas. City centers or downtowns are usually dominated by hardscapes (paving stones, walls, cement benches, and the like), and ornamental plants are grown in containers. These may be trees, shrubs, perennial **forbs**, or **annuals** of hardy species bred to withstand, at least for a season, the rigors of the urban environment. As pollution damages their appearance, they are replaced with new specimens bred in nurseries and greenhouses outside the city. In older city centers and in residential areas, trees may line the streets to provide shade. These trees must be able to withstand air pollution and often nutrient enrichment from the excreta of domestic dogs. Commonly, only a few species are planted in any given part of the world. The most pollution-resistant trees in England were found to be Manchester poplar (*Populus nigra* var. *betulifolia*) and London plane tree (*Platanus* x *acerifolia*) (Gilbert 1989). Plane trees have hard leaves that allow rainwater to easily wash off particulate matter, and the bark regularly sheds to remove soot from the trunk (Wheater 1999). In Western Europe, lime or linden trees (*Tilia* spp.) are quite popular as street trees; most are hybrid varieties better able to withstand air pollution than native wild species (Gilbert 1989). In some Chinese cities, hybrid varieties of plane trees (*Platanus* spp.) are also among those preferred.

In the United States, the American elm (*Ulmus americana*) was the most popular urban tree until the 1930s, when it began to succumb to Dutch elm disease, a fungal infection spread by insects and passed through the roots of diseased trees. The fungus invades the vascular tissues of elms and chokes off the trees' supply of water and nutrients. The fungus involved was an exotic species, *Ceratocystis ulmi*, from Eurasia. It eventually spread from New York throughout eastern North America, west to the Rocky Mountains and south to Texas. Its **dispersal** was aided by another exotic, the European elm bark beetle (*Scolytus multistriatus*), which moved westward ahead of the fungus. The bark beetle breeds under the bark of dying trees but may feed on the leaves of healthy ones; it carries with it any fungal spores picked up in a diseased tree (Schmid 1975). Although replacements for the elm vary regionally, people commonly use varieties of Norway maple (*Acer platanoides*); a spineless honey locust (*Gleditsia triacanthos*); and more recently, the Bradford flowering pear (*Pyrus calleryana* "Bradford") an introduction from China. In the warmer parts of the United States, jacarandas (*Jacaranda mimosifolia*), eucalpyts, or palms may line streets.

At the base of street trees, there may be a distinct canine zone of **epiphytes** nourished by the nitrogen in dog urine. Where studies have been done in Europe, the canine zone is dominated by a green alga (*Prasiola crispa*) that forms a dark velvety covering at the base of the trunk. Also present may be other algae such as *Hormidium flaccidum* and *Stichococcus* sp., the nitrogen-loving **lichen** *Lecanora dispersa*, and the moss *Bryum capillare*. Above the canine zone, the tree trunk may be covered with the lichen *Lecoanora conizacoides* or alga *Pluerococcus viridis* (Gilbert 1989).

In typical middle-class North American neighborhoods composed of single-family houses, plants are seemingly a requirement (Photo 16.2). Houses are set

Photo 16.2 Suburban gardens. Such habitat harbor; horticultural varieties of plants, weeds, and pets. (Photo by SLW.)

back from the street, and a well-kept lawn of Bermuda grass (*Cynodon dactylon*), fescues (*Festuca* spp.), or other grasses in the front yard is socially mandated. Shrubs—often pollution-tolerant, **needle-leaved** plants such as yews (*Taxus* spp.) and junipers (*Juniperus* spp.), or **broad-leaved** plants such as boxwood (*Buxus* spp.) and privet (*Ligustrum vulgare*)—may be planted in a more or less formal fashion along the foundation of the house or as hedges to indicate property lines. Shade trees or flowering trees may be scattered across the lawn. The plants used are mostly horticultural varieties bred from species collected from around the world; they are carefully manicured and supplied with water, fertilizer, and pesticides. Front yards tend to reflect public taste and neighborhood standards, as well as the status of the homeowner. Often they also record the fashions of the period when the neighborhood was first established. Backyards, however, are less formal and more exuberant in their plantings. There, flower gardens are common and reflect personal tastes more than landscaping in the front yard does. Vegetables also may be planted in the backyard—something almost never done in the front yard (Schmid 1975). Ethnic neighborhoods—especially those inhabited by recent immigrants from Asia, Africa, or the Caribbean—may display interesting divergences from this norm.

Yards are not a global characteristic of single-family residences. In southern Europe, Southwest Asia, and elsewhere, house fronts abut the street, and gardens are placed in inner courtyards. Houses may have little or no space between them. Containers filled with **herbs** or ornamentals are placed by the front door or below windows to soften the hardscape and individualize homes. In contrast, in Russian towns and cities, single-family homes are often surrounded by **kitchen gardens** that supply households with fruits and vegetables to preserve for winter consumption. Flowers are grown among the more utilitarian plants simply for their beauty or for cut flowers to enhance the homes' interiors. In the squatter settlements associated with rapidly growing cities in the developing countries of Latin America or Africa, plants have not been studied; they appear to be weeds similar to those found in heavily trodden, untended urban sites rather than ornamentals like those found in middle- and upper-class neighborhoods.

People living in apartment buildings use window sills, balconies, and rooftops for their gardens and use containers to grow a wide range of plants—forbs, vines, shrubs, even trees—from all over the world (Photo 16.3). They may be strictly ornamentals, or they may be fruits, vegetables, or medicinal herbs. Rooftop greenhouses may even be used to grow orchids and other tropical plants in cold-climate areas. In many cities, neighborhood residents have taken over vacant lots as community gardens. Today, a significant percentage of the urban food supply may be produced locally within urban gardens (Photo 16.4). In addition, lawns and other cover plantings are being encouraged on rooftops of all types of city buildings, at least in the United States, as a way to lower urban temperatures and reduce the amount of energy consumed for cooling.

Photo 16.3 An apartment house in Chongqing, China, with potted plants perched on windowsills and balconies. Often, caged birds are also placed outside on porches. (Photo by SLW.)

Large cemeteries are another well-managed habitat in many urban areas. In Europe, they began to replace churchyard graveyards around 1825 to accommodate a growing population. By 1850, after a series of cholera epidemics, almost every town had a cemetery. Cemeteries were initially laid out at the outskirts of the built-up area, so many included patches of natural vegetation, which in rare instances persist. Most were treated as parks or botanical gardens and were planted with what were felt to be appropriate trees and shrubs. In the nineteenth century, tall, upright trees such as Italian cypress (*Cupressus sempivirens*) and Irish yew (*Taxus baccata*) were planted to promote air circulation. Pendulant, weeping trees—the Victorians' "Trees of Sorrow"—such as weeping ash or weeping willows, also were used to suggest an atmosphere of mourning. Old varieties of ornamentals may be

Photo 16.4 A community garden in Seattle's International District. All over the world, an increasing amount of the city dweller's food supply is now produced in the city itself. (Photo © Kevin R. Morris/CORBIS.)

still be preserved in some cemeteries. However, as large, old trees die off, they are being replaced by small flowering trees such as cherries (*Prunus* spp.) and crabapples (*Malus* spp.), or removed altogether because it is much easier to maintain an open lawn (Gilbert 1989).

The unofficial wild places receive escapes from city gardens, weeds, and other plants that have entered the urban areas by railroad, canal, or highway. (In modern times, deliveries of topsoil and mulch from outside the city also bring seeds.) Little study has been done on these plant communities outside Western Europe, but examples from England and Germany can demonstrate what has or is happening elsewhere. Between 40 and 60 percent of urban plants are aliens introduced from other regions or continents. Some of these are garden plants that have become naturalized; examples in England include Michaelmas daisy or New England aster (*Aster novae-angliae*), goldenrod (*Solidago canadensis*), and evening primroses (*Oenothera* spp.). Oxford ragwort (*Senecio squalidus*), one of the first plants to invade a newly cleared site, escaped from a botanical garden in Oxford, England, in 1794 (Gilbert 1989). The most successful colonizers of disturbed and neglected urban habitats are tall herbs with small, wind-dispersed seeds. They are often **geophytes** able to spread vegetatively once they become established (Prach and Pysek 2000). Examples occurring in English cities include fireweed or rosebay willowherb (*Epilobium angustifolium*), butter-and-eggs (*Linaris vulgaris*), and coltsfoot (*Tussilago fafara*) (Gilbert 1989).

Old stone and brick walls, tile roofs, and areas paved with stone or brick offer special habitats for plants. The vertical surfaces of walls replicate seaside and inland cliffs and rock faces. Weathering and accompanying deterioration of the mortar is a necessary preparation for colonization, and these processes are accelerated in industrial cities. Pioneering lichens and mosses also contribute to the breakdown of the solid surface and allow the roots of vascular plants to gain hold. Walls are dry habitats and select for drought-adapted (xerophytic) plants that exhibit such traits as succulence, reflective leaf surfaces, hairy leaves, rolled leaves, reduced surface area and/or tolerance of extreme dessication. More than 1,200 species have been reported from walls in Europe. These include **cushion**-forming mosses (their shapes reduce the ratio of surface area to volume and thereby help to retain moisture); rustyback fern (*Cetarch offininarum*), which can dry out completely and then green up quickly in response to rainfall; **succulent** stonecrops (*Sedum* spp.); navelwort (*Umbilicus rupestris*); and grasses with rolled leaves such as narrow-leaved and flattened meadowgrass (*Poa angustifolium* and *P. compresa*, respectively). Nonxerophytes native to cliffs and rock faces and found growing in city walls include Pellitory-of-the-wall (*Parietaria judaica*), wall lettuce (*Mycelis muralis*), and thyme-leaved sandwort (*Arenaria serpyllifolia*). Among garden escapes found on walls throughout Europe are wall flower (*Cheiranthis cheiri*), snapdragon (*Antirrhinum majus*), and red valerian (*Centranthus ruber*). On the tops of walls and on small ledges, wall annuals such as common witlow grass (*Erophila verna*), rue-leaved saxifrage (*Saxifraga tridactylites*), and thale cress (*Arabidopsis thaliana*) can be found in late winter and early spring (Woodell 1979).

At the base of walls, windblown and downwashed debris collects and creates a moister, more nutrient-rich substrate than the wall itself. Pavings also typically have higher nutrient levels than walls and are affected by waterlogging beneath the stones and trampling on the surface. Common plants found growing in England at the foot of a wall, where there is little impact from foot or vehicular traffic, are nettles (*Utrica diocia*), chickweed (*Stellaria media*), wall barley (*Hordeum murinum*), mugwort (*Artemisia vulgaris*), and dandelion (*Taraxacum officinale*). Thyme (*Thymus* spp.) and chamomile (*Chamaemelum nobile*) grow between paving stones in less-trampled areas; in heavily trampled areas a prostrate or low-growth form is favored, and only plantain (*Plantago major*), knotweed (*Polygonum aviculare*), and pearlwort (*Sagina procumbens*) are found (Woodell 1979). In Hong Kong and other parts of China, a varied group of plants that includes some 30 tree species have colonized ancient walls; some individuals have grown up to 20 m (65 ft) tall. These are mostly members of the mulberry or fig family (Moraceae) (Jim 1998).

One of the key attributes of urban vegetation is a high percentage of alien or nonnative species. In North America, these plants usually have Eurasian origins; in Western and Central Europe, they mostly stem from North America or the warmer parts of Eurasia. Urbanized areas are collection points for aliens. Many arrive as horticultural varieties and then escape from gardens to invade the city's wild places. Other points of entry for plants into the city include industrial areas and transportation routes. Industries that process agricultural products are key sites for the establishment of new weeds. A brewery at Burton-on-Trent, England, was found to have 267 kinds of grains growing around it. The woolen industry in England imported seeds embedded in sheep's wool, mainly from Australia. By 1960, 529 different alien species had been identified from this source, but few other than pirri-pirri bur (*Acaena novaezalandica*) had become naturalized. The paper industry imported cotton rags contaminated with seeds from Egypt and elsewhere in North Africa. The oil-milling industry imported pigweed (*Amaranthus paniculatus*), sunflower (*Helianthus annuus*), wormwood (*Artemisia biennis*), and barley (*Hordeum jubatum*), mostly from North America as contaminants in industrial crops such as rapeseed, soybeans, and castor beans. Typically, alien plants first entered at ports and then spread in cargo along canal and railroad networks. They may have been jolted free at stations and sidings or purposefully sorted from the commodity in question at the industrial site; then they ended up in refuse areas and from there eventually invaded the city (Gilbert 1989).

The proportion of aliens in well-studied urban **floras** in Poland is 50–70 percent in cities, 40–50 percent in large towns, 35–40 percent in small towns, and 30 percent in villages (Sukopp et al. 1979). This demonstrates a common trend of increased dominance by nonnatives the more densely built up and disturbed an urban area is (Kent et al. 1999). A terminology has developed, particularly in Europe, to distinguish among different types of introduced plants. Those that do not reproduce in their new environment but maintain populations through continual reintroduction—mostly annuals from warmer climate regions—are called casuals or ephemerophytes. Aliens that have established reproducing populations but are restricted to man-made habitats are referred to as epoekophytes, and those established in natural or seminatural habitats are called naturalized aliens. If a species must depend upon humans for its dispersal, it is synanthropic or hemerochoric. Further distinction is made in the Old World between adventive plants that colonized urban areas before A.D. 1500—the archeophytes, and those that were introduced after A.D. 1500—the neophytes. Archeophytes consist mainly of field weeds with origins in the Mediterranean region (Gilbert 1989; Sukopp and Werner 1983; Kent et al. 1999). In his analysis of urban plants in Central Europe, Pysek (1998) found that the percentage of archeophytes decreased with increasing density of the human population, while the proportion of neophytes increased. He attributed the higher proportion of archeophytes in Polish cities compared to German cities to the fact that Polish cities are smaller and more rural than highly industrial German cities. Germany also has a much longer history of trade and communications with the rest of the world. Pysek concluded that neophytes, which increase both in numbers and percentage with city size, were the group of plants most closely associated with human-dominated sites.

The number of neophytes increased especially after 1840 as a consequence of the Industrial Revolution and increased global trade. Over time, each town or city has developed a distinctive plant **community** that reflects its history of imports, trading partners, and transportation technology. New introductions are relatively rare today, since modern industry receives most of its raw materials in containers and treats them, through one process or another, with seed-killing heat (Gilbert 1989; Sukopp and Werner 1983).

An interesting exception to the pattern of cities accumulating alien plant species occurs in Italy, where in five cities studied, only 12–26 percent of urban plants were nonnative. Native mediterranean species, it seems, were well adapted to disturbance, and they successfully invaded built-up areas and outcompeted aliens (Grapow and Blasi 1998). However, Chile, another urbanized mediterranean country, had no native annual weeds. There, alien plants are found concentrated along roads and in cities (Kalin Arroyo et al. 2000).

Distinctive plant communities develop in relation to the historical growth of an urban place. In their study of the spontaneously growing plants in the city of Plymouth in the United Kingdom, Kent et al. (1999) were able to match specific assemblages of plants to four habitat types: the city center, oldest residential areas, housing areas constructed between 1945 and 1960, and neighborhoods developed since 1960. This pattern suggests that urban plant communities are dynamic and change as a city ages. Older downtown areas undergo a reduction in overall plant diversity, while expanding areas at the margin of the city have a high species richness.

On a smaller scale, succession has been documented in some undisturbed wild places in urban settings. In England, the first couple of years are dominated by annual and short-lived perennial forbs, often **rosette** plants. In one study area, 41 species pioneered in the first eight months, and 13 more were added the next year. These included Oxford ragwort, coltsfoot, lamb's quarter (*Chenopodium album*), rosebay willowherb, knotgrass (*Polygonum viculare*) and rhizomatous grasses such as couchgrass (*Elymus repens*). (Most of these or their close relatives also grow in vacant lots in North America.) A tall herb stage comes to characterize the community three to six years after initial colonization. Rosebay willowherb dominates. Although originally dispersed by seed, it expands on the site by vegetative propagation. It is joined by other tall, leafy alien perennials such as Michaelmas daisy, goldenrod, lupine (*Lupinus polyphyllus*), feverfew (*Chrysanthemum parthenium*), and Shasta daisy (*Leucanthemum maximum*), and some native species such as thistles (*Cirsium* spp.) and goatsbeard (*Tragopogon pratensis*). Biennials such as mullein (*Verbascum thapus*), wild teasel (*Dipsacus fullonum*), and centaury are common on some sites. At 8 to 10 years, this flowering meadow gives way to a grassland stage in which couchgrass, *Festuca rubra*, and Yorkshire-fog (*Holcus lanatus*) are dominants. Thickets of bushy Japanese knotweed (*Reynoutria japonica*) may be overgrown with greater bindweed (*Calystegia sepium*). Ribwort plantain (*Plantago lanceolata*) and sheep sorrel (*Rumex acetosella*) are also often conspicuous. Mosses may occur as a ground cover and, where vegetation is thin and untrampled, lichens (*Cladonia* spp.) resistant to air pollution may also cover patches of ground. As succession continues and early pioneer shrub species such as goat willow (*Salix caprea*), butterfly bush (*Buddleja davidii*), and birches (*Betula* spp.) grow taller, the grassland stages changes into a **scrub** woodland. The early shrubs were species with light, windborne seeds that arrived when the plant cover was still open. Later arrivals are shrubs and trees with larger seeds, often dispersed by birds, and able to germinate under a **closed canopy**. Hawthorn (*Crataegus* spp.), elder (*Sambucus* spp.), domestic apple (*Malus sylvestris*), and Swedish whitebeam (*Sorbus intermedia*) are typical of the later arrivals, as are ash (*Fraxinus* spp.), sycamore (*Platanus* spp.), and broom (*Cytisus scoparius*). Finally, after 40 years or so, should a site be left alone that long, a closed woodland develops. Such woodlands described from Newcastle-upon-Tyne and Birmingham, England, contain a unique mix of native ash, hawthorn, willow, broom, and Guelden rose (*Viburnum opulus*) and alien apple, whitebeam, and garden privet (*Ligustrum vulgare*). The ground layer is composed of shade-tolerant weeds, ivies, and ferns (Gilbert 1989).

## ANIMAL LIFE

Urban animals, like plants, are a diverse group of species often introduced by people and adapted to life among humans. Like plants, urban animals could be subdivided into ornamentals (those deliberately introduced for some aesthetic or emotional reason), escapes from domestication, and weeds, both native and exotic. Pets constitute the ornamentals, but so do birds such as Starling (*Sturnus vulgaris*) and House Sparrow (*Passer domesticus*), which were deliberately introduced to the United States. The rapidly expanding House Finch (*Caropodacus mexicanus*), a bird from the western United States that was introduced into the eastern United States for its song and color, is another example. Feral animals such as free-roaming dogs, cats, cage birds, and pigeons, are among the escapes from domestication. And the weeds are the animals that are either exotics accidentally introduced (for example, the common rat, house mouse, cockroaches, and silverfish) or natives that arrived on their own, animals such as crows and ravens (*Corvus* spp.), Herring Gulls (*Larus argentatus*), raccoons (*Procyon lotor*), and coyotes (*Canis latrans*). These animals thrive independent of human care in the neglected habitats of vacant lots, sewers, and dump sites or the nooks and crannies of homes.

Pets, especially domestic dogs (*Canis familiaris*), cats (*Felis cattus*), and birds, are an important component wherever people live. Indeed, a survey of animals in the built environment would probably reveal a surprisingly high diversity of animals deliberately kept in houses and apartments. Pet dogs, usually leashed, affect urban vegetation through deposits of their excrement. Pet cats more often roam free and have retained their hunting instincts; they can have major effects on suburban wildlife, particularly shrews, rodents, and songbirds. Most of the cage birds in the United States are bred in the pet industry and remain inside, but in other parts of the world it is customary to have a wild-caught singing bird caged and displayed on a balcony or at the front door. In China, popular caged birds include the laughingthrush called Hwamei (*Garrulax canorus*), Mongolian Skylarks (*Melanocorypha mongolica*), and the Eurasian Skylark (*Alauda arvensis*). Each town and city has a bird market; and each morning, owners hang their caged birds from trees in city parks so the birds can sing to each other. In a few of the warmer United States cities escaped cage birds have established breeding populations. In Los Angeles, for example, one can find Ringed Turtledove (*Streptopelia risoria*), Budgerigar (*Melopsitticus undulatus*), Red-crowned Parrot (*Amazona viridigenalis*), and Yellow-headed Parrot (*A. oratrix*). They feed on the seeds and fruits of the alien tree species planted in yards and along city streets (Orians 2000).

For the most part, wildlife in the city defies human control. Habitats including gardens, waste sites, and hardscapes offer opportunities for those able to exploit them. The House Sparrow, native to Eurasia but an urban dweller in cities around the world, is representative of many urban vertebrates: it is a generalist in feeding behavior, it is flexible in its use of nesting sites, it is tolerant of noise and disturbance; and it is intelligent. In the United States, where it was introduced in 1850, it is one of the most common urban birds along with the Pigeon or Rock Dove (*Columba livia*), introduced in the early 1700s, and the Starling, introduced in 1890. In northern cities these three may constitute 95 percent of the bird population in winter (Adams 1994). The pigeon is an interesting example of the long relationships between many urban animals and people. The Pigeon was spread around the world by people as a domesticated animal raised for its meat. The Pigeon may actually have been the first bird to have become domesticated. Its origins trace to the eastern Mediterranean, where early grain farmers in Mesopotamia sometime between 10,000 and 5,000 years ago may have encouraged wild rock doves to breed in their towns and cities by building artificial nest sites. The native doves were year-round residents nesting in caves and on ledges and had breeding seasons longer than any other natives, which ensured a lengthy supply of meat. Dovecotes or pigeon houses became standard features of homes in farming villages and towns lasting into the Colonial Period in North America. During a time when it was difficult to impossible to overwinter larger livestock and all but the breeding stock were slaughtered in autumn, pigeons were a staple winter food. Genetically flexible, they were selected for a variety of characteristics such as high reproductive rates, tolerance of human activities, and a wide range of plumage colors and patterns that can be seen in populations that descend from birds that escaped from their owners. These domestics gone wild are called feral Pigeons; their long history with humans preadapted them to life in the city (Johnston and Janiga 1995).

Other birds that do well in cities include opportunistic feeders such as crows, ravens, and herring gulls. In urban settings, they have changed their diet to feed upon the intentional offerings of bread crusts, seeds, and the like that humans provide, as well as to scavenge in dumps. Birds that naturally roost and nest on cliffs can find suitable habitats in city centers where window ledges and cornices on high-rise buildings are adequate substitutes. Pigeons use these artificial cliffs, as do Peregrine Falcon (*Falco peregrinus*) and other hawks that feed upon pigeons and starlings. Since the 1970s, in the Prairie Provinces of Canada, merlins (*F. columbarius*) have been colonizing cities; these hawks of open vegetation feed on House Sparrows in urban areas. Common Nighthawks (*Chordeiles minor*) nest on flat rooftops and in the evening can be seen swooping beneath city street lamps in pursuit of insects (Gilbert 1989; Adams 1994).

In residential areas in the United States with carefully tended lawns, gardens, and trees, many native birds have disappeared, but others such as American Robin (*Turdus migratorius*), Gray Catbird (*Dumatella caroliniensis*), and Northern Mockingbird (*Mimus polyglottos*) have increased in numbers. Various finches, chickadees, and other small birds visit bird feeders, and hummingbirds seek nectar from flowers and feeders. Sharp-shinned Hawks (*Accipter striatus*) and Cooper's Hawk (*A. cooperii*) are often attracted to the abundance of songbirds in suburban yards, and the small Screech Owl (*Otus asio*) is a common nocturnal predator (Adams 1994). In drier parts of the United States and in cities in Brazil, the Burrowing Owl (*Speotyto cunicularia*) is frequently seen in parks and between runways at airports.

The most common waterfowl at urban ponds throughout the Northern Hemisphere are the Mallard (*Anas platyrhynchos*) and the Canada Goose (*Branta canadensis*) (Adams 1994). In many instances, domesticated ducks, including the red-wattled Muscovy Duck (*Cairina moschatus*), have been released onto these ponds and sometimes have hybridized with Mallards.

Small and medium-sized herbivores, especially those that eat seeds, are the most common mammals in urbanized areas. Among those found in North America are deer mice (*Peromyscus maniculatus*), meadow voles (*Microtus pennsylvanicus*), tree and ground squirrels, cottontails (*Sylvilagus floridanus*), and woodchucks (*Marmota monax*). Gray squirrel (*Sciurus caroliniensis*), red squirrel (*Tamiasciurus hudsonicus*), and fox squirrels (*Sciurus niger*)—depending upon the location—are abundant in yards and city parks. The highest reported density of gray squirrels, 49 per **hectare** (20 per acre) occurs in Lafayette Park next to the White House in Washington, D.C. Populations in nearby rural areas are 2–5 per hectare (1–2 per acre) (Adams 1994).

Larger herbivores, in particular members of the deer family, are becoming more common in U.S. towns and cities. White-tailed deer (*Odocoileus virginianus*) finds garden plants and shrubbery as delectable as wild plants and is becoming a year-round pest in the eastern United States, as suburbia continues to encroach upon the countryside. In the western United States, deer, elk (*Cervus canadensis*), and moose (*Alces alces*) generally are seen in towns only during severe winters or extended periods of drought (Adams 1994).

Aquatic herbivores such as muskrat (*Ondrata zithebeca*) and beaver (*Castor canadensis*) may be found in towns across the United States wherever streams or lakes occur. In Florida, manatee (*Trichechus manatus*) may swim in urban rivers or gather in winter near the warm water effluents of power plants (Adams 1994).

Among omnivores, raccoons can reach high numbers in urban settings, where they feed mostly on pet food and garbage. Coyotes have inhabited Los Angeles since the 1930s and are increasingly found in other cities across the United States. In town, these omnivores feed heavily on garbage, pet food, small mammals, and fruits, and also take house cats and small dogs (Adams 1994).

Small mammalian predators are abundant in towns and cities. Free-ranging domestic cats can be particularly destructive to ground-nesting birds in gardens and park areas. A study of feral cats in Brooklyn, New York, indicated that most urban cats receive supplemental feed from people and are more scavengers than hunters. The highest densities (4.88/ha or approximately 2/ac) correlated not with handouts, though, but with the abundance of shelter available in abandoned buildings. [Rural areas had densities considerably lower than 1/ha ( 0.4/ac)] (Calhoon and Haspel 1989). Other urban predatory mammals include red fox (*Vulpes fulva*) that feed on mice and squirrels, skunks that hunt grubs in lawns, shrews and moles that eat insects, and bats that consume flying insects by the thousands. One of the most common urban bats in the United States is the little brown bat (*Myotis lucifugus*) (Adams 1994). It eats midges and mosquitoes along with other insects, so it is sometimes attracted by landowners who put up special bat houses for them. In European cities, the most widespread of six urban bats is *Pipistrellus pipstrellus.* Greatest bat activity occurs where houses are widely spaced and there are many trees, and the least occurs in the city center. The paucity of pipistrelles in the downtown area is attributed to the scarcity of trees and prevalence of low-pressure sodium street lamps whose yellowish light does not attract insects. Where the white glow of mercury vapor street lamps attracts insects, bats are much more common (Glaisler et al. 1998).

Amphibians and reptiles tend to be sensitive to the urban environment and are less abundant there than in surrounding areas. Their populations decline with the destruction of groundcover, increased predation, pollution, and use of pesticides. Roads are often barriers to dispersal, which leaves them in small, fragmented populations highly susceptible to local extinction. A study of lizards in Tucson, Arizona, found that the tree lizard (*Urosaurus ornatus*) was an exception. These reptiles readily climb buildings and exotic trees, and their numbers actually were highest in areas with very high-density housing. The only other urban-dwelling lizard whose numbers were greatest where housing density was above average was the greater earless lizard (*Cophosaurus texanus*), which is found in moderately developed areas along with low numbers of desert spiny lizards (*Sceloporus magister*) and Clark's spiny lizard (*S. clarkii*). Highly sensitive species such as whiptails (*Cnemidophorus* spp.), zebra-tail (*Callisaurus draconoides*), and the lesser earless lizard (*Holbrookia maculata*)—though relatively common in the surrounding desert—were rare in town (Germaine and Wakeling 2001).

Among invertebrates thriving in gardens and lawns are earthworms, spiders, crickets, ants, beetles, butterflies, snails, and slugs.

Some animals have succeeded so well in the city that they are viewed as pest species by the human inhabitants, and many town governments have a special department for animal control. The white-tailed deer previously mentioned is only the most recent nuisance. The common rat (*Rattus norvegicus*) traveled the world in the days of sailing ships. As it spread from its place of origin in Asia to the cities of Europe in the 1300s, it carried with it bubonic plague, the Black Death that killed 25 percent of Europe's population between 1346 and 1352 (Diamond 1997). Plague is a bacterial disease that originally infected wild rodents, but it spread to the rats inhabiting the old market cities of Central Asia, India, and China. It is transmitted from animal to animal by fleas, and the fleas on urban rats could easily jump to city dwellers living in the squalor of medieval cities (Learmonth 1988). From ancient Asian centers of trade, the rat eventually spread to the port cities of the rest of the world. Today it lives in sewerage systems and garbage dumps and is considered a scourge in its own right, a symbol of filth.

Bats that roost in attics and pigeons and starlings that congregate in large numbers on city buildings or under bridges produce unsightly and unhealthful accumulations of droppings. The acids in the bird droppings can erode monuments and buildings, and the accumulated manure may harbor a disease-causing fungus (*Histoplasma capsulatum*) that can be transmitted to humans (Johnston and Janiga 1995). House Sparrows will displace favored native species from cavities and bird houses needed for nesting. Raccoons carry rabies (as do bats and skunks) and can be destructive if they get into buildings, as can squirrels and mice. Skunks, of course, are unpleasant when they are challenged by pet dogs and retaliate by spraying them. Coyotes are unwelcome when they take pet cats or small dogs.

## URBAN IMPACTS ON PLANT AND ANIMAL LIFE

Human-dominated towns and cities offered new habitats and new niches for those plants and animals that could colonize and adapt to the urban environment. Frequent disturbance and an abundance of artificial surfaces selected for those species that either could survive with constant nurturing by peoples (e.g., ornamental plants and pets) or were flexible generalists that could live independently but in the midst of human activity. In most urban places, this has meant a replacement of many of the native plants of presettlement vegetation with ornamentals and alien weeds. The ornamentals typically are **cultivars** and hybrids selected for odd but attractive features such as variegated or colored foliage, pendulous habit, or large and abundant flowers. Rarely are native plants brought into gardens, although native trees may remain in parks and other green spaces. The plants that usually dominate the urban wastelands—vacant lots, demolition sites, roadway gutters, and the like—tend to be either escapes from gardens or alien weeds distributed along the trade routes of the world (Gilbert 1989).

As pollution levels increase, the urban lichen community is simplified. The number of species increase, the biomass and cover decrease. New, often exotic, species such as *Lecanora conizaeoides* gain dominance. Lichens are visible enough that they have been useful indicators of the severity of air pollution. A similar response occurs among mosses in highly polluted areas, but since they tend to hide in cracks and crevices, they have not be used as pollution indicators. In English towns, a unique moss community dominated by only two species (*Bryum argenteum* and *Ceraton purpureus*) develops. In rural areas, the moss community may comprise as many as 30 species (Gilbert 1989).

The animal community has also been altered from that of the natural areas outside the urban place. Many natives do not succeed in proximity to human habitation, and these are lost. But other species, such as both the American Robin (*Turdus migratorius*) and the English Robin (*Erithacus rubecula*), thrive and have even increased in the lawns and gardens of residential areas. Indeed, gardens are the final refuge for animals from the vanished temperate broadleaf deciduous forest biome of the British Isles (see Chapter 7). Surviving natives are joined in urban places by species not seen in natural areas.

Some evolution has occurred among the life-forms of the city. Michaelmas daisy has become increasingly more variable in its traits in English cities. Hybridization has occurred to create uniquely urban species. In Great Britain, two species evolved as purely urban forms, the St. Kilda house mouse (*Mus muralis*) and London pride (*Saxifraga* x *urbinum*). The St. Kilda house mouse lived among settlements on a remote island west of the Outer Hebrides off Scotland. When the island's human inhabitants were removed, the mouse went extinct within two to three years. London pride, a hybrid plant, is known only from urban places in England (Gilbert 1989).

Speciation may have occurred among evening primroses (*Oenothera* spp.) in Europe. Originally introduced from North America 350 years ago, evening primroses are now represented on the continent by 15 species, only two of which are identical to North American species. Dandelions and buttercups limited to manmade habitats may also be the products of speciation within the city environments (Sukopp et al. 1979).

Adaptation to pollution has occurred in a few plants, most famously in the development of copper and zinc tolerance in creeping bentgrass (*Agrostis stolonifera*) and lead tolerance in ribwort plantain (*Plantago lanceolata*) (Pickett et al. 2001). A classic case of evolution by natural selection in animals is the peppered moth (*Biston betularia*), which derived from selection in a habitat in urban England blackened by soot. Moth populations in rural areas had a preponderance of white individuals, while those in the urban setting were mostly black—a response known as industrial melanism. The dark color afforded city moths protection; light-colored individuals were much more obvious and vulnerable to predation.

Among native species inhabiting urban environments, demographic and behavioral changes have been noted. Among birds, at least, species richness is often higher than in surrounding natural areas.

> The suburbanization of areas in lowland California may result in the displacement of certain bird species, whose environmental-support requirements are eliminated or altered. Yet the subsequent increases in species numbers and individuals in suburban over pre-suburban environments are striking phenomena, and these trends are contrary to the conventional wisdom of viewing human impacts on biotas as being solely detrimental to them. (Vale and Vale 1976, p. 165)

Animals live in higher densities in cities than in the wild. Squirrels in Washington, D.C., were mentioned earlier. White-tailed deer in Chicago are reported at densities of 26/km$^2$ (67/mi$^2$); in Winnipeg, Manitoba, at 30/km$^2$ (78/mi$^2$); and in a residential area in New Jersey at 65/km$^2$ (168/mi$^2$). In the wild, a high average is 10/km$^2$ (26 mi$^2$). Density of urban raccoons can be close to 1/ha (0.4/ac), while in the wild it is more like 0.16/ha (1/15–45 ac). Home ranges of raccoons in Toronto average 0.4 km$^2$ (0.16 mi$^2$), while in rural areas, home ranges vary from 10 to 26 km$^2$ (4–10 mi$^2$). Skunks also have much smaller home ranges in urban places, where the average is 0.62 km$^2$ (0.25 mi$^2$) compared to 3–5 km$^2$ (1.2–2.0 mi$^2$) in rural areas (Adams 1994).

Changes in social structure have been noted in suburban foxes in England. They form territorial pairs in wild situations, but in the city they run in groups of three to five and change location often. Part of this change is attributed to differences in the food supply and part to higher mortality rates due to traffic deaths that frequently break up pairs (Wheater 1999). Sex ratios in urban populations of mallard favor males (65:35), while in rural populations the ratio is more even (52:48). The cause is believed to be the high mortality of nesting hens and low survivorship of nestlings in cities due to predation. No such variation in sex ratios occurs among Canada Goose populations; geese strongly defend their own goslings and also accept orphans (Adams 1994).

## ORIGINS OF CITIES

Whether permanent settlements predated agriculture or became possible only after agriculture was adopted is debated. There is no question, however, that agricultural surpluses were a prerequisite to the establishment of cities. The earliest cities were confined to four locations in the Old World and two in the New World. The first arose by 4000 B.C. in Mesopotamia, where Ur may have achieved a population of 200,000 people. A thousand years later, the Nile Valley had cities. Thebes, in ancient Egypt, had grown to about 225,000 by 1600 B.C. Urban centers grew up in the valley of the Indus River by 2500 B.C. and on the Hwang He (Yellow River) of China by 2000 B.C. The earliest cities in the Americas, in Mexico and Peru, date to A.D. 500 (Brunn and Williams 1993).

Rome was the first city to reach a population of 1 million. This mark was met in A.D. 200s, but during the Dark Ages it declined to fewer than 200,000 people. London was the next European city to reach 1 million, a feat not accomplished until the early 1800s. In Asia, Xi'an, China, had a population greater than a million in the seventh century, and Kyoto, Japan, had reached a million by the middle of the eighteenth century. Most other Asian cities were much larger than their European counterparts before the nineteenth century (Brunn and Williams 1993).

Rapid population growth in Western cities coincided with industrialization. Before then, death rates were so high in cities because of contagious diseases and polluted drinking water that urban populations had to be maintained by a constant inflow of rural migrants. After about 1840, improved sanitation lowered death rates to levels lower than birth rates, and natural increase allowed both rural and urban populations to expand (White 1994). Between 1800 and the 1890s, London's population grew from 860,000 to 5 million, and New York went from 62,500 to 2.7 million. Raw materials extracted at great distances flowed to the urban factories.

Immigrants from the countryside and—in the United States—from abroad flooded into the cities. Tenements to house them filled in the empty spaces in the built-up area. By 1900, 13 cities had populations greater than a million—all but three in Europe or North America (Platt 1994a).

The picture changed dramatically after 1970. By 1980, 230 cities in the world had populations over a million, and most were in developing countries. Mexico City doubled in size between 1970 and 1980. Of the nine cities with populations greater than 9 million, seven were in Asia, Latin America, and Africa. This growth was fueled in large part by rapid population growth and the migration of the rural poor to the cities rather than by the expansion of industry. People found work in the service sector of the economy or in the so-called informal working sector composed of street hawkers, day laborers, and others who do not make it into official labor statistics. Others simply became unemployed. Accompanying this immigration was the creation of severe social, economic, and environmental conditions in the urban centers due to inadequate housing; inadequate infrastructure (water, sanitation, and transportation systems); and inadequate public health, welfare, and educational services. One-quarter to one-half of the population of many cities lived in squatter settlements or shantytowns (Platt 1994a; Brunn and Williams 1993).

Less than 3 percent of the world's people lived in urban places in 1800. Urbanization is often considered a measure of the level of economic development of a country: developed countries are also urbanized (at least according to the model of Europe and North America). In 1900, Great Britain became the first urbanized society, meaning more than 50 percent of its population lived in urban areas. The United States achieved this status in 1920. By 1955, there were 18 such urbanized countries, 30 in 1965, and 64 in 1990. Most such countries are in Europe and the Americas, very few are in Africa (Brunn and Williams 1993).

Urban land-use patterns developed differently in different cultures, but modern cities the world over bear the imprint of the European city. Many outside of Europe were created or greatly modified by Western colonialism and imperialism. Often, a new Western commercial city was appended to an existing traditional one, although in Latin America and Africa traditional cities were largely destroyed and then rebuilt by European powers. In Asia, remaining sections of the old cities add complexity and distinctiveness to the urban fabric (as well as habitats largely unexplored by scientists). Islamic cities in Southwest Asia, for example, developed from ancient Middle Eastern urban centers originally circular in layout and surrounded by a wall. Temples and public buildings were placed in the centers, and wide streets radiated outward from them. Residences were typically a compact mass of rectangular buildings with inner courtyards and windowless walls facing the alleyways. Streets and cul de sacs developed in a haphazard fashion to create an irregular pattern that reflects an Islamic ideal of equality: status is not advertised or recognized by neighborhood design or adornment of house facades. A central permanent marketplace, the bazaar or *suq*, is another distinctive feature of Moslem communities. The bazaar consists of a cluster of small stalls set next to each other, usually under a common canopy or dome. Those offering similar fresh agricultural products, manufactured wares, or services are grouped to together; a network of aisles winds among the various sectors. These neighborhoods are usually described as formless by Western observers used to grid-pattern street plans or otherwise controlled design of European and North American cities. A modern city of supermarkets and skyscrapers looking much like any in the West may be only a block away (Brunn and Williams 1993).

The point to be made here is that our knowledge of the urban biome derives from Western cities, particularly those of Western Europe and North America. Largely unexplored biotas await in the urban places of Latin America, Asia, and Africa. It is to be expected that there may be unique habitats and lingering native plants and animals, both domesticated and wild. However, it is also to be expected that these communities are rapidly changing and becoming more Europeanized with the globalization of trade and tourism and with rising economic levels in many developing countries where people are acquiring Western tastes.

## HISTORY OF SCIENTIFIC INVESTIGATION OF URBAN ECOLOGY

Scientific interest in the urban environment first arose in response to public health concerns. Sir Edwin Chadwick conducted pioneer work in the 1830s and 1840s in Sheffield, England, to correlate data on illness and infant mortality with environmental conditions in different parts of the city. His reports stimulated the development of urban planning in Great Britain, France, and the United States. This discipline and practice spread to other parts of the world with European colonialism. At about the same time, in the United States, George Perkins Marsh published his classic *Man and Nature, Physical Geography as Modified by Man.* While he did not focus directly on cities, Marsh pointed out the effects humans had on nature and initiated the conservation movement in America. Conservation blended with the romanticism then popular in painting and literature and gave rise to movements to bring nature into the city. Frank Law Olmstead, Sr., designed large public parks with this in mind, although his nature was very much controlled. Central Park in New York City (1853) was the first of the parks Olmstead designed for American cities. "City Beautiful" projects followed from the 1890s to 1950s and resulted in tree-lined streets and lawns outlined by hedges. Ebenezer Howard's Garden City concept, first published in 1898, promoted greenbelts around cities and open space, but the spaces were to be used for recreation and other human activities—not left to grow over in untamed vegetation. In the 1960s, the guide for modern urban planning was Ian McHarg's (1969) *Design with Nature*, a book that promoted taking natural processes into account and working with them rather than trying to deny their existence or correct them. None of these philosophies and resulting practices, however, addressed the ecological relations between plant and animal life and the urban environment (Platt 1994a, Platt 1994b), but urban planners and landscape architects have contributed importantly to our basic knowledge of urban plants and animals.

Urban ecology had largely fallen between the cracks of academic science until recently. Ecologists tended to seek out habitats as natural as possible for their studies. Botanists ignored ornamentals and scorned alien species. Sociologists interested in the urban environment focused on people and the city as habitat for humans. Urban geographers were largely interested in city morphology and growth and the development of cultural landscapes, without examining small-scale features of vegetation and animal life. Landscape architects focused on the grander parks and monuments of created habitats (Schmid 1975). Conservationists interested in urban environments have been interested mostly in improving the quality of life for the human inhabitants through cleaner air and cleaner water. This often involves establishing vegetation to trap pollutants and restoring wetlands and other habitats. Most recently, the focus has been on the effects of industrial cities on the global environment with such issues as acid precipitation, resource depletion, and global warming paramount.

True scientific exploration of the ecology of plants and animals in the urban areas originated with studies of the plants colonizing bombsites in Europe immediately after World War II. This early work gave rise to numerous case studies of urban floras in both western and eastern Europe and later led to an interest in urban ecology in North America, Latin America, and Japan (Kent et al. 1999). Forest Stearns (1971) was one of the first to encourage urban research in the United States (Pickett et al. 2001). Schmid's (1975) study of urban vegetation in Chicago was one of the first comprehensive looks at a city's plant life. The structure and composition of vegetation and the interaction between trees and the urban environment have been continuing focuses of urban forestry, and an extensive body of literature has developed.

Wildlife biologists took the lead in the study of urban animal life. They have concentrated on open spaces and, as befits their profession, have been most interested in the applied aspects of ecology related to habitat conservation and restoration. They have largely ignored built-up areas and the wild nooks and crannies of the city. A National Institute for Urban Wildlife was founded in Columbia, Maryland; it hosts conferences, publishes on urban wildlife management issues, and offers outreach programs for public education (Pickett et al. 2001).

Urban ecology is now a recognized subdiscipline of ecology and is just entering a stage that goes beyond inventory and description to engage in studies of ecological processes (Kent et al. 1999). A professional journal, *Urban Ecology*, is devoted to the publication of research results of ecologists, landscape architects, and urban planners. In 1997, the U.S. National Science Foundation added two urban sites to its program for Long Term Ecological Research (LTER), one in Phoenix, Arizona, and the other in Baltimore, Maryland. Germany, Taiwan, and Singapore have plans for similar projects (Kloor 1999). The urban biome is now seen as a laboratory for investigating dispersal and colonization processes, and plant and animal community responses to changes in species composition; climate; landscape structure; disturbance regimes; and cultural, economic, and political conditions. Humans are embraced as part of the **ecosystem** instead of being seen as an external factor separate from nature (McDonnell and Pickett 1990). Urban ecology is a young field and one with great growth potential as the world becomes increasingly urbanized.

## REFERENCES

Adams, Lowell W. 1994. *Urban Wildlife Habitats, A Landscape Perspective.* Minneapolis: University of Minnesota Press.

Brunn, Stanley D. and Jack F. Williams. 1993. *Cities of the World: World Regional Urban Development.* 2nd ed. New York: Harper Collins College Publishers.

Calhoon, Robert E. and Carol Haspel. 1989. "Urban cat populations compared by season, subhabitat, and supplemental feeding." *Journal of Animal Ecology* 58: 321–328.

Collins, J. P., A. Kinzig, N. B. Grimm, W. F. Fagan, D. Hope et al. 2000. "A new urban ecology." *American Scientist* 88: 416–425.

Diamond, Jared. 1997. *Guns, Germs, and Steel: The Fates of Human Societies.* New York: W. W. Norton & Company.

Germaine, S. S. and B. F. Wakeling. 2001. "Lizard species distribution and habitat occupation along an urban gradient in Tucson, Arizona, USA." *Biological Conservation* 97(2): 229–237.

Gilbert, O. L. 1989. *The Ecology of Urban Habitats.* London: Chapman & Hall.

Glaisler, J. J. Zukal, and Z. Rehak. 1998. "Habitat preferences and flight activity of bats in a city." *Journal of Zoology* 244(3): 439–445.

Grapow, L. C. and C. Blasi. 1998. "A comparison of urban flora in different phytoclimatic regions in Italy." *Global Ecology and Biogeography Letters* 7(5): 367–378.

Hough, Michael. 1994. "Design with city nature: an overview of some issues." Pp. 40–48 in Rutherford H. Platt, Rowan A. Rowntree, and Pamela C. Muick, eds., *The Ecological City: Preserving and Restoring Urban Biodiversity.* Amherst, MA: The University of Massachusetts Press.

Ioffe, Grigory and Tatyana Nefedova. 2000. *The Environs of Russian Cities.* Lewiston, NY: The Edwin Mellon Press.

Jim, C. Y. 1998. "Old stone walls as ecological habitat for urban trees in Hong Kong." *Landscape and Urban Planning* 42(1): 29–43.

Johnston, Richard F. and Marian Janiga. 1995. *Feral Pigeons.* Oxford: Oxford University Press.

Kalin Arroyo, Mary T., Clodomiro Marticorena, Oscar Matthei, and Lohengrin Cavieres. 2000. "Plant invasions in Chile: Present patterns and future predictions." Pp. 385–421 in Harold A. Mooney and Richard J. Hobbs, eds., *Invasive Species in a Changing World.* Washington, DC and Covelo, CA: Island Press.

Kent, M., R. A. Stevens and L. Zhang. 1999. "Urban plant ecology patterns and processes: a case study of the flora of the City of Plymouth, Devon, U.K." *Journal of Biogeography* 26(6): 1281–1298.

Kloor, Keith. 1999. "A surprising tale of life in the city." *Science* 286(5440): 63.

Learmonth, Andrew. 1988. *Disease Ecology.* Oxford: Basil Blackwell, Ltd.

McDonnell, M. J. and S. T. A. Pickett. 1990. "Ecosystem structure and function along urban-rural gradients: an unexploited opportunity for ecology." *Ecology* 71: 1232–1237.

Orians, G. H. 2000. "Site characteristics favoring invasions." Pp. 133–176 in Harold A. Mooney and Richard J. Hobbs, eds., *Invasive Species in a Changing World.* Washington, DC and Covelo, CA: Island Press.

Pickett, S. T. A., M. L. Cadenasso, J. M. Grove, C. H. Nilon, R. V. Pouyat, W. C. Zipperer, and R. Constanza. 2001. "Urban ecological systems: linking terrestrial ecological, physical and socioeconomic components of metropolitan areas." *Annual Review of Ecology and Systematics* 32: 127–157.

Platt, Rutherford H. 1994a. "The ecological city: introduction and overview." Pp. 1–17 in Rutherford H. Platt, Rowan A. Rowntree, and Pamela C. Muick, eds., *The Ecological City: Preserving and Restoring Urban Biodiversity.* Amherst, MA: The University of Massachusetts Press.

Platt, Rutherford H. 1994b. "From commons to commons: evolving concepts of open space in North American cities." Pp. 21–39 in Rutherford H. Platt, Rowan A. Rowntree, and Pamela C. Muick, eds., *The Ecological City: Preserving and Restoring Urban Biodiversity.* Amherst, MA: The University of Massachusetts Press.

Prach, K. and P. Pysek. 2000. "How do species dominating in succession differ from others?" *Journal of Vegetation Science* 10(3): 383–392.

Pysek, P. 1999. "Alien and native species in Central European urban floras: a quantitative comparison." *Journal of Biogeography* 25(1): 155–163.

Schmid, James A. 1975. *Urban Vegetation: A Review and Chicago Case Study. Research Paper No. 161.* Chicago: Department of Geography, University of Chicago.

Stearns, Forest. 1971. "Urban botany—an essay on survival." *University of Wisconsin Field Station Bulletin* 4: 1–6. (cited in Picketts et al. 2001).

Sukopp, Herbert, Hans-Peter Blume, and Wolfram Kunick. 1979. "The soil, flora, and vegetation of Berlin's waste lands." Pp. 115–132 in Ian Laurie, ed., *Nature in the City: The Natural Environment in the Design and Development of Urban Green Space.* New York: John Wiley & Sons.

Sukopp, H. and P. Werner. 1983. "Urban environments and vegetation." Pp. 247–260 in W. Holzner, M. J. A. Werger and I. Ikusima, eds., *Man's Impact on Vegetation.* The Hague: Dr. W. Junk Publishers.

Theobold, David M. 2001. "Land-use dynamics beyond the American urban fringe." *Geographical Review* 3: 544–564.

Vale, Thomas R. and Geraldine R. Vale, 1976. "Suburban bird populations in west-central California." *Journal of Biogeography* 3: 157–165.

Wheater, C. Philip. 1999. *Urban Habitats.* London: Routledge.

White, Rodney R. 1994. *Urban Environmental Management: Environmental Change and Urban Design.* New York: John Wiley & Sons.

Woodell, Stanley. 1979. "The flora of walls and pavings." Pp. 135–157 in Ian Laurie, ed., *Nature in the City: The Natural Environment in the Design and Development of Urban Green Space.* New York: John Wiley & Sons.

# GLOSSARY

**Affinity.** Relationship; used in plant and animal geography to indicate taxonomic similarities between the flora and fauna of different regions.

**Agroecosystem.** A group of plants, animals, and soils that interact with each other and their physical environment under human management in an agricultural setting.

**Alien (species).** Nonnative. A species living in an area outside its natural range. Also called introduced or exotic species. The term alien is usually used with plants.

**Anaerobic.** Living or being without free oxygen.

**Annual.** A plant that completes its life cycle in one year or one growing season.

**Arboreal.** Living in trees; adapted to life in the trees.

**Asexual reproduction.** The multiplication of individual organisms by processes that do not involve the combining of ova and pollen (plants) or egg and sperm (animals). Examples include budding and cloning. See also vegetative reproduction.

**Badlands.** An intricately eroded surface in desert or semiarid environments, usually without vegetation.

**Bajada.** Coalesced alluvial fans at the base of desert mountain ranges.

**Biodiversity.** The total variability of life contained in genes, in species, and in ecosystems.

**Biogeography.** The study of the distribution of life on Earth, past and present, and the processes that determine where plants, animals, and other organisms occur.

**Biome.** A large region with similar vegetation, animal life, and environmental conditions. One of the largest recognizable ecological units on Earth.

**Breed.** A group of domesticated animals exhibiting a narrow range of traits produced by selective mating controlled by humans. Individuals mated with other members of the group produce offspring possessing these characteristics.

**Broad-leaved.** Plants that have thin, flat leaves. Contrast with needle-leaved.

**Brownfield.** A former industrial or urban site now abandoned. May or may not be contaminated.

**Canopy.** The uppermost layer of foliage in vegetation.

**Cauliflory.** The condition of having flowers and fruits produced on the trunks of trees. Common among the trees of tropical rainforests.

**Climate.** The average weather (especially temperature and precipitation) patterns that occur during a normal year and are experienced over decades or centuries. Weather refers to the conditions of the atmosphere at any given moment.

**Climax community.** The stable, persistent plant community established in response to the regional climate. The theoretical end stage in ecological succession.

**Closed canopy.** Crowns of adjacent plants in the same layer of the vegetation intermingle and overlap to prevent much sunlight from penetrating to lower layers.

**Community.** The living organisms (i.e., species) assembled in a given area. Sometimes refers to only a subset of these species, such as the plant community, or the bird community, or the benthic community.

**Compound leaf.** A leaf composed of several distinct leaflets.

**Conifer.** A cone-bearing plant such as the needle-leaved pines and spruce.

**Cosmopolitan.** Worldwide in distribution.

**Cover.** The proportion of a surface on which vegetation occurs, usually measured as a percent.

**Crown.** The top mass of foliage on a tree or shrub.

**Crustose (lichen).** Crustlike. Compare to foliose and fruticose.

**Cultivar.** A particular variety of some domesticated plant species produced by selective breeding and maintained in the garden or agricultural plot.

**Cursorial.** Adapted to running.

**Cushion plant.** A low-growing, multistemmed plant that grows as a dense mound. This growth form is mostly associated with cold and dry climate regions.

**Cyanobacteria.** Also known as blue-green algae. These protistans are found in soil and water and are able to fix nitrogen and photosynthesize.

**Deciduous.** Refers to plants, usually trees, and shrubs, that shed their leaves during non-growing seasons.

**Dispersal.** The movement of an organism away from its place of origin. In biogeography, this refers to the movement and establishment of a species beyond the limits of its previous distribution area.

**Disturbance.** An event that disrupts an ecosystem and usually destroys at least part of the vegetation. May be of physical origin (e.g., floods, landslides, or fire) or of biological origin (e.g., disease, insect outbreaks, or overgrazing).

**Drift (glacial).** All materials deposited by processes associated with glaciation. Includes outwash and till.

**Ecology.** The study of the interrelationships between organisms and their environment, including the other organisms inhabiting the same area.

**Ecosystem.** A community of organisms and the physical environment with which it interacts. All living and nonliving elements of a given area that function together as a single ecological unit to maintain a flow of energy and the recycling of nutrients.

**Edaphic.** Refers to conditions of the substrate. Edaphic factors that may limit plant growth include waterlogged conditions, low-nutrient content of soils, and the development of a hardpan.

**Endemic.** Originating in and restricted to a particular geographic area.

**Ephemeral.** Refers to an annual plant that completes its life cycle in a extremely short period of time (usually a few weeks). Also refers to desert stream that flows only for a few hours during or immediately after a rainfall event.

**Epiphyte.** A plant that grows on the branches or stems of another plant and uses its host only for support. One of Raunkiaer's growth forms, also known as an air plant.

**Ericaceous.** Referring to a member of the Ericaceae, a plant family consisting of heathers, bilberries, labrador tea, rhododendrons, etc.

**Evapotranspiration.** The combined processes of adding water vapor to the atmosphere through evaporation from the soil and bodies of water and from the passage of water out of plants through the stomata in their leaves.

**Evergreen.** Refers to plants, usually trees or shrubs, which maintain leaves all year. Leaves may be replaced throughout the year or in a flush during a single season, but the plant is never without live foliage.

**Exotic (species).** Nonnative. See also alien. Usually used in reference to introduced animals.

**Exposure.** The direction toward which a slope faces. Also known as aspect.

**Fauna.** All the animal species in a given area.

**Feral.** Refers to domesticated species that have escaped from human control and are living and reproducing in a wild state.

**Fishery.** An area where commercial fishing takes place. The fishing industry, often as related to a specific type of fish.

**Flora.** All the plant species in a given area.

**Foliose (lichen).** Leaflike in appearance. Compare to crustose and fruticose.

**Forb.** A broad-leaved, green-stemmed plant. One type of herb.

**Fruticose (lichen).** Shrubby, upright lichens. Compare to crustose and foliose.

**Gallery forest.** A narrow strip of trees that extends along river banks in savannas or other nonforest vegetation types.

**Genus (plural = genera).** A taxonomic unit composed of one or more closely related species.

**Geophyte.** A perennial plant with its renewal organ protected well below the surface as a bulb or corm. One of Raunkiaer's life forms.

**Green Revolution.** The period marked by the development of improved, high-yielding varieties of major food crops that began in the 1950s. These new varieties depended upon large inputs of chemical fertilizers and pesticides but greatly increased the world's food supply at a time when the human population was growing rapidly.

**Growth form.** The appearance or morphology of a plant that is adapted to particular environmental conditions. Examples include tree, shrub, epiphyte.

**Habit (leaf).** The condition of being either deciduous or evergreen.

**Habitat.** The place where a species lives and the local environmental conditions of that place.

**Half shrub.** See subshrub.

**Heath.** Small-leaved shrubs such as heather and bilberries, which are members of the Ericaceae. Leaves are often drought adapted. Heaths are a widespread group of plants occurring in several different biomes.

**Hectare.** A unit of area in the metric system of measurement. Equivalent to 2.5 acres.

**Herb.** A nonwoody or soft, green-stemmed plant that dies down each year. May be annual or perennial. Broad-leaved herbs are called forbs. Grasses and sedges are called graminoids.

**Horticultural variety.** A plant selected for use in gardens or landscaping. Also called an ornamental.

**Horticulture.** The intensive cultivation associated with gardening.

**Humus.** Partially decayed plant and animal matter that occurs as a dark brown organic substance in soils and is important for conserving water and some plant nutrients.

**Industrial species.** Domestic livestock species or crop bred for maximum production of a single product. Usually individuals in a breed or variety are genetically identical and require high inputs of energy, fertilizer or high quality feed. Livestock must be kept in an environmentally controlled structure; crops often grown with irrigation.

**Intercropping.** The simultaneous planting of two or more crop species in the same agricultural plot. Plants of different species may be intermingled or may be planted in alternating rows.

**Introduced (species).** A species transported accidentally or deliberately by humans beyond its natural distribution area. See also alien, exotic.

**ITC.** Intertropical Convergence Zone. This is the location where the trade winds of the Northern and Southern hemispheres meet. The converging air masses are forced upward, and as they cool they release moisture. The ITC moves back and forth across the equator in sync with the annual migration of the direct rays of the sun. This gives rise to the alternating rainy and dry seasons commonly experienced in the tropics depending on whether the ITC is close or distant. Also ITCZ.

**Kitchen garden.** An intensively cultivated plot close to a home where a large number of crops are typically produced for the consumption of the home's occupants. The crops often include food and medicinal plants as well as ornamentals. Chickens and other animals may also be raised. Also called dooryard garden.

**Landrace.** A population of domesticated plants or animals whose traits are selected as much by natural environmental conditions as by human selection. Individuals are more genetically variable than those of plant varieties or animal breeds.

**Landscape.** A mosaic of interacting ecosystems, the spatial pattern of which is repeated over a wide area.

**Latitude.** Distance north or south of the equator, measured in degrees. The equator is 0° latitude. Low latitudes lie between 0° and 30° north and south; midlatitudes between 30° and 60°, and high latitudes between 60° and 90°.

**Leachate.** Dissolved organic compounds that seep from plant cells.

**Leaching.** The process by which dissolved compounds are removed from soil horizons by the downward movement of water.

**Legume.** Any plant that is a member of the families Mimosaceae, Caesalpinaceae, or Fabaceae. These families are sometimes lumped into a single family, the Leguminosae, the pea and bean family. Many of these plants have rhizobial bacteria occurring in nodules in their roots. These bacteria fix nitrogen and allow legumes to grow in otherwise nutrient-poor conditions.

**Liana.** A woody-stemmed vine or climber.

**Lichen.** A form of life composed of a fungus and an alga joined in a symbiotic relationship and classified as a single organism.

**Life form (Raunkiaer's).** A category of plant life based upon morphology and the position of the renewal bud. See "Introduction to Terrestrial Biomes" for more details.

**Limnology.** The study of bodies of still water, such as lakes, and the organisms that inhabit them.

**Loess.** Fine, wind-blown particles that accumulate in thick deposits.

**Microhabitat.** A tiny nook or habitat with specific environmental conditions that differ from that of the larger habitat in which it occurs.

**Monoculture.** The production of a single crop species over large areas. Commonly associated with plantation and other forms of commercial agriculture.

**Morphology.** Form and structure, size and shape of an organism. General appearance of an organism.

**Multicropping.** The production of two or more crops in the same plot. May be grown simultaneously or in sequence. See also intercropping.

**Myccorhiza.** The association between certain fungi and the root hairs of many plants.

**Myccorhizal fungi.** The fungi that are associated with plant roots and collect and prepare nutrients for uptake by the plant.

**Nebka.** A small mound of sand or soil held in place by a desert plant.

**Needle-leaved.** Refers to conifers with slender, pointed leaf shapes such as pines and spruce.

**Open canopy.** The crowns of adjacent plants in the same layer of the vegetation do not touch, and abundant sunlight reaches lower layers or the ground.

**Orographic.** Refers to the condition of air being forced to rise and cool as it passes over a land barrier, usually a mountain, or to the precipitation so produced.

**Outwash (glacial).** Sorted materials deposited by the meltwater from glaciers.

**Paludification.** The process by which acidic, waterlogged conditions are produced and bogs are expanded due to the growth of mosses. Associated with the Boreal Forest Biome.

**Parent material.** Unconsolidated material from which soil is developed.

**Perennial.** Refers to a plant that lives for several years and undergoes active growth each year.

**pH.** A measure of acidity or alkalinity. Represents the negative logarithm of the concentration of hydrogen ions in solution and uses a scale from 0 (extremely acidic) through 7 (neutral) to 14 (extremely alkaline).

**Photosynthesis.** The process by which green plants convert oxygen and carbon dioxide in the presence of sunlight to sugars. Electromagnetic energy in the form of visible light is transformed into chemical energy that can be used by living organisms.

**Phylogeny.** The evolutionary history or descent of a species or other taxon.

**Physiognomy.** The characteristic appearance of a plant community or of vegetation.

**Physiographic province.** A region defined according to its characteristic landforms and underlying geology.

**Physiology.** The metabolic functions and processes of organisms. The study of these functions and processes.

**Plate tectonics.** The movement of pieces of Earth's outermost layer and the rearrangement and deformation of the crust that results. Continental drift, ocean floor spreading, and mountain building are some of the processes involved that have significant effects on climate change and the distribution of organisms at a geological time scale.

**Reef.** Any mass of rock with a surface at or just below the low-tide mark.

**Rhizobial bacteria.** Bacteria associated with the roots of some plants. They are often enclosed in nodules and are important because they can fix free nitrogen into compounds that plants can use.

**Rhizome.** A horizontal root structure that lies just below the surface.

**Rosette.** A growth form characterized by a basal whorl of leaves around a central stem or renewal bud.

**Savanna.** A vegetation structure consisting of a continuous cover of grasses above which is an open canopy of trees or shrubs.

**Scrub.** A vegetation type characterized by sparse, small shrubs.

**Semishrub.** See subshrub.

**Sexual reproduction.** The formation of new individuals by the fusion of gametes (ova and pollen in plants; egg and sperm in animals).

**Soil.** The uppermost layer on land, composed of a mixture of mineral and organic materials and in which plants grow.

**Soil horizon.** A layer within the soil which is fairly distinct in terms of its chemistry and color.

**Species.** A group of sexually reproducing individuals that can produce viable offspring. The fundamental unit of classification in taxonomy.

**Stolon.** A horizontal stem that suns just above or just below the ground surface. Common in grasses.

**Stomata.** The leaf pores through which plants exchange gases with the atmosphere.

**Subshrub.** A shrub in which the upper branches die back during the unfavorable seasons. Also called half shrub and semishrub.

**Subsistence.** A type of farming or other economic activity in which foods and other products are consumed by the producer and his/her immediate family and not sold to others.

**Succession (ecological).** The theoretical development of a plant community over time beginning with a barren site. A series of communities come to replace each other on the site, each one altering the habitat to make it more favorable for the next group of species. Finally, a stable, persistent community (the climax) in equilibrium with the regional climate is reached.

**Succulent.** A growth-form that has specialized tissue in the stem, leaves, or an underground organ for the storage of water. Associated primarily with the Desert Biome but also found in highly saline terrestrial habitats.

**Taxon (plural = taxa).** Any taxonomic group at any level in the taxonomic hierarchy.

**Taxonomy.** The science of describing, classifying, and naming organisms.

**Temperate.** Mild. Refers generally to the temperature patterns of the midlatitudes where summers are warm to hot and winters are mild to cool.

**Temperate zone.** Those parts of Earth that lie between the tropics and the Arctic and Antarctic circles. Essentially corresponds to the midlatitudes.

**Thallus.** The body of a nonvascular plant that is not differentiated into stem, root, or leaf.

**Till (glacial).** Unsorted materials deposited directly by glaciers.

**Tropics.** The parts of Earth that lie between 23° 30′ N and 23° 30′ S or between the Tropic of Cancer and the Tropic of Capricorn.

**Tussock.** A growth form of grasses and sedges in which individuals grow as tufts or clumps and form conspicuous hummocks on the ground.

**Ungulate.** Any hoofed mammal.

**Variety.** A population of domesticated plants in which a narrow range of desirable traits have been perpetuated through selective breeding.

**Vascular plant.** Any plant with conducting vessels (phloem and xylem) that move nutrients water between roots and leaves. Includes ferns and the flowering plants.

**Vegeculture.** The cultivation of crops propagated by vegetative reproduction. Involves root crops such as potatoes, yam, and manioc; also such crops as bananas and sugar cane. Contrasts with seed agriculture.

**Vegetation.** The general plant cover of an area, defined according to the appearance of the plants rather than the actual species present.

**Vegetative reproduction.** The formation of new plants from pieces of the parent plant such as leaf fragments of root, stem, or rhizome.

**Weathering.** The physical and chemical processes by which solid rock is broken into finer particles and, ultimately, clays.

**Zonation.** The occurrence of particular forms of life in distinct belts. These may be determined by latitude, elevation, or depth below the surface of water.

# BIBLIOGRAPHY

Abd el Rahman, A. A. 1986. "The deserts of the Arabian peninsula." Pp. 29–54 in Michael Evenari, Imanuel Noy-Meir, and David W. Goodall, eds., *Hot Deserts and Arid Shrublands, B*. Ecosystems of the World, 12B. Amsterdam: Elsevier.

Abell, Robin A., et al. 2000. *Freshwater Ecoregions of North America: A Conservation Assessment.* Washington, DC: Island Press.

Abyzov, Sabit S. 1993. "Microorganisms in the Antarctic ice." Pp. 265–295 in E. Imre Friedmann, ed., *Antarctic Microbiology*. New York: Wiley-Liss, Inc.

Achitur, Yair and Zvy Dubinsky. 1990. "Evolution and zoogeography of coral reefs." Pp. 1–9 in Z. Dubinsky, ed., *Coral Reefs*. Ecosystems of the World, 25. Amsterdam: Elsevier.

Acocks, J. P. H. 1953. "Veld types of South Africa." *Memoirs of the Botanical Survey of South Africa* 28: 1–192.

Acton, D. F. 1992. "Grassland soils." Pp. 25–54 in R. T. Coupland, ed., *Natural Grasslands: Introduction and Western Hemisphere*. Ecosystems of the World, 8A. Amsterdam: Elsevier.

Adams, Lowell W. 1994. *Urban Wildlife Habitats, A Landscape Perspective*. Minneapolis: University of Minnesota Press.

Allan, J. David. 1995. *Stream Ecology: Structure and Function of Running Waters*. London: Chapman & Hall.

Alongi, Daniel M. 1998. *Coastal Ecosystem Processes*. Boca Raton: CRC Press.

American Philosophical Society. n.d. "Mark Catesby." http://www.amphilsoc.org/library/exhibits/nature/carolina.htm.

Andrade Lima, D. 1981. "The caatinga dominium." *Revista Brasileira de Botânica* 4: 149–1963.

Angel, Martin and Tegwyn Harris. 1977. *Animals of the Oceans, the Ecology of Marine Life*. New York: Two Continents Publishing Group.

Arianoutsou, M. and R. H. Groves, eds. 1994. *Plant-Animal Interactions in Mediterranean-type Ecosystems*. Dordrecht: Kluwer.

Aubréville, A. 1949. *Climats, Forêts et Désertification de l'Afrique Tropicale*. Paris: Larose.

Ayyad, Mohamed A. and Samir I. Ghabbour. 1986. "Hot deserts of Egypt and the Sudan." Pp. 149–202 in Michael Evenari, Imanuel Noy-Meir, and David W. Goodall, eds.,

*Hot Deserts and Arid Shrublands, B.* Ecosystems of the World, 12B. Amsterdam: Elsevier.

Axelrod, D. I. 1989. "Age and origin of chaparral." Pp. 7–19 in S.C. Keeley, ed., *The California Chaparral: Paradigms Revisted.* Sci. Ser. No. 34. Los Angeles: Natural History Museum of Los Angeles County.

Axelrod, D. I. and P. Raven. 1985. "Origins of the Cordilleran flora." *Journal of Biogeography* 12: 21–47.

Bahre, Conrad J. 1979. *Destruction of the Natural Vegetation of North-Central Chile.* Berkeley: University of California Press.

Barnes, Burton V. 1991. "Deciduous Forests of North America." Pp. 219–344 in E. Röhrig and B. Ulrich, eds., *Temperate Deciduous Forests.* Ecosystems of the World, 7. Amsterdam: Elsevier.

Barnes, D. J. and B. E. Chalker. 1990. "Calcification and photosynthesis in reef-building corals and algae." Pp. 109–131 in Z. Dubinsky, ed., *Coral Reefs.* Ecosystems of the World, 25. Amsterdam: Elsevier.

Barnes, R. S. K. and R. N. Hughes. 1999. *An Introduction to Marine Ecology.* 3rd edition Oxford: Blackwell Scientific.

Beard, J. S. 1953. "The savanna vegetation of northern tropical Amazonas." *Ecological Management* 23: 149–215.

Beard, J. S. 1955. "The classification of tropical American vegetation-types." *Ecology* 36: 89–100.

Bender, Lionel. 1982. *Reference Handbook on the Deserts of North America.* Westport, CT: Greenwood Press.

Bews, J. W. 1916. "An account of the chief types of vegetation in Southern Africa, with notes on plant succession." *Journal of Ecology* 4: 129–159.

Bixby, Donald E., Carolyn J. Christman, Cynthia J. Ehrman, and D. Phillip Sponenberg. 1994. *Taking Stock: The North American Livestock Census.* Blacksburg, VA: The McDonald & Woodward Publishing Company.

Blatter, E. and F. Hallberg. 1918. "The flora of the Indian desert (Jodhpur and Jaisaimer)." *Journal of the Bombay Natural History Society* 26: 218–246.

Bliss, L. C. 1997. "Arctic ecosystems of North America." Pp. 551–683 in F. E. Wielgolaski, ed., *Polar and Alpine Tundra.* Ecosystems of the World, 3. Amsterdam: Elsevier.

Block, William. 1997. "Arctic ecosystems of North America." Pp. 551–683 in F. E. Wielgolaski, ed., *Polar and Alpine Tundra.* Ecosystems of the World, 3. Amsterdam: Elsevier.

Blondel, Jacques and James Aronson. 1999. *Biology and Wildlife of the Mediterranean Region.* Oxford: Oxford University Press.

Bolshov, Alexandr A. 1999. "Main priorities for conservation of biodiversity of caspian region." Annex 6 in Caspian Environmental Programme, Biodiversity Caspian Regional Thematic Center, First Regional Workshop in Kokshetau, Kazakhstan. http://www.caspianenvironment.org/biodiversity/first/annex6.htm.

Bonan, G. B. and H. H. Shugart. 1989. "Environmental factors and ecological processes in boreal forests." *Annual Review in Ecology and Systematics* 20: 128.

Boughey, A. S. 1957a. "The physiognomic delimitation of West African vegetation types." *Journal of the West African Science Association* 3: 148–163.

Boughey, A. S. 1957b. "The vegetation types of the Federation." *Proceedings and Transactions of the Rhodesian Science Association* 45: 73–91.

Bourliere, F. and Hadley M. 1983. "Present day savannas: An overview." Pp. 1–17 in Francois Bourliere, ed., *Tropical Savannas.* Ecosystems of the World, 13. Amsterdam: Elsevier Scientific Publishing Company.

Bourliere, Francois. 1983. "Mammals as secondary consumers in savanna ecosystems." Pp. 463–475 in Francois Bourliere, ed., *Tropical Savannas.* Ecosystems of the World, 13. Amsterdam: Elsevier Scientific Publishing Company.

Boyden, Stephen. 1992. *Biohistory: The Interplay between Human Society and the Biosphere.* Park Ridge, NJ: Parthenon Publishing Group.

Braun, E. Lucy. 1950. *Deciduous Forests of Eastern North America.* Philadelphia: Blakiston.

Breckle, S. W. 1983. "Temperate deserts and semideserts of Afghanistan and Iran." Pp. 271–319 in Neil E. West, ed., *Temperate Deserts and Semi-deserts.* Ecosystems of the World, 5. Amsterdam: Elsevier Scientific Publishing Company.

Briggs, J. C. 1974. *Marine Zoogeography.* New York: McGraw-Hill.

Brönmark, Christer and Lars-Anders Hansson. 1998. *The Biology of Lakes and Ponds.* New York: Oxford University Press.

Brookfield, Harold. 2001. *Exploring Agrodiversity.* New York: Columbia University Press.

Brown, Dwight A. 1993. "Early nineteenth century grasslands of the midcontinent plains." *Annals of the Association of American Geographers* 83(4): 589–612.

Brunn, Stanley D. and Jack F. Williams. 1993. *Cities of the World: World Regional Urban Development.* 2nd ed. New York: Harper Collins College Publishers.

Bullock, Stephen H., Harold A. Mooney, and Ernesto Medina, eds. 1995. *Seasonally Dry Tropical Forests.* Cambridge: Cambridge University Press.

Burgis, Mary and Pat Morris. 1987. *The Natural History of Lakes.* Cambridge: Cambridge University Press.

Calhoon, Robert E. and Carol Haspel. 1989. "Urban cat populations compared by season, subhabitat, and supplemental feeding." *Journal of Animal Ecology* 58: 321–328.

Campbell, J. S. 1997. "North American alpine ecosystems." Pp. 211–261 in Wielgolaski, F. E., ed., *Polar and Alpine Tundra.* Ecosystems of the World, 3. Amsterdam: Elsevier.

Cannon, W. A. 1924. "General and physiological features of the vegetation of the more arid portions of South Africa with notes on the climatic environment." *Yearbook of the Carnegie Institution of Washington* 8(354): 1–159.

Carlton, J. T., J. B. Geller, M. L. Reaka-Kudla, and E. A. Norse. 1999. "Historical extinctions in the sea." *Annual Review of Ecology and Systematics* 30: 515–538.

Carpenter, J. R. 1940. "The grassland biome." *Ecological Monographs* 10: 617–684.

Castillo, Miris and Pier Luigi Nimis. 1997. "Diversity of lichens in Antarctic." Pp. 15–21 in B. Battaglia, J. Valencia, and D. W. H. Walton, eds., *Antarctic Communities: Species, Structure and Survival.* Cambridge: Cambridge University Press.

Ceballos, Gerardo. 1995. "Vertebrate diversity, ecology, and conservation in neotropical dry forests." Pp. 195–220 in Bullock, Stephen H., Harold A. Mooney, and Ernesto Medina, eds., *Seasonally Dry Tropical Forests.* Cambridge: Cambridge University Press.

Champion, H. G. 1936. "A preliminary survey of the forest types of India and Burma." *Indian Forest Records, New Series* 1: 1–286.

Chang, W. Y. B. 1987. "Fish Culture in China." *Fisheries* 12(3): 11–15.

Chernov, Y. I. and N. V. Mataveyeva. 1997. "Arctic ecosystems in Russia." Pp. 361–507 in Wielgolaski, F. E., ed., *Polar and Alpine Tundra.* Ecosystems of the World, 3. Amsterdam: Elsevier.

Chevrier, M., P. Vernon, and Y. Frenot. 1997. "Potential effects of two alien insects on a sub-Antarctic wingless fly in the Kerguelen Islands." Pp. 424–432 in B. Battaglia, J. Valencia, and D. W. H. Walton, eds., *Antarctic Communities: Species, Structure and Survival.* Cambridge: Cambridge University Press.

Childe, V. Gordon. 1951. *Man Makes Himself.* New York: The New American Library, Inc.

Ching, Kim K. 1991. "Temperate deciduous forests in East Asia." Pp. 539–555 in E. Röhrig and B. Ulrich, eds., *Temperate Deciduous Forests.* Ecosystems of the World, 7. Amsterdam: Elsevier.

Christopherson, Robert W. 1998. *Elemental Geosystems.* 2nd edition. Upper Saddle River, NJ: Prentice-Hall.

Clements, Frederic E. 1916. *Plant Succession: An Analysis of the Development of Vegetation.* Publication No. 242, Carnegie Institution of Washington. Washington, DC: Carnegie Institution of Washington.

Clements, F. E. 1920. *Plant Indicators: The Relation of Plant Community to Process and Practice.* Washington, DC: Carnegie Institute of Washington.

Clements, Frederic E. and Victor E. Shelford. 1939. *Bio-ecology.* New York: John Wiley.

Clutton-Brock, Juliet. 1981. *Domesticated Animals from Early Times.* Austin, TX: University of Texas Press.

Cohen, Mark Nathan. 1977. *The Food Crisis in Prehistory; Overpopulation and the Origins of Agriculture.* New Haven: Yale University Press.

Cole, M. M. 1986. *The Savannas: Biogeography and Geobotany.* New York: Academic Press.

Collins, J. P., A. Kinzig, N. B. Grimm, W. F. Fagan, D. Hope, et al. 2000. "A new urban ecology." *American Scientist* 88: 416–425.

Conacher, Arthur J. and Maria Sala, eds. 1998. *Land Degradation in Mediterranean Environments of the World: Nature and Extent, Causes and Solutions.* New York: John Wiley & Sons.

Connell, J. H. 1961a. "The influence of intra-specific competition and other factors on the distribution of the barnacle *Chthamalus stellatus.*" *Ecology* 42: 710–723.

Connell, J. H. 1961b. "Effects of competition, predation by *Thais lapillus,* and other factors on natural populations of the barnacle *Balanus balanoides.*" *Ecological Monographs* 31: 61–104.

Coupland, R. T. 1961. "A reconsideration of grassland classification in the northern Great Plains of North America." *Journal of Ecology* 49: 135–167.

Coupland, R. T. 1992a. "Mixed prairie." Pp. 150–182 in R. T. Coupland, ed., *Natural Grasslands: Introduction and Western Hemisphere.* Ecosystems of the World, 8A. Amsterdam: Elsevier.

Coupland, R. T. 1992b. "Overview of South American grasslands." Pp. 363–366 in R. T. Coupland, ed., *Natural Grasslands: Introduction and Western Hemisphere.* Ecosystems of the World, 8A. Amsterdam: Elsevier.

Cousteau, Jean-Michel. 2001. "The future of the ocean." Pp. 170–180 in Sylvia A. Earle, ed., *National Geographic Atlas of the Oceans.* Washington, DC: National Geographic Society.

Cowling, R. M., D. M. Richardson, and S. M. Pierce. 1997. *Vegetation of South Africa.* Cambridge: Cambridge University Press.

Crosby, Alfred W. 1972. *The Columbian Exchange: Biological Consequences of 1492.* Westport, CT: Greenwood Publishing Company.

Crosby, Alfred W. 1986. *Ecological Imperialism: The Biological Expansion of Europe, 900–1900.* Cambridge: Cambridge University Press.

Cuatrecasas, J. 1968. "Paramo vegetation and its life forms." *Colloquium geographicum* 9: 163–186.

Dallman, Peter. R. 1998. *Plant Life in the World's Mediterranean Climates.* Oxford: Oxford University Press.

Dansereau, Pierre. 1957. *Biogeography: An Ecological Perspective.* New York: Ronald Press.

Darwin, Charles R. 1842. *The Structure and Distribution of Coral Reefs.* London: Smith Elder and Co.

Daubenmire, Rexford. 1992. "Palouse prairie." Pp. 297–312 in R. T. Coupland, ed., *Natural Grasslands: Introduction and Western Hemisphere.* Ecosystems of the World, 8A. Amsterdam: Elsevier.

Davis, G. W. and D. M. Richardson, eds. 1995. *Mediterranean-Type Ecosystems: The Function of Biodiversity.* New York: Springer-Verlag.

Dean, Warren. 1995. *With Broadax and Firebrand, The Destruction of the Brazilian Atlantic Forest.* Berkeley: University of California Press.

Dejoux, Claude. 1992a. "The Mollusca." Pp. 311–336 in C. Dejoux and A. Iltis, eds., *Lake Titicaca, A Synthesis of Limnological Knowledge.* Dordrecht: Kluwer Academic Publishers.

Dejoux, Claude. 1992b. "The insects." Pp. 365–382 in C. Dejoux and A. Iltis, eds., *Lake Titicaca, A Synthesis of Limnological Knowledge.* Dordrecht: Kluwer Academic Publishers.

Dejoux, Claude. 1992c. "The avifauna." Pp. 460–469 in C. Dejoux and A. Iltis, eds., *Lake Titicaca, A Synthesis of Limnological Knowledge.* Dordrecht: Kluwer Academic Publishers.

Dejoux, Claude and André Iltis. 1992. "Introduction." Pp. xv–xx in C. Dejoux and A. Iltis, eds., *Lake Titicaca, A Synthesis of Limnological Knowledge.* Dordrecht: Kluwer Academic Publishers.

D'Elia, Christopher F. and William J. Wiebe. 1990. "Biogeochemical nutrient cycles in coral-reef ecosystems." Pp. 49–74 in Z. Dubinsky, ed., *Coral Reefs.* Ecosystems of the World, 25. Amsterdam: Elsevier.

Delaney, John. 2001. "Antarctica and the Europa connection." Pp. 150–151 in Sylvia A. Earle, ed., *National Geographic Atlas of the Oceans.* Washington, DC: National Geographic Society.

Denevan, William. 2001. *Cultivated Landscapes of Native Amazonia and the Andes.* New York: Oxford University Press.

Desanker, P. V., P. G. H. Frost, C. O. Frost, C. O. Justice, and R. J. Scholes, eds. 1997. The Miombo Network: Framework for a Terrestrial Transect Study of Land-Use and Land-Cover Change in the Miombo Ecosystems of Central Africa, IGBP Report 41. Stockholm: The International Geosphere-Biosphere programme. Also available at http://miombo.gecp.virginia.edu/igbp/index.html.

Diamond, Jared. 1997. *Guns, Germs, and Steel: The Fates of Human Societies.* New York: W.W. Norton & Company.

Diaz, A., J. E. Péfour, and P. Durant. 1997. "Ecology of South American páramos with emphasis on the fauna of the Venezuelan páramos." Pp. 263–310 in Wielgolaski, F. E., ed., *Polar and Alpine Tundra.* Ecosystems of the World, 3. Amsterdam: Elsevier.

di Castri, Francesco and Harold A. Mooney, eds. 1973. *Mediterranean-type Ecosystems: Origin and Structure.* New York: Springer-Verlag.

Dickenson, Ann. n.d. "Curator's notes." for Mark Catesby, The Natural History of Carolina, Florida, and the Bahama Islands: A Hypertext Exhibit. http://www.people.virginia.edu/~jad7e/catesby/contents.html.

Dillon, Michael O. n.d. "Lomas formations of the Atacama Desert, Northern Chile," *Centres of Plant Diversity, Vol 3: The Americas.* Washington, DC: National Museum of Natural History. http://www.nmnh.si.edu/botany/projects/cpd/sa/sa43.htm.

Di Prisco, Guido. 1997. "Physiological and biochemical adaptations in fish to a cold marine environment." Pp. 251–260 in B. Battaglia, J. Valencia, and D. W. Walton (eds.), *Antarctic Communities: Species, Structures and Survival.* Cambridge: Cambridge University Press.

Donoso, Claudio. 1996. "Ecology of *Nothofagus* forests in Central Chile." Pp. 271–292 in Thomas T. Veblen, Robert S. Hill, and Jennifer Reed, eds., *The Ecology and Biogeography of* Nothofagus *Forests.* New Haven: Yale University Press.

Drude, O. 1890. *Handbuch der Pflanzengeographie.* Stuttgart: Engelhorn.

Dubinsky, Z. 1990. "Preface." Pp. v–vi in Z. Dubinsky, ed., *Coral Reefs.* Ecosystems of the World, 25. Amsterdam: Elsevier.

Dugan, Patrick. 1993. *Wetlands in Danger, A World Conservation Atlas.* New York: Oxford University Press.

Edwards, M. V. 1950. "Burma forest types according to Champion's classification." *Indian Forest Records, New Series* 7: 135–173.

Eisma, D. 1988. "An introduction to the geology of continental shelves." Pp. 39–91 in H. Postma and J. J. Zijlstra, eds., *Continental Shelves.* Ecosystems of the World, 27. Amsterdam: Elsevier.

Ekman, S. 1953. *Zoogeography of the Sea.* London: Sidgwick and Jackson.

Elliott, S., J. F. Maxwell, and O. P. Beaver. 1989. "A transect survey of monsoon forest in Doi Suthep-Pui National Park." *Natural History Bulletin of the Siam Society* 37: 137–141.

Elliott-Fisk, D. L. 1983. "The stability of the northern Canadian tree limit." *Annals of the Association of American Geographers* 73: 560–576.

Elliott-Fisk, Deborah L. 2000. "The taiga and boreal forest." Pp. 41–73 in Michael G. Barbour and William Dwight Billings, eds., *North American Terrestrial Vegetation.* 2nd edition. Cambridge: Cambridge University Press.

Ellison, Aaron M. and Elizabeth J. Farnsworth. 2001. "Mangrove communities." Pp. 423–442 in Mark D. Bertness, Steven D. Gaines, and Mark E. Hay, eds., *Marine Community Ecology.* Sunderland, MA: Sinauer Associates, Inc.

Ellison, A. M., E. J. Farnsworth, and R. E. Merkt. 1999. "Origins of mangrove ecosystems and mangrove biodiversity anomalies." *Global Ecology and Biogeography Letters* 8: 9–115.

Elvebakk, Arve. 1997. "Tundra diversity and ecological characteristics of Svalbard." Pp. 347–359 in Wielgolaski, F. E., ed., *Polar and Alpine Tundra.* Ecosystems of the World, 3. Amsterdam: Elsevier.

EMBRAPA. 1996. *Atlas do Meio ambiente do Brasil* (Atlas of the Environment of Brazil). Brasilia, DF: EMBRAPA-SPI.

Esseen, Per-Anders, Bengt Ehnstrom, Lars Ericson, and Kjell Sjoberg. 1997. "Boreal forest." Pp. 16–47 in Lennart Hannson, ed., *Boreal Ecosystems and Landscapes: Structures, Processes, and Conservation of Biodiversity.* Copenhagen: Munksgaard International Publishers.

Etter, Ron J. and Lauren S. Mullineaux. 2001. "Deep-sea communities." Pp. 367–394 in Mark D. Bertness, Steven D. Gaines, and Mark E. Hay, eds., *Marine Community Ecology.* Sunderland, MA: Sinauer Associates, Inc.

Eyre, S. R. 1968. *Vegetation and Soils.* 2nd edition. Chicago: Aldine Publishing Company.

Falkowski, Paul G., Paul L. Jokiel, and Robert A. Kinzie III. 1990. "Irradiance and coral." Pp. 89–107 in Z. Dubinsky, ed., *Coral Reefs.* Ecosystems of the World, 25. Amsterdam: Elsevier.

Forbes, S. A. 1887. "The lake as a microcosm." *Peoria Science Association Bulletin.* Republished in 1920 in *Illinois Natural History Survey Bulletin* 15: 537–550.

Forestry Resource Assessment. 2000. "AN Boreal coniferous forest (Ba)." Food and Agriculture Organization of the United Nations. http://www.fao.org/forestry/fo/fra/index.html.

Foster, M. S. 1991. "Open coast intertidal and shallow subtidal ecosystems of the Northeast Pacific." Pp. 235–272 in A. C. Mathieson and P. H. Nienhuis, eds., *Intertidal and Littoral Ecosystems.* Ecosystems of the World 24. Amsterdam: Elsevier.

Fowler, Cary and Pat Mooney. 1990. *Shattering; Food, Politics, and the Loss of Genetic Diversity.* Tucson: The University of Arizona Press.

Fraser, William R. and Donna L. Patterson. 1997. "Human disturbance and long-term changes in Adélie penguin populations: A natural experiment at Palmer Station, Antarctic Peninsula." Pp. 445–452 in B. Battaglia, J. Valencia, and D. W. H. Walton, eds., *Antarctic Communities: Species, Structure and Survival.* Cambridge: Cambridge University Press.

Fry, C. H. 1983. "Birds in savanna ecosystems." Pp. 337–357 in Francois Bourliere, ed., *Tropical Savannas.* Ecosystems of the World, 13. Amsterdam: Elsevier Scientific Publishing Company.

Furley P. A. and J. A. Ratter. 1988. "Soil resources and plant communities of the central Brazilian cerrado and their development." *Journal of Biogeography* 15: 97–108.

Furley, P. A., J. Proctor, and J. A. Ratter, eds. 1992. *Nature and Dynamics of Forest-Savanna Boundaries.* London: Chapman and Hall.

Furley, P. A. and W. W. Newey. 1983. *Geography of the Biosphere, An Introduction to the Nature, Distribution and Evolution of the World's Life Zones.* London: Butterworths.

Fussell, L. K. 1992. "Semi-arid cereal and grazing systems of West Africa." Pp. 485–518 in C. J. Pearson, ed., *Field Crop Ecosystems*. Ecosystems of the World, 18. Amsterdam: Elsevier.

Gade, Daniel W. 1999. *Nature and Culture in the Andes*. Madison: University of Wisconsin Press.

Galera, Francisca M. and Lorenzo Ramella. n.d. "Gran Chaco data sheet." *Centres of Plant Diversity, Vol. 3. The Americas*. Washington, DC: National Museum of Natural History. http://www.nmnh.si.edu.

Garda, Eduardo Carlos, ed. 1996. *Atlas do Meio Ambiente do Brasil*. Brasilia: EMBRAPA-SPI.

Gates, David M. 1993. *Climate Change and Its Biological Consequences*. Sunderland, MA: Sinauer Associates, Inc.

Gentry, Alwyn H. 1995. "Diversity and floristic composition of neotropical dry forests." Pp. 146–194 in Bullock, Stephen H., Harold A. Mooney, and Ernesto Medina, eds., 1995. *Seasonally Dry Tropical Forests*. Cambridge: Cambridge University Press.

Germaine, S. S. and B. F. Wakeling. 2001. "Lizard species distribution and habitat occupation along an urban gradient in Tucson, Arizona, USA." *Biological Conservation* 97(2): 229–237.

Gilbert, O. L. 1989. *The Ecology of Urban Habitats*. London: Chapman & Hall.

Giller, Paul S. and Björn Malmqvist. 1998. *The Biology of Streams and Rivers*. Oxford: Oxford University Press.

Gillon, Dominique. 1983. "The fire problem in tropical savannas." Pp. 617–641 in Francois Bourliere, ed., *Tropical Savannas*. Ecosystems of the World, 13. Amsterdam: Elsevier Scientific Publishing Company.

Glaisler, J. J. Zukal, and Z. Rehak. 1998. "Habitat preferences and flight activity of bats in a city." *Journal of Zoology* 244(3): 439–445.

Gleason, H. A. 1926. "The individualistic concept of the plant association." *Torrey Botanical Club Bulletin* 53: 7–26.

Glynn, Peter W. 1990. "Feeding ecology of selected coral-reef macroconsumers: Patterns and effects on coral community structure." Pp. 365–400 in Z. Dubinsky, ed., *Coral Reefs*. Ecosystems of the World, 25. Amsterdam: Elsevier.

Good, Ronald. 1947. *The Geography of the Flowering Plants*. London: Longmans, Green and Company.

Goudie, Andrew. 1981. *The Human Impact: Man's Role in Environmental Change*. Cambridge, MA: The MIT Press.

Goulding, Michael. 1980. *The Fishes and the Forest: Explorations in Amazonian Natural History*. Berkeley: University of California Press.

Grabherr, Georg. 1997. "The high-mountain ecosystems of the Alps." Pp. 97–121 in Wielgolaski, F. E., ed., *Polar and Alpine Tundra*. Ecosystems of the World, 3. Amsterdam: Elsevier.

Gradwohl, Judith (curator). 1996. "Ocean Planet, A Smithsonian Traveling Exhibition." Washington, DC: National Museum of Natural History. http://seawifs.gsfc.nasa.gov/ocean_planet.html.

Graham, Alan and David Dilcher. 1995. "The Cenozoic record of tropical dry forest in northern Latin America and the southern United States." Pp. 124–145 in Bullock, Stephen H., Harold A. Mooney, and Ernesto Medina, eds., *Seasonally Dry Tropical Forests*. Cambridge: Cambridge University Press.

Grapow, L. C. and C. Blasi. 1998. "A comparison of urban flora in different phytoclimatic regions in Italy." *Global Ecology and Biogeography Letters* 7(5): 367–378.

Grassle, J. F. and N. J. Maciolek. 1992. "Deep-sea species richness: Regional and local diversity estimates from quantitative bottom samples." *American Naturalist* 139: 313–341.

Gremmen, Niek J. M. 1997. "Changes in the vegetation of sub-Antarctic Marion island resulting from introduced vascular plants." Pp. 417–423 in B. Battaglia, J. Valencia, and D. W. H. Walton, eds., *Antarctic Communities: Species, Structure and Survival.* Cambridge: Cambridge University Press.

Grigg, Richard W. and Steven J. Dollar. 1990. "Natural and anthropogenic disturbance on coral reefs." Pp. 439–467 in Z. Dubinsky, ed., *Coral Reefs.* Ecosystems of the World, 25. Amsterdam: Elsevier.

Grisebach, August H. R. 1872. *Die Vegetation der Erde nach Ihrer Klimatischen Anordnung.* Leipzig: Wilhelm Englemann.

Groves, R. H. and F. di Castri, eds. 1991. *Biogeography of Mediterranean Invasions.* Cambridge: Cambridge University Press.

Gupta, R. K. 1986. "The Thar Desert." Pp. 55–99 in Michael Evenari, Imanuel Noy-Meir, and David W. Goodall, eds., *Hot Deserts and Arid Shrublands, B.* Ecosystems of the World, 12B. Amsterdam: Elsevier.

Hagen, Joel. 1992. *An Entangled Bank: The Origins of Ecosystem Ecology.* New Brunswick, NJ: Rutgers University Press.

Hanbury-Tenison, Robin. 1989. "People of the forest." Pp. 189–213 in Lisa Silcock, ed., *The Rainforests: A Celebration.* San Francisco: Chronicle Books.

Hanfei, Dong. 1992. "Upland Rice systems." Pp. 183–203 in C. J. Pearson, ed., *Field Crop Ecosystems.* Ecosystems of the World, 18. Amsterdam: Elsevier.

Hansen, Michael J. and James W. Peck. n.d. "Lake Trout in the Great Lakes." http://biology.usgs.gov/s+t/m2130.htm.

Happold, D.C. D. 1983. "Rodents and lagomorphs." Pp. 363–400 in Francois Bourliere, ed., *Tropical Savannas.* Ecosystems of the World, 13. Amsterdam: Elsevier Scientific Publishing Company.

Hare, F. K. 1950. "Climate and zonal divisions of the boreal forest formation in eastern Canada." *Geographical Review* 40: 615–635.

Harlan, Jack R. 1995. *The Living Fields: Our Agricultural Heritage.* Cambridge: Cambridge University Press.

Harriott, Thomas. 1588. *Briefe and true report of the new found land of Virginia.* 1972 Reprint of 1590 version by Dover Publications, Inc., New York.

Harris, David R. 1976. "Traditional systems of plant food production and origins of agriculture in West Africa." Pp. 311–356 in Jack R. Harlan, Jan M. J. DeWet, and Ann B. Stemler, eds., *Origins of African Plant Domestication.* The Hague: Mouton Publishers.

Harris, David R. 1977. "Alternative pathways toward agriculture." Pp. 179–243 in Charles A. Reed, ed., *Origins of Agriculture.* The Hague: Mouton Publishers.

Harshberger, J. W. 1911. *Phytogeographic Survey of North America.* New York: Hafner Publishers.

Hart, John Fraser. 1998. *The Rural Landscape.* Baltimore: The Johns Hopkins University Press.

Hartshorn, G. S. 1983. "Plants." Pp. 118–157 in D. H. Janzen, ed., *Costa Rican Natural History.* Chicago: University of Chicago Press.

Head, Lesley. 2000. *Cultural Landscapes and Environmental Change.* New York: Oxford University Press.

Heady, H. F., J. W. Bartolome, M. D. Pitt, G. D. Savelle, and M. C. Stroud. 1992. "California prairie." Pp. 313–335 in R. T. Coupland, ed., *Natural Grasslands: Introduction and Western Hemisphere.* Ecosystems of the World, 8A. Amsterdam: Elsevier.

Hedberg, O. 1964. "Features of afroalpine plant ecology." *Acta Phytogeographica Suecia* 49: 1–144.

Hedberg, O. 1986. "Origins of the Afroalpine flora." Pp. 443–468 in Francois Vuilleumier and Maximina Monasterio, eds., *High Altitude Tropical Biogeography.* New York: Oxford University Press.

Hedberg, O. 1992. "Afroalpine vegetation compared to paramo: Convergent adaptations and divergent differentiation." Pp. 15–29 in H. Balslev and J. L. Luteyn, eds., *Paramo, An Andean Ecosystem under Human Influence.* New York: Academic Press.

Hill, Robert S. and Mary E. Dettmann. 1996. "Origins and diversification of the genus *Nothofagus*." Pp. 11–24 in Thomas T. Veblen, Robert S. Hill, and Jennifer Reed, eds., *The Ecology and Biogeography of* Nothofagus *Forests*. New Haven: Yale University Press.

Holdridge, L. R. 1947. "Determination of world plant formations from simple climatic data." *Science* 105: 367–368.

Hopkins, Brian. 1983. "Successional processes." Pp. 605–616 in Francois Bourliere, ed., 1983. *Tropical Savannas*. Ecosystems of the World, 13. Amsterdam: Elsevier Scientific Publishing Company.

Hough, Michael. 1994. "Design with city nature: an overview of some issues." Pp. 40–48 in Rutherford H. Platt, Rowan A. Rowntree, and Pamela C. Muick, eds., *The Ecological City: Preserving and Restoring Urban Biodiversity*. Amherst, MA: The University of Massachusetts Press.

Iltis, André and Phillippe Mourguiart. 1992. "Higher plants: distribution and biomass." Pp. 241–253 in C. Dejoux and A. Iltis, eds., *Lake Titicaca, A Synthesis of Limnological Knowledge*. Dordrecht: Kluwer Academic Publishers.

Ioffe, Grigory and Tatyana Nefedova. 2000. *The Environs of Russian Cities*. Lewiston, NY: The Edwin Mellon Press.

Jahn, Gisela. 1991. "Temperate deciduous forests of Europe." Pp. 377–502 in E. Röhrig and B. Ulrich, eds., *Temperate Deciduous Forests*. Ecosystems of the World, 7. Amsterdam: Elsevier.

Janzen, D. H. 1972. "Protection of *Barteria* (Passifloraceae) by *Pachysima* ants (Pseudomyrmecine) in a Nigerian rain forest." *Ecology* 53(5): 885–892.

Janzen, D. H. 1988. "Tropical dry forests, the most endangered major tropical ecosystem." Pp. 130–137 in E. O. Wilson, ed., *Biodiversity*. Washington, DC: National Academy Press.

Jeffries, Michael and Derek Mills. 1990. *Freshwater Ecology: Principals and Applications*. London: Belhaven Press.

Jenkins, Robert E. and Noel M. Burkhead. 1994. *Freshwater Fishes of Virginia*. Bethesda, MD: American Fisheries Society.

Jim, C. Y. 1998. "Old stone walls as ecological habitat for urban trees in Hong Kong." *Landscape and Urban Planning* 42(1): 29–43.

Johnson, Edward A. 1992. *Fire and Vegetation Dynamics. Studies from the North American Boreal Forest*. Cambridge: Cambridge University Press.

Johnston, Richard F. and Marian Janiga. 1995. *Feral Pigeons*. Oxford: Oxford University Press.

Josens, Guy. 1983. "The soil fauna of tropical savannas. III. The termites." Pp. 505–524 in Francois Bourliere, ed., *Tropical Savannas*. Ecosystems of the World, 13. Amsterdam: Elsevier Scientific Publishing Company.

Joshi, M. C. and R. K. Gupta. 1973. "Ecology of arid and semiarid zones in India." Pp. 111–153 in *Progress of Plant Ecology in India*. New Delhi: Today & Tomorrow.

Juo, A. S. R. and H. C. Ezumah. 1992. "Mixed root-crop systems in wet sub-Saharan Africa." Pp. 243–258 in C. J. Pearson, ed., *Field Crop Ecosystems*. Ecosystems of the World, 18. Amsterdam: Elsevier.

Jürgens, N., A. Burke, M. K. Seely, and K. M. Jacobson. 1997. "Desert." Pp. 189–214 in R. M. Cowling, D. M. Richardson, and S. M. Pierce, eds., *Vegetation of Southern Africa*. Cambridge: Cambridge University Press.

Kalin Arroyo, Mary T., P. H. Zedler, M. D. Fox, eds. 1995. *Ecology and Biogeography of Mediterranean Ecosystems in Chile, California, and Australia*. NY: Springer Verlag.

Kalin Arroyo, Mary T., Clodomiro Marticorena, Oscar Matthei, and Lohengrin Cavieres. 2000. "Plant invasions in Chile: "Present patterns and future predictions." Pp. 385–421 in Harold A. Mooney and Richard J. Hobbs, eds., *Invasive Species in a Changing World*. Washington, DC and Covelo, CA: Island Press.

Kanda, H. and V. Komárková. 1997. "Antarctic terrestrial ecosystems." Pp. 721–761, in Wielgolaski, F. E., ed., *Polar and Alpine Tundra*. Ecosystems of the World, 3. Amsterdam: Elsevier.

Kasischke, Eric S. and Brian J. Stocks. 2000. "Introduction." Pp. 1–6 in Eric S Kasischke and Brian J. Stocks, eds., *Fire, Climate Change and Carbon Cycling in the Boreal Forest*. Ecological Studies 138. New York: Springer.

Kassas, M. 1953. "Landforms and plant cover in the Egyptian desert." *Bull. Soc.Geogr. Egypt* 24: 193–205.

Keay, John, ed. 1991. *History of World Exploration*. New York: Mallard Press.

Kendeigh, S. Charles. 1961. *Animal Ecology*. Englewood Cliffs, NJ: Prentice-Hall.

Kent, M., R. A. Stevens, and L. Zhang. 1999. "Urban plant ecology patterns and processes: a case study of the flora of the City of Plymouth, Devon, U.K." *Journal of Biogeography* 26(6): 1281–1298.

King, G. 1879. "Sketches of the flora of Rajasthan." *Indian Forestry* 4: 226–236.

Kingdon, Jonathan. 1989. "Bigger, brighter, louder—Signals through the leaves." Pp. 127–151 in Lisa Silcock, ed., *The Rainforests: A Celebration*. San Francisco: Chronicle Books.

Kingdon, Jonathan. 1989. *Island Africa: The Evolution of Africa's Rare Animals and Plants*. Princeton, NJ: Princeton University Press.

Kitchings, J. Thomas and Barbara T. Walton. 1991. "Fauna of the North American temperate deciduous forest." Pp. 345–370 in E. Röhrig and B. Ulrich, eds., *Temperate Deciduous Forests*. Ecosystems of the World, 7. Amsterdam: Elsevier.

Klee, Gary A. 1999. *The Coastal Environment. Toward Integrated Coastal and Marine Sanctuary Management*. Upper Saddle River, NJ: Prentice Hall.

Klinger, Lee F. 1991. "Peatland formation and ice ages: A possible Gaian mechanism related to community succession." Pp. 247–255, in S. H. Schneider and P. J. Boston, eds., *Scientists on Gaia*. Cambridge: M. I. T. Press.

Kloor, Keith. 1999. "A surprising tale of life in the city." *Science* 286(5440): 63.

Knight, Dennis H. 1994. *Mountains and Plains, The Ecology of Wyoming Landscapes*. New Haven: Yale University Press.

Knowlton, Nancy and Jeremy B. C. Jackson. 2001. "The ecology of coral reefs." Pp. 395–422 in Mark D. Bertness, Steven D. Gaines, and Mark E. Hay, eds., *Marine Community Ecology*. Sunderland, MA: Sinauer Associates, Inc.

Köppen, W. 1923. *Grundiss der Klimakunde: Die Klimate die Erde*. Berlin: DeGruyter.

Körner, Christian. 1999. *Alpine Plant Life. Functional Plant Ecology of High Mountain Ecosystems*. Berlin: Springer.

Kucera, C. L. 1992. "Tall-grass prairie." Pp. 227–268 in R. T. Coupland, ed., *Natural Grasslands: Introduction and Western Hemisphere*. Ecosystems of the World, 8A. Amsterdam: Elsevier.

LaMotte, Maxime. 1983. "Amphibians in savanna ecosystems." Pp. 313–323 in Francois Bourliere, ed., *Tropical Savannas*. Ecosystems of the World, 13. Amsterdam: Elsevier Scientific Publishing Company.

Lancaster, Nicholas 1996. "Desert environments." Pp. 211–237 in William S. Adams, Andrew S. Goudie, and Antony R. Orme, eds., *The Physical Geography of Africa*. Oxford: Oxford University Press.

Larsen, James A. 1980. *The Boreal Ecosystem*. Physiological ecology series. New York: Academic Press.

Larsen, James A. 1982. *Ecology of Northern Lowland Bogs and Coniferous Forests*. New York: Academic Press.

Lauenroth, W. K. and D. G. Milchunas. 1992. "Short-grass steppe." Pp. 183–226 in R. T. Coupland, ed., *Natural Grasslands: Introduction and Western Hemisphere*. Ecosystems of the World, 8A. Amsterdam: Elsevier.

Lauzanne, Laurent. 1992. "Fish fauna: Native species." Pp. 405–419 in C. Dejoux and A. Iltis, eds., *Lake Titicaca, A Synthesis of Limnological Knowledge*. Dordrecht: Kluwer Academic Publishers.

Lavenu, Alain. 1992. "Origins." Pp. 3–15 in C. Dejoux and A. Iltis, eds., *Lake Titicaca, A Synthesis of Limnological Knowledge*. Dordrecht: Kluwer Academic Publishers.

Lavrenko, E. M. and Z. V. Karamysheva. 1993. "Steppes of the former Soviet Union and Mongolia." Pp. 3–59 in R. T. Coupland, ed., *Natural Grasslands: Eastern Hemisphere and Résumé*. Ecosystems of the World, 8B. Amsterdam: Elsevier.

Learmonth, Andrew. 1988. *Disease Ecology*. Oxford: Basil Blackwell, Ltd.

LeHouérou, H. N. 1959. *Recherces Écologiques et Floristiques sur la Vegetation de la Tunisie Méridionale*. Algiers: Institut des Recherches Sahariennes, University of Algiers.

LeHouérou, Henri Noel. 1986. "The desert and arid zones of northern Africa." Pp. 101–147 in Michael Evenari, Imanuel Noy-Meir, and David W. Goodall, eds., *Hot Deserts and Arid Shrublands, B*. Ecosystems of the World, 12B. Amsterdam: Elsevier.

Lenihan, Hunter S. and Fiorenza Micheli. 2001. "Soft-sediment communities." Pp. 253–288 in Mark D. Bertness, Steven D. Gaines, and Mark E. Hay, eds., *Marine Community Ecology*. Sunderland, MA: Sinauer Associates, Inc.

Lincoln, Roger, Geoff Boxshall, and Paul Clark. 1998. *A Dictionary of Ecology, Evolution and Systematics*. 2nd edition. Cambridge: Cambridge University Press.

Lindeman, Raymond. 1942. "The trophic dynamic aspect of ecology." *Ecology* 23: 399–418.

Linzey, Donald W. 1998. *The Mammals of Virginia*. Blacksburg, VA: The McDonald & Woodward Publishing Company.

Lippson, Alice Jane and Robert L. Lippson. 1984. *Life in the Chesapeake Bay*. Baltimore: The Johns Hopkins University Press.

Logan, R. F. 1960. *The Central Namib Desert, South West Africa*. National Research Council Publication 758. Washington, DC: National Academy of Science.

Longhurst, Alan. 1998. *Ecological Geography of the Sea*. New York: Academic Press.

Longley, W. H. and S. F. Hildebrand. 1941. "Systematic catalogue of the fishes of Tortugas, Florida, with observations on color, habits, and local distribution." *Papers of the Tortugas Laboratory* 3: 1–331.

López-Bellido, L. 1992. "Mediterranean cropping systems." Pp. 311–356 in C. J. Pearson, ed., *Field Crop Ecosystems*. Ecosystems of the World, 18. Amsterdam: Elsevier.

Loubens, Gérard. 1992. "Introduced species: *Salmo gairdneri* (Rainbow trout)." Pp. 420–426 in C. Dejoux and A. Iltis, eds., *Lake Titicaca, A Synthesis of Limnological Knowledge*. Dordrecht: Kluwer Academic Publishers.

Loubens, Gérard and Francisco Osorio. 1992. "Introduced species: *Basilichthys bonariensis* (the 'Pejerrey')." Pp. 427–448 in C. Dejoux and A. Iltis, eds., *Lake Titicaca, A Synthesis of Limnological Knowledge*. Dordrecht: Kluwer Academic Publishers.

Loucks, Orie. 1998. "In changing forests, a search for answers." Pp. 85–97 in Harvard Ayers, Jenny Hager, and Charles Little, eds., *An Appalachian Tragedy: Air Pollution and Tree Death in the Eastern Forests of North America*. San Francisco: Sierra Club.

Löve, A. and D. Löve. 1974. "Origin and evolution of the arctic and alpine floras." Pp. 571–603 in J. D. Ives and R. G. Barry, eds., *Arctic and Alpine Environments*. London: Methuen.

Lowe, Charles H., ed. 1964. *The Vertebrates of Arizona*. Tucson: The University of Arizona Press.

Lowe-McConnell, Rosemary. 1991. "Ecology of cichlids in South American and African waters, excluding the African Great Lakes." Pp. 60–85 in Miles H. A. Keenleyside, ed., *Cichlid Fishes: Behaviour, Ecology, and Evolution*. London: Chapman & Hall.

Lugo, A. E., J. A. Gonzales-Libby, B. Cintrón, and K. Dugger. 1978. "Structure, productivity, and transpiration in a subtropical dry forest in Puerto Rico." *Biotropica* 10: 278–291.

Lüning, K. and R. Asmus. 1991. "Physical characteristics of littoral ecosystems, with special reference to marine plants." Pp. 7–26 in A. C. Mathieson and P. H. Nienhuis, eds., *Intertidal and Littoral Ecosystems.* Ecosystems of the World, 24. Amsterdam: Elsevier.

Luteyn, J. L. 1992. "Paramos: Why study them?" Pp. 1–14 in H. Balslev and J. L. Luteyn, eds., *Paramo, An Andean Ecosystem under Human Influence.* New York: Academic Press.

MacDonald, Glen. 2003. *Biogeography: Introduction to Space, Time and Life.* New York: John Wiley & Sons.

MacDonald, Glen M., Julian M. Szeicz, Jane Claricoates, and Kursti A. Dale. 1998. "Response of the central Canadian treeline to Recent climatic changes." *Annals of the Association of American Geographers* 88: 183–208.

MacMahon, James A. 2000. "Warm Deserts." Pp. 285–322 in Michael G. Barbour and William Dwight Billings, eds., *North American Terrestrial Vegetation.* 2nd edition. Cambridge: Cambridge University Press.

Mandaville, S. M. 2001. "Saline lakes (The largest, highest and lowest lakes of the world!)." http:/www.chebucto.ns.ca/Sciences/SWCS/saline1.html.

Mares, M. A. 1992. "Neotropical mammals and the myth of Amazonian biodiversity." *Science* 255: 976–979.

Mares, M. A., W. F. Blair, F. A. Enders, D. Greegor, J. Hunt, A. C. Hulse, D. Otte, R. Sage, and C. Tomoff. 1977. "The strategies and community patterns of desert animals." Pp. 107–163 in G. H. Orians and O. T. Solbrig, eds., *Convergent Evolution in Warm Deserts.* US/IBP Synthesis Series 3. Stroudsburg, PA: Dowden, Hutchinson & Ross.

Mares, M. A., J. Morello, and G. Goldstein. 1985. "The Monte Desert and other subtropical semi-arid biomes of Argentina, with comments on their relation to North American arid areas." Pp. 203–237 in Michael Evenari, Imanuel Noy-Meir, and David W. Goodall, eds., *Hot Deserts and Arid Shrublands, A.* Ecosystems of the World, 12A. Amsterdam: Elsevier.

Mares, M. A., M. R. Willig, K. E. Steinlein, and T. E. Lacher, Jr. 1981. "The mammals of Northeastern Brazil: a preliminary assessment." *Annals of the Carnegie Museum* 50: 81–137.

Markgraf, Vera, Edgardo Romero, and Carolina Villagian. 1996. "History and paleogeography of South American *Nothofagus* forests." Pp. 354–386 in Thomas T. Veblen, Robert S. Hill, and Jennifer Reed, eds., *The Ecology and Biogeography of* Nothofagus *Forests.* New Haven: Yale University Press.

Marloth, Rudolf. 1908. *Das Kapland insonderheit das Reich der Kapflora, das Waldegebeit und die Karroo, pflanzengeographisch dargestellt.* Jena: Gustav Fischer.

Martin, Claude. 1991. *The Rainforests of West Africa, Ecology—Threats—Conservation.* Boston: Birkhäuser Verlag.

Martin, Paul S. and Richard G. Klein, eds. 1984. *Quaternary Extinctions: A Prehistoric Revolution.* Tucson: The University of Arizona Press.

Martin, P. S. and David A. Yetman. 2000. "Introduction and prospect: Secrets of a tropical deciduous forest." Pp. 3–18 in Robichaux, Robert H. and David A. Yetman, eds., *The Tropical Deciduous Forest of Alamos: Biodiversity of a Threatened Ecosystem in Mexico.* Tucson: The University of Arizona Press.

Mathieson, Arthur C., Clayton A. Penniman, and Larry G. Harris. 1991. "Northwest Atlantic rocky shore ecology." Pp. 109–191 in A. C. Mathieson and P. H. Nienhuis, eds., *Intertidal and Littoral Ecosystems.* Ecosystems of the World, 24. Amsterdam: Elsevier Science Publishers.

McDonnell, M. J. and S. T. A. Pickett. 1990. "Ecosystem structure and function along urban-rural gradients: an unexploited opportunity for ecology." *Ecology* 71: 1232–1237.

McLusky, D. S. and A. D. McIntyre. 1988. "Characteristics of the benthic fauna." Pp. 131–154 in H. Postma and J. J. Zijlstra, eds., *Continental Shelves.* Ecosystems of the World, 27. Amsterdam: Elsevier.

Menaut, J. C. 1983. "African savannas." Pp. 109–149 in Francois Bourliere, ed., *Tropical Savannas.* Ecosystems of the World, 13. Amsterdam: Elsevier Scientific Publishing Company.

Menaut, Jean-Claude, Michel Lepage, and Luc Abbadie. 1995. "Savannas, woodlands, and dry forests in Africa." Pp. 64–92 in Bullock, Stephen H., Harold A. Mooney, and Ernesto Medina, eds., *Seasonally Dry Tropical Forests.* Cambridge: Cambridge University Press.

Menge, Bruce A. and George M. Branch. 2001. "Rocky intertidal communities." Pp. 221–252 in Mark D. Bertness, Steven D. Gaines, and Mark E. Hay, eds., *Marine Community Ecology.* Sunderland, MA: Sinauer Associates, Inc.

Merriam, C. Hart. 1894. "Laws of temperature control of the geographic distribution of terrestrial animals and plants." *National Geographic* 6: 229–238.

Meyer, William B. 1996. *Human Impact on the Earth.* Cambridge: Cambridge University Press.

Miehe, Georg. 1997. "Alpine vegetation types in the central Himalaya." Pp. 161–183 in Wielgolaski, F. E., ed., *Polar and Alpine Tundra.* Ecosystems of the World, 3. Amsterdam: Elsevier.

Milius, Susan. 2001. "Torn to ribbons in the desert." *Science News,* 160: 266–268.

Milton, S. J., R. I. Yeaton, W. R. J. Dean, and J. H. J. Vlok. 1997. "Succulent Karoo." Pp. 131–166 in R. M. Cowling, D. M. Richardson, and S. M. Pierce, eds., *Vegetation of Southern Africa.* Cambridge: Cambridge University Press.

Mistry, J. 2000. *World Savannas: Ecology and Human Use.* Harlow, England: Pearson Education Ltd.

Mittermeier, Russell, Norman Myers, and Cristina Goettsch Mittermeir. 1999. *Hotspots: Earth's Biologically Richest and Most Endangered Terrestrial Ecosystems.* Mexico City: CEMEX, S.A. andWashington, DC: Conservation International.

Monasterio, Maximina and Francois Vuilleumier. 1986. "Introduction: High tropical mountain biota of the world." Pp. 3–7 in Francois Vuilleumier and Maximina Monasterio, eds., *High Altitude Tropical Biogeography.* New York: Oxford University Press.

Monod, T. 1954. "Modes contracté et diffus de la végétation saharienne." Pp. 35–37 in J. L. Cloudsley-Thompson, ed., *Biology of Deserts.* London: Tavistock House.

Monod, T. 1963. "Après Yangambi (1956): notes de phytogéographie africaine." *Bulletin de l'IFAN, Série A* 24: 594–657.

Montgomery, R. F. and G. P. Askew. 1983. "Soils of tropical savannas." Pp. 63–78 in Francois Bourliere, ed., *Tropical Savannas,* Ecosystems of the World, 13. Amsterdam: Elsevier Scientific Publishing Company.

Montgomery, W. Linn 1990. "Zoogeography, behavior, and ecology of coral-reef fishes." Pp. 329–364 in Z. Dubinsky, ed., *Coral Reefs.* Ecosystems of the World, 25. Amsterdam: Elsevier.

Mooney, Harold A., Stephen H. Bullock and Ernesto Medina. 1995. "Introduction." Pp. 1–8 in Stephen H. Bullock, Harold A. Mooney, and Ernesto Median, eds., *Seasonally Dry Tropical Forests.* Cambridge: Cambridge University Press.

Moreau, R. E. 1966. The Bird Faunas of Africa and Its Islands. New York: Academic Press.

Morello, J. 1958. "La provincia fitogeografica del monte." *Opera Lilloana* 2: 1–155.

Moreno, Jose M. and Walter C. Oechel, eds. 1994. *The Role of Fire in Mediterraenan-Type Ecosystems.* New York: Springer Verlag.

Moreno, Jose M. and Walter C. Oechel, eds. 1995. *Global Change and Mediterranean-Type Ecosystems.* New York: Springer Verlag.

Mortimer, Clifford 1979. "Props and actors on a massive stage." Pp. 1–12 in John Rousmaniere, ed., *The Enduring Great Lakes.* New York: W. W. Norton & Company.

Moss, Brian. 1998. *Ecology of Fresh Waters; Man and Medium, Past to Future.* 3rd edition. Oxford: Blackwell Scientific.

Muller, C. H. 1939. "Relations of the vegetation and climate types in Nuevo Leon, Mexico." *American Midland Naturalist* 2: 687–728.

Muller, C. H. 1947. "Vegetation and climate of Coahuila, Mexico." *Madroño* 9: 33–57.

Murphy, P. G. and A. E. Lugo. 1986a. "Ecology of tropical dry forest." *Annual Review of Ecology and Systematics* 17: 67–88.

Murphy, P. G. and A. E. Lugo. 1986b. "Structure and biomass of a subtropical dry forest in Puerto Rico." *Biotropica* 18: 89–96.

Murphy, Peter G. and Ariel E. Lugo. 1995. "Dry forests of Central America and the Caribbean." Pp. 9–34 in Bullock, Stephen H., Harold A. Mooney, and Ernesto Medina, eds., *Seasonally Dry Tropical Forests.* Cambridge: Cambridge University Press.

Nienhuis, P. H. and A. C. Mathieson. 1991. "Introduction." Pp. 1–5 in A. C. Mathieson and P. H. Nienhuis, eds., *Intertidal and Littoral Ecosystems.* Ecosystems of the World, 24. Amsterdam: Elsevier Science Publishers.

Nix, H. A. 1983. "Climate of tropical savannas." Pp. 37–62 in Francois Bourliere, ed., *Tropical Savannas.* Ecosystems of the World, 13. Amsterdam: Elsevier Scientific Publishing Company.

Nowak, Ronald M. 1991. *Walker's Mammals of the World.* 5th Edition. Vol. II. Baltimore: The Johns Hopkins University Press.

O'Connor, T. G. and G. J. Bredenkamp. 1997. "Grassland." Pp. 215–257 in R. M. Cowling, D. M. Richardson, and J. M. Pierce, eds., *Vegetation of Southern Africa.* Cambridge: Cambridge University Press.

Odum, E. P. and A. de la Cruz. 1967. "Particulate organic detritus in a Georgia salt marsh-estuarine system." Pp. 383–385 in G. H. Lauff, ed., *Estuaries.* AAAS Publication 83. Washington, DC: AAAS.

Odum, Howard T. 1957. "Trophic structure and productivity of Silver Springs." *Ecological Monographs* 27: 55–112.

Odum, H. T. and E. P. Odum. 1955. "Trophic structure and productivity of windward coral reef community on Eniwetok Atoll." *Ecological Monographs* 25: 292–320.

Orians, G. H. 2000. "Site characteristics favoring invasions." Pp. 133–176 in Harold A. Mooney and Richard J. Hobbs, eds., *Invasive Species in a Changing World.* Washington, DC and Covelo, CA: Island Press.

Orians, G. H. and O. T. Solbrig, eds. 1977. *Convergent Evolution in Warm Deserts.* US/IBP Synthesis Series 3. Stroudsburg, PA: Dowden, Hutchinson & Ross.

Ormond, Rupert F. G., John D. Gage, and Martin V. Angel. 1997. "Foreword, The value of biodiversity." Pp. xiii–xxii in Rupert F. G. Ormond, John D. Gage, and Martin V. Angel, eds., *Marine Biodiversity, Patterns and Processes.* Cambridge: Cambridge University Press.

Orshan, G. 1986. "The deserts of the Middle East." Pp. 1–26 in Michael Evenari, Imanuel Noy-Meir, and David W. Goodall, eds., *Hot Deserts and Arid Shrublands, B.* Ecosystems of the World, 12A. Amsterdam: Elsevier.

Paine, R. T. 1969. "The *Pisaster-Tegula* interaction: Prey patches, predator food preferences and intertidal community structure." *Ecology* 50: 950–961.

Palumbi, Stephen R. 2001. "The ecology of marine protected areas." Pp. 509–530 in Mark D. Bertness. Steven D. Gaines, and Mark E. Hay, eds., *Marine Community Ecology.* Sunderland, MA: Sinauer Associates, Inc.

Pastor, J., R. H. Gardner, V. H. Dale, and W. M. Post. 1987. "Successional changes in nitrogen availability as a potential factor contributing to spruce decline in boreal North America." *Canadian Journal of Forest Research* 17(11): 1394–1400.

Payne, A. I. 1986. *The Ecology of Tropical Lake and Rivers.* New York: Wiley & Sons Ltd.

Pennings, Steven C. and Mark D. Bertness. 2001. "Salt marsh communities." Pp. 289–316 in Mark D. Bertness, Steven D. Gaines, and Mark E. Hay, eds., *Marine Community Ecology.* Sunderland, MA: Sinauer Associates, Inc.

Peres, Francisco L. 1998. "Human impact on the high paramo landscape of the Venezuelan Andes." Pp. 147–183 in Zimmerer, K. S. and K. R. Young, eds., *Nature's Geography, New Lessons for Conservation in Developing Countries.* Madison: The University of Wisconsin Press.

Perry, William L., Jeffrey L. Feder, and David M. Lodge. 2001. "Implications of hybridization between introduced and resident *Orconectes* crayfishes." *Conservation Biology* 15: 1656–1666.

Petersen, C. G. J. 1918. "The sea bottom and its production of fish food. A survey of the work done in connection with valuation of the Danish waters from 1883–1917." *Rap. Danish Biol. Stat.* 25: 1–62.

Peterson, Charles H. and James A. Estes. 2001. "Conservation and management of marine communities." Pp. 469–507 in Mark D. Bertness, Steven D. Gaines, and Mark E. Hay, eds., *Marine Community Ecology.* Sunderland, MA: Sinauer Associates, Inc.

Petrov, M. P. 1966–67. *Deserts of Central Asia.* 2 vols (in Russian). Leningrad: Nauka.

Pickett, S. T. A., M. L. Cadenasso, J. M. Grove, C. H. Nilon, R. V. Pouyat, W. C. Zipperer, and R. Constanza. 2001. "Urban ecological systems: linking terrestrial ecological, physical and socioeconomic components of metropolitan areas." *Annual Review of Ecology and Systematics* 32: 127–157.

Pielou. E. C. 1988. *The World of Northern Evergreens.* Ithaca: Comstock Publishing Associates.

Pielou, E. C. 1991. *After the Ice Age, the Return of Life to Glaciated North America.* Chicago: University of Chicago Press.

Pivello, Vania R. 2000. "Cerrado." http://www.brazil.org.uk: Brazilian Embassy in London.

Platt, Rutherford H. 1994a. "The ecological city: introduction and overview." Pp. 1–17 in Rutherford H. Platt, Rowan A. Rowntree, and Pamela C. Muick, eds., *The Ecological City: Preserving and Restoring Urban Biodiversity.* Amherst, MA: The University of Massachusetts Press.

Platt, Rutherford H. 1994b. "From commons to commons: evolving concepts of open space in North American cities." Pp. 21–39 in Rutherford H. Platt, Rowan A. Rowntree, and Pamela C. Muick, eds., *The Ecological City: Preserving and Restoring Urban Biodiversity.* Amherst, MA: The University of Massachusetts Press.

Potsma, H. 1988. "Physical and chemical oceanographic aspects of continental shelves." Pp. 5–37 in H. Postma and J. J. Zijlstra, eds., *Continental Shelves.* Ecosystems of the World, 27. Amsterdam: Elsevier.

Potsma, H. and J. J. Zijlstra. 1988. "Introduction." Pp. 1–4 in H. Postma and J. J. Zijlstra, eds., *Continental Shelves.* Ecosystems of the World, 27. Amsterdam: Elsevier.

Prach, K. and P. Pysek. 2000. "How do species dominating in succession differ from others?" *Journal of Vegetation Science* 10(3): 383–392.

Pysek, P. 1999. "Alien and native species in Central European urban floras: a quantitative comparison." *Journal of Biogeography* 25(1): 155–163.

RADAMBRASIL. 1983. *Projecto Radambrasil. Levantamento dos Recursos Naturais.* Vol. Folhas SF23/24. Rio de Janeiro: IBGE.

Raffaelli, David and Stephen Hawkins. 1996. *Intertidal Ecol*ogy. London: Chapman & Hall.

Ratter, J. A., J. F. Ribeiro, and S. Bridgewater. 1997. "The Brazilian cerrado vegetation and threats to its biodiversity." *Annals of Botany* 80: 223–230.

Rauh, Walter. 1985. "The Peruvian-Chilean deserts." Pp. 239–267 in Michael Evenari, Imanuel Noy-Meir, and David W. Goodall, eds., *Hot Deserts and Arid Shrublands, A.* Ecosystems of the World, 12A. Amsterdam: Elsevier.

Raunkiaer, C. 1934. *The Life Forms of Plants and Statistical Plant Geography.* Oxford; Clarendon Press.

Ray, G. C., B. P. Hayden, M. G. McCormick-Ray, and T. M. Smith. 1997. "Land-seascape diversity of USA East Coast coastal zone with particular reference to estuaries." Pp.

337–371 in Rupert F. G. Ormond, John D. Gage, and Martin V. Angel, eds., *Marine Biodiversity, Patterns and Processes.* Cambridge: Cambridge University Press.

Raynal-Roques, Aline. 1992. "Macrophytes: The higher plants." Pp. 223–231 in C. Dejoux and A. Iltis, eds., *Lake Titicaca, A Synthesis of Limnological Knowledge.* Dordrecht: Kluwer Academic Publishers.

Reed, Charles A. 1969. "The pattern of animal domestication in the prehistoric Near East." Pp. 361–380 in Peter J. Ucko and G. W. Dimbleby, eds., *The Domestication and Exploitation of Plants and Animals.* Chicago: Aldine Publishing Company.

Reed, Charles A., ed. 1977. *Origins of Agriculture.* The Hague: Mouton Publishers.

Richards, P. W. 1996. *The Tropical Rain Forest.* 2nd Edition. Cambridge: Cambridge University Press.

Ricketts, Taylor H., Eric Dinerstein, David M. Olson, Colby J. Loucks, et al. 1999. *Terrestrial Ecoregions of North America, A Conservation Assessment.* Washington, DC: Island Press.

Riemer, Donald N. 1993. *An Introduction to Freshwater Vegetation.* Malabar, FL: Krieger.

Rindos, David. 1984. *The Origins of Agriculture, An Evolutionary Perspective.* New York: Academic Press.

Robinson, Bruce. 2001. "Water column life." Pp. 98–103 in Sylvia A. Earle, ed., *National Geographic Atlas of the Oceans.* Washington, DC: National Geographic Society.

Röhrig, Ernst. 1991a. "Introduction." Pp. 1–5 in E. Röhrig and B. Ulrich, eds., *Temperate Deciduous Forests.* Ecosystems of the World, 7. Amsterdam: Elsevier.

Röhrig, Ernst. 1991b. "Seasonality." Pp. 25–33 in E. Röhrig and B. Ulrich, eds., *Temperate Deciduous Forests.* Ecosystems of the World, 7. Amsterdam: Elsevier.

Röhrig, Ernst. 1991c. "Floral composition and its evolutionary development." Pp. 17–23 in E. Röhrig andB. Ulrich, eds., *Temperate Deciduous Forests.* Ecosystems of the World, 7. Amsterdam: Elsevier.

Round, P. D. 1988. *Resident Forest Birds in Thailand: their status and conservation.* Monograph No. 2. Cambridge: International Council for Bird Preservation.

Rowe, J. S. 1972. *Forest Regions of Canada.* Publication No. 1300. Ottawa: Canadian Forestry Service, Department of the Environment.

Rundel, Philip W. and Kansri Boonpragob. 1995. "Dry forest ecosystems of Thailand." Pp. 93–123 in Bullock, Stephen H., Harold A. Mooney, and Ernesto Medina, eds., *Seasonally Dry Tropical Forests.* Cambridge: Cambridge University Press.

Rundel, P. W., G. Montenegro, and F. M. Jaksic, eds. 1998. *Landscape Disturbance and Biodiversity in Mediterranean-type Ecosystems.* Berlin: Springer.

Russell, G. 1991. "Vertical distribution." Pp. 43–65 in A. C. Mathieson and P. H. Nienhuis, eds., *Intertidal and Littoral Ecosystems.* Ecosystems of the World, 24. Amsterdam: Elsevier Science Publishers.

Russell, Stephen M. 2000. "Birds of the tropical deciduous forest of the Alamos, Sonora, area." Pp. 200–244 in Robichaux, Robert H. and David A. Yetman, eds., *The Tropical Deciduous Forest of Alamos: Biodiversity of a Threatened Ecosystem in Mexico.* Tucson: The University of Arizona Press.

Rysin, L. P. and M. V. Nadeshdina, eds. 1968. *Long-term Biogeocenotic Investigations in the Southern Taiga Subzone.* Jerusalem: Israel Program for Scientific Translations.

Sampaio, Everardo V. S. B. 1995. "Overview of the Brazilian caatinga." Pp. 35–63 in Bullock, Stephen H., Harold A. Mooney, and Ernesto Medina, eds., *Seasonally Dry Tropical Forests.* Cambridge: CambridgeUniversity Press.

Sarmiento, G. 1983. "The savannas of tropical America." Pp. 245–288 in Francois Bourliere, ed., *Tropical Savannas.* Ecosystems of the World, 13. Amsterdam: Elsevier Scientific Publishing Company.

Sarmiento, Guillermo. 1986. "Ecological features of climate in high tropical mountains." Pp. 11–45 in Francois Vuilleumier and Maximina Monasterio, eds., *High Altitude Tropical Biogeography.* New York: Oxford University Press.

Sarmiento, G. and M. Monasterio. 1983. "Life forms and phenology." Pp. 79–108 in Francois Bourliere, ed., *Tropical Savannas.* Ecosystems of the World, 13. Amsterdam: Elsevier Scientific Publishing Company.

Sauer, C. O. 1975. "Man's dominance by use of fire." Pp. 1–13 in R. H. Kesel, ed., *Geoscience and Man 10.* Baton Rouge: Louisiana State University.

Sauer, Jonathan D. 1993. *Historical Geography of Crop Plants, A Select Roster.* Boca Raton, FL: CRC Press.

Schaller, George B. 1998. *Wildlife of the Tibetan Steppe.* Chicago: University of Chicago Press.

Schimper, A. F. W. 1898. *Pflanzen-Geographie auf physiologischer Grundlage.* Jena: Fischer. English translation: 1930 *Plant Geography Under a Physiological Basis.* Oxford: Clarendon Press.

Schmaltz, Jürgen. 1991. "Deciduous forests of southern South America." Pp. 557–578 in E. Röhrig and B. Ulrich, eds., *Temperate Deciduous Forests.* Ecosystems of the World, 7. Amsterdam: Elsevier.

Schmid, James A. 1975. *Urban Vegetation: A Review and Chicago Case Study. Research Paper No. 161.* Chicago: Department of Geography, University of Chicago.

Schmida, A. 1985. "Biogeography of the desert flora." Pp. 23–77 in Michael Evernari, Imanual Noy-Meir, and David W. Goodall, eds., *Hot Deserts and Arid Shrublands, A.* Ecosystems of the World 12A. Amsterdam: Elsevier.

Schmutz, E. M., E. L. Smith, P. R. Ogden, M. L. Cox, J. O. Klemmedson, J. J. Norris, and L. C. Fierro. 1992."Desert grassland." Pp. 337–362 in R. T. Coupland, ed., *Natural Grasslands: Introduction and Western Hemisphere.* Ecosystems of the World, 8A. Amsterdam: Elsevier.

Schultz, J. 1995. *The Ecozones of the World: The Ecological Divisions of the Geosphere.* Berlin: Springer-Verlag.

Schwable, Cecil R. and Charles H. Lowe. 2000. "Amphibians and reptiles of the Sierra de Alamos." Pp. 172–199 in Robichaux, Robert H. and David A. Yetman, eds., *The Tropical Deciduous Forest of Alamos: Biodiversity of a Threatened Ecosystem in Mexico.* Tucson: The University of Arizona Press.

Schweinfurth, U. 1983. "Man's impact on vegetation and landscape in the Himalayas." Pp. 297–309 in W. Holzner, M. J. A. Werger, and I. Ikusima, eds., *Man's Impact of Vegetation.* The Hague: Dr. W. Junk Publishers.

Shaeffer, M. 1991. "Fauna of the European temperate deciduous forest." Pp. 503–525 in E. Röhrig and B. Ulrich, eds., *Temperate Deciduous Forests.* Ecosystems of the World, 7. Amsterdam: Elsevier.

Sharp, G. D. 1988. "Fish populations and fisheries. Their perturbations, natural and man-induced." Pp. 155–202 in H. Postma and J. J. Zijlstra, eds., *Continental Shelves.* Ecosystems of the World, 27. Amsterdam: Elsevier.

Shramm, W. 1991. "Chemical characteristics of marine littoral ecosystems." Pp. 27–42 in A. C. Mathieson and P. H. Nienhuis, eds., *Intertidal and Littoral Ecosystems.* Ecosystems of the World, 24. Amsterdam: Elsevier Science Publishers.

Shreve, Forrest. 1910. "The establishment of the giant cactus." *Plant World* 13: 235–240.

Shreve, Forrest. 1917. "The establishment of desert perennials." *Journal of Ecology* 5: 210–216.

Shreve, Forrest. 1942. "The desert vegetation of North America." *Botanical Review* 8(4): 195–246.

Shreve, Forrest. 1951. "Vegetation of the Sonoran Desert." *Carnegie Institution Washington Publication 591.*

Sick, Helmut. 1993. *Birds in Brazil, A Natural History.* Princeton, NJ: Princeton University Press.

Siegert, Martin J. 1999. "Antarctica's Lake Vostok." *American Scientist* 87: 511–517.

Simmons, I. G. 1989. *Changing the Face of the Earth: Culture, Environment, History.* Oxford: Basil Blackwell.
Simmons, I. G. 1996. *Changing the Face of the Earth.* 2nd edition. Oxford: Blackwell Publishers.
Sinclair, A. R. E. 1983. "The adaptations of African ungulates and their effects on community function." Pp. 401–426 in Francois Bourliere, ed., *Tropical Savannas.* Ecosystems of the World, 13. Amsterdam: Elsevier Scientific Publishing Company.
Sirenian International. 2000–2002. "Amazonian manatee." http://sirenian.org/amazonian.htm.
Sirois, L., G. B. Bonan, and H. H. Shugart. 1994. "Development of a simulation model of the forest-tundra transition zone of northeastern Canada." *Canadian Journal of Forest Research* 24: 697–706.
Smetacek, Victor. 1988. "Planktonic characteristics." Pp. 93–130 in H. Postma and J. J. Zijlstra, eds., *Continental Shelves.* Ecosystems of the World, 27. Amsterdam: Elsevier.
Smith, Nigel J. F. 1999. *The Amazon River Forest, A Natural History of Plants, Animals, and People.* New York: Oxford University Press.
Smith, Stanford. 1979. "Pushed toward extinction: The salmon and the trout." Pp. 33–46 in John Rousmaniere, ed., *The Enduring Great Lakes.* New York: W. W. Norton & Company.
Solbrig, O. T. 1996. "The diversity of the savanna ecosystem." Pp. 1–27 in O. T. Solbrig, E. Medina, and J. F. Silva, eds., *Biodiversity and Savanna Ecosystem Processes. A Global Perspective.* Berlin: Springer.
Solbrig, O. T. and G. H. Orians. 1977. "The adaptive characteristics of desert plants." *American Scientist* 65: 412–421.
Solomon, Derek. 1997. "Harare." Durban, South Africa: Southern African Birding. http://www.sabirding.co.za/birdspot/120408.asp.
Somme, L. 1997. "Adaptation to the alpine environment in insects and other terrestrial arthropods." Pp. 11–26 in Wielgolaski, F. E., ed., *Polar and Alpine Tundra.* Ecosystems of the World, 3. Amsterdam: Elsevier.
Songqiao, Zhao. 1986. *Physical Geography of China.* New York: Wiley & Sons, Inc.
Soriano, Alberto. 1983. "Deserts and semi-deserts of Patagonia." Pp. 423–460 in Neil E. West, ed., *Temperate Deserts and Semi-deserts.* Ecosystems of the World, 5. Amsterdam: Elsevier Scientific Publishing Company.
Soriano, Alberto. 1992. "Rio de la Plata grasslands." Pp. 367–407 in R. T. Coupland, ed., *Natural Grasslands: Introduction and Western Hemisphere.* Ecosystems of the World, 8A. Amsterdam: Elsevier.
Stearns, Forest. 1971. "Urban botany—an essay on survival." *University of Wisconsin Field Station Bulletin* 4: 1–6. (cited in Picketts et al. 2001).
Steneck, Robert S. and James T. Carlton. 2001. "Human alterations of marine communities: Students beware!" Pp. 445–468 in Mark D. Bertness, Steven D. Gaines, and Mark E. Hay, eds., *Marine Community Ecology.* Sunderland, MA: Sinauer Associates, Inc.
Stephenson, T. A. and A. Stephenson. 1949. "The universal features of zonation between tidemarks on rocky coasts." *Journal of Ecology* 38: 289–305.
Stephenson, T. A. and A. Stephenson. 1972. *Life between Tidemarks on Rocky Shores.* San Francisco: W. H. Freeman.
Stewart, O. C. 1956. "Fire as the first great force employed by man." Pp. 115–133 in W. L. Thomas, ed., *Man's Role in Changing the Face of the Earth.* Chicago: University of Chicago Press.
Stoermer, Eugene F. 1979. "Bloom and crash: Algae in the Lakes." Pp. 13–20 in John Rousmaniere, ed., *The Enduring Great Lakes.* New York: W. W. Norton & Company.
Sukopp, Herbert, Hans-Peter Blume, and Wolfram Kunick. 1979. "The soil, flora, and vegetation of Berlin's waste lands." Pp. 115–132 in Ian Laurie, ed., *Nature in the City: The*

*Natural Environment in the Design and Development of Urban Green Space.* New York: John Wiley & Sons.
Sukopp, H. and P. Werner. 1983. "Urban environments and vegetation." Pp. 247–260 in W. Holzner, M. J. A. Werger and I. Ikusima, eds., *Man's Impact on Vegetation.* The Hague: Dr. W. Junk Publishers.
Tainton, N. M. and B. H. Walker. 1993. "Grasslands of southern Africa." Pp. 265–290 in R. T. Coupland, ed., *Natural Grasslands: Eastern Hemisphere and Résumé.* Ecosystems of the World, 8B. Amsterdam: Elsevier.
Taylor, David. 1996. "Mountains." Pp. 287–306 in Adams, W. M., A. S. Goudie, and A. R. Orme, eds., *The Physical Geography of Africa.* Oxford: Oxford University Press.
Teal, J. M. 1962. "Energy flow in the salt marsh ecosystem of Georgia." *Ecology* 43: 614–624.
Tharp, Marie. 2001. "Ocean cartography." Pp. 32–44 in Sylvia A. Earle, ed., *National Geographic Atlas of the Oceans.* Washington, DC: National Geographic Society.
Theobold, David M. 2001. "Land-use dynamics beyond the American urban fringe." *Geographical Review* 3: 544–564.
Thrower, N.J.W. and D. E. Bradbury. 1977. *Chile-California Mediterranean Scrub Atlas, A Comparative Analysis.* Stroudsburg, PA: Dowden, Hutchinson,& Ross, Inc.
Ting-Chen, Zhu. 1993. "Grasslands of China." Pp. 61–82 in R. T. Coupland, ed., *Natural Grasslands: Eastern Hemisphere and Résumé.* Ecosystems of the World, 8B. Amsterdam: Elsevier.
Tirdád. n.d. "Fishes of northern Iran, Southern Caspian sea basin." Gorgán homepage: http://medlem.spray.se/davidgorgan/fishes.html.
Tivy, Joy. 1990. *Agricultural Ecology.* Essex, England: Longman Scientific & Technical.
Troll, C. 1930. "Die trophischen Andenländer: Bolivien, Peru, Ecuador, Columbien und Venezuela." Pp. 309–490 in F. Klute, ed., *Handbuch der Geographischen Wissenschaften,* Band Südamerika. Potsdam: Athenaion Wildpark.
U.S. Geological Survey (USGS). 2001. "Great Salt Lake, Utah." http://ut.water.usgs.gov/greatsaltlake/index.html.
Vale, Thomas R. and Geraldine R. Vale. 1976. "Suburban bird populations in west-central California." *Journal of Biogeography* 3: 157–165.
van der Hammen, T. 1983. "The paleoecology and paleogeography of savannas." Pp. 19–35 in Francois Bourliere, ed., *Tropical Savannas.* Ecosystems of the World, 13. Amsterdam: Elsevier Scientific Publishing Company.
van der Hammen, Thomas and Antoine M. Cleef. 1986. "Development of high Andean paramo flora and vegetation." Pp. 153–201 in Francois Vuilleumier and Maximina Monasterio, eds., *High Altitude Tropical Biogeography.* New York: Oxford University Press.
Van Devender, Thomas R., Andrew C. Sanders, Rebecca K. Wilson, and Stephanie A. Meyer. 2000. "Vegetation, flora, and seasons of the Rio Cuchujaqui, a tropical deciduous forest near Alamos, Sonora." Pp. 36–101 in Robichaux, Robert H. and David A. Yetman. eds., *The Tropical Deciduous Forest of Alamos: Biodiversity of a Threatened Ecosystem in Mexico.* Tucson: The University of Arizona Press.
Van Dover, Cindy Lee. 2000. *The Ecology of Deep-Sea Hydrothermal Vents.* Princeton: Princeton University Press.
Van Dover, C. L., S. E. Humphris, D. Fornari, C. M. Cavanaugh, et al. 2001. "Biogeography and ecological setting of Indian Ocean hydrothermal vents." *Science* 294(5543): 818–823.
Van Gelder, Richard G. 1982. *Mammals of the National Parks.* Baltimore: The Johns Hopkins University Press.
Vankat, John L. 1979. *The Natural Vegetation of North America, an Introduction.* New York: John Wiley & Sons.

Vannote, R. L., G. W. Minshall, K. W. Cummings, J. R. Sedell, and C. E. Cushing. 1980. "The river continuum concept." *Canadian Journal of Fisheries and Aquatic Science* 37: 120–137.

Vasey, Daniel E. 1992. *An Ecological History of Agriculture, 10,000 B.C.–A.D. 10,000.* Ames, IA: Iowa State University Press.

Veblen, Thomas T., Claudio Donoso, Thomas Kitberger, and Alan J. Rebertus. 1996. "Ecology of southern Chilean and Argentinean *Nothofagus* forests. Pp. 293–353 in Thomas T. Veblen, Robert S.Hill, and Jennifer Reed, eds., *The Ecology and Biogeography of* Nothofagus *Forests.* New Haven: Yale University Press.

Vellard, Jean. 1992. "Associated animal communities: the Amphibia." Pp. 449–457 in C. Dejoux and A. Iltis, eds., *Lake Titicaca, A Synthesis of Limnological Knowledge.* Dordrecht: Kluwer Academic Publishers.

Vermeij, Geerat J. 2001. "Community assembly in the sea. Geologic history of the living shore biota." Pp. 39–60 in Mark D. Bertness, Steven D. Gaines, and Mark E. Hay, eds., *Marine Community Ecology.* Sunderland, MA: Sinauer Associates, Inc.

Vitousek, Peter M., Harold A. Mooney, Jane Lubchenko, and Jerry M. Mellilo. 1997. "Human domination of Earth's ecosystems." *Science* 277(5325): 494–499.

Walker, B. H., W. L. Steffen, and J. Landridge. 1999. "Interactive and integrated effects of global change on terrestrial ecosystems." Pp. 329–375 in B. Walker, B. Steffen, J. Canadell, and J. Ingram, eds., *The Terrestrial Biosphere and Global Change: Implications for Natural and Managed Ecosystems.* Cambridge: Cambridge University Press.

Wallace, Alfred Russel. 1853. *A Narrative of Travels on the Amazon and Rio Negro.* London: Reeve (cited in Goulding 1980).

Walter, H. 1936. "Die ökologischen Verhältnisse in der Namib-Nebelwüste (Südwestafrika) unter Auswertung der Aufzeichnungen des Dr. G. Boss (Swakopmund)." *Jahrbuch der Wissenschaften Botanische* 84: 58–222.

Walter, H. 1974. *Die vegetation Osteuropas, Nord- und Zentralasiens.* Band VII, Vegetationsmonographien der einzelnen Grossräume. Stuttgart: Fischer-Verlag.

Walter, Heinrich. 1985. *Vegetation of the Earth and Ecological Systems of the Geo-biosphere.* 3rd edition. New York: Springer-Verlag.

Walter, H. 1986. "The Namib Desert." Pp. 245–282 in Michael Evenari, Imanuel Noy-Meir, and David W. Goodall, eds., *Hot Deserts and Arid Shrublands, B.* Ecosystems of the World, 12B. Amsterdam: Elsevier.

Walter, H. and E. O. Box. 1983a. "Overview of Eurasian continental deserts and semi-deserts." Pp. 3–7 in Neil E. West, ed., *Temperate Deserts and Semi-deserts.* Ecosystems of the World, 5. Amsterdam: Elsevier Scientific Publishing Company.

Walter, H. and E. O. Box. 1983b. "The Karakum Desert, an example of a well-studied eu-biome." Pp. 105–159 in Neil E. West, ed., *Temperate Deserts and Semi-deserts.* Ecosystems of the World, 5. Amsterdam: Elsevier Scientific Publishing Company.

Walter, H. and E. O. Box. 1983c. "The deserts of Central Asia." Pp. 193–236 in Neil E. West, ed., *Temperate Deserts and Semi-deserts.* Ecosystems of the World, 5. Amsterdam: Elsevier Scientific Publishing Company.

Walter, H. and E. O. Box. 1983d. "Middle Asia deserts." in Neil E. West, ed., *Temperate Deserts and Semi-deserts.* Ecosystems of the World, 5. Amsterdam: Elsevier Scientific Publishing Company.

Weaver, J. E. 1954. *North American Prairie.* Lincoln, NE: Johnsen.

Weberbauer, A. 1911. "Die Pflanzenwelt der peruanischen Anden." In Vegetation der Erde, 12. Leipzig: Englemann (cited Rauh 1985).

Werger, M. J. A. 1986. "The Karoo and Southern Kalahari," Pp. 283–359 in Michael Evenari, Imanuel Noy-Meir, and David W. Goodall, eds., *Hot Deserts and Arid Shrublands, B.* Ecosystems of the World, 12B. Amsterdam: Elsevier.

West, Neil E. 1983a. "Overview of North American temperate deserts and semi-deserts." Pp. 321–330 in Neil E. West, ed., *Temperate Deserts and Semi-deserts.* Ecosystems of the World, 5. Amsterdam: Elsevier Scientific Publishing Company.

West, Neil E. 1983b. "Great Basin-Colorado sagebrush semi-desert." Pp. 331–349 in Neil E. West, ed., *Temperate Deserts and Semi-deserts*. Ecosystems of the World, 5. Amsterdam: Elsevier Scientific Publishing Company.

West, Neil E. 1983c. "Western intermontane sagebrush steppe." Pp. 351–374 in Neil E. West, ed., *Temperate Deserts and Semi-deserts*. Ecosystems of the World, 5. Amsterdam: Elsevier Scientific Publishing Company.

West, Neil E. 1983d. "Intermontane salt-desert shrubland." Pp. 375–397 in Neil E. West, ed., *Temperate Deserts and Semi-deserts*. Ecosystems of the World, 5. Amsterdam: Elsevier Scientific Publishing Company.

West, Neil E. 1983e. "Comparisons and contrasts between the temperate deserts and semi-deserts of three continents." Pp. 461–472 in Neil E. West, ed., *Temperate Deserts and Semi-deserts*. Ecosystems of the World, 5. Amsterdam: Elsevier Scientific Publishing Company.

West, N. E. and James A. Young. 2000. "Intermountain valleys and lower mountain slopes." Pp. 255–284 in Michael G. Barbour and Dwight Billings, eds., *North American Terrestrial Vegetation*. 2nd edition. Cambridge: Cambridge University Press.

Wheater, C. Philip. 1999. *Urban Habitats*. London: Routledge.

White, F. 1983. *The Vegetation of Africa: A Descriptive Memoir to Accompany the UNESCO/AEFAT/UNSO Vegetation Map of Africa*. Paris: UNESCO.

White, Mary E. 1990. *The Flowering of Gondwana: The 400 Million Year Story of Australian Plants*. Princeton, NJ: Princeton University Press.

White, Rodney R. 1994. *Urban Environmental Management: Environmental Change and Urban Design*. New York: John Wiley & Sons.

Whitmore, T. C. 1984. *Tropical Rain Forests of the Far East*. 2nd edition. Oxford: Clarendon Press.

Whitmore, T. C. 1990. *An Introduction to Tropical Rain Forests*. New York: Oxford University Press.

Whitney, Gordon G. 1994. *From Coastal Wilderness to Fruited Plain. A History of Environmental Change in Temperate North America 1500 to the Present*. New York: Cambridge University Press.

Wieboldt, Thomas F. 1989 "Early botanical exploration and plant notes from the New River, Virginia section." *Proceedings of the Eighth Annual New River Symposium:* 149–159. Oak Hill, West Virginia: National Park Service.

Wielgolaski, F. E. 1997a. "Introduction." Pp. 1–5, in Wielgolaski, F. E., ed., *Polar and Alpine Tundra*. Ecosystems of the World, 3. Amsterdam: Elsevier.

Wielgolaski, F. E. 1997b. "Adaptation in plants." Pp. 7–10, in Wielgolaski, F. E., ed., *Polar and Alpine Tundra*. Ecosystems of the World, 3. Amsterdam: Elsevier.

Williams, Michael. 1989. *Americans and Their Forests: A Historical Geography*. New York: Cambridge University Press.

Williams, O. B. and J. H. Calaby. 1985. "The hot deserts of Australia." Pp. 269–312 in Michael Evenari, Imanuel Noy-Meir, and David W. Goodall, eds., *Hot Deserts and Arid Shrublands, A*. Ecosystems of the World, 12A. Amsterdam: Elsevier.

Williams, Susan L. and Kenneth L. Heck, Jr. 2001. "Sea grass community ecology." Pp. 317–337 in Mark D. Bertness, Steven D. Gaines, and Mark E. Hay, eds., *Marine Community Ecology*. Sunderland, MA: Sinauer Associates, Inc.

Witman, Jon D. and Paul K. Dayton. 2001. "Rocky subtidal communities." Pp. 339–366 in Mark D. Bertness, Steven D. Gaines, and Mark E. Hay, eds., *Marine Community Ecology*. Sunderland, MA: Sinauer Associates, Inc.

Wolfe, Art and Ghillean T. Prance. 1998. *Rainforests of the World: Water, Fire, Earth, and Air.* New York: Crown Publishers, Inc.

Wood, Stanley, Kate Sebastian, and Sara J. Scherr. 2000. *Pilot Analysis of Global Ecosystems: Agroecosystems*. Washington, DC: International Food Policy Research Institute and World Resources Institute. Also available at http://www.wri.org/wr2000.

Woodell, Stanley. 1979. "The flora of walls and pavings." Pp. 135–157 in Ian Laurie, ed., *Nature in the City: The Natural Environment in the Design and Development of Urban Green Space.* New York: John Wiley & Sons.

Woodward, F. I. 1995. "Ecophysiological controls of conifer distributions." Pp. 79–94 in William K. Smith and Thomas M. Hinckley, eds., *Ecophysiology of Coniferous Forests.* San Diego: Academic Press.

Woodward, Susan L. 1976. "Feral Burros of the Chemehuevi Mountains, California: The Biogeography of a Feral Exotic." Ph.D. dissertation, The University of California, Los Angeles, 178 p.

Zlotin, R. J. 1997. "Geography and organization of high-mountain ecosystems in the former U.S.S.R." Pp. 133–159 in Wielgolaski, F. E., ed., *Polar and Alpine Tundra.* Ecosystems of the World, 3. Amsterdam: Elsevier.

Zohary, M. 1940. "Geobotanical analysis of the Syrian Desert." *Palestine Journal of Botany* 2: 46–96.

Zohary, M. and G. Orshan. 1956. "Ecological studies in the vegetation of the Near East deserts II. Wadi Araba." *Vegetatio* 71: 15–37.

Zoltai, S. C. and F. C. Pollett. 1983. "Wetlands in Canada: their classification, distribution, and use." Pp. 245–268 in A. J. P. Gore, ed., *Mires—Swamp, Bog, Fen, and Mire.* Ecosystems of the World, 4B. Amsterdam: Elsevier Scientific Publishing Company.

# INDEX

**About the Author**

SUSAN L. WOODWARD is Professor of Geography at Radford University, where she teaches courses in biogeography, physical geography, and human ecology. Dr. Woodward received her Ph.D. in geography from the University of California, Los Angeles, and has studied biomes in North America, Brazil, Chile, Ecuador, Great Britain, Russia, and China.